Periodic Table of the Elements with the Gmelin System Numbers

1	2	3	4	5	6	7	8	9	10	11	12	13	14	15	16	17	18
1 H 2																	2 He 1
3 Li 20	4 Be 26											5 B 13	6 C 14	7 N 4	8 O 3	9 F 5	10 Ne 1
11 Na 21	12 Mg 27											13 Al 35	14 Si 15	15 P 16	16 S 9	17 Cl 6	18 Ar 1
19* K 22	20 Ca 28	21 Sc 39	22 Ti 41	23 V 48	24 Cr 52	25 Mn 56	26 Fe 59	27 Co 58	28 Ni 57	29 Cu 60	30 Zn 32	31 Ga 36	32 Ge 45	33 As 17	34 Se 10	35 Br 7	36 Kr 1
37 Rb 24	38 Sr 29	39 Y 39	40 Zr 42	41 Nb 49	42 Mo 53	43 Tc 69	44 Ru 63	45 Rh 64	46 Pd 65	47 Ag 61	48 Cd 33	49 In 37	50 Sn 46	51 Sb 18	52 Te 11	53 I 8	54 Xe 1
55 Cs 25	56 Ba 30	57** La 39	72 Hf 43	73 Ta 50	74 W 54	75 Re 70	76 Os 66	77 Ir 67	78 Pt 68	79 Au 62	80 Hg 34	81 Tl 38	82 Pb 47	83 Bi 19	84 Po 12	85 At 8a	86 Rn 1
87 Fr 25a	88 Ra 31	89*** Ac 40	104 71	105 71													

*Lanthanides 39

58 Ce	59 Pr	60 Nd	61 Pm	62 Sm	63 Eu	64 Gd	65 Tb	66 Dy	67 Ho	68 Er	69 Tm	70 Yb	71 Lu

***Actinides

90 Th 44	91 Pa 51	92 U 55	93 Np 71	94 Pu 71	95 Am 71	96 Cm 71	97 Bk 71	98 Cf 71	99 Es 71	100 Fm 71	101 Md 71	102 No 71	103 Lr 71

* NH_4 23

A Key to the Gmelin System is given on the Inside Back Cover

Gmelin Handbook of Inorganic Chemistry

8th Edition

Gmelin Handbook of Inorganic Chemistry

8th Edition

Gmelin Handbuch der Anorganischen Chemie

Achte, völlig neu bearbeitete Auflage

Prepared and issued by	Gmelin-Institut für Anorganische Chemie der Max-Planck-Gesellschaft zur Förderung der Wissenschaften Director: Ekkehard Fluck

Founded by	Leopold Gmelin
8th Edition	8th Edition begun under the auspices of the Deutsche Chemische Gesellschaft by R. J. Meyer
Continued by	E. H. E. Pietsch and A. Kotowski, and by Margot Becke-Goehring

Springer-Verlag Berlin Heidelberg GmbH 1989

Gmelin-Institut für Anorganische Chemie
der Max-Planck-Gesellschaft zur Förderung der Wissenschaften

Volumes published on Platinum Metals

Ir * Iridium Main Volume – 1939
      * Iridium Suppl. Vol. 1 (Metal, Alloys) – 1978
      ° Iridium Suppl. Vol. 2 (Compounds) – 1978

Os * Osmium Main Volume – 1939
      Osmium Suppl. Vol. 1 – 1980

Pd * Palladium 1 (Element) – 1941
      * Palladium 2 (Compounds) – 1942
      Palladium Suppl. Vol. B 2 (Compounds) – 1989 (present volume)

Pt * Platinum A 1 (History, Occurrence) – 1938
      * Platinum A 2 (Occurrence) – 1939
      * Platinum A 3 (Preparation of Platinum Metals) – 1939
      * Platinum A 4 (Detection and Determination of the Platinum Metals) – 1940
      * Platinum A 5 (Alloys of Platinum Metals: Ru, Rh, Pd) – 1949
      * Platinum A 6 (Alloys of Platinum Metals: Os, Ir, Pt) – 1951
      * Platinum B 1 (Physical Properties of the Metal) – 1939
      * Platinum B 2 (Physical Properties of the Metal) – 1939
      * Platinum B 3 (Electrochemical Behavior of the Metal) – 1939
      * Platinum B 4 (Electrochemical Behavior and Chemical Reactions of the Metal) – 1942
      * Platinum C 1 (Compounds up to Platinum and Bismuth) – 1939
      * Platinum C 2 (Compounds up to Platinum and Caesium) – 1940
      * Platinum C 3 (Compounds up to Platinum and Iridium) – 1940
      * Platinum D (Complex Compounds of Platinum with Neutral Ligands) – 1957
      Platinum Suppl. Vol. A 1 (Technology of Platinum-Group Metals) – 1986
      Platinum Suppl. Vol. A 2 – 1989

Rh * Rhodium Main Volume – 1938
      Rhodium Suppl. Vol. B 1 (Compounds) – 1982
      Rhodium Suppl. Vol. B 2 (Coordination Compounds) – 1984
      Rhodium Suppl. Vol. B 3 (Coordination Compounds) – 1984

Ru * Ruthenium Main Volume – 1938
      * Ruthenium Suppl. Vol. – 1970

* Completely or ° partially in German

Gmelin Handbook
of Inorganic Chemistry

8th Edition

Pd
Palladium

Supplement Volume B 2

Palladium Compounds

With 87 illustrations

AUTHORS

William P. Griffith, Imperial College of Science and
Technology, London, Great Britain

Stephen D. Robinson, Kings College, London, Great Britain

Kurt Swars, Gelnhausen, Bundesrepublik Deutschland

EDITORS

William P. Griffith, Imperial College of Science and
Technology, London, Great Britain

Kurt Swars, Gelnhausen, Bundesrepublik Deutschland

System Number 65

Springer-Verlag Berlin Heidelberg GmbH 1989

LITERATURE CLOSING DATE: END OF 1986
IN MANY CASES MORE RECENT DATA ARE CONSIDERED

Library of Congress Catalog Card Number: Agr 25-1383

ISBN 978-3-662-09190-6   ISBN 978-3-662-09188-3 (eBook)
DOI 10.1007/978-3-662-09188-3

© by Springer-Verlag Berlin Heidelberg 1989
Originally published by Springer-Verlag Berlin Heidelberg New York London Paris Tokyo in 1989
Softcover reprint of the hardcover 8th edition 1989

Preface

With platinum and rhodium, palladium is one of the most important members of the platinum metal group. The last Gmelin treatment of it was in 1942, and knowledge of its properties and chemistry has made enormous strides since then.

This volume is primarily concerned with binary compounds and with the coordination complexes derived from them. Although it is a member of the nickel-palladium-platinum triad, it more closely resembles platinum in its binary and coordination chemistry, though being a second-row transition element it displays less tendency than does platinum to assume higher oxidation states. In heterogeneous and homogeneous catalysis, referred to at appropriate points, palladium and its complexes are of great importance in bulk and fine chemicals production, effecting a wide variety of organic transformations.

The arrangement of material in this volume follows the traditional Gmelin arrangement. Within each category of compounds or complexes the material is arranged, as usual, in order of ascending metal oxidation states (e.g., palladium(II) precedes palladium(IV)). The chemistry of the palladium-hydrogen system is so large that it merits a separate volume, so this book starts with the binary oxides and oxopalladates followed by hydroxides, hydroxo complexes and aquo complexes.

Then nitrides and nitrates are treated. They are followed by the large chapters on halides and their complexes (172 pages). The largest single chapter in this volume (110 pages) deals with chlorides, chloropalladates and other chloro complexes.

The palladium-sulphur chapter contains sulphides, sulphito complexes, sulphates and sulphato complexes, some substituted sulphonato complexes, and thiosulphato complexes. Compounds with selenium, tellurium and boron follow.

The next 50 pages deal with palladium-carbon compounds, i.e., mainly with cyano-palladates containing Pd^{II} and Pd^{IV}, and with thiocyanates and thiocyanato complexes.

As this volume is not supposed to report on technological procedures the treatment of crystalline and amorphous Pd–Si phases is limited to the description of the binary system and the properties of defined silicides.

Twenty pages are devoted to the palladium-phosphorus chapter, mainly to binary and polynary phosphides and to complexes containing thiocyanate and phosphines or tertiary phosphites.

Binary and polynary arsines as well as complexes containing thiocyanate and tertiary or chelating arsines are listed in the Pd–As chapter. As the Pd–Sb system is to be treated in the Gmelin supplement A series, only PdSbS and substituted thiocyanato complexes with tertiary and chelating stibines are described on one page.

London/Gelnhausen, April 1989   William P. Griffith, Kurt Swars

Table of Contents

	Page
Compounds of Palladium	1
1 Palladium and Oxygen	1
1.1 Adsorption of Oxygen on Palladium Surfaces	1
1.2 Oxygen Interactions at Palladium Anodes	3
1.3 Palladium Oxides	5
1.3.1 Oxides PdO_{1-x}	6
1.3.2 Palladium Monoxide PdO	6
1.3.3 Palladium Dioxide PdO_2	16
1.3.4 Palladium Trioxide PdO_3	17
1.4 Palladium Dioxygen Complexes	17
1.5 Palladium Aquo Complexes	18
1.6 Palladium Hydroxides	20
1.7 Hydroxopalladates	22
1.7.1 Palladium Hydroperoxide $Pd(OOH)_2$	23
1.7.2 Sodium Hydroxopalladates	23
1.7.3 Potassium Hydroxopalladates	24
1.7.4 Rubidium and Caesium Hydroxopalladates	24
1.7.5 Calcium Hydroxopalladate	25
1.7.6 Strontium Hydroxopalladates	25
1.7.7 Barium Hydroxopalladates	26
1.8 Oxopalladates	26
1.8.1 Tetraoxo Anion $[PdO_4]^-$ (?)	27
1.8.2 Bismuth Palladates	27
1.8.3 Lithium Palladates	28
1.8.4 Sodium Palladates	29
1.8.5 Potassium Palladates	30
1.8.6 Rubidium and Caesium Palladates	31
1.8.7 Magnesium Palladates	31
1.8.8 Calcium Palladate $CaPd_3O_4$	31
1.8.9 Strontium Palladates	32
1.8.10 Barium Palladates	33
1.8.11 Zinc and Cadmium Palladates	33
1.8.12 Indium Palladate $In_2Pd_2O_7$	33
1.8.13 Thallium Palladate $TlPd_3O_4$	34
1.8.14 Scandium and Yttrium Palladates	34
1.8.15 Rare Earth Palladates	34
1.8.16 Titanium Palladates	37
1.8.17 Lead Palladate $PbPdO_2$	37

Page

1.8.18 Vanadium Palladates 38

1.8.19 Chromium Palladate CrPdO$_2$ 38

1.8.20 Cobalt Palladate CoPdO$_2$ 38

1.8.21 Copper Palladates 39

1.8.22 Rhodium Palladates 39

1.8.23 Palladium Palladate Pd$_{0.5}$Pd$_3$O$_4$ 39

2 Palladium and Nitrogen 40

2.1 The Pd–N System 40

2.2 Ternary Nitrides 40

2.3 Palladium Nitrates 40

2.3.1 Palladium(II) Nitrates 41

2.3.2 Palladium(IV) Nitrates 43

3 Palladium and Fluorine 44

3.1 Palladium Monofluoride PdF 44

3.2 Palladium Difluoride PdF$_2$ 44

3.3 Palladium Trifluoride PdF$_3$ 46

3.4 Palladium Tetrafluoride PdF$_4$ 47

3.5 Palladium Pentafluoride PdF$_5$ 48

3.6 Palladium Hexafluoride PdF$_6$ 48

3.7 Fluoropalladates, Polynary Pd Fluorides, and Adducts 49

3.7.1 Fluoropalladates of Nonmetals 50

3.7.2 Alkali-Metal Fluoropalladates 53

3.7.3 Alkaline-Earth Fluoropalladates 59

3.7.4 Fluoropalladates of Zinc, Cadmium and Mercury 61

3.7.5 Miscellaneous Metal Fluoropalladates 62

4 Palladium and Chlorine 66

4.1 The Palladium-Chlorine System 66

4.2 Palladium Dichloride PdCl$_2$ 66

4.2.1 Preparation 67

4.2.2 Physical Properties 68

4.2.3 Chemical Reactions 72

4.2.4 Palladium Dichloride in Organic Synthesis 80

4.2.5 Applications of PdCl$_2$ 83

4.3 Palladium Tetrachloride PdCl$_4$(?) 86

4.4 Palladium Oxochloride Pd$_2$OCl$_2$ 86

4.5 Chloropalladates 87

4.5.1 Chloropalladates(I) 87

Page

4.5.2 Tetrachloropalladates [PdCl₄]²⁻ 87
Vibrational Spectra 87
Electronic Spectra 89
Molecular Orbital (MO) Calculations and Studies on [PdCl₄]²⁻ 92
Stability Constants 92
Thermodynamic Data on [PdCl₄]²⁻ 98
Electrochemical Behaviour 99
Extraction of [PdCl₄]²⁻ 102
Chemical Reactions 105
Biological Effects of [PdCl₄]²⁻ 115
Applications of [PdCl₄]²⁻ 116
Miscellaneous Data for [PdCl₄]²⁻ 117
Tetrachloropalladates of Nonmetals 117
Tetrachloropalladates of Alkali Metals and Ammonium 128
Heavy-Metal Chloropalladates 140

4.5.3 Chloropalladates with Coordination Numbers >4 146

4.5.4 Hexachloropalladates(II) [PdCl₆]⁴⁻ 146

4.5.5 Chloropalladates(III) 147

4.5.6 Hexachloropalladates(IV) [PdCl₆]²⁻ 147

4.5.7 Hexachloropalladate(V) [PdCl₆]⁻ 158

4.5.8 Hexachloropalladate(VI) PdCl₆(?) 158

4.5.9 Binuclear Chloropalladates 159
The [Pd₂Cl₄]²⁻ Ion 159
Salts of [Pd₂Cl₆]²⁻ 159
Vibrational Spectra of [Pd₂Cl₆]²⁻ 159
Electronic Spectra 160
Existence of [Pd₂Cl₆]²⁻ in Solutions 161
Equilibrium Constants 161
Extraction of [Pd₂Cl₆]²⁻ 162
Reactions of [Pd₂Cl₆]²⁻ 162
Salts of [Pd₂Cl₆]²⁻ with Nonmetallic Cations 162
Alkali-Metal Salts 164
Other Binuclear Species 165

4.6 Miscellaneous Multicomponent Systems and Compounds 165

4.7 Hydrido-chloro Complexes 171

4.8 Chloro-aquo Complexes 172

4.9 Chloro-hydroxo Complexes 175

5 Palladium and Bromine 176

5.1 Palladium Dibromide, PdBr₂ 176

5.2 Bromopalladates 178

5.2.1 Bromopalladates(I) 179

5.2.2 Tetrabromopalladates, [PdBr₄]²⁻ 179
Bromopalladates of Nonmetals 187
Tetrabromopalladates of Alkali Metals and Ammonium 191

Page

5.2.3 Bromopalladates(II) with Coordination Number >4 195
5.2.4 Bromopalladates(III) 195
5.2.5 Hexabromopalladates $[PdBr_6]^{2-}$ 196
5.2.6 Binuclear Bromopalladates 198
5.2.7 Ternary Bromopalladates 202
5.2.8 Aquo-bromo Complexes, $[PdBr_n(H_2O)_{4-n}]^{2-n}$ 202
5.2.9 Chloro-bromo Complexes 203

6 Palladium and Iodine 205
6.1 Palladium Iodides 205
6.1.1 Palladium Di-iodide PdI_2 205
6.1.2 Palladium Tetra-iodide PdI_4(?) 208
6.2 Iodopalladates 208
6.2.1 Tetra-iodopalladates 208
6.2.2 Higher Iodopalladates 212
6.2.3 Binuclear Iodopalladates 212
6.2.4 Ternary Iodopalladates 214
6.2.5 Iodo-aquopalladates 215
6.3 Mixed Halo-iodo Complexes 215

7 Palladium and Sulphur 217
7.1 Phase Diagram 217
7.2 Binary Palladium Sulphides 217
7.3 Ternary Palladium Sulphides 222
7.4 Quaternary Palladium Sulphides 228
7.5 Palladium Sulphoxylate Pd_2SO_2(?) 228
7.6 Palladium Sulphito Complexes 228
7.6.1 Unsubstituted Complexes 228
7.6.2 Substituted Sulphito Complexes 230
7.7 Palladium Sulphates and Sulphato Complexes 233
7.7.1 Palladium Sulphate 233
7.7.2 Sulphato Complexes 236
Unsubstituted Species 236
Substituted Sulphato Complexes 237
7.8 Fluorosulphonato (FSO_3^-) Complexes 238
7.9 Chlorosulphonato ($ClSO_3^-$) Complex $Na_2[Pd(SO_3Cl)_4]$ 239
7.10 Thiosulphato ($S_2O_3^{2-}$) Complexes 239

8 Palladium and Selenium 240
8.1 Phase Diagram 240
8.2 Binary Palladium Selenides 240

Page

8.3   Ternary Palladium Selenides 244
   8.3.1   Alkali-Metal Selenides 245
   8.3.2   Other Ternary Selenides 245
8.4   Quaternary Palladium Selenides 247

9   Palladium and Tellurium 248
9.1   Phase Diagram 248
9.2   Binary Palladium Tellurides 249
9.3   Ternary Palladium Tellurides 254

10   Palladium and Boron 256
10.1   Phase Diagram 256
10.2   Binary Palladium Borides 257
10.3   Pd Borides with Other Metals 259
10.4   Palladoboranes 260
10.5   Miscellaneous Compounds 260

11   Palladium and Carbon 262
11.1   Solid Solution PdC_x (x = 0.10 to 0.15) 262
11.2   Palladium Carbides 262
11.3   Cyanides and Cyano Complexes 264
   11.3.1   Palladium Dicyanide $Pd(CN)_2$ 264
11.3.2   Cyano Complexes 265
   Palladium(0) Complexes 265
   Palladium(I) Complexes 266
   Palladium(II) Complexes 266
      Tetracyanopalladates 266
      Palladium(II) Complexes of Coordination Number > 4 280
   Hexacyanopalladates(IV) 281
   Salts of $[Pd(CN)_6]^{2-}$ 282
   Substituted Complexes 285
   Polynuclear Species 287
11.4   Palladium Cyanate(?) and Cyanato Complexes 288
11.5   Thiocyanates 288
   11.5.1   Palladium Thiocyanate $Pd(SCN)_2$ 289
   11.5.2   Palladium Thiocyanato Complexes 289
   The $[Pd(SCN)_4]^{2-}$ Ion 289
   Free Acid and Salts with Organic Cations "$H_2[Pd(SCN)_4]$" 295
   Salts with Alkali Metal Cations 297
   Substituted Complexes 301
      Polymeric Heavy-Metal Thiocyanato Complexes 310

Page

12 Palladium and Silicon 312

12.1   The Pd–Si System 312

    12.1.1   Diffusion 312

    12.1.2   Phase Diagram 313

    12.1.3   Formation and Preparation 315

    12.1.4   Reactions of Pd–Silicides 316

12.2   Dipalladium Monosilicide Pd_2Si 317

    12.2.1   Formation and Preparation 317

    12.2.2   Physical Properties 319

12.3   Monopalladium Monosilicide PdSi 320

    12.3.1   Formation and Preparation 320

    12.3.2   Physical Properties 321

12.4   Other Palladium Silicides 321

13 Palladium and Phosphorus 323

13.1   Pd–P Phase Diagram 323

13.2   Binary Compounds 324

13.3   Polynary Compounds 327

13.4   Palladium Phosphite $Pd(PO_3)_2$ 331

13.5   Palladium Phosphate and Phosphato Complexes 331

    13.5.1   Palladium Phosphate(?) 331

    13.5.2   Phosphato Complexes 331

13.6   Palladium Pyrophosphate and Pyrophosphato Complexes 332

    13.6.1   Palladium Pyrophosphate $Pd_2P_2O_7$(?) 332

    13.6.2   Pyrophosphato Complexes 332

13.7   Miscellaneous Phosphato Compound $PdP_5Cr_{1.5}O_{15.8}$ 333

13.8   Complexes Containing Thiocyanate and Tertiary Phosphines 333

13.9   Complexes Containing Thiocyanate and Chelating Phosphines 336

13.10   Complexes with Thiocyanate and Phosphorus-Containing Chelates 339

    Phosphorus-Nitrogen Donors 339

    Phosphorus-Sulphur Donors 340

13.11   Substituted Thiocyanato Complexes with Tertiary Phosphites 341

14 Palladium and Arsenic 343

14.1   The Pd–As System 343

14.2   Binary Compounds 344

14.3   Ternary Compounds 347

		Page
14.4	Complexes Containing Thiocyanate and Tertiary Arsines	348
14.5	Complexes Containing Thiocyanate and Chelating Arsines	350
15	**Palladium and Antimony**	352
	Substituted Thiocyanato Complexes with Tertiary and Chelating Stibines	352
	Physical Constants and Conversion Factors	353

1 Palladium and Oxygen

This chapter first deals with adsorption of oxygen on palladium surfaces and some reactions of oxygen at palladium anodes, then proceeds to cover palladium oxides, peroxides or dioxygen adducts, aqua ions, hydroxides, hydroxopalladates and finally palladates.

The systems of Pd with **noble gases** and **hydrogen** will be treated in a separate volume.

1.1 Adsorption of Oxygen on Palladium Surfaces

Isobars for adsorption of oxygen on evaporated films of palladium and other metals have been measured [1]. Adsorption on palladium stops short of forming a complete monolayer and the heat of adsorption (-280 kJ/mol O_2) is considerably greater than the heat of PdO formation (-176 kJ/mol O_2). These results indicate that true chemisorption rather than oxide formation occurs [2]. Identifications based on Auger and He^I UPS studies of Pd(100), PdO and Pd$_4$Si suggest that three types of oxide – PdO, "collapsed" or decomposed PdO, and Si-stabilized oxides – persist in the subsurface region of Pd and are responsible for the known surface property modifications induced by oxygen pretreatment of palladium. Auger and He^I UPS spectra are reproduced [3]. Thermal, chemical, structural and electronic data point to the presence of at least three types of oxygen atoms in the surface region of Pd(111). The initial sticking coefficient is 0.3 and the adsorption energy is given as 230 kJ/mol. High temperature ($\sim$1000 K) treatment with oxygen causes formation of a more tightly bound surface species, characterised by a 2×2 LEED pattern, which is chemically stable and is a transition state to PdO. Interaction with NO at 1000 K affords PdO. LEED patterns, thermal desorption spectra (700 to 1100 K) and UV photoelectron spectra are reproduced in the paper [4]. The texture of palladium and its surface oxide has been analysed by electron diffraction and discussed in terms of ordered chemisorption of oxygen [5]. X-ray data for quenched samples provide evidence that the interaction of sub-μ palladium black with oxygen at 180 to 300°C leads to generation of a three-dimensional surface compound ("Pd–O") before formation of palladium oxide (PdO) [6]. Above 250°C a significant amount of oxygen is incorporated into bulk, polycrystalline palladium at pressures far below the dissociation pressure of PdO. Pressure independence of the uptake rate suggests that oxygen penetration into the bulk palladium occurs from an oxygen-saturated surface and is limited by the rate at which chemisorbed oxygen moves out of the surface layer toward the interior of the palladium. The temperature dependence of the uptake rate in this pressure-independent region gave an Arrhenius pre-exponential factor of 5.2×10^{20} atoms $\cdot$ cm$^{-2} \cdot$ min^{-1} and an activation energy of 71 kJ/mol [7]. Oxygen uptake by evaporated palladium films has been shown to occur quite rapidly but equilibrium was never attained and consequently data on rate and extent of adsorption are not available [8]. Oxygen-palladium (100) interactions have been studied by LEED, AES and electron energy loss spectroscopy (ELS) to characterise initial oxidation and elucidate the relation between surface-oxide reconstructions and bulk oxidation. Five oxygen related LEED patterns were identified and classified into three types:

I)   chemisorption structure p(2×2) replaced by c(2×2) upon increased exposure to oxygen at room temperature;

II) surface oxide reconstruction p(5×5) which formed >200°C and was replaced by a ($\sqrt{5} \times \sqrt{5}$) R27° at >300°C;

III) a multilayer oxide, distinct from PdO, formed above 800°C.

LEED pattern and energy loss spectra are reproduced [9].

Thin films of PdO prepared by oxidation of evaporated palladium films at 973 K for 24 h display p-type conduction. Heating cycles (77 to 500 K) in He, N_2 or vacuum caused decrease in conductivity, in contrast heating to ~300 K in H_2 led to rapid increase in conductivity arising from reduction of oxide to palladium metal. The conductivity at >500 K was almost the same for all specimens indicating intrinsic conduction. The energy gap was estimated as ~1.5 eV. Hall coefficient and Hall mobility measured at 300 K in a magnetic field of 3.5 kG were ~5 cm^3/C and ~17 $cm^3 \cdot V^{-1} \cdot s^{-1}$, respectively. Plots of reciprocal temperature versus conductivity are given in the paper [10]. The solubility limit of oxygen in palladium at 1200°C has been given as >0.4 at% [11]. Formation of oxygen chemisorbed palladium atoms (PdO_{ads}) and palladium oxide (PdO) on palladium surfaces exposed to air at 600 to 900°C has been detected by ESCA measurements [12]. The contact resistance and mechanical properties of surface oxide films on palladium have been investigated [13].

Oxidation of palladium deposited on Al_2O_3 has been studied by electron reflection spectroscopy [14]. The interaction of oxygen with palladium black or palladium supported on Al_2O_3 or SiO_2 involves surface chemisorption of oxygen (300 to 470 K), incorporation of oxygen into bulk palladium (470 to 770 K) and complete oxidation of PdO (>770 K) [15]. Oxidation of palladium particles supported on SiO_2 at temperatures in the range 300 to 600°C gives a product of stoichiometry PdO. Gas uptake plots and X-ray diffraction patterns are given in paper [16]. Adsorption of oxygen by reduced Pd/MgO catalysts at 20°C leads to formation of Pd^+ and O_2^- ions, surface Pd^{3+} ions are also generated. The O_2^- ions appear to be associated with the magnesium cations [17].

In contrast O_2^- ions formed under similar conditions on Pd/γ-Al_2O_3 are apparently stabilised by the palladium ions [18]. ESR spectra are reproduced in both papers [17, 18]. Reactions of CO [19 to 21], H_2 [19] and C_2H_4 [19] with oxygen adsorbed on palladium films have been reported.

The heats of chemisorption of oxygen on palladium black and on palladium-silver alloys have been measured in the temperature range 500 to 700 K and for fraction of surface coverage, Θ, between 1×10^{-4} and 0.7. Values (tabulated in paper) range from −102.9 kJ/mol PdO for pure palladium to −225.9 kJ/mol Ag_2O for pure silver [22]. The bond energy for oxygen bound to palladium surfaces has been calculated by a kinetic method [23]. Adsorption of oxygen on palladium film leads to increased resistance, at room temperature and $P_{O_2} = 10^{-2}$ Torr increase is ~8 to 10% [21].

The adsorption of oxygen on polycrystalline palladium and the kinetics of the reaction between adsorbed oxygen and carbon monoxide have been studied by Auger electron spectroscopy. Oxygen adsorption curves (413 to 783 K) and Auger spectra data are reproduced in the paper [24]. A LEED Debye-Waller analysis of vibrational hybridisation for p(2×2)-O/Pd(100) has been reported [25]. The temperature dependence of the 45 eV specular LEED beam intensity has been measured between ~100 to 300°C and ~50 to 200°C for clean p(1×1) and quarter monolayer O-covered p(2×2) Pd(100), respectively [25].

References:

[1] Lanyon, M. A. H.; Trapnell, B. M. W. (Proc. Roy. Soc. [London] A **227** [1955] 387/99).

[2] Brennan, D.; Hayward, D. O.; Trapnell, B. M. W. (Proc. Roy. Soc. [London] A **256** [1960] 81/105).

[3] Bader, S. D.; Richter, L.; Orent, T. W. (Surf. Sci. **115** [1982] 501/12).

[4] Conrad, H.; Ertl, G.; Küppers, J.; Latta, E. E. (Surf. Sci. **65** [1977] 235/44, 245/60).

[5] Shishakov, N. A.; Andrushchenko, N. K.; Asanov, U. A. (Izv. Akad. Nauk SSSR Otd. Khim. Nauk **1961** 1234/40; C. A. **58** [1963] 1191).

[6] Guiot, J. M. (J. Appl. Phys. **39** [1968] 3509/11).

[7] Campbell, C. T.; Foyt, D. C.; White, J. M. (J. Phys. Chem. **81** [1977] 491/4).

[8] Dogoberto, S. M. (Rev. Fac. Quim. Univ. Nac. Mayor San Marcos **17** [1965] 34/48; C. A. **65** [1966] 9778).

[9] Orent, T. W.; Bader, S. D. (Surf. Sci. **115** [1982] 323/34).

[10] Okamoto, H.; Aso, T. (Japan. J. Appl. Phys. **6** [1967] 779; C. A. **67** [1967] No. 85983).

[11] Raub, E.; Plate, W. (Z. Metallk. **48** [1957] 529/39).

[12] Kim, K. S.; Gossmann, A. F.; Winograd, N. (Anal. Chem. **46** [1974] 197/200).

[13] Wilson, R. W. (Proc. Phys. Soc. [London] B **68** [1955] 625/41).

[14] Kravchuk, L. S.; Kupcha, L. A.; Zaretskii, M. V.; Markevich, S. V. (Kinetika Kataliz **17** [1976] 1353; Kinet. Catal. [USSR] **17** [1976] 1173).

[15] Paryjczak, T.; Jóźwiak, W. K.; Góralski, J. (J. Chromatog. **155** [1978] 9/17).

[16] Lam, Y. L.; Boudart, M. (J. Catal. **47** [1977] 393/8).

[17] Sass, A. S.; Shvets, V. A.; Savel'eva, G. A.; Popova, N. M.; Kazanskii, V. B. (Kinetika Kataliz **24** [1983] 1167/72; Kinet. Catal. [USSR] **24** [1983] 994/9).

[18] Shubin, V. E.; Shvets, V. A.; Savel'eva, G. A.; Popova, N. M.; Kazanskii, V. B. (Kinetika Kataliz **23** [1982] 1153/60; Kinet. Catal. [USSR] **23** [1982] 982/8).

[19] Stephens, S. J. (J. Phys. Chem. **63** [1959] 188/194).

[20] Alexander, E. G.; Russell, W. W. (J. Phys. Chem. **68** [1964] 1614/8).

[21] Kawasaki, K.; Sugita, T.; Ebisawa, S. (J. Chem. Phys. **44** [1966] 2313/6).

[22] Bortner, M. H.; Parravano, G. (Advan. Catal. **9** [1957] 424/33).

[23] Kiperman, S. L.; Balandin, A. A. (Kinetika Kataliz Akad. Nauk SSSR Sb. Statei **1960** 159/68; C. A. **57** [1962] 7967).

[24] Matsushima, T.; White, J. M. (Surf. Sci. **67** [1977] 122/38).

[25] Bader, S. D. (Surf. Sci. **99** [1980] 392/404).

1.2 Oxygen Interactions at Palladium Anodes

Anodic passivation of palladium has been reviewed [1]. The anodic passivation of palladium in neutral buffer solutions of 0.5 M NaCl (pH = 6.55, potential range 600 to 1200 mV vs. SCE) has been studied by cathodic charging techniques. Data indicate that compound responsible for initial passivation of palladium under these conditions is Pd^{II} oxide or hydroxide [2]. Coloured films formed on palladium by anodic treatment in aqueous solution, at potentials above that of the evolution of oxygen, contain PdO, PdO_2 and in some cases PdO_3 [3]. The corrosion and passivation of a palladium electrode at 20 to 80°C and at different pH values in the absence of complexing ions has been investigated. Anodic potentiodynamic polarisation curves have been plotted and the activation energy of the process $Pd \rightarrow Pd^{2+} + 2e$ has been calculated: 27.6 kJ/mol at pH = 0; 30.5 kJ/mol at pH = 1. At alkaline pH and high temperature the formation of PdO_2 is favoured over PdO [4]. Formation of oxygen multilayers at a smooth palladium electrode has been demonstrated by the method of charging curves (0.2 to 12 N H_2SO_4 or N alkali, potential range 2 to 4 V vs. SHE, 0 to 60°C) [5]. There is evidence from charging curves that two oxides, PdO and PdO_2, may be formed electrochemically on palladium electrodes [6 to 12]. Equilibrium potentials of palladium oxides in 0.5 M H_2SO_4, phosphate buffer and M NaOH have been measured and compared with estimated values [7]. Oxides on palladium in 10.6 M KOH have been studied by cathodic charging curve method, after preliminary polarisation at different potentials. At −30°C a layer of adsorbed oxygen began to form at $\psi_a = 1.05$ V and increased with ψ_a, a second oxide, probably PdO_2, began to form at $\psi_a = 1.6$ V [10]. The exchange current of chemisorbed oxygen on palladium has been determined and its dependence on the potential, fractional coverage, and pH has been discussed [13].

Adsorption of oxygen on oxide-free palladium (and other noble metal) electrodes immersed in 0.5 M H_2SO_4 has been correlated with the d-electron configuration of the metals concerned. The fraction of the palladium surface covered by oxygen monolayer is given as 0.22 [14]. Anodic passivation layers deposited on palladised Pt electrode suspended in 0.5 M H_2SO_4 solution contain $PdO_2 \cdot x H_2O$ which is progressively reduced to PdO and then Pd by 0.5 M H_2SO_4 containing 1 to 10 M HCO_2H [15]. Structural changes in surface oxide layers on palladium during anodic formation and cathodic reduction have been investigated in per-chloric acid solutions using cyclic voltammetry and linear sweep polarisation. Cyclic voltam-mograms and plots of cathode charge amounts, Δq, as a function of E_a, waiting time t_w at different E_a, and the reduction peak potential E_p are reproduced in the paper [16].

Oxidation of a palladium anode during electrochemical evolution of oxygen in 0.1, 1.0 and 10 M KOH has been studied by an oscillographic method. Data indicate that PdO is significant-ly more stable than PdO_2. The stability of both oxides increases with KOH concentration [17]. X-ray photoelectron spectroscopic studies of oxidised palladium electrodes in 1 N H_2SO_4 reveal PdO and PdO_2 at potentials above +0.90 V S.H.E. Oxide coatings range from ~5 Å at 0.9 V to >40 Å at 1.7 V. Spectra indicate the presence of adsorbed H_2O, hydrated PdO, PdO_2 and $Pd(OH)_2$ or $Pd(OH)_4$ [18]. The chemisorption of oxygen on Pd electrodes in 0.1 N H_2SO_4 and in KOH has been studied as a function of time at different potentials and degrees of surface coverage [19]. Rates of reduction of oxygen on Pd and PdO electrodes in KOH and 1 M H_2SO_4 solutions have been compared. The rate of oxygen reduction on the PdO electrode, with account being taken of the surface, is lower than on a palladium metal electrode. An anode of PdO has a low anodic potential and high corrosion resistance in 2 N H_2SO_4 solution [20]. The process of molecular oxygen ionisation on palladium has been studied, cathode curves of oxygen ionisation were measured up to $\psi = 0.2$ to 0.3 V and a two stage mechanism proposed: 1) $O_2 \rightarrow H_2O_2$; 2) $H_2O_2 \rightarrow H_2O + OH^-$. Surface Pd oxides slow down the rate of 1) but increase the rate of 2) [21]. Coatings formed on palladium by galvanostatic anodisation have been shown by reflection electron diffraction technique to consist of poorly crystalline PdO [22]. Anodic polarisation curves for PdO and other transition metal oxides have been measured in acid solution at 10 to 50°C. The activation energy for oxygen evolution from PdO increases linearly with overpotential [23]. The fundamental properties of p-PdO for photoelectrochemi-cal and transport data rationalised the observation that the preferred Faradaic route for photogenerated electrons at the surface of PdO is the generation of Pd rather than hydrogen cal and transport data rationalized the observation that the preferred faradaic route for photogenerated electrons at the surface of PdO is the generation of Pd rather than hydrogen [24]. The electron exit value for oxygen-bearing palladium surfaces has been determined as 5.97 to 6.32 V [25]. The electrode system Pd/PdO, O^{2-} in fused LiCl/KCl eutectic behaves reversibly, data for the potential of the Pd/PdO electrode at various concentrations of added O^{2-} are tabulated [26].

References:

[1] Genesca, J.; Victori, L. (Rev. Coat. Corros. **4** [1981] 325/48).
[2] Tordesillas, I. M.; Victori, L.; Chao, J. M. (Anales Quim. **71** [1975] 642/6; C.A. **84** [1976] No. 157164).
[3] Genesca, J.; Victori, L. (Afinidad **38** [1981] 205/8; C.A. **95** [1981] No. 140685).
[4] Fortuny, R.; Genesca, J.; Victori, L. (Afinidad **38** [1981] 91/5; C.A. **95** [1981] No. 122799).
[5] Flerov, V. N.; Battalova, Yu. V.; Tyurin, Yu. M. (Elektrokhimiya **10** [1974] 502; Soviet Electrochem. **10** [1974] 488).
[6] El Wakkad, S. E. S.; Shams El Din, A. M. (J. Chem. Soc. **1954** 3094/8).
[7] Hickling, A.; Vrjosek, G. G. (Trans. Faraday Soc. **57** [1961] 123/9).
[8] Blackburn, T. R.; Lingane, J. J. (J. Electroanal. Chem. **5** [1963] 216/35).
[9] Will, F. G.; Knorr, C. A. (Z. Elektrochem. **64** [1960] 270/5).
[10] Mazitov, Yu. A.; Rosental', K. I.; Veselovskii, V. I. (Zh. Fiz. Khim. **38** [1964] 151/5; Russ. J. Phys. Chem. **38** [1964] 79/81).

[11] Vetter, K. J.; Berndt, D. (Z. Elektrochem. **62** [1958] 378/86).
[12] Burshtein, R. Kh.; Tarasevich, M. R.; Vilinskaya, V. S. (Elektrokhimiya **3** [1967] 349/55; Soviet Electrochem. **3** [1967] 301/6).
[13] Tarasevich, M. R.; Bogdanovskaya, V. A.; Vilinskaya, V. S. (Elektrokhimiya **7** [1971] 1684/9; Soviet Electrochem. **7** [1971] 1626/9).
[14] Rao, M. L. B.; Damjanovic, A.; Bockris, J. O'M. (J. Phys. Chem. **67** [1963] 2508/9).
[15] Genin, G.; Capel-Boute, C.; Decroly, C. (Compt. Rend. **259** [1964] 4660/2).
[16] Chierchie, T.; Mayer, C.; Lorenz, W. J. (J. Electroanal. Chem. Interfacial. Electrochem. **135** [1982] 211/20).
[17] Gorodetskii, Yu. S.; Chesalov, Yu. F. (Reaktsionnosposobn. Koord. Polim. Soedin. **1982** 123/7; C.A. **100** [1984] No. 199683).
[18] Kim, K. S.; Gossmann, A. F.; Winograd, N. (Anal. Chem. **46** [1974] 197/200).
[19] Tarasevich, M. R.; Bogdanovskaya, V. A.; Vilinskaya, V. S. (Elektrokhimiya **8** [1972] 89/93; Soviet Electrochem. **8** [1972] 87/90).
[20] Razina, N. F.; Gur'eva, L. N. (Izv. Akad. Nauk Kaz. SSR Ser. Khim. **1983** 21/3; C.A. **99** [1983] No. 29840).

[21] Sobol, V. V.; Khrushcheva, E. I.; Dagaeva, V. A. (Elektrokhimiya **1** [1965] 1332/8; Soviet Electrochem. **1** [1965] 1193/8).
[22] Masae, S.; Shigeo, S. (Denki Kagaku Oyobi Kogyo Butsuri Kagaku **44** [1976] 425/6; C.A. **86** [1977] No. 35645).
[23] Inai, M.; Iwakura, C.; Hideo, T. (Denki Kagaku Oyobi Kogyo Butsuri Kagaku **48** [1980] 173/9; C.A. **93** [1980] No. 83502).
[24] Dare-Edwards, M. P.; Goodenough, J. B.; Hamnett, A.; Katty, A.; Ramsden, M. H.; Trevellick, P. R. (Sol. Energy R & D Eur. Community D **2** [1983] 100/9; C.A. **99** [1983] No. 125611).
[25] Giner, J.; Lange, E. (Naturwissenschaften **40** [1953] 506).
[26] Laitinen, H. A.; Bhatia, B. B. (J. Electrochem. Soc. **107** [1960] 705/10).

1.3 Palladium Oxides

The most important of these is the monoxide PdO, the dioxide PdO_2 is also firmly established and a trioxide PdO_3 has been mentioned. Earlier reports concerning a sesquioxide $Pd_2O_3 \cdot x H_2O$ remain unconfirmed.

General Reference:

Arai, H.; Tsuda, N.; Ruthenium, Rhodium, Palladium, Osmium, Iridium, Platinum Oxides, Kinzoku Sankabutsu to Fukugo Sankabutsu **1978** 270/90.

1.3.1 Oxides PdO_{1-x}

Catalyst particles of oxygen-deficient stoichiometry have been obtained by tribochemical reactions between Pd and PdO. For $x = 0$ the material is an insulator but for $x > 0$ semiconductor properties are displayed. Electron diffraction data and photographs are presented in the paper. PdO_{1-x} is an efficient catalyst for hydrocarbon oxidation; catalytic activity for gasoline combustion has been determined for different temperatures and periods of time [1]. Formation of nonstoichiometric PdO_{1-x} on palladium sheet heated to 800°C has been detected by electron diffraction, data and photographs are included in the paper [2].

References:

[1] Yamaguchi, S. (Mater. Chem. **5** [1980] 257/66).
[2] Yamaguchi, S. (Z. Anal. Chem. **303** [1980] 409/11).

1.3.2 Palladium Monoxide PdO (see "Palladium" 1942, p. 259)

Samples of PdO free from metallic palladium have been obtained by heating finely divided $PdCl_2$ in a $NaNO_3$ melt at 520°C [1]. Thermolysis of $Pd(NO_3)_2 \cdot 2H_2O$ affords PdO as the final product [2]. A process for the preparation of PdO by electrolysis of molten $NaNO_3$/LiCl melts using a palladium anode has been reported in the patent literature [3, 4]. Heating finely divided palladium metal (obtained from $Pd(NO_3)_2$/HCHO) in air at 600°C for 10 h [5] or 700 to 750°C for 15 to 20 h [6] affords PdO. Vaporisation of palladium metal in a direct carbon arc at 50 V and 3000°C produces a mist of PdO [7]. Palladium oxide films of good optical quality have been obtained by oxidation of sputtered palladium films [8]. Preparation of PdO by heating palladium foil or powder in oxygen (160 Torr) at 400 to 700°C [9] and growth of PdO crystals by vapour transport have been described [10]. Chemical and electrochemical formation of PdO films on palladium surfaces is covered in Sections 1.1 and 1.2.

A hydrated form **PdO·xH₂O** is produced when aqueous $Pd(OH)_2$ or $Pd(NO_3)_2$ solutions are boiled; dehydration to PdO occurs at 90°C [11].

Two types of PdO crystal have been described: glistening dark green crystals and crystals with a black tinge (possibly Pd-black contamination). In addition PdO specimens obtained by oxidation of metallic palladium are reported to be blue-green in colour [12].

References:

[1] Waser, J.; Levy, H. A.; Peterson, S. W. (Acta Cryst. **6** [1953] 661/3).
[2] Shorokhov, N. A.; Vashman, A. A.; Samsonov, V. E.; Chuklinov, R. N.; Shmidt, V. S. (Zh. Neorgan. Khim. **27** [1982] 3137/40; Russ. J. Inorg. Chem. **27** [1982] 1773/6).
[3] Johnson Matthey and Co. Ltd. (Neth. Appl. 65-06719; C.A. **64** [1965] 13767).
[4] Steele, R.; J. Bishop and Co. Platinum Works (U.S. 3357904; C.A. **68** [1968] No. 55996).
[5] Shaplygin, I. S.; Bromberg, A. V.; Sokol, V. A. (Zh. Neorgan. Khim. **15** [1970] 2305; Russ. J. Inorg. Chem. **15** [1970] 1195).
[6] Goncharenko, G. I.; Lazarev, V. B.; Shaplygin, I. S. (Zh. Neorgan. Khim. **30** [1985] 3032/7; Russ. J. Inorg. Chem. **30** [1985] 1723/6).

[7] Amick, J. A.; Turkevich, J. (Ultrafine Particles **1963** 146/55).
[8] Rey, E.; Kamal, M. R.; Miles, R. B.; Royce, B. S. H. (J. Mater. Sci. **13** [1978] 812/6).
[9] Tardy, M.; Bozon-Verduraz, F. (Compt. Rend. C **280** [1975] 317/20).
[10] Rogers, D. B.; Shannon, R. D.; Gillson, J. L. (J. Solid State Chem. **3** [1971] 314/6).

[11] Glemser, O.; Peuschel, G. (Z. Anorg. Allgem. Chem. **281** [1955] 44/53).
[12] Gagarin, S. G.; Teterin, Yu. A.; Kovtun, A. P.; Gubskii, A. L. (Zh. Neorgan. Khim. **28** [1983] 2750/5; Russ. J. Inorg. Chem. **28** [1983] 1561/4).

1.3.2.1 Structural Properties

The tetragonal structure of PdO has been confirmed by X-ray and neutron diffraction data, crystals belong to the space group D_{4h}^9-$P4_2/mmc$ with $a = 3.03 \pm 0.01$, $c = 5.33 \pm 0.02$ Å, $Z = 2$; atomic positions are $2 \times$ Pd at 0, 0, 0 and $\frac{1}{2}$, $\frac{1}{2}$, $\frac{1}{2}$ and $2 \times$ O at $\frac{1}{2}$, 0, $\frac{1}{4}$ and $\frac{1}{2}$, 0, $\frac{3}{4}$ [1]. Neutron and X-ray diffraction patterns are reproduced and spacing and intensity data are tabulated [1]. There is evidence that two tetragonal forms of PdO may exist – PtS type and PbO type – with very similar cell constants [2]. Changes in the lattice constants with temperature (0 to 800°C) have been plotted and tabulated [3, 8]. The Madelung constant for the PdO structure has been computed as 1.604935 [4]. Madelung potentials for PdO obtained from calculations within the point charge approximation (22.613, -23.228 eV) and from an empirical expression (23.341, -23.341 eV) have been reported [5]. Dissolution of CuO in PdO occurs with retention of the basic lattice but contraction of unit cell dimensions, plots of cell dimensions versus composition are reproduced in the paper [6]. The X-ray diffraction pattern of hydrated palladium oxide PdO·xH$_2$O is similar to that of PdO, but with larger spacings, values of cell constants as a function of xH$_2$O range from $a = 3.036$, $c = 5.327$ Å (0% H$_2$O) to $a = 2.994$, $c = 5.403$ Å (8.8% H$_2$O), all ± 0.005 Å [7].

References:

[1] Waser, J.; Levy, H. A.; Peterson, S. W. (Acta Cryst. **6** [1953] 661/3).
[2] Gagarin, S. G.; Teterin, Yu. A.; Kovtun, A. P.; Gubskii, A. L. (Zh. Neorgan. Khim. **28** [1983] 2750/5; Russ. J. Inorg. Chem. **28** [1983] 1561/4).
[3] Bayer, G.; Wiedemann, H. G. (Thermochim. Acta **11** [1975] 79/88).
[4] Sakamoto, Y. (J. Chem. Phys. **28** [1958] 164/5, errata 733, 1253).
[5] Broughton, J. Q.; Bagus, P. S. (J. Electron Spectrosc. Relat. Phenom. **20** [1980] 261/80).
[6] Schmahl, N. G.; Eikerling, G. F. (Z. Physik. Chem. [Frankfurt] **62** [1968] 268/79).
[7] Glemser, O.; Peuschel, G. (Z. Anorg. Allgem. Chem. **281** [1955] 44/53).
[8] Bayer, G.; Wiedemann, H. G. (Arch. Hutn. **22** [1977] 3/13).

1.3.2.2 Spectroscopic Properties

The electronic spectrum of PdO, measured by diffuse reflection technique, contains absorption bands at $\lambda = 435$, 335 and 260 nm attributed to $^1A_{2g} \leftarrow {}^1A_{1g}$, $^1E_{1g} \leftarrow {}^1A_{1g}$ and $^1B_{1g} \leftarrow {}^1A_{1g}$, respectively. Spectra (200 to 2000 nm) taken at various temperatures are reproduced in the paper [1]. The far infrared spectrum of PdO (800 to 200 cm^{-1}, reproduced in the paper) shows strong bands at 660, 600 and 170 cm^{-1}, the first two of which are each split into two components with a separation of 5 to 8 cm^{-1} [2]. As part of a study of the effect of the chemical bonding on the L$\alpha_{1,2}$, X-ray lines of palladium in chemical compounds spectra have been obtained for PdO [3]. The effects of chemical combination on the X-ray L$_{III}$ absorption limit of palladium have been studied (spectra reproduced in paper), wavelength and energy of L_{III} absorption limit for PdO are 3907.5 mÅ and 3172.9 eV, respectively [4].

A comparison of X-ray photoelectron spectra for samples of PdO prepared by different methods with theoretical results on models of different structural types (PtS, PbO) has indicated the existence of two forms of PdO. Calculated and experimental spectra are reproduced in [5].

See also following section for references to use of X-ray and UV photoelectron spectra in determination of PdO electronic structure.

References:

[1] Tardy, M.; Bozon-Verduraz, F. (Compt. Rend. C **280** [1975] 317/20).
[2] Goncharenko, G. I.; Lazarev, V. B.; Shaplygin, I. S. (Zh. Neorgan. Khim. **30** [1985] 3032/7; Russ. J. Inorg. Chem. **30** [1985] 1723/6).
[3] Genchev, D. (Dokl. Bolg. Akad. Nauk **30** [1977] 821/3; C.A. **87** [1977] No. 159499).
[4] Kawata, S.; Mafda, K. (J. Phys. F **3** [1973] 167/78).
[5] Gagarin, S. G.; Teterin, Yu. A.; Kovtun, A. P.; Gubskii, A. L. (Zh. Neorgan. Khim. **28** [1983] 2750/5; Russ. J. Inorg. Chem. **28** [1983] 1561/4).

1.3.2.3 Electronic Structure

The electronic structure of PdO has been studied by calculations on a $Pd_6O_8^{4-}$ model within the Xα scattered wave method with finite boundary conditions [1, 2]. Results together with X-ray photoelectron spectra indicate existence of two forms of PdO (PtS and PbO types). A molecular orbital diagram for the clusters $Pd_6O_8^{4-}$ modelling the structures of the two forms is reproduced in the paper, together with calculated and experimental X-ray electron spectra for both forms [1]. The electronic structure of palladium oxide has been investigated by photo-emission (XPS and UPS) studies on polycrystalline surfaces of PdO. The spectra (reproduced in paper) show that the energy bands of PdO are narrow (~ 2 to 3 eV) and that the energy shift of the core levels ($|\Delta E_B^F| = 2$ eV) is important. These results suggest that the correlation between the d-electrons may be important in PdO. The valence band of PdO, which differs significantly from that of Pd, can be built up by four structures located at 0.5, 2.2, 4.5 and 6.5 eV below E_F [3]. The local electronic structure of PdO crystal and PdO catalyst supported on SiO_2 and γ-Al_2O_3 has been investigated by X-ray absorption near edge structure (XANES) using synchrotron radiation. The local structure of PdO on catalyst supports is different from that in PdO crystals and evidence of structural disorder in the former is reported. L_3 and L_1 XANES spectra of PdO and Pd are reproduced in the paper together with a scheme of core levels of PdO and Pd measured by XPS [4].

References:

[1] Gagarin, S. G.; Teterin, Yu. A.; Kovtun, A. P.; Gubskii, A. L. (Zh. Neorgan. Khim. **28** [1983] 2750/5; Russ. J. Inorg. Chem. **28** [1983] 1561/4).
[2] Gagarin, S. G.; Gubskii, A. L.; Kovtun, A. P.; Krichko, A. A.; Sachenko, V. P. (Kinetika Kataliz **23** [1982] 585/90; Kinet. Catal. [USSR] **23** [1982] 491/6).
[3] Holl, Y.; Krill, G.; Amamou, A.; Legare, P.; Hilaire, L.; Maire, G. (Solid State Commun. **32** [1979] 1189/92).
[4] Davoli, I.; Stizza, S.; Bianconi, A.; Benfatto, M.; Furlani, C.; Sessa, V. (Solid State Commun. **48** [1983] 475/8).

1.3.2.4 Kinetic Data

The decomposition kinetics of polydisperse PdO have been studied by the isothermal variant of the gas volumetric method at 780 to 850°C and a residual pressure of 53 Torr. The temperature variation of the rate constant is described by the equation $\log k[\min^{-1}] = -(10193 \pm 611)/T + (8.7344 \pm 0.5613)$ and the activation energy is given as (195.11 ± 11.69) kJ/mol PdO [1]. Plots of extent of reaction against time for various temperatures in the range 780 to 850°C are reproduced in the paper [1]. A later paper by the same authors reports an activation energy of 201.25 ± 13.0 kJ/mol PdO [2]. The kinetics of palladium black oxidation in air and PdO reduction in hydrogen have been investigated over the temperature range 848 to 923 K, activation energies are 74.0 and 208.0 kJ/mol, respectively [3].

An earlier report gives a value of 24.2 kcal/mol (101.3 kJ/mol) for the activation energy of palladium oxidation in air at 800 to 1400°C [4]. The kinetics of exchange between PdO and $^{18}O_2$ indicate surface layer exchange by a molecular reaction, the activation energy is 84 kJ/mol [5]. The reaction of PdO with HCl to form $PdCl_2$ proceeds at temperatures below 500°C and has an activation energy of 100.3 kJ/mol at 400.75°C, above 500°C Pd metal forms [6]. The kinetics of reduction of PdO by hydrogen have been investigated [7]. The kinetics of chlorate formation at a dimensionally stable PdO anode in 4 M NaCl solution have been followed by ultraviolet spectroscopy [8].

References:

[1] Kislyakova, G. V.; Komlev, G. A.; Tomskikh, I. V. (Zh. Fiz. Khim. **53** [1979] 494; Russ. J. Phys. Chem. **53** [1979] 273).

[2] Kislyakova, G. V.; Komlev, G. A.; Tomskikh, I. V. (Izv. Akad. Nauk SSSR Metally **1980** No. 6, pp. 70/1; Russ. Met. **1980** No. 6, pp. 66/7; C. A. **94** [1981] No. 72255).

[3] Coombs, P. G.; Munir, Z. A. (Precious Met. Min. Extr. Process Proc. Intern. Symp., Los Angeles 1984, pp. 567/81; C. A. **101** [1984] No. 11173).

[4] Phillips, W. L. (Am. Soc. Metals Trans. Quart. **57** [1964] 33/7; C. A. **60** [1964] 12979).

[5] Winter, E. R. S. (J. Chem. Soc. A **1968** 2889/902).

[6] Ivashentsev, Ya. I.; Ryumin, A. I. (Izv. Vysshikh Uchebn. Zavedenii Tsvetn. Metall. **1978** No. 5, pp. 70/4; C. A. **90** [1979] No. 27744).

[7] Verhoeven, W.; Delmon, B. (Bull. Soc. Chim. France **1966** 3065/73).

[8] Tasaka, A.; Tojo, T.; Kaya, M.; Yamashita, M. (Denki Kagaku Oyobi Kogyo Butsuri Kagaku **51** [1983] 197/8; C. A. **98** [1983] No. 224168).

1.3.2.5 Thermodynamic Data

Thermodynamic data for formation of PdO are collected in the table on p. 10. The discrepancy between values of $^f\Delta H^\circ_{298}$ PdO(c) quoted in the older literature (ca. -88 kJ·mol^{-1}) and those obtained in more recent work (ca. -121 kJ·mol^{-1}) has been attributed to the existence of two slightly different structural forms of PdO (see p. 8) [1].

Thermodynamic data for PdO(g) include $^f\Delta H^\circ_{298} = 349 \pm 12$ kJ/mol, $^f\Delta S^\circ_{298} = 78.7 \pm 10.5$ J·mol^{-1}·deg^{-1} [13] and $^f\Delta H^\circ_{298} = 334.7$ kJ/mol [14]. A recent survey gives for the dissociation energy and standard enthalpy of formation of gaseous PdO estimated values of 376 ± 84 kJ/mol and 245 ± 84 kJ/mol, respectively. Gibbs energy and enthalpy data for gaseous PdO over the temperature range 100 to 4000 K have been tabulated [17]. Another source gives values of 293 and 3625 kJ/mol, respectively, for homolytic $(Pd + \frac{1}{2}O_2)$ and heterolytic $(Pd^{2+} + O^{2-})$ dissociation of gaseous PdO [14].

Expressions for the standard free energy of formation of PdO(s) are listed below.

ΔG_f° (in J/mol) $= -112789.8 + 99.81T \pm 274$ [7]

ΔG_f° (in J/mol) $= -(113900 \pm 750) + (99.9 \pm 1.5)T$ [6]

ΔG_f° (in J/mol) $= -114893 + 100T \pm 1170$ [4]

ΔG_f° (in J/mol) $= -(109500 \pm 880) + (96.02 \pm 0.79)T$ [10]

The free energy function of PdO(s) has been given as $-40.81 \, J \cdot K^{-1} \cdot mol^{-1}$ (500 K) and $-58.48 \, J \cdot K^{-1} \cdot mol^{-1}$ (1000 K). Plots of ΔG_f° and ΔH_f° versus temperature (700 to 1100 K) are reproduced in the paper [7]. Thermodynamic parameters for the reaction $Pd + \frac{1}{2}O_2 = PdO$ are tabulated for temperatures in the range 830 to 1120 K [10].

The heat capacity, C_p, for PdO is estimated to be $35.35 + (1.26 \times 10^{-3})T - (3.47 \times 10^5) T^{-2} \, J \cdot K^{-1} \cdot mol^{-1}$ [13].

Theoretical E° values for solid or molten PdO range from 0.312 V (25°C) to -0.138 V (1000°C) [15]. Measurements of the thermal stability of PdO films in the presence of a low partial pressure of oxygen indicate a dissociation enthalpy of -1.17 ± 0.06 eV [16]. A Pourbaix-Ellingham diagram for the formation of PdO is reproduced [6].

Thermodynamic data for PdO:

method	$\Delta_f H_T^\circ$ in kJ/mol	$\Delta_f S_T^\circ$ in $J \cdot mol^{-1} \cdot deg^{-1}$	T in K	S_{298}° in $J \cdot mol^{-1} \cdot deg^{-1}$	Ref.
transpiration data	−108	−94	950	—	[2]
transpiration data	−112±8.4	−98.7±8.4	298	41.4±8.4	[2]
static methods	−118.6±1.6	—	298	36±1.2	[3]
EMF	−118±2.1	—	298	36.8±2	[4]
EMF	−115	−100	1050	—	[4]
O_2 partial pressure	−112.5	—	298	41.5	[5]
O_2 partial pressure	−106.7	−93.1	1000	—	[5]
impedance dispersion anal.	−121.01±0.8	—	298	33.5±1.5	[6]
EMF	−116.25±0.4	—	298	—	[7]
	−85.4	—	298	—	[8]
	−87.9±8	−84.9	298	—	[9]
EMF	−109.2±0.89	−95.69±0.84	930	—	[10]
EMF	−109.75±0.80	−96.27±0.67	930	—	[10]
	−117.2	—	298	—	[11]
O_2 partial pressure	−114.5	−100.4	298	—	[12, 18]
O_2 partial pressure	−107.3±3.6	−93.3±3.2	1050	—	[12]

References:

[1] Gagarin, S. G.; Teterin, Yu. A.; Kovtun, A. P.; Gubskii, A. L. (Zh. Neorgan. Khim. **28** [1983] 2750/5; Russ. J. Inorg. Chem. **28** [1983] 1561/4).

[2] Bell, W. E.; Inyard, R. E.; Tagami, M. (J. Phys. Chem. **70** [1966] 3735/6).

[3] Warner, J. S. (J. Electrochem. Soc. **114** [1967] 68/71).

[4] Kleykamp, H. (Z. Physik. Chem. [Frankfurt] **71** [1970] 142/8).

[5] Schmahl, N. G.; Minzl, E. (Z. Physik. Chem. [Frankfurt] **47** [1965] 142/63).

[6] De Bruin, H. J.; Badwal, S. P. S. (J. Solid State Chem. **34** [1980] 133/5).

[7] Mallika, C.; Sreedharan, O. M.; Gnanamoorthy, J. B. (J. Less-Common Metals **95** [1983] 213/20).
[8] Long, L. H. (Quart. Rev. [London] **7** [1953] 134/74).
[9] Brewer, L. (Chem. Rev. **52** [1953] 1/75).
[10] Levitskii, V. A.; Narchuk, P. B.; Kovba, M. L.; Skolis, Yu. Ya. (Zh. Fiz. Khim. **56** [1982] 2405/11; Russ. J. Phys. Chem. **56** [1982] 1474/9).

[11] Tagirov, V. K.; Chizhikov, D. M.; Kazenas, E. K.; Shubochkin, L. K. (Zh. Neorgan. Khim. **21** [1976] 2565/7; Russ. J. Inorg. Chem. **21** [1976] 1411/2).
[12] Bayer, G.; Wiedemann, H. G. (Thermochim. Acta **11** [1975] 79/88).
[13] Norman, J. H.; Staley, H. G.; Bell, W. E. (J. Phys. Chem. **69** [1965] 1373/6).
[14] Glidewell, C. (Inorg. Chim. Acta **24** [1977] 149/57).
[15] Hamer, W. J. (J. Electroanal. Chem. **10** [1965] 140/50).
[16] Rey, E.; Kamal, M. R.; Miles, R. B.; Royce, B. S. H. (J. Mater. Sci. **13** [1978] 812/6).
[17] Pedley, J. B.; Marshall, E. M. (J. Phys. Chem. Ref. Data **12** [1983] 967/1031).
[18] Bayer, G.; Wiedemann, H. G. (Arch. Hutn. **22** [1977] 3/13).

1.3.2.6 Dissociation Temperatures and Pressures

Least squares analysis of data for $\Delta_f G°$ versus temperature obtained from various sources give dissociation temperatures in the ranges 1061 to 1097 K and 1129 to 1149 K for PdO in air and pure oxygen, respectively [1]. Thermogravimetric data give a dissociation range 1035 to 1101 K with T_{max} at 1079 K for PdO under 100 Torr pressure of oxygen [2, 8]. Plots of dissociation pressure versus reciprocal temperature (830 to 1110°C) are reproduced in [2, 3]. Dissociation and partial pressure data for PdO are discussed in a report on high-temperature chemistry of fission product elements [4]. Oxygen pressures generated by PdO dissociation over the temperature range 717 to 797 K have been tabulated and compared with published data [5]. The equilibrium pressure of oxygen for the reaction $2PdO(solid) \rightleftharpoons 2Pd(solid) + O_2(gas)$ has been measured over the temperature range 910 to 1145 K and the following working equation given: $\log P_{O_2} = 31.905 - 6.29 \log T - 14510/T \pm 0.006$ [6].

The equilibrium of PdO decomposition has been measured for oxygen pressures ≦600 atm. By measurement of the depression of the Pd melting point with increase in oxygen pressure the liquidus line representing equilibrium between solid metal-liquid Pd–O solutions has been evaluated. Extrapolation of this liquidus line and of the pressure-temperature relationship for decomposition of the oxide gave a eutectic point for the system at 1400° and 1195 atm. [7].

References:

[1] Mallika, C.; Sreedharan, O. M.; Gnanamoorthy, J. B. (J. Less-Common Metals **95** [1983] 213/20).
[2] Bayer, G.; Wiedemann, H. G. (Thermochim. Acta **11** [1975] 79/88).
[3] Bell, W. E.; Inyard, R. E.; Tagami, M. (J. Phys. Chem. **70** [1966] 3735/6).
[4] Bell, W. E.; Norman, J. H. (GA-5611 [1964]; N.S.A. **21** [1967] No. 46; C.A. **66** [1967] No. 90832).
[5] Tagirov, V. K.; Chizhikov, D. M.; Kazenas, E. K.; Shubochkin, L. K. (Zh. Neorgan. Khim. **21** [1976] 2565/7; Russ. J. Inorg. Chem. **21** [1976] 1411/2).
[6] Warner, J. S. (J. Electrochem. Soc. **114** [1967] 68/71).
[7] Baker, E. H. (Trans. Inst. Min. Metall. C **93** [1984] 64/8; C.A. **101** [1984] No. 233718).
[8] Bayer, G.; Wiedemann, H. G. (Arch. Hütn. **22** [1977] 3/13).

1.3.2.7 Thermogravimetric Data

Thermogravimetric curves for PdO in air [1, 2] and in pure oxygen [2] are reproduced in papers. TG and DTA curves for oxidation of Pd to PdO and subsequent dissociation back to Pd are reproduced in [3].

References:

[1] Bayer, G.; Wiedemann, H. G. (Thermochim. Acta **11** [1975] 79/88).
[2] Mallika, C.; Sreedharan, O. M.; Gnanamoorthy, J. B. (J. Less-Common Metals **95** [1983] 213/20).
[3] Bayer, G.; Wiedemann, H. G. (Arch. Hütn. **22** [1977] 3/13).

1.3.2.8 Electrical and Optical Properties

Studies on thin films [1] and single crystals [2] have established semiconductor character for PdO. Resistivities of PdO films have been measured over the temperature range 77 to 560 K. A transition from extrinsic p-type conduction to intrinsic behaviour occurs as the temperature rises through ~500 K [1]. Electrical resistivity measurements on single crystals over the temperature range 4.2 to 300 K have confirmed the same conductivity of PdO, typical resistivity values at 298 K are 10 to 1000 $\Omega \cdot cm$. Seebeck coefficients are all positive indicating that the predominant current carriers are of the p-type. However, high levels of impurities noted for the PdO samples used may have affected the data collected. A schematic one-electron energy diagram for PdO, consistent with the observed semiconductor character, is reproduced in the paper [2]. Another source reports a resistivity of 0.04 $\Omega \cdot cm$ [3]. Hall [1, 4] and Verdet [4] coefficients of PdO have been reported.

The optical transmittance of thin PdO films has been measured at 0.5 to 5.4 eV. The real and imaginary parts of the refractive index and dielectric function, together with the optical conductivity have been determined by Kramers-Kronig analysis of the results. A small energy gap (0.8 eV) and a high dielectric constant (8.0) are indicated [5]. The value for the energy gap confirms an earlier result (0.6 eV) obtained from diffuse reflectance measurements [6]. However, optical absorption and photoconductivity measurements indicate an extrapolated band gap of 2.13 ± 0.03 eV and 2.67 ± 0.03 eV, respectively [3]. Finally, a value of 1.5 eV has been deduced from electrical conductivity measurements [1] and very low values (0.04 to 0.10 eV) have been attributed to acceptor levels in the band gap caused by various impurities [2].

References:

[1] Okamoto, H.; Aso, T. (Japan. J. Appl. Phys. **6** [1967] 779).
[2] Rogers, D. B.; Shannon, R. D.; Gillson, J. L. (J. Solid State Chem. **3** [1971] 314/6).
[3] Rey, E.; Kamal, M. R.; Miles, R. B.; Royce, B. S. H. (J. Mater. Sci. **13** [1978] 812/6).
[4] Kogutyuk, I. P.; Skripnik, F. V. (Fiz. Elektron. [L'vov] No. 19 [1979] 146/50; C.A. **93** [1980] No. 59042).
[5] Nilsson, P. O.; Shivaraman, M. S. (J. Phys. C **12** [1979] 1423/7).
[6] Hulliger, F. J. (J. Phys. Chem. Solids **26** [1965] 639/45).

1.3.2.9 Solubility

The solubility of PdO in molten LiCl–KCl eutectic has been determined as $(11.1\pm0.5)\times10^{-3}$ M at 485°C [1, 2] and as 9.4×10^{-3} M at 450°C [3] by potentiometric measurements. Results indicate complete dissociation of PdO in the melt [2]. The solubility of PdO in molten $NaNO_3$–KNO_3 eutectic has been given as $<10^{-6}$ M at 254°C [4].

References:

[1] Scrosati, B. (Anal. Chem. **38** [1966] 1588/9).
[2] Scrosati, B. (Ric. Sci. **37** [1967] 829/34).
[3] Laitinen, H. A.; Bhatia, B. B. (J. Electrochem. Soc. **107** [1960] 705/10).
[4] Inman, D. (Electrochim. Acta **10** [1965] 11/20).

1.3.2.10 Uses of PdO

In addition to its widespread use in catalysis of organic reactions, PdO finds extensive applications in metal glaze resistors, gas sensors and specialised electrodes.

Metal Glaze Resistors

The fabrication and characterisation of palladium oxide/silver metal glaze resistors has been extensively described in patents [1, 2, 3] and in the open literature [4 to 7]. The influence of PdO [5], layer composition and firing conditions [4] on the long-term resistance stability of PdO/Ag glazes have been investigated. Resistivities ranging from tens to thousands of ohms can be obtained by changing the Ag/Pd ratio and glaze content [4]. The structure and conduction mechanism of thick resistive PdO/Pd–Ag films have been examined [8, 9]. X-ray phase analysis of thick film resistor structures, including PdO + glaze and PdO + Ag + glaze, has been reported [10].

References:

[1] Boyd, J. R.; Mones, A. H.; Schottmiller, J. C. (U.S. 3372058 [1968]; C.A. **69** [1968] No. 39733).
[2] N.V. Philips Gloeilampenfabrieken (Neth. Appl. 81-02809 [1981/83]; C.A. **98** [1983] No. 136067).
[3] Miller, L. F. (U.S. 3414641 [1968]; C.A. **70** [1969] No. 32691).
[4] Kalita, W.; Rzasa, B. (Pr. Nauk. Inst. Technol. Elektron. Politech. Wroclaw No. 2 [1970] 49/62; C.A. **75** [1971] No. 145049).
[5] Kusy, A. (Pr. Nauk. Inst. Technol. Elektron. Politech. Wroclaw No. 2 [1970] 41/8; C.A. **75** [1971] No. 145046).
[6] Minarsky, E. (Elektrotech. Casopis **34** [1983] 917/28; C.A. **100** [1984] No. 112947).
[7] Kahan, G. J. (IBM J. Res. Develop. **15** [1971] 313/7; C.A. **75** [1971] No. 113639).
[8] Kusy, A. (Thin Solid Films **37** [1976] 281/302).
[9] Kusy, A. (Thin Solid Films **17** [1973] 345/61).
[10] Kalita, W.; Kusy, A. (Elektronika **11** [1970] 521/4; C.A. **75** [1971] No. 26652).

Gas Detectors

The use of PdO and other platinum-group metal oxides in combination with In_2O_3 and SnO_2 to prepare sintered elements for the detection of combustible gases has been described in a series of patents and reports. Preparative details are given in a series of patents [1 to 12]. Gases detected include H_2, CH_4, C_4H_{10}, C_2H_5OH and fuel gases [1, 3 to 6, 10, 12]. Industrial experience with metal-metal-oxide sensors, notably Pd/PdO, for measuring oxygen partial pressure in combustion gases has been discussed [13]. Measurements of changes in both electrical conductance and catalytic activity of PdO at 200 and 500°C in the presence of low concentrations of CO and methane have been reported [14]. The feasibility of using PdO/Pd reference electrodes in Y_2O_3–ZrO_2 solid-electrolyte cells for determining oxygen pressures (10^{-3} to 10^{-10} Pa) in gas mixtures by potentiometry has been studied [15].

References:

[1] Matsushita Electric Works Ltd. (Japan. Kokai Tokkyo Koho 58179347 [1982/83]; C.A. **100** [1984] No. 126321).

[2] Matsushita Electric Works Ltd. (Japan. Kokai Tokkyo Koho 58210558 [1982/83]; C.A. **100** [1984] No. 79149).

[3] Matsushita Electric Works Ltd. (Japan. Kokai Tokkyo Koho 58198751 [1982/83]; C.A. **100** [1984] No. 95768).

[4] Matsushita Electric Works Ltd. (Japan. Kokai Tokkyo Koho 5997047 [1982/84]; C.A. **101** [1984] No. 203541).

[5] Matsushita Electric Works Ltd. (Japan. Kokai Tokkyo Koho 5997048 [1982/84]; C.A. **101** [1984] No. 203540).

[6] Matsushita Electric Works Ltd. (Japan. Kokai Tokkyo Koho 59119251 [1982/84]; C.A. **101** [1984] No. 239548).

[7] Matsushita Electric Works Ltd. (Japan. Kokai Tokkyo Koho 59109851 [1982/84]; C.A. **101** [1984] No. 163013).

[8] Matsushita Electric Works Ltd. (Japan. Kokai Tokkyo Koho 5950354 [1982/84]; C.A. **101** [1984] No. 16474).

[9] Matsushita Electric Works Ltd. (Japan. Kokai Tokkyo Koho 5948648 [1982/84]; C.A. **101** [1984] No. 16475).

[10] Matsushita Electric Works Ltd. (Japan. Kokai Tokkyo Koho 5950353 [1982/84]; C.A. **101** [1984] No. 16546).

[11] Matsushita Electric Works Ltd. (Japan. Kokai Tokkyo Koho 57144452 [1981/82]; C.A. **98** [1983] No. 118810).

[12] Matsushita Electric Works Ltd. (Neth. Appl. 81-04198 [1980/81]; C.A. **97** [1982] No. 94540).

[13] Roettenbacher, R.; Schmidberger, R.; Mattes, G. (VDI-Ber. **509** [1984] 281/3).

[14] Gentry, S. J.; Jones, T. A. (Sens. Actuators **4** [1983] 581/6; C.A. **101** [1984] No. 136006).

[15] Levitskii, V. A.; Narchuk, P. B.; Kovba, M. L.; Skolis, Yu. Ya. (Zh. Fiz. Khim. **57** [1983] 16/22; Russ. J. Phys. Chem. **57** [1983] 8/13).

Miscellaneous Uses

Use of PdO thin films as neutron dosimeters has been suggested following the discovery that fast neutron irradiation drastically lowers resistivity by causing H_2 evolution which reduces PdO to palladium metal [1]. Palladium-based electric contacts with little or no catalytic activity are prepared by forming a composite of PdO and palladium metal through electroplating onto a nickel base from a plating solution containing Pd oxide particles [2]. Use of PdO (and/or PdO_2) to catalyse combination of H_2 and O_2 in sealed battery cells and thus prevent pressure accumulation has been described [3]. Solutions of $Pd(NO_3)_2$ sprayed onto sheet glass at 1200°C produce a PdO-coated glass suitable for use in mirrors [4].

References:

[1] Childers, H. M. (Rev. Sci. Instr. **29** [1958] 1008/10; C.A. **1960** 13883).
[2] Matsushita Electric Works Ltd. (Japan. Kokai Tokkyo Koho 5925999 [1982/84]; C.A. **101** [1984] No. 47266).
[3] Hennigan, T. J.; Donnelly, P. C.; Palandati, C. F. (U.S. 2278174 [1963/66] C.A. **66** [1967] No. 34284; see also correction U.S. 3287174 [1963/66]; C.A. **66** [1967] No. 71950).
[4] Raymond, R. F. (U.S. 2691323 [1954]; C.A. **1955** 2698).

Electrode Fabrication

PdO is extensively used in construction of specialised electrodes. A general fabrication method for M/MO$_x$ electrodes including Pd/PdO has been described [1]. Production of Pd/PdO electrodes suitable for use in pH measurements [2, 3] and in hydrogen-generating cells [4] has been reported. Ti/PdO electrodes, prepared by thermal decomposition of Pd and Ti chlorides in an oxygen containing atmosphere or by oxidation of electrodeposited palladium either with atmospheric oxygen or in a KNO$_2$–KNO$_3$ melt, showed a low overpotential and a high current efficiency for chlorine evolution [5]. The Ti-supported electrodes (Ti/MnO$_x$–PdO$_x$) have relatively low catalytic activities for chlorine evolution, due to their low electrical conductivity, as compared with Ti/PdO$_x$ electrodes [6]. A galvanostatic polarisation curve for Ti/PdO anode in 1M KOH is reproduced in [12].

Fabrication of a nonconsumable PdO-based electrode for brine electrolysis has been described [7]. Electrodes prepared by coating porous platinum with iridium oxide (30 to 99) and palladium oxide (70 to 1 mol%) have excellent mechanical strength and corrosion resistance, and show low overvoltage in electrolysing aqueous solutions containing H$_2$SO$_4$ or low concentrations of chloride [8]. Electrodes consisting of PdO-doped SnO$_2$ on a titanium base have been fabricated and their electrochemical activity assessed [9, 10]. Use of PdO as an electrode catalyst in the chlor-alkali industry has been explored, catalyst suppresses evolution of oxygen, prolongs cell life and reduces voltage necessary [11].

References:

[1] Every, R. L. (J. Electrochem. Soc. **112** [1965] 524/6).
[2] Grubb, W. T.; King, L. H. (Anal. Chem. **52** [1980] 270/3).
[3] Liu, C.-C.; Bocchicchio, B. C.; Overmyer, P. A.; Neuman, M. R. (Science **207** [1980] 188/9).
[4] Lu, W-T. P.; Western Electric Co. Inc. (Braz. Pedido 8004939 [1981]; C.A. **95** [1981] No. 88249).
[5] Bondar, R. U.; Kalinovskii, E. A. (Elektrokhimiya **16** [1980] 1492/6; Soviet Electrochem. **16** [1980] 1223/6).
[6] Morita, M.; Iwakura, C.; Tamura, H. (Denki Kagaku Oyobi Kogyo Butsuri Kagaku **48** [1980] 12/5; C.A. **93** [1980] No. 56838).
[7] Nakamura, A.; Saito, S.; TDK Electronics Co. Ltd. (Japan. Kokai Tokkyo Koho 79-20968 [1977/79]; C.A. **90** [1979] No. 212322).
[8] Ishifuku Metal Industry Co. Ltd. (Japan. Kokai Tokkyo Koho 59116388 [1982/84]; C.A. **101** [1984] No. 237381).
[9] Bondar, R. U.; Sorokendya, V. S.; Kalinovskii, E. A. (Elektrokhimiya **20** [1984] 1369/12; Soviet Electrochem. **20** [1984] 1268/70).
[10] TDK Electronics Co Ltd. (Japan. Kokai Tokkyo Koho 80-97486 [1979/80]; C.A. **93** [1980] No. 227608).

[11] Saito, S. (CEW Chem. Eng. World **15** [1980] 31/4; C.A. **93** [1980] No. 33865).
[12] Tamura, H.; Iwakura, C. (Int. J. Hydrogen Energy **7** [1982] 857/65, 863).

1.3.3 Palladium Dioxide PdO$_2$ (see "Palladium" 1942, p. 266)

There are published data indicating that PdO$_2$ is unstable [1]. However, synthesis has been achieved by heating a 0.5 M mixture of powdered PdO and KClO$_3$ at 950 ± 20°C and 65 kbar pressure for 15 min, then quenching rapidly to room temperature [2]. There are also numerous reports of PdO$_2$ formation by anodic oxidation of palladium [3 to 7] or as an intermediate in the electrochemical oxidation of oxalic or formic acids using palladium electrodes [6, 8]. Palladium dioxide is a black powder which begins to lose oxygen at 65 to 70°C forming an oxygen-deficient phase PdO$_{2-x}$, stable as far as x ≈ 0.2 to 0.24. Oxygen loss also occurs on grinding in the dry state but not when triturated under acetone. On further heating complete decomposition to PdO and oxygen occurs. Thermal analysis curves (TGA, DTA and DTGA) are reproduced in the paper [2]. The stability of anodically deposited PdO$_2$ is increased with increasing pH (KOH concentration) of the electrolysis solution [3, 4].

Palladium dioxide has a rutile structure, space group P4$_2$/mmm with a = 4.483 ± 0.004, c = 3.101 ± 0.003 Å, Z = 2, D$_{pyk}$ = 7.44 and D$_{X-ray}$ = 7.732 g/cm^3, indexing data are tabulated in the paper [2]. The infrared spectrum (1000 to 200 cm^{-1}) of PdO$_2$ is reproduced in papers, ν(PdO) occurs at 715 and 640 cm^{-1}, δ(OPdO) at 310 and 230 cm^{-1} [2, 9]. Palladium dioxide is a semiconductor, specific resistance 6.2 × 10^2 Ω^{-1}·cm^{-1} at 295 K, the temperature dependence of the specific resistance (80 to 300 K) is plotted and the data tabulated in the paper [2].

Palladium dioxide is reported to be chemically stable, it does not dissolve in cold concentrated HNO$_3$ or H$_2$SO$_4$ and is unaffected by solutions of alkalies or ammonia [2]. However, the hydrated form PdO$_2$·nH$_2$O is amphoteric and dissolves in NaOH solution to form [Pd(OH)$_5$]$^-$ species. The solubility isotherm is reproduced, the maximum occurs at a base concentration of 10.2 M for a PdO$_2$·nH$_2$O concentration of 1.607 × 10^{-2} M [10].

The estimated value of the free energy function $-F°-H_{298}°/T$ for PdO$_2$ ranges from 261.5 (298.15 K) to 348.1 J·deg^{-1}·mol^{-1} (3000 K) [11]. The dissociation energy of gaseous PdO$_2$ has been calculated to be not greater than 652.7 kJ/mol [11, 12]. The donor-acceptor properties of PdO$_2$–Al$_2$O$_3$ and PdO$_2$–SiO$_2$ catalysts have been studied by ESR [13].

References:

[1] Sleight, A. W. (Mater. Res. Bull. **3** [1968] 699/704).

[2] Shaplygin, I. S.; Aparnikov, G. L.; Lazarev, V. B. (Zh. Neorgan. Khim. **23** [1978] 884/7; Russ. J. Inorg. Chem. **23** [1978] 488/90).

[3] Gorodetskii, Yu. S.; Chesalov, Yu. F. (Reaktsionnosposobn. Koord. Polim. Soedin **1982** 123/7; C.A. **100** [1984] No. 199683).

[4] Fortuny, R.; Genesca, J.; Victori, L. (Afinidad **38** [1981] 91/5).

[5] Kim, K. S.; Gossmann, A. F.; Winograd, N. (Anal. Chem. **46** [1974] 197/200).

[6] Genin, G.; Capel-Boute, C.; Decroly, C. (Compt. Rend. **259** [1964] 4660/2).

[7] Genesca, J.; Victori, L. (Afinidad **38** [1981] 205/8).

[8] Blackburn, T. R.; Campbell, P. C. (J. Electroanal. Chem. **8** [1964] 145/50).

[9] Goncharenko, G. I.; Lazarev, V. B.; Shaplygin, I. S. (Zh. Neorgan. Khim. **30** [1985] 3032/7; Russ. J. Inorg. Chem. **30** [1985] 1723/6).

[10] Ivanov-Emin, B. N.; Borzova, L. D.; Egorov, A. M.; Sudzhben, D. (Zh. Neorgan. Khim. **19** [1974] 1570/2; Russ. J. Inorg. Chem. **19** [1974] 855/6).

[11] Brewer, L.; Rosenblatt, G. M. (Chem. Rev. **61** [1961] 257/63).

[12] Alcock, C. B.; Hooper, G. W. (Proc. Roy. Soc. [London] A **254** [1960] 551/61).

[13] Bodrikov, I. V.; Khulbe, K. C.; Mann, R. S. (Geterog. Katal. **3** [1975] 474/9; C.A. **90** [1979] No. 21806).

1.3.4 Palladium Trioxide PdO₃

Little is known about this highly unstable oxide of palladium(VI). Electrochemical measurements on Pd in water at 25°C have permitted calculation of the free energy of formation of PdO_3 (100.4 kJ/mol) [1]. Formation of PdO_3 in oxide films during anodic treatment of palladium in aqueous solutions at high potentials has been reported [2].

References:

[1] de Zoubov, N.; van Muylder, J.; Pourbaix, M. (Rappt. Tech. Centre Belge Etude Corros. No. 60 [1957] 1/12; C.A. **1958** 18012).
[2] Genesca, J.; Victori, L. (Afinidad **38** [1981] 205/8).

1.4 Palladium Dioxygen Complexes

Pd(O₂). This species has been trapped in a low-temperature matrix by co-condensing palladium atoms with a dilute $^{16}O_2 : {}^{18}O_2 : Ar = 2:1:600$ mixture at 4.2 to 10 K, and has been detected by infrared spectroscopy. Frequencies are $\nu(^{16}O-^{16}O) = 1023.5$; $\nu(^{18}O-^{18}O) = 966.5$ cm⁻¹. Use of a dilute $^{16}O_2 : {}^{16}O^{18}O, {}^{18}O_2 : Ar = 1:2:1:800$ mixture gave a product displaying three infrared bands, 1023.5, 966.5 and $\nu(^{16}O-^{18}O) = 995.5$ cm⁻¹ with the multiplicity of bands indicating "side-on" rather than "end-on" coordination of dioxygen [1, 2]. Spectra are reproduced in a paper [1]. However, in a later report the infrared frequencies are interpreted in favour of an "end-on" bonded structure with an O–O bond order of ~1.5 [3]. $\nu(Pd-O)$ is reported to occur at 427 cm⁻¹ and the force constants k_{O-O} and k_{Pd-O} are given as 4.15 and 1.35 mdyn/Å, respectively [4]. A normal coordinate analysis of $Pd(O_2)$ has been reported [1].

Pd(O₂)₂. This species has been trapped in a low-temperature matrix by co-condensing Pd atoms with a dilute $^{16}O_2 : {}^{18}O_2 : Ar = 2:1:600$ mixture at 4.2 to 10 K and has been characterised by infrared spectroscopy. On the basis of the multiplicity of infrared absorption bands recorded (see table below) a D_{2h} "spiro" structure has been proposed [1, 2]. Spectra are reproduced and a normal coordinate analysis has been performed [1].

Observed (and calculated) frequencies for isotopic species of the bis(dioxygen) complex $(O_2)Pd(O_2)$:

molecule	frequency in cm⁻¹	assignment
$(^{16}O_2)Pd(^{16}O_2)$	1111.5(1111.1)	ν(O–O)
$(^{16}O_2)Pd(^{16}O-^{18}O)$	1092.2(1092.4)	ν(O–O)
$(^{16}O-^{18}O)Pd(^{16}O-^{18}O)$	1080.5(1080.1)	ν(O–O)
$(^{16}O_2)Pd(^{18}O_2)$	1067.2(1067.3)	ν(O–O)
$(^{16}O-^{18}O)Pd(^{18}O_2)$	1060.5(1060.5)	ν(O–O)
$(^{18}O_2)Pd(^{18}O_2)$	1048.5(1047.9)	ν(O–O)
$(^{16}O_2)Pd(^{16}O_2)$	504.0 (503.7)	ν(Pd–O)

Force constants for $Pd(O_2)_2$ have been evaluated as $k_{O-O} = 5.50$, $k_{Pd-O} = 1.84$ mdyn/Å [1].

The electronic spectrum with $\lambda_{max} \approx 280$ nm and weak shoulder at 320 nm has been recorded (600 to 200 nm) and is reproduced in the paper. Extended Hückel MO calculations have been reported and an energy level diagram for $Pd(O_2)_2$ and the analogous rhodium species is reproduced in the paper [4]. The charge-transfer electronic spectra of a range of metal dioxygen complexes $M(O_2)_2$ including $Pd(O_2)_2$ have been examined [3, 5]. Data recorded for $Pd(O_2)_2$ are ligand-to-metal charge transfer (LMCT) = 4.46 eV or 4.59 eV, electron affinity 19.4 eV and $\nu(O-O) = 1.112$ kK [3]. A plot of LMCT band energies versus electron affinities of M(II) for a range of $M(O_2)_2$ complexes including $Pd(O_2)_2$ is reproduced [5].

References:

[1] Huber, H.; Klotzbücher, W.; Ozin, G. A.; Vander Voet, A. (Can. J. Chem. **51** [1973] 2722/36).
[2] Ozin, G. A.; Vander Voet, A. (Accounts Chem. Res. **6** [1973] 313/8).
[3] Lever, A. B. P. (J. Mol. Struct. **59** [1980] 123/35).
[4] Hanlan, A. J. L.; Ozin, G. A. (Inorg. Chem. **16** [1977] 2848/57).
[5] Lever, A. B. P.; Ozin, G. A.; Gray, H. B. (Inorg. Chem. **19** [1980] 1823/4).

1.5 Palladium Aquo Complexes

"$Pd(H_2O)_2$". A theoretical investigation of the addition of dihydrogen to "$Pd(H_2O)_2$" has been undertaken using multireference C. I. calculations. The presence of the H_2O ligands lowers the atomic d^9s configuration relative to the ground state and thus facilitates sd hybridisation. This in turn leads to formation of covalent Pd–H bonds and complete splitting of the H–H bond in the product, "cis-$PdH_2(H_2O)_2$" [1].

$[Pd(H_2O)_4]^{2+}$. This cation, which is presumed to be present in aqueous perchloric acid solutions of palladium(II), has been isolated as the perchlorate salt by dissolving palladium sponge in concentrated nitric acid, adding perchloric acid (72%) and heating the solution until it fumes strongly, then cooling. It forms deliquescent brown needles. The hydrated cation is stable in aqueous solution only at low pH and in the absence of any coordinating ligands. The pH values of 10^{-2} and 10^{-3} M solutions are 1.8 and 2.8, respectively. The absorption spectrum (1M $HClO_4$) shows a maximum at 382 mμ and a minimum at 310 mμ. On standing for some hours the solution darkens and a new maximum at 268 mμ appears [2]. The cation $[Pd(H_2O)_4]^{2+}$ is also the final product formed when $[Pd(NH_3)_4]^{2+}$ [3] or $[PdF_6]^{2-}$ [4] is treated with aqueous $HClO_4$. The electronic spectrum of $[Pd(H_2O)_4]^{2+}$ (500 to 260 nm) is reproduced in the paper [4].

Oxygen-17 Fourier transform NMR line widths for the coordinated H_2O resonance ($\delta_{H_2O} = -131.8$ ppm) measured at 27.11 and 48.78 MHz have been used to study H_2O exchange at square-planar $[Pd(H_2O)_4]^{2+}$ as a function of temperature (240 to 345 K) and pressure (0.1 to 260 MPa at ~324 K; plot reproduced in paper). An associative activation mechanism is proposed, exchange parameters are $k_{ex}^{298} = (560 \pm 40)\,s^{-1}$, $\Delta H^* = (49.5 \pm 1.9)$ kJ/mol, $\Delta S^* = -(26 \pm 6)$ J·K^{-1}·mol^{-1} and $\Delta V^* = -(2.2 \pm 0.2)$ cm^3/mol for aqueous perchlorate solutions with ionic strength 2.0 to 2.6 M and perchloric acid concentration 0.8 to 1.7 M. The exchange rate for $[Pd(H_2O)_4]^{2+}$ is 1.4×10^6 times faster than that for $[Pt(H_2O)_4]^{2+}$ at 298 K. NMR relaxation rate data are tabulated and presented in graphical form [5]. Stability constant data have been recorded for the species $[Pd(H_2O)_{(4-n)}Cl_n]^{(2-n)+}$ with n = 0 to 4 [6, 7, 8], $[Pd(H_2O)_{(4-n)}(O_2CMe)_n]^{(2-n)+}$ with n = 1 to 3 [9] and $[Pd(H_2O)_3(H_2NC_6H_4X)]^{2+}$ (X = H, m- or p-Me, p-OMe, m-Br, m- or p-NO$_2$) [10]. For details see sections devoted to palladium chloro, carboxylato and amine complexes, respectively.

Rate constants and activation parameters at 25°C for reactions between $[Pd(H_2O)_4]^{2+}$ and various entering ligands, X, and for the hydrolysis of various complexes $[Pd(H_2O)_3X]^{n+}$ (n = 1 or 2) have been reported and given in tables below [5].

Rate constants and activation parameters at 25°C for reaction $Pd(H_2O)_4^{2+} + X^{n-} \xrightarrow{k_x} Pd(H_2O)_3X^{(2-n)+} + H_2O$:

X	k_x^{298} in $M^{-1} \cdot s^{-1}$	k_x/k_{H_2O}	ΔH^* in kJ/mol	ΔS^* in $J \cdot K^{-1} \, mol^{-1}$	ΔV^* in cm^3/mol
$(CH_3)_2SO$	2.6	0.06	58	−44	−10.4
H_2O	$41^{a)}$	1	50	$−48^{a)}$	−2.2
CH_3CN	130	3	43	−61	—
Cl^-	1.83×10^4	4×10^2	42	−25	—
Br^-	9.2×10^4	2×10^3	42	−13	—
I^-	1.1×10^6	3×10^4	12	−75	—

[a)] Rate constant and ΔS^* for exchange of one of the four coordinated water molecules recalculated to second-order units, i.e., $4 k_{ex}^{298}/55$.

Rate constants and activation parameters at 25°C for acid hydrolysis of various complexes $PdX(H_2O)_3^{(2-n)+}$, where X is the leaving ligand, reaction $Pd(H_2O)_3X^{(2-n)+} + H_2O \xrightarrow{k_x} Pd(H_2O)_4^{2+} + X^{n-}$:

complex	X	k_x^{298} in s^{-1}	ΔH^* in kJ/mol	ΔS^* in $J \cdot K^{-1} \, mol^{-1}$	ΔV^* in cm^3/mol
$Pd(H_2O)_3Cl^+$	Cl^-	0.8	59	−55	—
$Pd(H_2O)_3Br^+$	Br^-	0.8	59	−55	—
$Pd(H_2O)_3I^+$	I^-	0.8	55	−50	—
$Pd(H_2O)_3CH_3CN^{2+}$	CH_3CN	24	54	−65	—
$Pd(H_2O)_4^{2+}$	H_2O	560	50	−26	−2.2

The salt $[Pd(H_2O)_4][ClO_4]_2$ reacts with 1,4-bis(2-aminoethyl) 1,4-diazacycloheptane (L) in aqueous ethanol to afford $[PdL][ClO_4]_2$ [11]. Reduction of $[Pd(H_2O)_4][ClO_4]_2$ to metallic palladium by molecular hydrogen is autocatalytic and involves unstable Pd^{II}-hydride intermediates; the activation energy is 81.5 kJ/mol [12]. Additional interactions between alkyl halides and Pd^{2+}, not taken into account by an ordinary $M^+ - S_n2$ model of transition states, are postulated for the hydrolysis of alkyl halides in aqueous $Pd(ClO_4)_2$ solution [13], $[Pd(H_2O)_4]^{2+}$ is rapidly reduced by C_2H_4 to form "$Pd(C_2H_4)$" and "$Pd(C_2H_4)_2$" which protonate to $[PdH(C_2H_4)]^+$ and $[PdH(C_2H_4)_2]^+$, respectively, in aqueous acid solution [14], metallic palladium precipitates if C_2H_4 is pumped off [15]. The catalytic effect of $[Pd(H_2O)_4][ClO_4]_2$ in the oxidation of C_2H_4 to CH_3CHO increases when 1,2-naphthoquinone-4-sulphonic acid is added to the reaction [16]. Electrochemical corrosion of palladium in 1M $HClO_4$ solution at 25°C to yield $[Pd(H_2O)_4]^{2+}$ has been studied, corrosion rates and current factors at different potentials (750 to 1700 mV) are tabulated and the stationary potentiostatic polarisation curve for Pd in 1M $HClO_4$ at 25°C is reproduced [17].

References:

[1] Brandemark, U. B.; Blomberg, M. R. A.; Pettersson, L. G. M.; Siegbahn, P. E. M. (J. Phys. Chem. **88** [1984] 4617/21).
[2] Livingstone, S. E. (J. Chem. Soc. **1957** 5091/2).
[3] De Berry, W. J.; Reinhardt, R. A. (Inorg. Chem. **11** [1972] 2401/4).
[4] Shipachev, V. A.; Zemskov, S. V.; Al't, L. Ya. (Koord. Khim. **6** [1980] 932/5; Soviet J. Coord. Chem. **6** [1980] 472/5).
[5] Helm, L.; Elding, L. I.; Merbach, A. E. (Helv. Chim. Acta **67** [1984] 1453/60).
[6] Kragten, J. (Talanta **27** [1980] 375/7).

[7] Burke, R. L. (AD-758684 [1972] 1/61; C.A. **79** [1973] No. 84027).

[8] Sundaram, A. K.; Sandell, E. B. (J. Am. Chem. Soc. **77** [1955] 855/7).

[9] Yatsimirskii, A. K.; Ryabov, A. D.; Berezin, I. V. (Izv. Akad. Nauk SSSR Ser. Khim. **1976** 1490/3; Bull. Acad. Sci. USSR Div. Chem. Sci. **1976** 1426/9).

[10] Vargaftik, M. N.; German, E. D.; Dogonadze, R. R.; Syrkin, Ya. K. (Dokl. Akad. Nauk SSSR **206** [1972] 370/3; C.A. **78** [1973] No. 8494).

[11] Phillip, A. T. (Australian J. Chem. **22** [1969] 259/62).

[12] Fasman, A. B.; Golodov, V. A.; Pustyl'Nikov, L. M.; Luk'Yanov, A. T.; Baranovskii, B. P.; Darinskii, Yu. V. (Izv. Sibirsk. Otd. Akad. Nauk SSSR Ser. Khim. Nauk **1968** 144/8; C.A. **70** [1969] No. 31962).

[13] Zamashchikov, V. V.; Litvinenko, S. L.; Rudakov, E. S. (Org. React. [Tartu] **16** [1979] 95/102; C.A. **92** [1980] No. 21730).

[14] Shitova, N. B.; Matveev, K. I.; Obynochnyi, A. A. (Kinetika Kataliz **12** [1971] 1417/25; Kinet. Catal. [USSR] **12** [1971] 1258/64).

[15] Shitova, N. B.; Matveev, K. I.; Kuznetsova, L. I. (Izv. Sibirsk. Otd. Akad. Nauk SSSR Ser. Khim. Nauk **1973** 25/30; C.A. **78** [1973] No. 152113).

[16] Matveev, K. I.; Shitova, N. B. (Kinetika Kataliz **10** [1969] 717/21; Kinet. Catal. [USSR] **10** [1969] 586/9).

[17] Llopis, J. F.; Gamboa, J. M.; Victori, L. (Electrochim. Acta **17** [1972] 2225/30).

1.6 Palladium Hydroxides

Pd(OH). Palladium(I) hydroxide has been postulated as an intermediate in the electrochemical reduction of PdO under acid or alkaline conditions [1].

$Pd(OH)_2$ has been prepared by hydrolysis of sodium palladate (Na_2PdO_2) [2] and by adding alkali (to pH = 7 to 8) to $PdCl_2$/HCl solutions [3]. Formation of $Pd(OH)_2$ monolayers by anodic oxidation of palladium in solutions of varying pH values at low current density has been reported [4]. Precipitation and co-precipitation with $Zr(OH)_4$ of $Pd(OH)_2$ from 1M KNO_3 solution using KOH has been studied by ^{109}Pd tracer methods [5]. The solubility product of $Pd(OH)_2$ has been given as 1.1×10^{-29} [3]. The mechanism of dissolution of $Pd(OH)_2$ in 0.1M Na(Cl, ClO_4) has been studied at 0.003 to 0.1M [Cl^-], 4×10^{-6} to 0.001M Pd^{II} and pH = 1 to 4. At pH $\leq$ 4.5 the dissolution can be described as $Pd(OH)_2 + nCl^- = PdCl_n^{(2-n)+} + 2(OH^-)$ where n = 3 and 4 and the equilibrium constants are 3.9×10^{14} and 3.6×10^{13}, respectively. At pH = 5.6 to 6.0, penta- or hexameric hydrolysed Pd^{II} complexes are formed in solution. At pH > 11, $Pd(OH)_2$ dissolves as $[Pd(OH)_n]^{(2-n)+}$ (n = 3 or 4). Diagrams expressing the solubility of $Pd(OH)_2$ as a function of pH and Cl^- concentration are reproduced [6, 7]. The effect of pH on the solubility of $Pd(OH)_2$ in ClO_4^-, Cl^- and SO_4^{2-} solutions has been investigated [3, 8, 9, 10], and the results presented in graphical form [3, 8]. The solubility curves pass through a minimum ($\sim 4.0 \times 10^{-6}$ mol/l) at pH = 2.5 to 11 (ClO_4^- solutions), and at pH = 7.5 to 11 (Cl^- solutions) or 2.5 to 7.5 (SO_4^{2-} solutions) indicative of amphoteric behaviour [9].

Stability constant data reported for $Pd(OH)_2$ are $k_1 = 1.9 \times 10^{-12}$, $k_2 = 1.4 \times 10^{-12}$ [3] and $\log \beta_1^\circ = 13.0 \pm 0.4$, $\log \beta_2^\circ = 25.8 \pm 0.4$ (pH titration), or $\log \beta_1^\circ = 12.4 \pm 0.6$, $\log \beta_2^\circ = 26.5 \pm 0.7$ (spectrophotometric) at 25°C and zero ionic strength [11]. The concentrations of various ligands (NH_3, NO_2^-, Cl^-, Br^-, NCS^-, oxalate, citrate, tartrate and acetate) necessary to effect the dissolution of $Pd(OH)_2$ in 0.1M $NaClO_4$ at various pH values have been measured and used to determine relative complexing strengths [12, 13].

Calculated values for the Pd–OH bond energy and bond distance are 205.7 kJ/mol [14] and 1.94 Å [15]. The calculated and experimental heats of atomisation of $Pd(OH)_2$ are 1668.5 and 1695.8 kJ/mol [16]. The lattice energy is reported to be 2813 kJ/mol [17].

Reactions of organic acids with $Pd(OH)_2$ give palladium(II) complexes, examples include ethylcysteinate (Et-cyst-H) and propylene diamine tetraacetic acid $(PdtaH_4)$ which yield $Pd(Et-cyst)_2$ [18] and $Pd(PdtaH_2)$ [19]. Oxidative dehydrogenation of piperazine to pyrazine with $Pd(OH)_2$ in aqueous media has been described [20].

The use of $Pd(OH)_2$ to prepare metallopolymers based on poly(pyromellitimide) [21], poly(vinylchloride) [22] and bisphenol A epoxy resin ED-5 [23] has been reported.

$Pd(OOH)_2$. For palladium(II) hydroperoxide see p. 23.

$Pd(OH)_4$. Formation of a $Pd(OH)_4$ monolayer and its subsequent decomposition to $Pd(OH)_2$ during the anodic oxidation of palladium in solutions of varying pH at low current density has been reported [4]. The Pd–OH bond energy has been calculated to be 219.0 kJ/mol [14].

References:

[1] Tarasevich, M. R.; Vilinskaya, V. S.; Burshtein, R. K. (Elektrokhimiya **7** [1971] 1200; Soviet Electrochem. **7** [1971] 1152/5).

[2] Glemser, O.; Peuschel, G. (Z. Anorg. Allgem. Chem. **281** [1955] 44/53).

[3] Nabivanets, B. I.; Kalabina, L. V. (Zh. Neorgan. Khim. **15** [1970] 1595/600; Russ. J. Inorg. Chem. **15** [1970] 818/21).

[4] El Wakkad, S. E. S.; Shams El Din, A. M. (J. Chem. Soc. **1954** 3094/8).

[5] Chursin, G. P.; Shevelev, G. A.; Kochetkov, V. L.; Plotnikov, V. I.; Pikalova, L. I. (Izv. Akad. Nauk Kaz.SSR Ser. Fiz. Mat. **13** [1975] 27/31).

[6] Kalabina, L. V. (Tezisy Dokl. 10th Vses. Soveshch. Khim. Anal. Tekhnol. Blagorodn. Met.; Novosibirsk, USSR, 1976, Vol. 1, p. 31; C.A. **90** [1979] No. 44621).

[7] Nabivanets, B. I.; Kalabina, L. V. (Ukr. Khim. Zh. **41** [1975] 1094/8; C.A. **84** [1976] No. 22746).

[8] Nabivanets, B. I.; Kalabina, L. V.; Kudritskaya, L. N. (Zh. Neorgan. Khim. **16** [1971] 3281/4; Russ. J. Inorg. Chem. **16** [1971] 1736/8).

[9] Nabivanets, B. I.; Kalabina, L. V.; Kudritskaya, L. N. (Izv. Sibirsk. Otd. Akad. Nauk SSSR Ser. Khim. Nauk **1970** 51/3; C.A. **74** [1971] No. 68421).

[10] Nabivanets, B. I.; Kalabina, L. V.; Masloi, N. N. (Tezisy Dokl. 10th Vses. Soveshch. Khim. Anal. Tekhnol. Blagorodn. Met.; Novosibirsk. USSR 1976, Vol. 1, p. 22; C.A. **90** [1979] No. 44620).

[11] Izatt, R. M.; Eatough, D.; Christensen, J. J. (J. Chem. Soc. A **1967** 1301/4).

[12] Nabivanets, B. I.; Kalabina, L. V. (Ukr. Khim. Zh. **35** [1969] 889/91; C.A. **72** [1970] No. 16169).

[13] Nabivanets, B. I.; Kalabina, L. V. (Ukr. Khim. Zh. **36** [1970] 1294/5; C.A. **74** [1971] No. 131179).

[14] Rüetschi, P.; Delahay, P. (J. Chem. Phys. **23** [1955] 556/60).

[15] Bäckvall, J-E.; Bjorkman, E. E.; Pettersson, L.; Siegbahn, P. (J. Am. Chem. Soc. **107** [1985] 7265/7).

[16] Sobyanin, V. A.; Bulgakov, N. N.; Gorodetskii, V. V. (React. Kinet. Catal. Letters **6** [1977] 125/32).

[17] Karapet'yants, M. Kh. (Zh. Fiz. Khim. **28** [1954] 1136/52; C.A. **1955** 7917).

[18] Chandrasekharan, M.; Udupa, M. R.; Aravamudan, G. (J. Inorg. Nucl. Chem. **36** [1974] 1417/8).

[19] Gonzalez Garcia, S.; Gonzalez Vilchez, F. (Anales Quim. **66** [1970] 875/90; C.A. **74** [1971] No. 82636).

[20] Rudakova, R. I.; Rudakov, E. S.; Sheinkman, A. K. (Khim. Geterotsikl. Soedin. **1976** 713/4; C.A. **85** [1976] No. 94312).

[21] Korshak, V. V.; Danilenko, E. E.; Bryk, M. T.; Tseitlin, G. M.; Ustinova, M. S. (Ukr. Khim. Zh. **43** [1977] 625/30; Soviet Progr. Chem. **43** No. 6 [1977] 66/70).

[22] Khimchenko, Yu. I.; Radkevich, L. S. (Plasticheskie Massy **1975** No. 1, pp. 53/4; C.A. **82** [1975] No. 157126).

[23] Khimchenko, Yu. I.; Radkevich, L. S. (Khim. Tekhnol. [Kiev] **1981** No. 3, pp. 38/40; C.A. **95** [1981] No. 98461).

1.7 Hydroxopalladates

"**[Pd(OH)$_3$]$^-$**" has been mentioned as one of the species present when Pd(OH)$_2$ dissolves in aqueous solution at pH >11 [1, 2]. The instability constant ($k_3 = $ [Pd(OH)$_2$][OH$^-$]/[Pd(OH)$_3^-$]) is estimated to be ~0.014 [2].

[Pd(OH)$_4$]$^{2-}$. Several papers describe formation of [Pd(OH)$_4$]$^{2-}$ anions in solution by dissolution of PdO·nH$_2$O [3], [PdCl$_4$]$^{2-}$ [4, 5] or Pd(OH)$_2$ [1, 2] in aqueous alkali. Formation of [Pd(OH)$_4$]$^{2-}$ was confirmed by voltammetry and spectrophotometry [4, 5]. The instability constant ($k_4 = $ [Pd(OH)$_3^-$][OH$^-$]/[Pd(OH)$_4$]$^{2-}$) has been estimated to be ~0.1 [2]. Polarographic curves of Pd^{II} in >1M NaOH show a 2-electron diffusion wave for the irreversible reduction of [Pd(OH)$_4$]$^{2-}$ to metallic palladium with a half-wave potential of −0.46 V [5]. The electronic spectra [500 to 300 nm] of PdO·nH$_2$O/NaOH solutions are reproduced, λ_{max} ranges from 362 [20.5 M NaOH] to 372 nm [2M NaOH], a second band occurs at 335 nm [3]; $\lambda_{max} = 226$ nm [5]. Solubility isotherms of PdO·nH$_2$O in alkali solutions, MOH (M = Na, K, Rb, Cs) at 25°C are reproduced [3]. The stability constant, K, for [Pd(OH)$_4$]$^{2-}$ is reported to be 3.6×10^{-29} and the diffusion constant, D, is given as 6.14×10^{-6} cm^2/s [6]. The basic kinetic principles for the cathodic reduction of [Pd(OH)$_4$]$^{2-}$ in alkali baths have been studied in connection with the electrodeposition of palladium [7].

[Pd(OH)$_6$]$^{4-}$. Formation of this palladium(II) species from PdO·nH$_2$O and OH$^-$ at high alkali concentrations has been detected by electronic spectroscopy, distorted octahedral symmetry is proposed [3].

[Pd(OH)$_5$]$^-$. This palladium(IV) species is formed when PdO$_2$·nH$_2$O is dissolved in aqueous NaOH, the calculated concentration formation constant is 0.41. A solubility isotherm and tables of solubility data for PdO$_2$·nH$_2$O in aqueous NaOH are reproduced in the paper [8].

[Pd(OH)$_6$]$^{2-}$. Detection of this palladium(IV) species on palladium surfaces during electrochemical studies of monolayer and multilayer oxide films has been reported [9].

References:

[1] Kalabina, L. V. (Tezisy. Dokl. 10th Vses. Soveshch. Khim. Anal. Tekhnol. Blagorodn. Met., Novosibirsk, USSR, 1976, Vol. 1, p. 31; C.A. **90** [1979] No. 44621).

[2] Kalabina, L. V.; Nabivanets, B. I. (Zh. Neorgan. Khim. **15** [1970] 1595/600; Russ. J. Inorg. Chem. **15** [1970] 818/21).

[3] Ivanov-Emin, B. N.; Zaitsev, B. E.; Petrishcheva, L. P. (Zh. Neorgan. Khim. **30** [1985] 3144/7; Russ. J. Inorg. Chem. **30** [1985] 1786/8).

[4] Bardin, M. B.; Ketrush, P. M. (Izv. Sibirsk. Otd. Akad. Nauk SSSR Ser. Khim. Nauk **1974** 70/6; C.A. **82** [1975] No. 49219).

[5] Bardin, M. B.; Ketrush, P. M. (Teor. Prakt. Polyarogr. Metodov Anal. **1973** 62/8; C.A. **83** [1975] No. 21648).

[6] Visomirskis, R.; Morgenstern, Ya. L. (Lietuvos TSR Mokslu Akad. Darbai B **1966** No. 1, pp. 49/60; C.A. **66** [1967] 15895).

[7] Kuzhakova, G. M.; Krasikov, B. S. (Zh. Prikl. Khim. **52** [1979] 76/81; J. Appl. Chem. [USSR] **52** [1979] 64/8).

[8] Ivanov-Emin, B. N.; Borzova, L. D.; Egorov, A. M.; Sujben, D. (Zh. Neorgan. Khim. **19** [1974] 1570/2; Russ. J. Inorg. Chem. **19** [1974] 855/6).

[9] Burke, L. D.; Roche, M. B. C. (J. Electroanal. Chem. Interfacial Electrochem. **186** [1985] 139/54).

$[(C_2H_5)SnPd(OH)_p]^{(p-5)-}$. $[(C_2H_5)SnPd_2(OH)_q]^{(q-7)-}$. Complexes of these general formulas have been obtained from $SnCl_3(C_2H_5)$ and $PdCl_2$ in basic aqueous media. Exchange of the ethyl group between the two metallic atoms has been observed in both cases and the kinetics of the reaction have been investigated. The system has been analysed by polarography. Polarographic and kinetic data are plotted in the paper. Proposed structures are shown below.

Reference:

Joisson, D.; Devaud, M. (J. Organometal. Chem. **93** [1975] 95/106).

1.7.1 Palladium Hydroperoxide $Pd(OOH)_2$

A Pd–OOH bond distance of 1.94 Å has been calculated for this compound.

Reference:

Bäckvall, J.-E.; Björkman, E. E.; Pettersson, L.; Siegbahn, P. (J. Am. Chem. Soc. **107** [1985] 7265/7).

1.7.2 Sodium Hydroxopalladates

$Na_2[Pd(OH)_4]$ is prepared by heating freshly prepared $PdO \cdot nH_2O$ with 13.5M NaOH on a steam bath for 8 to 10 min, then allowing the filtered solution to evaporate (desiccator NaOH/ P_2O_5). It forms rather coarse yellow rectangular crystals. These, which have tetragonal or possibly orthorhombic symmetry, are pleochroic (yellow along n_p, colourless along n_g). The X-ray diffraction pattern is reproduced and interplanar spacings and intensities are recorded. The infrared spectrum (4000 to 400 cm^{-1}) is reproduced, ν(OH), δ(O–H) and ν(Pd–OH) occur at 3500 to 2800, 1110 and 790, and 515 cm^{-1}, respectively. The thermogram (reproduced in the paper) shows a single endothermic event at 268°C corresponding to loss of water and conversion into PdO and NaOH. A square-planar structure is proposed for the $[Pd(OH)_4]^{2-}$ anion. The crystals are readily soluble in water and in acids, and freely absorb atmospheric moisture and carbon dioxide [1].

Na$_2$[Pd(OH)$_6$] is prepared by adding a 50% excess of 10 M NaOH solution to a freshly prepared solution of H$_2$[PdCl$_6$] and is separated as reddish brown crystals by allowing the yellow-green solution to evaporate in a desiccator over KOH pellets [2]. The crystals are trigonal, space group P$\overline{3}$1m-D$_{3d}^1$ with a = 5.80(1), c = 4.720(6) Å, D$_{exp}$ = 2.85(2), D$_{calc}$ = 3.070(5) g/cm^3 for Z = 1. Comparison of line X-ray diffraction diagrams (reproduced in [3]) reveal that Na$_2$Pd(OH)$_6$ is isostructural with its platinum analogue [2, 3]. The infrared spectrum of Na$_2$-[Pd(OH)$_6$] (4000 to 400 cm^{-1}) is reproduced, assignments are ν(OH) = 3470, δ(PdO–H) = 1100, γ(PdO–H) = 650 and ν(Pd–OH) = 530 cm^{-1} [2].

According to the thermogravigram thermal decomposition to Na$_2$PdO$_3$ and H$_2$O commences at 170°C and is complete at 254°C [2].

References:

[1] Ivanov-Emin, B. N.; Borzova, L. D.; Sudzhben, D.; Ivanova, N. N.; Ezhov, A. I. (Zh. Neorgan. Khim. **19** [1974] 1880/3; Russ. J. Inorg. Chem. **19** [1974] 1026/8).

[2] Ivanov-Emin, B. N.; Venskovskii, N. U.; Lin'Ko, I. V.; Zaitsev, B. E.; Borzova, L. D. (Koord. Khim. **6** [1980] 928/31; Soviet J. Coord. Chem. **6** [1980] 469/72).

[3] Venskovskii, N. U.; Ivanov-Emin, B. N.; Lin'Ko, I. V. (Koord. Khim. **10** [1984] 1263/5; C. A. **101** [1984] No. 202004).

1.7.3 Potassium Hydroxopalladates

K$_2$[Pd(OH)$_4$]. A solubility isotherm of PdO·xH$_2$O in KOH solution at 25°C is reproduced in [1]. Solid K$_2$[Pd(OH)$_4$] has been obtained by concentrating PdO·xH$_2$O solutions over solid KOH/P$_2$O$_5$. The crystals are monoclinic, space group P2$_1$-C$_2^2$ with a = 5.561(3), b = 6.947(3), c = 6.763(2) Å and β = 102.13(4)°. K$_2$[Pd(OH)$_4$] is not isostructural with Na$_2$[Pd(OH)$_4$] [2, 5]. Infrared and thermal analysis data are reported, thermal decomposition to PdO, KOH and H$_2$O occurs [2].

K$_2$[Pd(OH)$_6$] is prepared from H$_2$PdCl$_6$ and 6 M KOH solution [3]. The crystals are trigonal, space group R$\overline{3}$-C$_{3i}^2$ with a = 6.38(1), c = 12.74(1) Å, D$_{exp}$ = 2.90(2), D$_{calc}$ = 3.180(5) g/cm^3 for Z = 3 [4], and are isomorphous with those of K$_2$[Pt(OH)$_6$] [3]. Infrared data have been recorded. Decomposition of K$_2$[Pd(OH)$_6$] to K$_2$PdO$_3$ occurs at 210 to 350°C [3].

References:

[1] Ivanov-Emin, B. N.; Zaitsev, B. E.; Petrishcheva, L. P. (Zh. Neorgan. Khim. **30** [1985] 3144/7; Russ. J. Inorg. Chem. **30** [1985] 1786/8).

[2] Ivanov-Emin, B. N.; Petrishcheva, L. P.; Zaitsev, B. E. (Koord. Khim. **12** [1986] 537/9; C. A. **105** [1986] No. 17120).

[3] Venskovskii, N. U.; Ivanov-Emin, B. N.; Lin'ko, I. V. (Izv. Vysshikh Uchebn. Zavedenii Khim. Khim. Tekhnol. **24** [1981] 531/4; C. A. **95** [1981] No. 125328).

[4] Venskovskii, N. U.; Ivanov-Emin, B. N.; Lin'ko, I. V. (Koord. Khim. **10** [1984] 1263/5; C. A. **101** [1984] No. 202004).

[5] Il'inets, A. M.; Ivanov-Emin, B. N.; Petrishcheva, L. P.; Izmailovich, A. S. (Koord. Khim. **13** [1984] 1660/1; C. A. **108** [1988] No. 104292).

1.7.4 Rubidium and Caesium Hydroxopalladates

Rb$_2$[Pd(OH)$_4$]. A solubility isotherm of PdO·nH$_2$O in RbOH solution at 25°C is reproduced in [1].

Rb$_2$[Pd(OH)$_6$] is prepared by treatment of H$_2$PdCl$_6$(?) (abstract specifies H$_2$PdCl$_4$) with 6M RbOH solution, and has been characterised by infrared spectroscopy [2].

Cs$_2$[Pd(OH)$_4$]. A solubility isotherm of PdO·nH$_2$O in CsOH solution at 25°C is reproduced in [1].

References:

[1] Ivanov-Emin, B. N.; Zaitsev, B. E.; Petrishcheva, L. P. (Zh. Neorgan. Khim. **30** [1985] 3144/7; Russ. J. Inorg. Chem. **30** [1985] 1786/8).

[2] Venskovskii, N. U.; Ivanov-Emin, B. N.; Lin'ko, I. V. (Izv. Vysshikh Uchebn. Zavedenii Khim. Khim. Tekhnol. **24** [1981] 531/4; C.A. **95** [1981] No. 125328).

1.7.5 Calcium Hydroxopalladate

Ca[Pd(OH)$_6$]. This sparingly soluble salt is prepared by treating a 0.2N solution of Na$_2$[Pd(OH)$_6$] with the stoichiometric amount of 2N CaCl$_2$ solution, and precipitates as yellow trigonal crystals. The salt is isostructural with the corresponding hydroxoplatinate [1]. The crystals belong to the trigonal space group P$\bar{3}$1c-D$_{3d}^2$ with a = 5.66, c = 9.23 Å, D$_{pyk}$ = 3.09, D$_{X-ray}$ = 3.22 g/cm^3, Z = 2. Line X-ray diffraction patterns for Ca[Pd(OH)$_6$] and Ca[Pt(OH)$_6$] are reproduced in the paper together with tabulated intensity and d-spacing data for the former [2]. The infrared spectrum (4000 to 500 cm^{-1}) is reproduced and salient absorptions are tabulated; ν(OH) = 3370; δ(PdO–H) = 1150, 1110, 1070, 980; γ(PdO–H) = 695; ν(Pd–OH) = 560 cm^{-1}. Thermal analysis curves (DTGA, DTA and TGA) are reproduced, decomposition occurs in the range 260 to 440°C to yield CaPdO$_3$. The binding energy of the d-electrons is 340.8 eV [1].

References:

[1] Ivanov-Emin, B. N.; Venskovskii, N. U.; Zaitsev, B. E.; Lin'ko, I. V. (Zh. Neorgan. Khim. **26** [1981] 2195/9; Russ. J. Inorg. Chem. **26** [1981] 1181/3).

[2] Venskovskii, N. U.; Ivanov-Emin, B. N.; Lin'ko, I. V. (Zh. Neorgan. Khim. **28** [1983] 1065/7; Russ. J. Inorg. Chem. **28** [1983] 605/6).

1.7.6 Strontium Hydroxopalladates

Sr[Pd(OH)$_4$] is prepared by adding a solution of Sr(OH)$_2$ to a freshly prepared solution of Na$_2$[Pd(OH)$_4$] and crystallises as a hydrate which dehydrates at 180°C. The crystals are almost colourless, transparent, anisotropic lamina with refractive indices n$_g$ = 1.740 ± 0.005, n$_p$ = 1.659 ± 0.005. X-ray data (intensities and d-spacings) are tabulated, the crystals are orthorhombic with a = 5.42, b = 15.43, c = 6.23 Å, D$_{pyk}$ = 3.52, D$_{X-ray}$ = 3.53 g/cm^3. Infrared data are tabulated and thermal analysis curves (DTGA, DTA, TGA) are reproduced in the paper. The thermolysis products are SrPdO$_2$ and H$_2$O [1]. Cell dimensions for a **monohydrate** Sr[Pd(OH)$_4$]·H$_2$O are a = 5.532(5), b = 16.072(5), c = 6.387(8) Å, D$_{exp}$ = 3.52, D$_{X-ray}$ = 3.52 g/cm^3 [2].

Sr[Pd(OH)$_6$]. This sparingly soluble salt is prepared by treating a 0.2N solution of Na$_2$[Pd(OH)$_6$] with a stoichiometric amount of 2N SrCl$_2$ solution, and precipitates as yellow crystals. The infrared spectrum (4000 to 500 cm^{-1}) is reproduced and bands assigned: ν(O–H) = 3390; δ(PdO–H) = 1105, 1065, 975; γ(PdO–H) = 790, 680, ν(Pd–OH) = 555 cm^{-1}. Thermal analysis curves (DTGA, TGA, DTA) are reproduced, decomposition occurs over the range 200 to 410°C to yield SrPdO$_3$ [3].

References:

[1] Ivanov-Emin, B. N.; Petrishcheva, L. P.; Zaitsev, B. E.; Ivlieva, V. I.; Izmailovich, A. S.; Dolganev, V. P. (Zh. Neorgan. Khim. **29** [1984] 2046/50; Russ. J. Inorg. Chem. **29** [1984] 1169/71).

[2] Petrishcheva, L. P.; Ivanov-Emin, B. N.; Zaitsev, B. E.; Ivlieva, V. I. (Deposited Doc. VINITI-1316-84 Pt. 2 [1983] 93/6; C.A. **102** [1985] No. 71691).

[3] Ivanov-Emin, B. N.; Venskovskii, N. U.; Zaitsev, B. E.; Lin'ko, I. V. (Zh. Neorgan. Khim. **26** [1981] 2195/9; Russ. J. Inorg. Chem. **26** [1981] 1181/3).

1.7.7 Barium Hydroxopalladates

Ba[Pd(OH)$_4$] is prepared from Na$_2$[Pd(OH)$_4$] and Ba(OH)$_2$, and crystallises as a monohydrate which dehydrates at 180°C. The transparent, lemon-yellow tabular crystals are anisotropic and display a distinct pleochroism. Refractive indices are $n_g = 1.745$, $n_m = 1.698$, $n_p = 1.650$ (all ± 0.005). The crystals are orthorhombic with a = 5.532(5), b = 16.072(5), c = 6.387(8) Å, $D_{exp} = 3.83$, $D_{calc} = 3.86$ g/cm^3. The infrared spectra (4000 to 500 cm^{-1}) of Ba[Pd(OH)$_4$] and Ba[Pd(OD)$_4$] are reproduced and assignments are given. Thermal analysis curves (DTGA, DTA, TGA) are reproduced, thermal decomposition gives BaPdO$_2$ and H$_2$O [1]. Preparation, infrared, thermal analysis and X-ray data have been reported for the **monohydrate** Ba[Pd(OH)$_4$]·H$_2$O. Cell dimensions (orthorhombic) are a = 5.415, b = 15.430, c = 6.229 Å [2].

Ba[Pd(OH)$_6$]. This sparingly soluble compound is prepared by treating a 0.2 N solution of Na$_2$[Pd(OH)$_6$] with the stoichiometric amount of 2 N BaCl$_2$ solution and precipitates as yellow crystals. The infrared spectrum (4000 to 500 cm^{-1}) is reproduced and bands are assigned: ν(O–H) = 3430, 3400, 3360; δ(PdO–H) = 1110, 1065, 960; γ(PdO–H) = 790, 680; ν(Pd–OH) 555 cm^{-1}. Decomposition of Ba[Pd(OH)$_6$] to yield BaPdO$_3$ occurs at 180 to 380°C [3].

References:

[1] Ivanov-Emin, B. N.; Petrishcheva, L. P.; Zaitsev, B. E.; Ivlieva, V. I.; Izmailovich, A. S.; Dolganev, V. P. (Zh. Neorgan. Khim. **29** [1984] 2046/50; Russ. J. Inorg. Chem. **29** [1984] 1169/71).

[2] Petrishcheva, L. P.; Ivanov-Emin, B. N.; Zaitsev, B. E.; Ivlieva, V. I. (Deposited Doc. VINITI 1316-84 Pt. 2 [1983] 93/6; C.A. **102** [1985] No. 71691).

[3] Ivanov-Emin, B. N.; Venskovskii, N. U.; Zaitsev, B. E.; Lin'ko, I. V. (Zh. Neorgan. Khim. **26** [1981] 2195/9; Russ. J. Inorg. Chem. **26** [1981] 1181/3).

1.8 Oxopalladates

Palladium forms ternary or mixed oxides ("palladates") with many s, p, d and f block metals; in so doing it displays four formal oxidation states I, II, III and IV. The oxidation state Pd^I is found in ternary oxides MPdO$_2$ (M = Cr, Co, Rh) of the delafossite structural type. As might be anticipated palladium(II) mixed oxides are the most numerous, examples include derivatives of the alkali metals (notably M$_2$PdO$_2$ and M$_2$Pd$_3$O$_4$), the alkaline earths (MPdO$_2$ and MPd$_3$O$_4$) and the lanthanides (M$_2$PdO$_4$, M$_2$Pd$_2$O$_5$ and M$_4$PdO$_7$). Palladium(III) is encountered along with palladium(II) in bronzes of the form Na$_x$Pd$_3$O$_4$ (x = 0.8 to 1.0). Finally, palladium(IV) is present in alkali and alkaline earth metal mixed oxides of the form M$_2$PdO$_3$ and MPdO$_3$, respectively, and in the pyrochlores M$_2$Pd$_2$O$_7$ (M = In, Sc, Y, Gd, Dy, Er and Yb).

General References:

Sirchenko, T. D.; Volkov, V. I.; Mixed Oxides of Palladium and Non-Precious Metals, Khim. Tekhnol. Materialov dlya Nov. Tekhn. M. **1980** 92/7; C.A. **95** [1981] No. 72252.

Lazarev, V. B.; Shaplygin, I. S.; Electrical Conductivity of Platinum Metal-Nonplatinum Metal Double Oxides, Mater. Res. Bull. **13** [1978] 229/35.

Hagenmüller, P.; Oxygenated Bronzes, Progr. Solid State Chem. **5** [1971] 71/144.

1.8.1 Tetraoxo Anion [PdO$_4$]$^-$(?)

The electronic structures of 3d and 4d metal tetraoxo anions including the hypothetical [PdO$_4^-$] anion have been calculated using the discrete variational Xα method. One-electron energies (in eV) obtained in the corresponding Slater transition states, electron affinities (in eV) and Mulliken charges on atoms (in e) are tabulated in the paper.

Reference:

Gutsev, G. L.; Boldyrev, A. I. (Chem. Phys. Letters **108** [1984] 255/8).

1.8.2 Bismuth Palladates

A tentative equilibrium diagram showing the Bi$_2$O$_3$-rich corner (0 to 1.0 mol% PdO) of the PdO–Bi$_2$O$_3$ system is reproduced. A eutectic with 0.05 mol% PdO is found to melt at 805 $\pm$ 5°C [1]. For data on Bi$_2$PdO$_4$ see below.

Bi$_2$PdO$_4$ is prepared from PdO and Bi$_2$O$_3$ at 500°C [2, 3], 600 to 750°C [4] or 700 to 750°C [1]. The reddish brown compound [3] melts at 855 $\pm$ 5°C [1] or 835 $\pm$ 10°C [3] decomposing above this temperature into PdO or Pd and Bi$_2$O$_3$ [1, 3]. It is chemically stable, and is insoluble in water and in dilute or concentrated acids or alkalies [3].

Several X-ray diffraction studies have been reported, the crystals have been assigned to the tetragonal space group P4/ncc-D$_{4h}^8$; a = 8.614, c = 5.892 Å, D$_{exp}$ = 8.81, D$_{calc}$ = 8.937 g/cm^3 [1], a = 8.622, c = 5.907 Å, D$_{exp}$ = 8.79, D$_{calc}$ = 8.90 g/cm^3, Z = 4 [4], a = 8.621, c = 5.889 Å, D$_{exp}$ = 8.81, D$_{calc}$ = 8.93 g/cm^3, Z = 4 [3]. The space group C$_{4v}^{10}$-I4 cm with a = 8.623 and c = 5.909 Å has been reported [2] but this assignment is said to be erroneous [1]. The crystal structure (atomic projection reproduced in the paper) contains square-planar PdO$_4$ units linked by bridging bismuth atoms and is isostructural with Bi$_2$CuO$_4$ [2]. The infrared spectrum (1000 to 200 cm^{-1}) and thermal analysis (DTA-TG) curves are reproduced [1]. Bi$_2$PdO$_4$ is a semiconductor [1, 3] with a resistivity of (3 to 5) $\times 10^2$ $\Omega \cdot$cm at room temperature [1] or (3 to 9) $\times 10^2$ $\Omega \cdot$cm [3]. A plot of electrical resistivity versus mol% PdO for the PdO/Bi$_2$O$_3$ system is reproduced [3].

Bi$_2$Pd$_x$Cu$_{(1-x)}$O$_4$. Properties of solid solutions of Bi$_2$PdO$_4$ and Bi$_2$CuO$_4$ have been investigated by thermal, thermogravimetric and X-ray diffraction analysis, infrared spectroscopy and static magnetic susceptibility measurements. There is a continuous series of solid solutions Bi$_2$Pd$_x$Cu$_{(1-x)}$O$_4$ (x = 0 to 1) with anomalous behaviour indicative of possible ordering in the Cu/Pd sublattice for x = 0.5. Plots of melting point, lattice parameters (a, c and c/a) and magnetic moments, μ_{eff}, against composition (x) are reproduced together with infrared spectra (700 to 100 cm^{-1}) for different values of x [5].

References:

[1] Kakhan, B. G.; Lazarev, V. B.; Shaplygin, I. S. (Zh. Neorgan. Khim. **24** [1979] 1663/8; Russ. J. Inorg. Chem. **24** [1979] 922/5).

[2] Arpe, R.; Müller-Buschbaum, H. (Z. Naturforsch. **31b** [1976] 1708/9).

[3] Lazarev, V. B.; Shaplygin, I. S. (Zh. Neorgan. Khim. **19** [1974] 2388/90; Russ. J. Inorg. Chem. **19** [1974] 1305/6).

[4] Boivin, J.-C.; Conflant, P.; Thomas, D. (Compt. Rend. C **282** [1976] 749/51).

[5] Kakhan, B. G.; Lazarev, V. B.; Shaplygin, I. S.; Ellert, O. G. (Zh. Neorgan. Khim. **26** [1981] 232/7; Russ. J. Inorg. Chem. **26** [1981] 124/7).

1.8.3 Lithium Palladates

Li–Pd–O System. A tentative phase diagram is shown in **Fig. 1** [1].

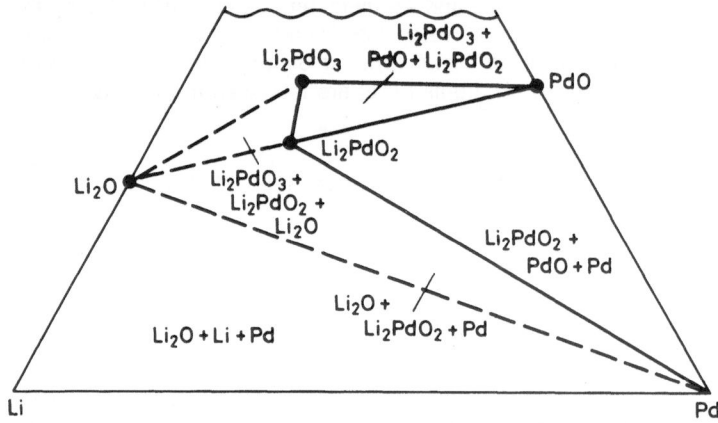

Fig. 1. Tentative partial diagram of the Li–Pd–O system, showing phases which can coexist in equilibrium at 600°C. Continuous lines represent joins that have been confirmed experimentally; broken lines show joins whose existence is inferred [1].

Li_2PdO_2 has been prepared in impure form by heating a 1.5:1 mixture of Li_2CO_3 and $Pd(O_2CCH_3)_2$ in a palladium foil boat at 700 to 800°C for 24 h in air. The product varies in colour from greenish yellow (excess lithium) to black (excess PdO), pure material is reported to be yellow-brown. Colour changes probably reflect deviations from ideal stoichiometry but X-ray powder d-spacings are unchanged [1]. A more recent paper describes the formation of Li_2PdO_2 as ruby crystals from the reaction of a palladium tube with the mixture of Na_2O, Li_2O_2 and "Tb_4O_7" contained therein over a period of 7 d at 1100°C [2]. According to X-ray powder diffraction measurements the needle-shaped crystals of Li_2PdO_2 are body-centred orthorhombic, space group $Immm-D_{2h}^{25}$ with $a = 3.74_0$, $b = 2.97_5$, $c = 9.35_4$ Å. X-ray powder diffraction data and projections of the crystal structure on (100) and (010) planes are reproduced in the paper. The crystal structure comprises sheets of tetrahedrally coordinated lithium and oxygen atoms, parallel to (001) and held together by ribbons of $(PdO_2)_\infty$ in which each palladium is coordinated by four oxygens in a rectangular-planar arrangement. Li_2PdO_2 is isostructural with Li_2CuO_2 with which it is completely miscible at ~750°C. A plot of variations in the unit cell dimensions of Li_2PdO_2, Li_2CuO_2 and their solid solutions with composition is reproduced [1]. A more recent single-crystal study confirms the Li_2CuO_2 type structure, space group $Immm-D_{2h}^{25}$, but gives significantly different cell parameters, $a = 3.7534(5)$, $b = 2.9818(4)$, $c = 9.3158(10)$ Å; $D_{X-ray} = 4.85$ g/cm^3. The Madelung part of the lattice energy (MAPLE) for Li_2PdO_2 (1884.3 kcal/mol) corresponds to the sum of the MAPLE values of the component binary oxides (1895.4 kcal/mol) [2]. Pure Li_2PdO_2 does not melt or decompose at temperatures below 1050°C [1].

Li$_2$PdO$_3$. Two papers mention this compound but give few details [3, 4]. A later paper describes preparation by heating Li$_2$PdO$_2$ in a stream of oxygen at 600°C for 10 d. The compound is thought to be isostructural with Li$_2$TiO$_3$. X-ray powder data are tabulated but further crystallographic characterisation was not attempted [1].

References:

[1] Dubey, B. L.; Gard, J. A.; Glasser, F. P.; West, A. R. (J. Solid State Chem. **6** [1973] 329/34).
[2] Wolf, R.; Hoppe, R. (Z. Anorg. Allgem. Chem. **536** [1986] 77/80).
[3] Scheer, J. J.; van Arkel, A. E.; Heyding, R. D. (Can. J. Chem. **33** [1955] 683/6).
[4] Dorrian, J. F.; Newnham, R. E. (Mater. Res. Bull. **4** [1969] 179).

1.8.4 Sodium Palladates

Na$_2$PdO$_2$ is obtained as a yellow diamagnetic powder by heating equivalent amounts of Na$_2$O and PdO at 650°C for 2 to 4 h under argon [1]. It is also prepared as yellow rod-like crystals by heating Na$_2$O and PdO (Na:Pd = 8:1) in a palladium bomb at 750°C (3 d) or 1050°C (2 d). The crystals are orthorhombic, space group Cmc2$_1$-C$_{2v}^{12}$ or Cmcm-D$_{2h}^{17}$ with a = 3.07$_7$, b = 10.35$_9$, c = 8.35$_1$ Å, D$_{exp}$ = 4.63, D$_{calc}$ = 4.60 g/cm^3; Z = 4. X-ray (Guinier-Simon absorption) data are tabulated in the paper [2].

Na$_2$Pd$_3$O$_4$ is prepared by heating Na$_2$O$_2$ or Na$_2$O with PdO in dry Ar at 900°C. Single crystals have been obtained by thermal decomposition of Na$_2$PdO$_2$ in a sealed bomb at 1100°C for $\leqq$ 24 h [3]. Preparation by heating PdO with Na$_2$CO$_3$ in air at 800°C has also been described and a decomposition temperature of 950 ± 20°C has been recorded [4]. It forms black rod-like orthorhombic crystals, space group Immm-D$_{2h}^{25}$ with a = 3.043, b = 5.446, c = 13.175 Å, D$_{exp}$ = 6.43, D$_{calc}$ = 6.50 g/cm^3 for Z = 2 [3] or a = 3.041, b = 5.470, c = 13.188 Å, D$_{exp}$ = 6.36, D$_{calc}$ = 6.49 g/cm^3 [4]. The structure shown in **Fig. 2** contains Pd$_3$O$_4$ double layers normal to

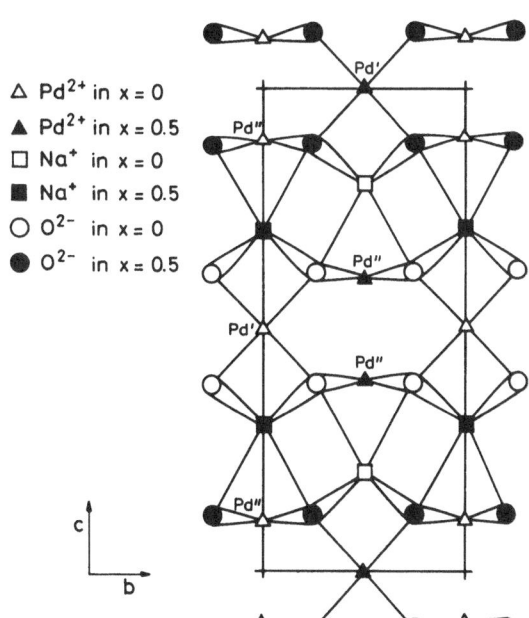

△ Pd^{2+} in x = 0
▲ Pd^{2+} in x = 0.5
□ Na$^+$ in x = 0
■ Na$^+$ in x = 0.5
○ O^{2-} in x = 0
● O^{2-} in x = 0.5

Fig. 2. Arrangement of atoms in Na$_2$Pd$_3$O$_4$; projection on (100).

(001) connected by Na atoms in trigonal-prismatic coordination [3]. $Na_2Pd_3O_4$ is a semiconductor, resistivity is reported to be 72(295 K) or 57(393 K) $\Omega \cdot cm$ [4].

$Na_xPd_3O_4$ (x = 0.8 to 1.0). These bronzes are obtained by heating together palladium and sodium nitrates in air at 750 to 1000°C [5] or by heating finely divided palladium metal with sodium carbonate in a stream of oxygen [6]. The crystals are cubic, space group $Pm3n\text{-}O_h^3$ with a = 5.654 ± 0.002 Å [5, 7] or 5.64 Å [6]. Debye-Scherrer data are tabulated in [6]. Thermolysis of $NaPd_3O_4$ commences at 844°C [5]. The bronzes are assigned the general formula $Na_xPd_{1+x}^{II}\text{-}Pd_{2-x}^{III}O_4$ [6].

Na_2PdO_3. This black, diamagnetic palladium(IV) compound is obtained from Na_2PdO_2 and dry air at 450°C over a period of 350 h [1]. The crystals are hexagonal ($\alpha\text{-}NaFeO_3$ type), with a = 3.15, c = 15.76 Å, $D_{X\text{-}ray}$ = 4.92 g/cm³, Z = 3 [1]. Another reddish brown form of Na_2PdO_3 is isostructural with Li_2MnO_3, the crystals are monoclinic, space group $C/2c\text{-}C_{2h}^6$ with a = 5.37$_4$, b = 9.30$_9$, c = 10.78$_9$ Å, β = 99.5°, D_{exp} = 4.7$_3$, D_{calc} = 5.00 g/cm³ for Z = 8 [2]. Thermal degradation of Na_2PdO_3 at 650°C and 10^{-4} Torr for 6 to 8 h affords Na_2PdO_2 [1].

References:

[1] Sabrowsky, H.; Hoppe, R. (Naturwissenschaften **53** [1966] 501/2).
[2] Wilhelm, M.; Hoppe, R. (Z. Anorg. Allgem. Chem. **424** [1976] 5/12).
[3] Wilhelm, M.; Hoppe, R. (Z. Anorg. Allgem. Chem. **409** [1974] 60/8).
[4] Lazarev, V. B.; Shaplygin, I. S. (Zh. Neorgan. Khim. **23** [1978] 1456/60; Russ. J. Inorg. Chem. **23** [1978] 802/4).
[5] Bergner, D.; Kohlhaas, R. (Z. Anorg. Allgem. Chem. **401** [1973] 15/20).
[6] Scheer, J. J.; van Arkel, A. E.; Heyding, R. D. (Can. J. Chem. **33** [1955] 683/6).
[7] Lazarev, V. B.; Shaplygin, I. S. (Zh. Neorgan. Khim. **23** [1978] 2902/6; Russ. J. Inorg. Chem. **23** [1978] 1610/2).

1.8.5 Potassium Palladates

K_2PdO_2 is prepared by heating PdO with equivalent amounts of K_2O (600°C, 4 h) [1, 2] or K_2CO_3 (800°C) [3] and is described as a yellow solid, decomposition temperature 880 ± 15°C [3]. The crystals are orthorhombic, space group $Immm\text{-}D_{2h}^{25}$ with a = 8.52$_3$, b = 6.08$_9$, c = 3.11$_9$ Å, D_{exp} = 4.2, D_{calc} = 4.44 g/cm³ and Z = 2 [2] or a = 8.523, b = 6.089, c = 3.119 Å, D_{exp} = 4.38, D_{calc} = 4.436 g/cm³ [3]. A chain type structure corresponding to that of K_2PtS_2 is proposed and a projection along the c-axis is reproduced in the paper [2]. Oxidation by oxygen gives K_2PdO_3 [1, 2]. K_2PdO_2 is a semiconductor, resistivities are 3.7×10^3 (295 K) and 8.9×10^2 (393 K) $\Omega \cdot cm$ [3]. The crystals are readily hydrolysed by moist air and decompose in water over a period of hours [3].

K_6PdO_4 is obtained as transparent orange, square prisms by heating a mixture of $KO_{0.54}$ and PdO (K:Pd = 6 to 8:1) in a silver bomb for 1 d at 500°C and 2 d at 600°C. The crystals are orthorhombic (pseudo-tetragonal), space group $P2_12_12_1\text{-}D_2^4$ with a = b = 12.36$_8$, c = 13.59$_3$ Å; Z = 8 [4].

K_2PdO_3. This dark green compound is prepared by heating K_2PdO_2 in dry oxygen for 17 h at 450°C and 800 Torr pressure [1, 2]. A less pure product is obtained by heating 2:1 mixtures of $KO_x\text{-}Pd$ or $KO_x\text{-}PdO$ under oxygen at 700°C for 12 h [1]. Single crystals of K_2PdO_3 have been obtained by a similar procedure [4]. The black, square-prismatic crystals are orthorhombic, with a = 6.20$_2$, b = 9.21$_9$, c = 4 × 11.37$_1$ = 45.4$_8$ Å, $D_{X\text{-}ray}$ = 3.60, D_{exp} = 3.5$_0$ g/cm³; Z = 24. X-ray data are tabulated, and a plot of $1/\psi$ versus temperature (0 to 300 K) is reproduced in the paper [4].

References:

[1] Sabrowsky, H.; Hoppe, R. (Naturwissenschaften **53** [1966] 501/2).
[2] Sabrowsky, H.; Bronger, W.; Shmitz, D. (Z. Naturforsch. **29 b** [1974] 10/12).
[3] Lazarev, V. B.; Shaplygin, I. S. (Zh. Neorgan. Khim. **23** [1978] 1456/60; Russ. J. Inorg. Chem. **23** [1978] 802/4).
[4] Wilhelm, M.; Hoppe, R. (Z. Anorg. Allgem. Chem. **424** [1976] 5/12).

1.8.6 Rubidium and Caesium Palladates

Rb_2PdO_2 has been mentioned as a hygroscopic semiconducting solid [1].

Rb_2PdO_3. This black powder is obtained by heating a mixture of $RbO_{1.60}$ and PdO [Rb:Pd = 2.2:1] for 1 d at 500°C and 12 d at 700°C in a gold tube. The crystals are isostructural with K_2PdO_3 and have a = 6.35_4, b = 9.60_5 (or 9.30_9), c = 11.58_6 Å, $D_{X\text{-ray}}$ = 4.58, D_{exp} = 4.4_7 g/cm³. X-ray data are tabulated in the paper [2].

Cs_2PdO_2 has been mentioned as a hygroscopic, semiconducting solid, no other details are given in the paper [1].

References:

[1] Lazarev, V. B.; Shaplygin, I. S. (Mater. Res. Bull. **13** [1978] 229/35).
[2] Wilhelm, M.; Hoppe, R. (Z. Anorg. Allgem. Chem. **424** [1976] 5/12).

1.8.7 Magnesium Palladates

$MgPd_3O_4$. A phase of this composition with the $Na_xPt_3O_4$ structure has been mentioned but no details given [1].

Mg_2PdO_4 is obtained when $Mg(OH)_2$/Pd-black mixtures are heated at 900°C and 200 atm oxygen pressure. Impure samples adopt an inverse spinel structure with Pd^{4+} ions in octahedral sites, a = 8.501 Å [2].

References:

[1] Lazarev, V. B.; Shaplygin, I. S. (Mater. Res. Bull. **13** [1978] 229/35).
[2] Müller, O.; Roy, R. (Advan. Chem. Ser. No. 98 [1971] 28/38).

1.8.8 Calcium Palladate $CaPd_3O_4$

This is one of a number of products obtained when $CaCO_3$ and PdO are mixed and repeatedly heated to 800°C in an air furnace. It forms a black powder with a greenish cast. Analytical data, X-ray powder patterns (cubic unit cell, a = 5.746 ± 0.001 Å) and similarity to $Na_xPt_3O_4$ support the proposed stoichiometry. Resistivity of 0.2 Ω·cm decreasing with temperature indicates p-type semiconductor character [1]. Later papers confirm formation of $CaPd_3O_4$ [2 to 6]. Data recorded are space group Pm3n-O_h^3 [4, 6], lattice parameter a = 5.747 Å [3, 6], decomposition temperature 1010 to 1040°C [3], 1012 to 1063°C [4], resistivity $\varrho_{298 K}$ = 4.0 Ω·cm [3] and thermogravimetric (TGA, DTA) results [4].

32

References:

[1] Wnuk, R. C.; Touw, T. R.; Post, B. (IBM J. Res. Develop. **8** [1964] 185/6).
[2] Wasel-Nielen, H. D.; Hoppe, R. (Z. Anorg. Allgem. Chem. **375** [1970] 209/13).
[3] Lazarev, V. B.; Shaplygin, I. S. (Zh. Neorgan. Khim. **23** [1978] 2902/6; Russ. J. Inorg. Chem. **23** [1978] 1610/2).
[4] Bergner, D.; Kohlhaas, R. (Z. Anorg. Allgem. Chem. **401** [1973] 15/20).
[5] Lazarev, V. B.; Shaplygin, I. S. (Mater. Res. Bull. **13** [1978] 229/35).
[6] Müller, O.; Roy, R. (Advan. Chem. Ser. No. 98 [1971] 28/38).

1.8.9 Strontium Palladates

$SrPdO_2$ is the final thermolysis product of $Sr[Pd(OH)_4]$, no other details are reported [1].

Sr_2PdO_3 is prepared by heating a mixture of SrO or $Sr(NO_3)_2$ and PdO ($Sr:Pd = 2.2:1.0$) under dry oxygen at 1250 to 1300°C for 24 h [2]. The brown crystals are orthorhombic, space group Immm-D_{2h}^{25} with $a = 3.977$, $b = 3.530$, $c = 12.82$Å, and $Z = 2$ [2] or $a = 3.970$, $b = 3.544$, $c = 12.84$ Å. X-ray data and a diagram of the structure, see **Fig. 3**, are reproduced in the paper [2]. Powder data have been tabulated [6].

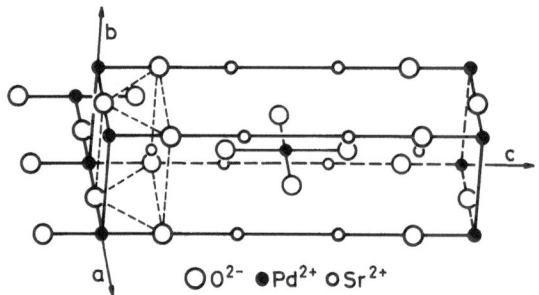

$\bigcirc$ O^{2-} $\bullet$ Pd^{2+} $\circ$ Sr^{2+}

Fig. 3.   Unit cell of Sr_2PdO_3.

It has also been obtained by heating $Sr(OH)_2 \cdot 8H_2O$ and Pd-black in air at 950°C for several days. Analyses correspond to $Sr_{1.93}PdO_{3+x}$ ($x = 0$ to 0.6) [6].

The compound is thermally stable to 950°C but dissolves rapidly in dilute acids and hydrates in the presence of water [6].

$SrPd_3O_4$ is prepared by heating a mixture of SrO or $Sr(NO_3)_2$ and PdO ($Sr:Pd = 1:3$) at temperatures between 800 and 1000°C [2], by heating a mixture of $SrCO_3$ and PdO under argon [3] or by heating a mixture of $Sr(OH)_2 \cdot 8H_2O$ and palladium black at ~950°C [6]. The black crystals are cubic ($NaPt_3O_4$ type), space group Pm3n-O_h^3 [4] with $a = 5.826$ [2, 4, 6] or 5.825 Å [3]. Interatomic distances are Pd–O 2.059; Sr–O 2.523; Pd–Pd 2.913 Å [6]. Other data reported are decomposition temperature (1020 to 1042°C) [4] or 960 to 990°C [3], resistivity $\varrho_{298\,K} = 48\ \Omega \cdot cm$ [3] and thermogravimetric (TGA and DTA) results [3, 4].

$SrPdO_3$ is the final product obtained on thermolysis of $Sr[Pd(OH)_6]$, no other details are reported [5].

References:

[1] Ivanov-Emin, B. N.; Petrishcheva, L. P.; Zaitsev, B. E.; Ivlieva, V. I.; Izmailovich, A. S.; Dolganev, V. P. (Zh. Neorgan. Khim **29** [1984] 2046/50; Russ. J. Inorg. Chem. **29** [1984] 1169/71).

[2] Wasèl-Nielen, H. D.; Hoppe, R. (Z. Anorg. Allgem. Chem. **375** [1970] 209/13).

[3] Lazarev, V. B.; Shaplygin, I. S. (Zh. Neorgan. Khim. **23** [1978] 2902/6; Russ. J. Inorg. Chem. **23** [1978] 1610/2).

[4] Bergner, D.; Kohlhaas, R. (Z. Anorg. Allgem. Chem. **401** [1973] 15/20).

[5] Ivanov-Emin, B. N.; Venskovskii, N. U.; Zaitsev, B. E.; Lin'ko, I. V. (Zh. Neorgan. Khim. **26** [1981] 2195/9; Russ. J. Inorg. Chem. **26** [1981] 1181/3).

[6] Müller, O.; Roy, R. (Advan. Chem. Ser. No. 98 [1971] 28/38).

1.8.10 Barium Palladates

BaPdO$_2$ is mentioned as the final thermolysis product obtained from Ba[Pd(OH)$_4$] [1].

BaPd$_3$O$_4$ has been mentioned but no details given [2]. An attempted synthesis from BaO/ PdO fused in PbCl$_2$ gave "Pd$_{0.5}$Pd$_3$O$_4$" [3].

BaPdO$_3$ is reported to be the final product obtained on thermolysis of Ba[Pd(OH)$_6$] [4].

References:

[1] Ivanov-Emin, B. N.; Petrishcheva, L. P.; Zaitsev, B. E.; Ivlieva, V. I.; Izmailovich, A. S.; Dolganev, V. P. (Zh. Neorgan. Khim. **29** [1984] 2046/50; Russ. J. Inorg. Chem. **29** [1984] 1169/71).

[2] Lazarev, V. B.; Shaplygin, I. S. (Mater. Res. Bull. **13** [1978] 229/35).

[3] Meyer, H.-J.; Müller-Buschbaum, H.-K. (Z. Naturforsch. **34 b** [1979] 1661/2).

[4] Ivanov-Emin, B. N.; Venskovskii, N. U.; Zaitsev, B. E.; Lin'ko, I. V. (Zh. Neorgan. Khim. **26** [1981] 2195/9; Russ. J. Inorg. Chem. **26** [1981] 1181/3).

1.8.11 Zinc and Cadmium Palladates

Zn$_2$PdO$_4$. This brick-red compound is prepared by heating a mixture of ZnO, PdO and KClO$_3$ (ratio 2:1:1.5) at 920°C and 70 kbar for 30 min. It adopts an inverse spinel structure with a cubic unit cell, a = 8.509 ± 0.005 Å. Magnetic moment measurements confirm low spin (t$_{2g}^6$e$_g^0$) palladium(IV) except for a small amount which occupies tetrahedral lattice sites. A plot of reciprocal magnetic susceptibility versus temperature (0 to 300 K) is reproduced [1].

CdPd$_3$O$_4$. This compound is prepared by heating mixtures of Pd(NO$_3$)$_2$/Cd(NO$_3$)$_2$ (ratio 3:1) [2, 4], PdO/CdCO$_3$ [2] or CdO/palladium black [3]. The crystals are cubic (Na$_x$Pt$_3$O$_4$ type), space group Pm3n-O$_h^3$ with a = 5.741 ± 0.002 [4] or 5.742 Å [3]. Interatomic distances are Pd–O = 2.030, Cd–O = 2.486, shortest Pd–Pd = 2.871 Å [3]; resistivity $\varrho_{298\,K}$ = 2.1 Ω·cm, decomposition interval 820 to 860°C [2].

References:

[1] Demazeau, G.; Omeran, I.; Pouchard, M.; Hagenmüller, P. (Mater. Res. Bull. **11** [1976] 1449/52).

[2] Lazarev, V. B.; Shaplygin, I. S. (Zh. Neorgan. Khim. **23** [1978] 2902/6; Russ. J. Inorg. Chem. **23** [1978] 1610/2).

[3] Müller, O.; Roy, R. (Advan. Chem. Ser. No. 98 [1971] 28/38).

[4] Bergner, D.; Kohlhaas, R. (Z. Anorg. Allgem. Chem. **401** [1973] 15/20).

1.8.12 Indium Palladate In$_2$Pd$_2$O$_7$

This is prepared by heating a 1:2:1 molar mixture of In$_2$O$_3$, PdO and KClO$_3$ at 1000°C and 65 kbar for 4 h or 700°C and 3 kbar for 8 h. It has the pyrochlore structure, space group Fd3m-O$_h^7$, with a = 9.919 ± 0.002 Å.

Reference:

Sleight, A. W. (Mater. Res. Bull. **3** [1968] 699/704).

1.8.13 Thallium Palladate TlPd₃O₄

This is prepared by heating PdO with $TlNO_3$ or Tl_2O_3 at 550°C. The structure has been determined by X-ray and neutron diffraction methods. The crystals are cubic, space group Fm3m-O_h^5 [1, 2] with a = 9.596(2) Å, D_{exp} = 8.74(8), D_{calc} = 8.83 g/cm³ for Z = 8 [1] or a = 9.5807(7) [2]. The compound can be formulated as a mixed-valence Tl^{III}-Tl^{I} oxopalladate(II). The structure contains square-planar PdO_4 units joined in a 3D framework by triply bonding oxygen atoms [1, 2].

References:

[1] Zöllner, C.; Thiele, G.; Müllner, M. (Z. Anorg. Allgem. Chem. **443** [1978] 11/8).
[2] Müllner, M.; Thiele, G.; Zöllner, C. (Z. Anorg. Allgem. Chem. **443** [1978] 19/22).

1.8.14 Scandium and Yttrium Palladates

Sc₂Pd₂O₇. This pyrochlore is prepared by heating a 1 : 2 : 1 molar mixture of Sc_2O_3, PdO and $KClO_3$ at 1000°C and 65 kbar pressure for 4 h or at 700°C and 3 kbar pressure for 8 h. The crystals are cubic, space group Fd3m-O_h^7 with a = 9.804 ± 0.002 Å.

Y₂Pd₂O₇ is prepared by heating a 1 : 2 : 1 molar mixture of Y_2O_3, PdO and $KClO_3$ at 1000°C and 65 kbar pressure for 4 h or at 700°C and 3 kbar pressure for 8 h. The pyrochlore structure has space group Fd3m-O_h^7 with a = 10.126 ± 0.002 Å.

Reference:

Sleight, A. W. (Mater. Res. Bull. **3** [1968] 699/704).

1.8.15 Rare Earth Palladates

Two major papers survey data on palladates formed with a wide range of rare earths, other publications deal with palladates of specific metals, notably La and Nd.

Phase relations between PdO and various rare earth sesquioxides, M_2O_3 (M = La, Nd, Sm, Eu, Gd, Dy, Ho, Er, Tm, Yb, Lu) in air have been examined and phase diagrams (see **Fig. 4**, **Fig. 5**, p. 35, and **Fig. 6**, p. 36) published. All syntheses were performed by preheating oxide mixtures at 770 to 780°C for ≧18 h, then reheating at 780 to 1310°C for periods of 2 to 168 h. Details of reagent ratios, heating temperatures and times used, and products obtained (determined by X-ray analysis) are extensively tabulated in the paper [1].

Rare earth palladates obtained from M_2O_3/PdO mixtures, with dissociation temperatures (in °C) [1]:

M_2O_3/PdO	1 : 2	1 : 1	2 : 1
M = La	$La_2Pd_2O_5$ 1105	—	La_4PdO_7 1190
Nd	$Nd_2Pd_2O_5$ 1085	Nd_2PdO_4 (860)[1)]	Nd_4PdO_7 1135

M_2O_3/PdO	1:2	1:1	2:1
Sm	$Sm_2Pd_2O_5$ 1060	Sm_2PdO_4 (940)	Sm_4PdO_7 1115
Eu	$Eu_2Pd_2O_5$ 1045	Eu_2PdO_4 (985)	Eu_4PdO_7 1085
Gd	$Gd_2Pd_2O_5$ 1025	Gd_2PdO_4 (1005)	[2]
Dy	[2]	Dy_2PdO_4 (1030)	[2]

[1] Values in parentheses: decomposition temperatures of metastable 1/1 compounds. – [2] No detectable reaction between M_2O_3 and PdO, no reaction with oxides of Ho, Er, Tm, Yb and Lu.

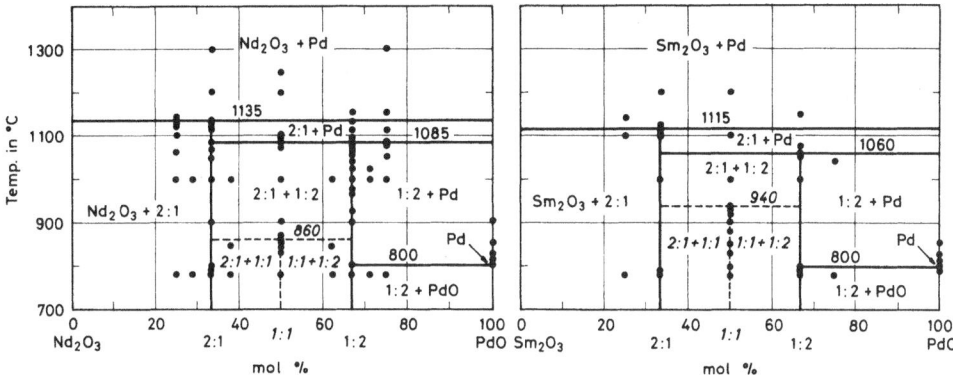

Fig. 4. Phase equilibrium diagram for the Nd_2O_3–PdO system in air.

Fig. 5. Phase equilibrium diagram for the Sm_2O_3–PdO system in air.

La_2PdO_4, La_2PdO_5, $La_4Pd_2O_7$. These ternary oxides have been prepared by heating mixtures of lanthanum oxalate and palladium dimethylglyoximate [Pd(dmg)$_2$] in air at temperatures in the range 500 to 600°C for 2 to 3 h, then annealing the product at higher temperatures. Annealing for 200 to 250 h at 790 to 810°C gives brown **La_2PdO_4**, space group I4/mmm-D_{4h}^{17}, a = 4.055, c = 12.62 Å, D_{X-ray} = 7.177 g/cm^3, Z = 2. Raising the annealing temperature to 920 to 940°C gives bright yellow **$La_2Pd_2O_5$**, space group P4$_2$/m-C_{4h}^2, a = 6.686, c = 5.616 Å, D_{X-ray} = 7.555 g/cm^3, Z = 2, and light brown **La_4PdO_7** [2]. The compound La_4PdO_7 has also been obtained by heating La_2O_3, $La_2(C_2O_4)_3$ or La(OH)$_3$ with PdO, Pd(dmg)$_2$ or [Pd(NH$_3$)$_4$][PdCl$_4$], or by calcining La(NO$_3$)$_3$ with Pd(NO$_3$)$_2$ [3]. Indexed X-ray powder data for all three compounds and infrared spectra for $La_2Pd_2O_5$ and La_4PdO_7 are reproduced in the paper. Electrical resistivities at room temperature are 4×10^{-6} Ω·cm for $La_2Pd_2O_5$ and $>10^{11}$ Ω·cm for La_4PdO_7. The solidus part of the La_2O_3/PdO equilibrium diagram is reproduced in the paper [2].

$La_2Pd_2O_6$ has been obtained by heating La_2O_3, $La_2(C_2O_4)_3$ or La(OH)$_3$ with PdO, Pd(dmg)$_2$ or [Pd(NH$_3$)$_4$][PdCl$_4$], or by calcining a mixture of La(NO$_3$)$_3$ and Pd(NO$_3$)$_2$. It is described as air-stable with a decomposition temperature >1100°C [3].

Nd_2PdO_4, $Nd_2Pd_2O_5$, Nd_4PdO_7. These ternary oxides have been prepared by heating mixtures of neodymium oxalate and palladium dimethylglyoxime in air at 790 to 940°C for long periods. Annealing a mixture of the appropriate composition for 200 to 220 h at 790 to 810°C gave **Nd_2PdO_4** as a dark brown solid, space group I4/mmm-D_{4h}^{17}, a = 4.002, c = 12.44 Å, D_{X-ray} = 7.654/cm^3, Z = 2. Annealing for 15 to 20 h at 900 to 940°C gave **$Nd_2Pd_2O_5$** as a bright yellow solid, space group P4$_2$/m-C_{4h}^2, a = 6.562, c = 5.557 Å, D_{X-ray} = 8.074 g/cm^3, Z = 2, and **Nd_4PdO_7** as a pale brown solid [4]. Indexed X-ray powder data [1, 4] and infrared spectra (800 to

400 cm^{-1}) [4] are reproduced for all three compounds. Thermal analysis curves (DTA and DTG) are reproduced for $Nd_2Pd_2O_5$, decomposition to Nd_4PdO_7, Pd and O_2 begins at 1090 ± 5°C, Nd_4PdO_7 further decomposes to Nd_2O_3, Pd and O_2 at 1145 ± 10°C.

Electrical resistivities (in $\Omega \cdot cm$) are 4.6×10^5 for Nd_2PdO_4, 3.8×10^9 for $Nd_2Pd_2O_5$ and $> 10^{12}$ for Nd_4PdO_7. The solidus part of the Nd_2O_3/PdO equilibrium diagram is reproduced in the paper [4].

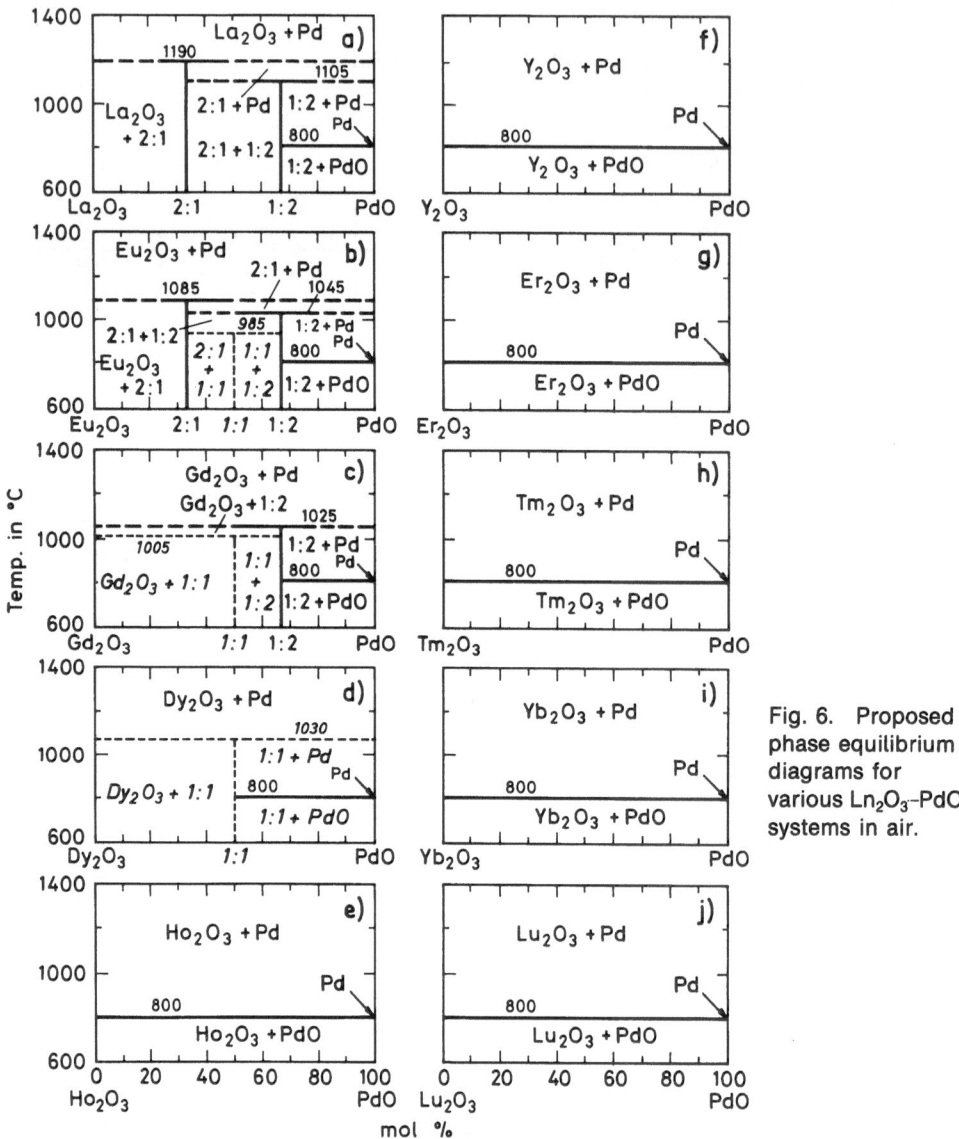

Fig. 6. Proposed phase equilibrium diagrams for various Ln_2O_3–PdO systems in air.

$M_2Pd_2O_7$ (M = Gd, Dy, Er, Yb). A series of ternary oxides $M_2Pd_2O_7$ with the pyrochlore structure has been prepared by heating 1:2:1 molar mixtures of the rare earth sesquioxide M_2O_3, palladium oxide and potassium chlorate at 1000°C under 65 kbar pressure for 4 h (platinum container) or at 700°C and 3 kbar pressure for 8 h (gold container). The compounds are cubic, space group Fd3m-O_h^7. Cell constants a in Å: $Gd_2Pd_2O_7$ 10.236, $Dy_2Pd_2O_7$ 10.155, $Er_2Pd_2O_7$ 10.103, $Yb_2Pd_2O_7$ 10.049, all ± 0.002 Å. Attempts to prepare $La_2Pd_2O_7$ failed [5].

References:

[1] McDaniel, C. L.; Schneider, S. J. (J. Res. Nat. Bur. Std. A **72** [1968] 27/37).
[2] Kakhan, B. G.; Lazarev, V. B.; Shaplygin, I. S. (Zh. Neorgan. Khim. **27** [1982] 2090/3; Russ. J. Inorg. Chem. **27** [1982] 1180/2).
[3] Kakhan, B. G.; Lazarev, V. B.; Shaplygin, I. S. (Term. Anal. Tezisy. Dokl. 7th Vses. Soveshch., Riga 1979, Vol. 1, pp. 243/4; C.A. **93** [1980] No. 229975).
[4] Kakhan, B. G.; Lazarev, V. B.; Shaplygin, I. S. (Zh. Neorgan. Khim. **27** [1982] 2395/2401; Russ. J. Inorg. Chem. **27** [1982] 1352/6).
[5] Sleight, A. W. (Mater. Res. Bull. **3** [1968] 699/704).

1.8.16 Titanium Palladates

The reaction between PdO (0.3) and TiO_2 (0.3 to 1.0) when diluted with Al_2O_3 (99.4 to 98.7 wt%) at 500 to 800°C in air has been followed by diffuse reflectance spectra. Growth of bands at 300 to 330 and 480 to 490 nm confirmed the reaction of PdO with TiO_2 and the reduction of Ti^{IV} to a lower oxidation state [1]. Black products with differing PdO/TiO_2 ratios, prepared by impregnating TiO_2 with $Pd(NO_3)_2$, then drying in air for 3 to 4 h at 80 to 90°C under an infrared lamp, have been studied by X-ray diffraction, differential thermography, conducto-metric measurements, ultraviolet spectroscopy and electron spin resonance spectroscopy. PdO dissolves in the TiO_2 lattice and a Pd–Ti interaction is observed. Thermal analysis curves, reflectance spectra and plots of resistivity versus temperature (500 to 200 K) are reproduced in the paper [2]. Mixtures of PdO/TiO_2 afford anticorrosion protection for titanium surfaces [3]. A semiconductor compound $PdTiO_{3+x}$ has been observed in the system PdO/TiO_2 under pressure but no details have been reported [4].

References:

[1] Kravchuk, L. S.; Makarova, E. A.; Ivashchenko, N. I.; Markevich, S. V.; Zaretskii, M. V. (Izv. Akad. Nauk SSSR Neorgan. Materialy **16** [1980] 1037/9; Inorg. Materials **16** [1980] 717/9).
[2] Kravchuk, L. S.; Ivashchenko, N. I.; Bogush, A. K.; Zaretskii, M. V.; Akimova, N. E. (Zh. Neorgan. Khim. **26** [1981] 2630/4; Russ. J. Inorg. Chem. **26** [1981] 1409/11).
[3] Fukuzuka, T.; Shimogori, K.; Satoh, H.; Kamikubo, F. (R & D Res. Develop. Kobe Steel Ltd. **32** No. 1 [1982] 32/5; C.A. **96** [1982] No. 166966).
[4] Lazarev, V. B.; Shaplygin, I. S. (Mater. Res. Bull. **13** [1978] 229/35).

1.8.17 Lead Palladate $PbPdO_2$

This is prepared by heating a 1:1 PbO palladium-black mixture in air at ~600 to 700°C for several days. The black powder, which is unstable with respect to PbO and $PbPd_3$ above ~820°C, gives an X-ray diffraction pattern which has been tentatively indexed on the basis of an hexagonal cell (possibly only a pseudo cell) with a = 10.902, c = 4.654 Å [1]. A preparation from PdO and PbO at 400 to 900°C has been mentioned, and decomposition to PbO, Pd and O_2 at 830 ± 10°C is reported. Interplanar spacings are tabulated [2].

References:

[1] Müller, O.; Roy, R. (Advan. Chem. Ser. No. 98 [1971] 28/38).
[2] Shaplygin, I. S.; Bromberg, A. V.; Sokol, V. A. (Zh. Neorgan. Khim. **15** [1970] 2305; Russ. J. Inorg. Chem. **15** [1970] 1195).

1.8.18 Vanadium Palladates

SnO_2/VO_x-PdO ceramics display a double action switching below 40°C which disappears above 60°C, oscillation is irregular with an amplitude of 30 to 40 V at room temperature.

Reference:

Furugochi, Y.; Kanefusa, S.; Teketa, Y.; Nitta, M.; Haradome, M. (Nihon Daigaku Seisankoga-kubu Hokoku A **11** [1978] 167/85; C.A. **91** [1979] No. 25994).

1.8.19 Chromium Palladate $CrPdO_2$

This delafossite type ternary oxide is obtained as a black powder by heating a mixture of Pd, $PdCl_2$ and $LiCrO_2$ at 800°C and is thermally stable to 925°C [1]. Cell dimensions a = 2.9230 ± 0.0003, c = 18.087 ± 0.003 Å, powder diffraction data are tabulated in the paper [1]. The compound is formally regarded as a chromium(III)-palladium(I) mixed oxide [1]. Thin film electrodes of $CrPdO_2$ electrocatalyse the reduction of O_2, and O_2 evolution in NaOH. Plots of resistivity versus temperature (0 to 300 K) and O_2 evolution current versus potential for $CrPdO_2$ thin films are reproduced in the paper [2].

References:

[1] Shannon, R. D.; Rogers, D. B.; Prewitt, C. T. (Inorg. Chem. **10** [1971] 713/8).
[2] Carcia, P. F.; Shannon, R. D.; Bierstedt, P. E.; Flippen, R. B. (J. Electrochem. Soc. **127** [1980] 1974/8).

1.8.20 Cobalt Palladate $CoPdO_2$

This delafossite type ternary oxide is obtained as hexagonal metallic platelets by heating a mixture of $PdCl_2$ and CoO at 700°C. It decomposes at 900°C to palladium, CoO and oxygen [1]. Cell dimensions a = 2.8300 ± 0.0003, c = 17.743 ± 0.002 Å, powder diffraction data are tabulated in the paper [1, 2], the space group is thought to be $R\bar{3}m$-D_{3d}^5 [2]. Structural evidence for the presence of low-spin Co^{III} and the very low non-Curie-Weiss paramagnetic susceptibility, $\psi_g \approx (94/T + 0.93) \times 10^{-6}$ emu/g, point to the presence of Pd^I [1, 2]. Interatomic distances and bond angles are tabulated and a diagram of the delafossite structure is reproduced [2]. Room temperature resistivities, measured parallel and perpendicular to the c-axis are 2.1×10^{-3} and 2×10^{-6} $\Omega \cdot$ cm, respectively, and support the presence of Pd–Pd bonds, Pd–Pd = 2.83 Å [3]. Thin film electrodes of $CoPdO_2$ electrocatalyse the reduction of O_2, and O_2 evolution from NaOH [4]. Plots of resistivity versus temperature (0 to 300 K) and O_2 evolution current versus potential for $CoPdO_2$ thin films are reproduced [4].

References:

[1] Shannon, R. D.; Rogers, D. B.; Prewitt, C. T. (Inorg. Chem. **10** [1971] 713/8).
[2] Prewitt, C. T.; Shannon, R. D.; Rogers, D. B. (Inorg. Chem. **10** [1971] 719/23).

[3] Rogers, D. B.; Shannon, R. D.; Prewitt, C. T.; Gillson, J. L. (Inorg. Chem. **10** [1971] 723/7).
[4] Carcia, P. F.; Shannon, R. D.; Bierstedt, P. E.; Flippen, R. B. (J. Electrochem. Soc. **127** [1980] 1974/8).

1.8.21 Copper Palladates

Equilibrium measurements have been conducted on the CuO–PdO system at 900 and 1000°C, results are given in the form of decomposition isotherms for numerous CuO–PdO mixtures at 900 and 1000°C. For these temperatures, phase triangles have been constructed which contain tie lines for the bivariant range (CuO, PdO) mixed crystals in equilibrium with Cu–Pd alloys. The triangles show that the dissolution of the monoclinic CuO in the tetragonal PdO gives a tetragonal phase over the mole fraction range $N_{PdO} = 1$ to 0.155.

Reference:

Schmahl, N. G.; Minzl, E. (Z. Physik. Chem. [Frankfurt] **47** [1965] 142/63).

1.8.22 Rhodium Palladates

RhPdO. A product obtained on heating high purity Pd-5 at% Rh alloy in a pure oxygen atmosphere at 900°C has been assigned this formulation on the basis of vacuum-fusion oxygen analysis and X-ray fluorescence analysis. A more appropriate formulation may be $(Pd_xRh_{(1-x)})_2O$ with $x = 0.4 \pm 0.1$ [1].

RhPdO$_2$. This delafossite type ternary oxide is obtained as a black powder on heating a mixture of palladium metal, $PdCl_2$ and $LiRhO_2$ at 800°C, and is thermally stable to 800°C. Cell dimensions $a = 3.0209 \pm 0.0002$, $c = 18.083 \pm 0.002$ Å, powder diffraction data are tabulated. Formulation as a Rh^{III}/Pd^I oxide is proposed [2]. Thin film electrodes of $RhPdO_2$ electrocatalyse the reduction of O_2 and evolution of O_2 in NaOH. Plots of resistivity versus temperature (0 to 300 K) and O_2 evolution current versus potential for $RhPdO_2$ thin film electrodes are reproduced [3].

References:

[1] Seybolt, A. U. (Trans. AIME **233** [1965] 248).
[2] Shannon, R. D.; Rogers, D. B.; Prewitt, C. T. (Inorg. Chem. **10** [1971] 713/8).
[3] Carcia, P. F.; Shannon, R. D.; Bierstedt, P. E.; Flippen, R. B. (J. Electrochem. Soc. **127** [1980] 1974/8).

1.8.23 Palladium Palladate Pd$_{0.5}$Pd$_3$O$_4$

This palladium(II)-palladium(IV) mixed oxide has been obtained by fusing an intimate mixture of PdO and BaO with $PbCl_2$ at 600°C for 3 h and 680°C for 10 d. It forms black crystals of the $Na_xPt_3O_4$-structure type, space group Pm3n-O_h^3 with $a = 5.756$ Å. The structure has been compared to that of $CaPd_3O_4$ with Pd^{4+} ions occupying half of the interstitial (calcium) sites in a Pd_3O_4 array containing square-planar palladium(II). An alternative formulation $Pd_{0.5}^{2+}Pd_{0.5}^{4+}Pd_{2.5}^{2+}O_4$ in which Pd^{2+} ions occupy half of the interstitial sites was considered but rejected on the grounds that it would imply the presence of square-planar coordinated Pd^{4+}.

Reference:

Meyer, H.-J.; Müller-Buschbaum, H. (Z. Naturforsch. **34 b** [1979] 1661/2).

2 Palladium and Nitrogen

See "Palladium" 1942, p. 268.

2.1 The Pd–N System

There seems to be no evidence for the existence of binary palladium nitrides. Calculated heats of formation for Pd_xN_y are +10 kJ/mol for $Pd_{0.67}N_{0.33}$, +22 kJ/mol for $Pd_{0.5}N_{0.5}$ and +39 kJ/mol for $Pd_{0.4}N_{0.6}$, based on chemical potential and electron density data. Thus the non-existence of binary nitrides was predicted [1].

Adsorption of N_2. Infrared spectroscopy has been used to study adsorption of N_2 on Pd surfaces at −90°C and 20°C; a N–N stretch was observed at 2260 cm^{-1} (spectra reproduced in paper from 2000 to 2500 cm^{-1}) [2]. For adsorption of various gases including N_2 on Pd, see [3].

The theory of bonding of transition metals to non-transition metals, including Pd to N, has been discussed [4].

References:

[1] Bouten, P. C. P.; Niedema, A. R. (J. Less-Common Metals **65** [1979] 217/28, 223).
[2] van Hardeveld, R.; van Montfoort, A. (Surf. Sci. **4** [1966] 396/430, 409).
[3] Kavtaradze, N. N. (J. Res. Inst. Catal. Hokkaido Univ. **13** [1965] 196/208; C.A. **65** [1966] 1418).
[4] Gelatt, C. D.; Williams, A. R.; Moruzzi, V. L. (Phys. Rev. [3] B **27** [1983] 2005/13).

2.2 Ternary Nitrides

Pd_2Zr_4N. No preparative details were given for this material, but it has the partially completed Ti_2Ni structure with a = 12.40 Å [1].

$PdMn_3N$. Reaction of Mn_3N with Pd at 750°C gave this compound [2, 3]. The structure is of the perovskite type with a = 3.979 Å [3].

References:

[1] Holleck, H.; Thümmler, F. (Monatsh. Chem. **98** [1967] 133/4).
[2] Samson, C.; Bouchaud, J.-P.; Fruchart, R. (Compt. Rend. **259** [1964] 392/3).
[3] Madar, R.; Gilles, L.; Rouault, A.; Bouchaud, J.-P.; Fruchart, E.; Lorthioir, G.; Fruchart, R. (Compt. Rend. C **264** [1967] 308/11).

2.3 Palladium Nitrates (see "Palladium" 1942, p. 269)

General References:

Critchlow, P. B.; Robinson, S. D.; A Review of Platinum Metal Nitrato Complexes, Coord. Chem. Rev. **25** [1978] 69/102, 81.

Addison, C. C.; Sutton, D.; Complexes Containing the Nitrate Ion. Prog. Inorg. Chem. **8** [1967] 195/286.

Field, B. O.; Hardy, C. J.; Inorganic Nitrates and Nitrato Compounds, Quart. Rev. [London] **18** [1964] 361/88.

2.3.1 Palladium(II) Nitrates

A volatile, moisture-sensitive anhydrous form of palladous nitrate is known but most papers refer to the dihydrate $Pd(NO_3)_2(H_2O)_2$ or aqueous solutions of palladium nitrate. Salts of the $[Pd(NO_3)_4]^{2-}$ anion are also well established.

$Pd(NO_3)_2$. The anhydrous salt is obtained by treatment of the dihydrate with liquid N_2O_5 and is isolated as a brown, volatile, moisture-sensitive solid which sublimes at 80°C and 3×10^{-3} Torr, and is readily hydrated. Infrared absorptions occur at 1630(m), 1506(s), 1275(s), 1170(vs), 1000(m), 775(m), 762(m), 734(m) and 647(m) cm^{-1}; bands at 1630 and 1170 cm^{-1} are tentatively attributed to bridging nitrate [1]. The standard enthalpy of formation, $-\Delta H^\circ_{298}$, and Gibbs free energy, $-\Delta G^\circ_{298}$, of $Pd(NO_3)_2$ are given as 272 ± 4 and 84 kJ/mol, respectively [2]. Anhydrous $Pd(NO_3)_2$ reacts with carbon monoxide at 20 to 120°C and 1 atm pressure to give Pd, CO_2 and NO_2 [3], and with olefins $RCH=CH_2$ (R = H or Me) in anhydrous dioxane or C_6F_6 at 25 to 60°C to give $O_2NOCHRCH_2ONO_2$ and palladium metal [4]. With toluene in acetic acid cresyl acetates and nitrotoluenes are formed [5].

$Pd(NO_3)_2 \cdot 2N_2O_4$. This adduct has been mentioned but no details given [6].

$Pd(NO_3)_2 \cdot N_2O_5$. An alternative formulation for $Pd(NO_3)_4$ considered but discounted on basis of chemical evidence [6].

$Pd(NO_3)_2(H_2O)_2$ is prepared by dissolving palladium metal in concentrated nitric acid and evaporating the solution to small volume. It forms light brown [7] or yellow-brown lamellar [8] crystals which can be pressed almost dry, then dried in vacuo over sodium hydroxide [7] or washed with diethyl ether and dried in vacuo [8]. A claim that this hydrate is really a palladium(IV) derivative $Pd(NO_3)_2(OH)_2$ was made [6] but was later refuted [8]. Chloride-free aqueous solutions of $Pd(NO_3)_2$ can be obtained by dissolving palladium metal in warm nitric acid with simultaneous addition of <2% $[Pd(NH_3)_4][NO_3]_2$ [9]. The kinetics of solution of metallic palladium in nitric acid (1.8 to 5.7M) have been studied, acid concentration and stirring rate have no noticeable effect on the rate of dissolution but the temperature coefficient (20%/°C) is unusually high [10]. Palladium nitrate dihydrate is readily soluble in acetone, appreciably soluble in diethyl ether and methyl isobutyl ketone, and dissolves in ethanol with precipitation of palladium black [8].

The infrared spectrum of $Pd(NO_3)_2(H_2O)_2$ shows absorption bands at 1502(s), 1362(s), 1307(s), 1274(s), 988(vs), 797(sh) and 784(s) cm^{-1}; assignments are given [7,11]. The ^{1}H NMR spectrum (reproduced in the paper) has been used to refute the alternative formulation $Pd(NO_3)_2(OH)_2$ [8]. The absorption and emission maxima and phosphorescent lifetime of $Pd(NO_3)_2$ have been recorded [12]. The effect of added pyridine on the polarographic reduction waves of palladium nitrate has been investigated [13]. Hydrolysis of $Pd(NO_3)_2(H_2O)_2$ gives polymeric "hydroxide nitrates" [8], $Pd_3O(NO_3)_2(OH)_2(H_2O)_2$ [14] or, in boiling dilute nitric acid, $PdO \cdot xH_2O$ [15]. Thermal decomposition of $Pd(NO_3)_2(H_2O)_2$ via $Pd(NO_3)(OH)(H_2O)$ and $Pd(OH)_2$ to PdO has been reported [14] and a thermal analysis curve is reproduced [8].

Ethylene is oxidised by $Pd(NO_3)_2 \cdot xH_2O$ (x = 1, 2 or 2.5) to $HOCH_2CH_2ONO_2$ [4].

For palladous nitrate adducts with oxygen, nitrogen, sulphur and phosphorus donor ligands see "Palladium" Suppl. Vol. B 3 (to be published).

$[Pd(NO_3)(H_2O)_3]^+$. Spectroscopic evidence for the formation of this cation in aqueous HNO_3 solution has been presented and an extrapolated spectrum (500 to 300 nm) is reproduced, $K_1 = 1.2 \pm 0.4$ mol^{-1} [16]. An earlier paper mentions shifts in the absorption spectrum of $[Pd(H_2O)_4][ClO_4]_2$ on addition of nitric acid [17].

Pd(NO₃)(OH)(H₂O) has been mentioned as an intermediate species in the thermal decomposition of $Pd(NO_3)_2(H_2O)_2$ [14].

Pd₃O(NO₃)₂(OH)₂(H₂O)₂ has been isolated from the hydrolysis of $Pd(NO_3)_2(H_2O)_2$. A proton NMR spectrum and thermal analysis curves are reproduced in the paper [14].

Pd(NO)₂(NO₃)₂. A black diamagnetic product obtained by treatment of palladous nitrate with NO and methanol vapour has been tentatively assigned this formulation. Infrared bands are $\nu(NO) = 1810$, 1780 cm⁻¹; $\nu(NO_3) = 1500$, 1300, 990, 820 cm⁻¹ [18].

K₂[Pd(NO)(NO₃)(NO₂)₄]·2H₂O is prepared by treatment of $K_2[Pd(NO_2)_4]$ with 8M HNO_3 at 80°C for 2 h and forms red crystals from 0.1M HNO_3. Infrared bands are $\nu(NO) = 1720$ cm⁻¹; $\nu(NO_3) = 1500$, 1280, 1010, 995 and 790 cm⁻¹; $\nu(NO_2) = 1300$ and 800 cm⁻¹ [18].

K₂[Pd(NO₃)₄] is prepared by evaporating a solution of $K_2[Pd(NO_2)_4]$ and concentrated HNO_3 [1, 2] or by adding a stoichiometric amount of KNO_3 to a solution of palladium in concentrated nitric acid [19, 20, 21]. It forms orange-red air-stable but moisture-sensitive crystals [19, 20]. X-ray investigations have shown the crystals to be monoclinic, space group $P2_1/c–C_{2h}^5$ with a = 7.940(2), b = 15.469(4), c = 9.453(2)Å, $\beta = 91.10(3)°$, $D_{exp} = 2.48(1)$, $D_{calc} = 2.479$ g/cm³, Z = 4. The complex anion has a basket-shaped structure with square-planar coordinated Pd^{II}; Pd–O = 1.995 to 2.010Å, see **Fig. 7** [21].

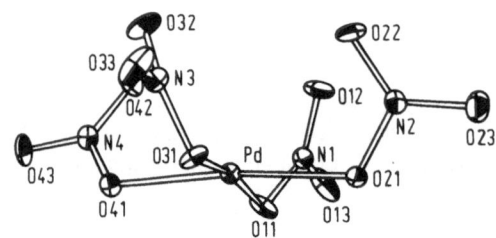

Fig. 7. Perspective view of the $[Pd^{II}(NO_3)_4]^{2-}$ ion [21].

The infrared spectrum is reproduced; $\nu(NO_3) = 1500$, 1370, 1270, 1235, 1010, 997, 802 and 796 cm⁻¹ [20]. Photoelectron and Auger line energy data show Ols BE = 532.3, KVV KE = 509.4 eV [22].

[A]₂[Pd(NO₃)₄] (A = tetramethyl-, methyltrioctyl-, tetraoctyl-, tributyl-, trioctyl-, mono-octyl- and monolauryl-ammonium, and benzoquinolinium cations). These salts are prepared by treatment of $Pd(NO_3)_2(H_2O)_2$ with the appropriate organic nitrate [A][NO₃] in aqueous 1M HNO_3 or acetone (for hydrophobic organic nitrates) solution. Colours range from ruby to pale yellow depending upon the nature of the cation and the particle size. Infrared spectra (3500 to 750 cm⁻¹) are tabulated and assignments given. Electronic reflectance (A = tetramethylammonium and benzoquinolinium) or absorption (A = tetraoctyl-, methyltrioctyl- and trioctyl-ammonium) spectra (520 to 360 nm) are reproduced in the paper. Thermal analysis curves (DTGA, DTA and TGA) are reproduced (A = tetramethyl- or tetraoctyl-ammonium and benzoquinolinium) [23].

References:

[1] Field, B. O.; Hardy, C. J. (J. Chem. Soc. **1964** 4428/34).
[2] Chukurov, P. M.; Drakin, S. I.; Karapet'yants, M. Kh. (Zh. Fiz. Khim. **50** [1976] 2989/90; Russ. J. Phys. Chem. **50** [1976] 1781).

[3] Moiseev, I. I.; Vargaftik, M. N.; Gentosh, O. I.; Zhavoronkov, N. M.; Kalechits, I. V.; Pazderskii, Yu. A.; Stromnova, T. A.; Shcherbakova, L. S. (Dokl. Akad. Nauk. SSSR. **237**
[1977] 645/7; C. A. **88** [1978] No. 42195).

[4] Likholobov, V. A.; Ermakov, Yu. I.; Burylin, S. M. (Kinetika Kataliz **15** [1974] 1092; Kinet.
Catal. [USSR] **15** [1974] 977).

[5] Tanaka, M.; Yoneyama, T.; Yamashita, J.; Hashimoto, H. (Technol. Rept. Tohoku. Univ. **43**
[1978] 105/10; C.A. **89** [1978] No. 196587).

[6] Addison, C. C.; Ward, B. G. (Chem. Commun. **1966** 155).

[7] Gatehouse, B. M.; Livingstone, S. E.; Nyholm, R. S. (J. Chem. Soc. **1957** 4222/5).

[8] Shmidt, V. S.; Shorokhov, N. A.; Vashman, A. A.; Samsonov, V. E. (Zh. Neorgan. Khim. **27**
[1982] 1254/6; Russ. J. Inorg. Chem. **27** [1982] 703/5).

[9] Funk, R. (Ger. [East] 114389 [1975]; C.A. **85** [1976] No. 35194).

[10] Mueller, R. L.; Afanas'eva-Potepun, E. Ya. (Zh. Neorgan. Khim. **2** [1957] 1306/16; Russ. J.
Inorg. Chem. **2** No. 6 [1957] 148/64; C.A. **1958** 3488).

[11] Gatehouse, B. M.; Livingstone, S. E.; Nyholm, R. S. (J. Inorg. Nucl. Chem. **8** [1958] 75/8).

[12] Maria, H. J.; Srinivasan, B. N.; McGlynn, S. P. (Mol. Lumin. Intern. Conf., Chicago 1968
[1969], pp. 787/800; C.A. **71** [1969] No. 26304).

[13] Magee, R. J.; Douglas, W. H. (J. Electroanal. Chem. **6** [1963] 261/6).

[14] Shorokhov, N. A.; Vashman, A. A.; Samsonov, V. E.; Chuklinov, R. N.; Shmidt, V. S. (Zh.
Neorgan. Khim. **27** [1982] 3137/40; Russ. J. Inorg. Chem. **27** [1982] 1773/6).

[15] Glemser, O.; Peuschel, G. (Z. Anorg. Allgem. Chem. **281** [1955] 44/53).

[16] Jørgensen, C. K.; Parthasarathy, V. (Acta Chem. Scand. A **32** [1978] 957/62).

[17] Sundaram, A. K.; Sandell, E. B. (J. Am. Chem. Soc. **77** [1955] 855/7).

[18] Griffith, W. P.; Lewis, J.; Wilkinson, G. (J. Chem. Soc. **1961** 775/82).

[19] Eskenazi, R.; Raskovan, J.; Levitus, R. (Chem. Ind. [London] **1962** 1327/8).

[20] Eskenazi, R.; Raskovan, J.; Levitus, R. (Anales Asoc. Quim. Arg. **50** [1962] 9/17).

[21] Elding, L. I.; Norén, B.; Oskarsson, A. (Inorg. Chim. Acta **114** [1986] 71/4).

[22] Wagner, C. D.; Zatko, D. Å.; Raymond, R. H. (Anal. Chem. **52** [1980] 1445/51).

[23] Shorokhov, N. A.; Teterin, E. G.; Chuklinov, R. N.; Shmidt, V. S. (Zh. Neorgan. Khim. **29**
[1984] 1255/60; Russ. J. Inorg. Chem. **29** [1984] 718/22).

2.3.2 Palladium(IV) Nitrates

Pd(NO$_3$)$_4$ is obtained as diamagnetic brown plate-like crystals by mixing Pd(NO$_3$)$_2$(H$_2$O)$_2$
with N$_2$O$_5$ at −78°C, then allowing to warm to room temperature. It forms a yellow acidic
solution in water which releases iodide quantitatively from KI but does not oxidise ferrous ions.

Pd(NO$_3$)$_2$(OH)$_2$. This formulation was erroneously assigned to the hydrate Pd(NO$_3$)$_2$(H$_2$O)$_2$.

Reference:

Addison, C. C.; Ward, B. G. (Chem. Commun. **1966** 155).

3 Palladium and Fluorine

General Remarks. The chemistry of the Pd/F system is extensive. The palladium oxidation states II, IV and VI are found in the well-characterised fluorides PdF_2, PdF_4 and PdF_6, respectively. $Pd^{III}F_3$ does not exist as such except possibly under high pressure, a claim to Pd^VF_5 has not been substantiated. However, all of these oxidation states are encountered amongst the many well-established complex fluoro palladates.

General References:

Müller, G. B.; Fluoride mit Kupfer, Gold und Palladium, Angew. Chem. **99** [1987] 1120/35.

Hagenmüller, P.; Inorganic Solid Fluorides, Academic Press 1985.

Edwards, A. J.; Solid-State Structures of the Binary Fluorides of the Transition Metals, Advan. Inorg. Chem. Radiochem. **27** [1983] 83/112.

Woolf, A. A.; Thermochemistry of Inorganic Fluorine Compounds, Advan. Inorg. Chem. Radiochem. **24** [1981] 1/55.

Zemskov, S. V.; Gabuda, S. P.; Structural Features of Hexafluoro Complexes of Noble Metals, Zh. Strukt. Khim. **17** [1976] 904/21; J. Struct. Chem. [USSR] **17** [1976] 772/89.

Allen, G. C.; Warren, K. D.; Electronic Spectra of the Hexafluoro Complexes of the Transition Metals, Coord. Chem. Rev. **16** [1975] 227/44.

Allen, G. C.; Warren, K. D.; Electronic Spectra of the Hexafluoro Complexes of the Second and Third Transition Series, Struct. Bonding [Berlin] **19** [1974] 105/65.

O'Donnell, T. A.; Chemical Reactivity of Higher Fluorides of the Transition Metals, Rev. Pure Appl. Chem. **20** [1970] 159/74.

Canterford, J. H.; Colton, R.; Halides of the Second and Third Row Transition Metals, Halides of the Transition Elements, Interscience, New York 1968, pp. 1/409.

Thiele, G.; Brodersen, K.; Zur Strukturchemie der Halogenide der Platinmetalle, Fortschr. Chem. Forsch. **10** [1968] 631/72.

3.1 Palladium Monofluoride PdF

Calculations on the heat of formation of the hypothetical compound "PdF" are included in a theoretical paper on bonding between transition metals and nontransition elements.

Reference:

Gelatt, C. D.; Williams, A. R.; Moruzzi, V. L. (Phys. Rev. [3] B **27** [1983] 2005/13).

3.2 Palladium Difluoride PdF₂ (see "Palladium" 1942, p. 270)

Palladium difluoride exists in two crystallographic forms – tetragonal (rutile) and cubic (fluorite).

Preparation. The pale violet tetragonal form is made by reduction of "PdF_3" or PdF_3BrF_3 with SF_4 [1] or SeF_4 [1, 2], by heating Pd/"PdF_3" mixtures in an argon/fluorine atmosphere at 500 to 750°C [3], in a sealed tube at 600°C [4] or 980°C [5] or by heating Pd/PdF_4 mixtures in a sealed tube at 980°C [5]. It has also been obtained by heating PdS with excess SF_4 at 150 to 350°C and 5 to 50 atm pressure [6]. It is rapidly hydrolysed in moist air and reacts with water to form

Pd(OH)$_2$ [1]. The violet-brown cubic form is obtained by compressing the tetragonal form at pressures up to 90 kbar (temperatures not specified) [4, 7, 8]. Complete conversion is achieved at 400°C and 50 kbar pressure [4, 8]. It is stable at 60 kbar and metastable at atmospheric pressure [7]. The structural transformation rutile → fluorite has been observed to occur for PdF$_2$ at room temperature under pressure [9] and has been followed by pressure vs. resistivity measurements [8].

Crystallographic Properties. The tetragonal form has space group P4$_2$/mnm-D$_{4h}^{14}$, a = 4.956 ± 0.002, c = 3.389 ± 0.002 Å, and is isostructural with MnF$_2$, FeF$_2$, CoF$_2$, NiF$_2$ and ZnF$_2$. Each palladium atom is surrounded by 6 fluorine atoms (Pd–F, 2 at 2.171, 4 at 2.155 Å) in an almost regular octahedral arrangement with the octahedra sharing corners [10]. An earlier report [1] described PdF$_2$ as a pale violet compound with an X-ray pattern corresponding to a tetragonal unit cell of the rutile type with a = 4.96 and c = 3.39 Å. The neutron diffraction patterns of tetragonal PdF$_2$ at 4.2 and 300 K are reproduced and the data tabulated [12].

The cubic form of PdF$_2$ belongs to the space group Pa3-T$_h^6$ [4, 7] or, more likely P2$_1$3-T^4 [4] with a = 5.32 [7] or 5.322 ± 0.002 Å (at 20°C by neutron diffraction) [4]. The structure is derived from the fluorite type by rhombohedral distortion of the cubic environment of the cations [4]. The palladium is surrounded by 6 fluorines at 2.178 [4] or 2.16 Å [8] and 2 at 3.167 Å [4]. Neutron diffraction patterns (20, 160 and 300 K) of the cubic form of PdF$_2$ [4] and a diagram of the structure [4, 15] are reproduced in papers. Relationships between the cubic form of PdF$_2$ and other MX$_2$ or M$_2$X$_4$ structural types have been discussed [15].

Thermodynamic Data. The standard enthalpy of formation of cubic PdF$_2$ has been given as −495 kJ/mol [11].

Magnetic Properties. PdF$_2$ is a rare example of a paramagnetic d^8 palladium(II) compound. Magnetic measurements on the tetragonal (rutile) form indicate a magnetic moment of 1.88 [2], 1.90 [7], 1.84 [10] or 1.97 BM [14] at room temperature, significantly below the spin only value of 2.83 BM expected for two unpaired electrons. Magnetic data collected over the temperature range 1.4 to 300 K include values of 217 K for T$_N$ [13] and 220 K for T$_c$ [7]. Plots of magnetisation and reciprocal susceptibility against temperature (0 to 300 K) [7, 13] and of reciprocal susceptibility against temperature (0 to 700 K) at different field strengths (0.25 to 1.74 T) [12] are reproduced. The orientation of spins in tetragonal PdF$_2$ at 4.2 K is perpendicular to the tetragonal c-axis and, therefore, similar to that found in NiF$_2$ [12]. Magnetic data have been explained in terms of the Moriya theory of single-ion magnetocrystalline anisotropy [13].

The cubic (fluorite) form of PdF$_2$ is also paramagnetic (μ$_{eff}$ = 2.0 ± 0.1 BM) and has a Néel temperature of 190 ± 5 K [4] and a Curie temperature of 160 K [7]. Plots of reciprocal susceptibility vs. temperature [7, 8] and variation of (100) and (110) magnetic lines with temperature [4] are reproduced. Diagrams of magnetic structure of cubic PdF$_2$ are reproduced [4].

Miscellaneous Physical Properties. The resistivity of cubic PdF$_2$ at room temperature and pressure (10^5 Ω·cm) is lower than that of the tetragonal form by a factor of 10^5. The tetragonal-cubic transition has been followed by resistivity vs. pressure measurements and the plot is reproduced [8].

Thermal decomposition of PdF$_2$ begins above 350°C and is complete at 800°C [8].

Theoretical papers on crystal chemistry of group VIII metal complexes [16], determination of ionic radii in rutile and fluorite structures [17, 18], approximate calculations of lattice energies [19] and the determination of upper valence bands of rutile type difluorides using the Slater-Koster simplified LCAO method for the periodic potential problem [20] all include PdF$_2$ amongst the compounds considered.

X-ray photoelectron spectra for PdF_2 according to [21]:

form	$Pd3d_{3/2}$[a]	$Pd3d_{5/2}$[a]	spin-orbit[b] coupling	F1s[a] components 1 and 2	
tetragonal	343.1	337.7	5.4	684.6	687.0
	346.0[c]	340.1[c]			
cubic	343.2	337.9	5.3	684.7	686.9
	345.7[c]	340.6[c]			

[a] Values in eV (± 0.4 eV). — [b] Values in eV. — [c] Satellites.

References:

[1] Bartlett, N.; Hepworth, M. A. (Chem. Ind. [London] **1956** 1425/6).
[2] Bartlett, N.; Quail, J. W. (J. Chem. Soc. **1961** 3728/32).
[3] Müller, B.; Hoppe, R. (Naturwissenschaften **58** [1971] 268).
[4] Tressaud, A.; Soubeyroux, J. L.; Touhara, H.; Demazeau, G.; Langlais, F. (Mater. Res. Bull. **16** [1981] 207/14).
[5] Müller, B.; Hoppe, R. (Mater. Res. Bull. **7** [1972] 1297/306).
[6] Smith, W. C. (U.S. 2952514 [1960]).
[7] Müller, B. G. (Naturwissenschaften **66** [1979] 519/20).
[8] Tressaud, A.; Langlais, F.; Demazeau, G.; Hagenmüller, P. (Mater. Res. Bull. **14** [1979] 1147/53).
[9] Demazeau, G.; Langlais, F.; Portier, J.; Tressaud, A.; Hagenmüller, P. (High Pressure Sci. Technol. Proc. 7th Intern. AIRAPT Conf., Le Creusot, France, 1979 [1980], Vol. 1, pp. 579/82; C.A. **95** [1981] No. 142509).
[10] Bartlett, N.; Maitland, R. (Acta Cryst. **11** [1958] 747/8).

[11] Hopkins, K. G. G.; Nelson, P. G. (J. Chem. Soc. Dalton Trans. **1984** 1393/9).
[12] Paus, D.; Hoppe, R. (Z. Anorg. Allgem. Chem. **431** [1977] 207/16).
[13] Rao, R. P.; Sherwood, R. C.; Bartlett, N. (J. Chem. Phys. **49** [1968] 3728/30).
[14] Paus, D.; Müller, B.; Hoppe, R. (Tr. Mezhdunar Konf. Magn., Moscow 1973 [1974], Vol. 3, pp. 32/5; C.A. **84** [1976] No. 37985).
[15] Tressaud, A.; Demazeau, G. (High-Temp. — High Pressures **16** [1984] 303/8).
[16] Bokii, G. B.; Porai-Koshits, M. A. (Kristallografiya **5** [1960] 605/19; Soviet Phys.-Cryst. **5** [1960] 580/93).
[17] Sasvári, K. (Acta Phys. [Budapest] **11** [1960] 333/51).
[18] Sasvári, K. (Acta Phys. [Budapest] **11** [1960] 353/90).
[19] Karapet'Yants, M. Kh. (Zh. Fiz. Khim. **28** [1954] 1136/52; C.A. **1955** 7917).
[20] Bazhenov, V. K.; Markolenko, Yu. K.; Timofeenko, V. V. (Ukr. Fiz. Zh. **26** [1981] 677/9; C.A. **95** [1981] No. 30636).

[21] Tressaud, A.; Khairoun, S.; Touhara, H.; Watanabe, N. (Z. Anorg. Allgem. Chem. **540/541** [1986] 291/9).

3.3 Palladium Trifluoride PdF_3 (see "Palladium" 1942, pp. 270/2)

The compound originally formulated as $Pd^{III}F_3$ has now been shown to be the palladium(II)/palladium(IV) complex $Pd^{II}[Pd^{IV}F_6]$ [1] and is discussed under this heading on p. 64. However, at high temperatures and pressures true PdF_3 may exist, changes in resistivity under these conditions have been attributed to the reversible reaction $Pd^{II} + Pd^{IV} \rightleftharpoons 2Pd^{III}$ [2, 3].

$PdF_3 \cdot BrF_3$. For this "adduct" see $[BrF_2][PdF_4]$, p. 52.

References:

[1] Bartlett, N.; Rao, P. R. (Proc. Chem. Soc. **1964** 393/4).
[2] Langlais, F.; Demazeau, G.; Portier, J.; Tressaud, A.; Hagenmüller, P. (Solid State Commun. **29** [1979] 473/6).
[3] Demazeau, G.; Langlais, F.; Portier, J.; Tressaud, A.; Hagenmüller, P. (High Pressure Sci. Technol. Proc. 7th Intern. AIRAPT Conf., Le Creusot, France, 1979 [1980], Vol. 1, pp. 579/82; C.A. **95** [1981] No. 142509).

3.4 Palladium Tetrafluoride PdF_4

Preparation. Impure PdF_4 was first prepared by fluorination of $Pd[PdF_6]$ at 300°C and ~7 atm pressure of fluorine [1], slow fluorination over a period of days [2] or use of $Pd[SnF_6]$ or $Pd[GeF_6]$ gives a purer product free from $Pd[PdF_6]$ [1, 2, 3]. Fluorination of palladium metal by KrF_2 in BrF_5 or HF solution has also been used to prepare PdF_4 [4]. It is a diamagnetic brick-red [1] or pink solid [2, 3] which is rapidly hydrolysed by moisture in air and reacts violently with water [1]. It has powerful oxidising properties [2] and undergoes rapid decomposition above 350°C but shows only slight conversion to $Pd[PdF_6]$ after 3 h under vacuum at 200°C [1].

Crystallographic Properties. The first report, based on an impure sample, described a tetragonal unit cell, space group $I4/amd\text{-}D_{4h}^{19}$ with $a = 6.585 \pm 0.0005$ Å, $c = 5.835 \pm 0.005$ Å, $Z = 4$ [1]. A later neutron powder diffraction study (pattern reproduced in paper) reported an orthorhombic space group $Fdd2\text{-}C_{2v}^{19}$ with $a = 9.339(5)$, $b = 9.240(7)$, $c = 5.828(3)$ Å and $Z = 8$ [5] and a revised X-ray diffraction study gave similar results, $a = 9.37(1)$, $b = 9.24(1)$, $c = 5.84(1)$ Å [2]. The structure consists of almost regular PdF_6 octahedra with 4 bridging and 2 (cis) terminal fluoride ligands (Pd–F = 1.91(1) to 2.00(5) and 1.94(5) Å, respectively). The Pd–F–Pd bridging angle $134 \pm 2.4°$ agrees well with that required for ideal hexagonal close packing of octahedra, see **Fig. 8** [5].

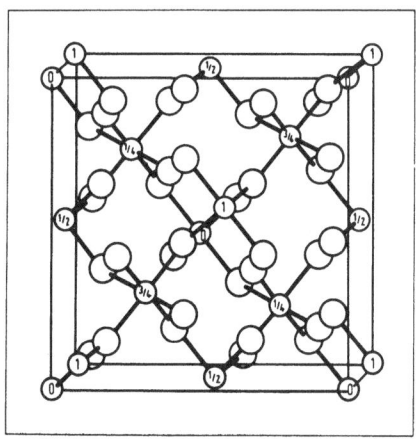

Fig. 8. Unit cell of PdF_4 projected on the a-b plane (viewed obliquely): Pd, small circles marked with the fractional z coordinate along c; F, large circles.

Magnetic Properties. PdF_4 is essentially diamagnetic (low-spin d^6 Pd^{IV}), weak paramagnetism (χ_g(298 K)=1.23×10^{-6} cgs units) observed for samples prepared at 300°C is attributed to contamination by $Pd[PdF_6]$ [1].

Vibrational Spectrum. The Raman spectrum (700 to 200 cm^{-1}) of impure PdF_4 is reproduced in a paper [3].

References:

[1] Bartlett, N.; Rao, P. R. (Proc. Chem. Soc. **1964** 393/4).
[2] Rao, P. R.; Tressaud, A.; Bartlett, N. (Inorg. Nucl. Chem. H. H. Hyman Mem. Vol. **1976** 23/8; C. A. **85** [1976] No. 153193).
[3] Bartlett, N.; Žemva, B.; Graham, L. (J. Fluorine Chem. **7** [1976] 301/20).
[4] Sokolov, V. B.; Drobyshevskii, Yu. V.; Prusakov, V. N.; Ryzhkov, A. V.; Khoroshev, S. S. (Dokl. Akad. Nauk SSSR **229** [1976] 641/4; Dokl. Chem. Proc. Acad. Sci. USSR **226/231** [1976] 503/5).
[5] Wright, A. F.; Fender, B. E. F.; Bartlett, N.; Leary, K. (Inorg. Chem. **17** [1978] 748/9).

3.5 Palladium Pentafluoride PdF_5

It has been suggested that PdF_5 is formed from metallic palladium and KrF_2 in anhydrous HF or BrF_5 solution. However, this conclusion has yet to be confirmed.

Reference:

Sokolov, V. B.; Drobyshevskii, Yu. V.; Prusakov, V. N.; Ryzhkov, A. V.; Khoroshev, S. S. (Dokl. Akad. Nauk SSSR **229** [1976] 641/4; Dokl. Chem. Proc. Acad. Sci. USSR **226/231** [1976] 503/5).

3.6 Palladium Hexafluoride PdF_6

Fluorination of palladium powder with atomic fluorine (900 to 1700 Pa) is reported to yield PdF_6 as a thermally unstable dark red solid which decomposes appreciably even at 273 K to form PdF_4.

Palladium hexafluoride is a very powerful oxidising agent, it oxidises dioxygen to give an orange solid and reacts with water to liberate O_2 and HF, and leave a red precipitate. Alkali metal fluorides, MF, react to form complex palladium(V) fluorides $M[PdF_6]$. Infrared data ($\nu_3 = 711 \, cm^{-1}$) are reported [1].

Prior to the discovery of PdF_6 several authors speculated on its existence [2] and performed theoretical calculations on its physical properties, notably infrared frequencies [3], force constants [4, 5] and Coriolis interaction constants [6].

References:

[1] Timakov, A. A.; Prusakov, V. N.; Drobyshevskii, Yu. V. (Zh. Neorgan. Khim. **27** [1982] 3007/9; Russ. J. Inorg. Chem. **27** [1982] 1704/5).
[2] Weinstock, B. (Chem. Eng. News. **42** No. 38 [1964] 89).
[3] Weinstock, B.; Goodman, G. L. (Advan. Chem. Phys. **9** [1965] 169).
[4] Labonville, P.; Ferraro, J. R.; Wall, M. C.; Basile, L. J. (Coord. Chem. Rev. **7** [1972] 257/87).
[5] Pandey, A. N.; Sharma, D. K.; Verma, U. P. (Acta Phys. Polon. A **51** [1977] 475/85).
[6] Timoshinin, V. S.; Godnev, I. N. (Dokl. Nauchn. Tekhn. Konf. Ivanovsk. Khim. Tekhnol. Inst., Ivanovo, USSR, 1971, pp. 3/5; C. A. **78** [1973] No. 153209).

3.7 Fluoropalladates, Polynary Pd Fluorides, and Adducts

The dominant fluoropalladate anions are square planar $[PdF_4]^{2-}$ and octahedral $[PdF_6]^{3-}$, $[PdF_6]^{2-}$ and $[PdF_6]^-$ representing the palladium oxidation states, II, III, IV and V, respectively.

$[PdF_4]^{2-}$. The relaxation energies, E_R, during ionisation of valence orbitals of $[PdF_4]^{2-}$ have been calculated within the 2nd-order perturbation energy. The electronic structure of the ground state was calculated within the framework of the semi-empirical INDO method and the contributions of various orbitals to E_R were obtained. The use of these results in the interpretation of the X-ray emission spectrum of $[PdF_4]^{2-}$ has been discussed [1]. The Madelung part of lattice energy (MAPLE) for salts $M^{II}[PdF_4]$ has been discussed [2].

$[PdF_6]^{3-}$. This anion is commonly found in Elpasolite type compounds of the form $A_2B[PdF_6]$ (A and B are alkali metals). Apart from $K_2Li[PdF_6]$ which is cubic (space group Fm3m-O_h^5) all Elpasolites of the form $A_2B[PdF_6]$ possess tetragonal symmetry (space group F4/mmm-D_{4h}^{17}) at room temperature. A diagram of the unit cell is reproduced, see **Fig. 9** [3]. Between 200 and 500 K every $A_2B[PdF_6]$ compound undergoes a tetragonal → cubic transition [3]. ESR spectra provide unequivocal evidence of Pd^{III} with a low-spin d^7 ($t_{2g}^6 e_g^1$) configuration which is stabilised by Jahn-Teller splitting of the ^2E_g state. ESR results also confirm that in most instances the $[PdF_6]^{3-}$ octahedra show significant tetragonal elongation [3, 4]. The occurrence of Pd^{III} in a low-spin state is corroborated by magnetic measurements [3] and by X-ray photoelectron spectra [17]. Strong axial distortion stabilises Pd^{3+} in elpasolite structures [4].

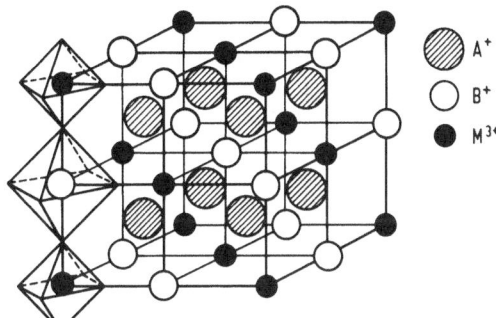

Fig. 9. A_2BMF_6 Elpasolite structure.

$[PdF_6]^{2-}$. The $[PdF_6]^{2-}$ anion is diamagnetic as expected for the $^1A_{1g}$ (t_{2g}^6) low-spin state. The electronic spectrum shows two well-marked peaks of moderate intensity at 25.0 kK ($^1A_{1g} \rightarrow {}^1T_{1g}$) and 30.0 kK ($^1A_{1g} \rightarrow {}^1T_{2g}$) together with a weaker broad shoulder at 21.0 kK ($^1A_{1g} \rightarrow {}^3T_{1g}$). Calculations established that the other expected spin forbidden band $^1A_{1g} \rightarrow {}^3T_{2g}$ would be obscured by the 25 kK absorption. Data were satisfactorily fitted with $D_q = 2600$ cm^{-1} and $B = 400$ cm^{-1}, using $C/B = 5.0$ [5, 6]. The electronic structure has been obtained from MO calculations by the CNDO method [7]. The effect of outer-sphere cations on the electronic structure of the $[PdF_6]^{2-}$ anion has been studied by X-ray (F K_α) spectroscopy, the spectrum is reproduced in paper [8]. The ^{19}F NMR spectrum of $[PdF_6]^{2-}$ in anhydrous HF solution (+284 ppm relative to the doublet signal of external $CF_2Cl \cdot CFCl_2$) displays a large temperature-independent diamagnetic shift, narrow line widths indicate that $[PdF_6]^{2-}$ is not labile [9]. From ^{19}F NMR data it has been concluded that the covalency of the Pd–F bonds increases as the counter cation changes $K \rightarrow Rb \rightarrow Cs$ [10]. Solid state ^{19}F NMR show separate peaks (ratio 1:2) for axial and equatorial fluoride ligands [10]. The infrared active vibration ν_3 appears at 602 cm^{-1} [11]. The characteristic features of noble metal hexafluoride anions including $[PdF_6]^{2-}$ have been reviewed, the radius of the Pd^{4+} cation in $[PdF_6]^{2-}$ is given as 0.64 Å [12]. Compounds containing the $[PdF_6]^{2-}$ anion are air- and moisture-sensitive [13, 14], treatment with concentrated HCl immediately affords $[PdCl_6]^{2-}$ anions [13]. The palladium(IV) species $M^{II}PdF_6$ are included in a review article on structure, bonding and, in particular, phase transitions in mixed metal fluorides $M^{II}M^{IV}F_6$ [15]. The force field for $[PdF_6]^{2-}$ has been calculated on the basis of a theoretical analysis of vibrational frequencies [18].

$[PdF_6]^-$. This anion exists in $O_2^+[PdF_6]^-$ and $Na^+[PdF_6]^-$ salts. The electronic structures of $[MF_6]^-$ anions including $[PdF_6]^-$ have been calculated using the discrete variational X_α-method, one-electron energies, atomic orbital populations, charges on M and F atoms and overlap populations are tabulated [16, 19].

References:

[1] Voityuk, A. A.; Mazalov, L. N. (Zh. Strukt. Khim. **23** [1982] 61/6; J. Struct. Chem. [USSR] **23** [1982] 50/4).

[2] Müller, B.; Hoppe, R. (Mater. Res. Bull. **7** [1972] 1297/306).

[3] Tressaud, A.; Khairoun, S.; Dance, J. M.; Hagenmüller, P. (Z. Anorg. Allgem. Chem. **517** [1984] 43/58).

[4] Khairoun, S.; Dance, J. M.; Grannec, J.; Demazeau, G.; Tressaud, A. (Rev. Chim. Minerale **20** [1983] 871/6).

[5] Brown, D. H.; Russell, D. R.; Sharp, D. W. A. (J. Chem. Soc. A **1966** 18/20).

[6] Allen, G. C.; Warren, K. D. (Struct. Bonding [Berlin] **19** [1974] 105/65 (121)).

[7] Sakaki, S.; Kato, H. (Bull. Chem. Soc. Japan **46** [1973] 2227/9).

[8] Mazalov, L. N.; Kravtsova, E. A.; Zemskov, S. V.; Nikonorov, Yu. I. (Zh. Strukt. Khim. **18** [1977] 565/72; J. Struct. Chem. [USSR] **18** [1977] 453/9).

[9] Matwiyoff, N. A.; Asprey, L. B.; Wageman, W. E.; Reisfeld, M. J.; Fukushima, E. (Inorg. Chem. **8** [1969] 750/3).

[10] Zemskov, S. V.; Gabuda, S. P.; Nikonorov, Yu. I.; Selezneva, V. A. (Zh. Strukt. Khim. **15** [1974] 933/4; J. Struct. Chem. [USSR] **15** [1974] 824/5).

[11] Peacock, R. D.; Sharp, D. W. A. (J. Chem. Soc. **1959** 2762/7).

[12] Zemskov, S. V.; Gabuda, S. P. (Zh. Strukt. Khim. **17** [1976] 904/21; J. Struct. Chem. [USSR] **17** [1976] 772/89).

[13] Sharpe, A. G. (J. Chem. Soc. **1953** 197/9).

[14] Shipachev, V. A.; Zemskov, S. V.; Al't, L. Ya. (Koord. Khim. **6** [1980] 932/5; Soviet J. Coord. Chem. **6** [1980] 472/5).

[15] Reinen, D.; Steffens, F. (Z. Anorg. Allgem. Chem. **441** [1978] 63/82).

[16] Gutsev, G. L.; Boldyrev, A. I. (Mol. Phys. **53** [1984] 23/31).

[17] Tressaud, A.; Khairoun, S.; Touhara, H.; Watanabe, N. (Z. Anorg. Allgem. Chem. **540/541** [1986] 291/9).

[18] Rogalevich, N. L.; Bobkova, E. Yu.; Novitskii, G. G. (Dokl. Akad. Nauk SSSR **29** [1985] 906/9; C.A. **103** [1985] No. 202972).

[19] Gutsev, G. L.; Boldyrev, A. I. (Koord. Khim. **11** [1985] 435; Soviet J. Coord. Chem. **11** [1985] 245/51).

3.7.1 Fluoropalladates of Nonmetals (see also adducts, p. 52)

[XeF][PdF₅] ($XeF_2 \cdot PdF_4$). This yellow diamagnetic solid is obtained by treatment of $Pd[PdF_6]$ or PdF_4 with liquid XeF_2 at 140 to 150°C. It is thermally stable at room temperature but loses XeF_2 at 140 to 150°C to form $[XeF][Pd_2F_9]$. The complex is amorphous to X-rays, however, an ionic structure with a tetrameric F-bridged $[PdF_5]_4^{4-}$ anion similar to $[RhF_5]_4$ is proposed [1, 2]. Raman and infrared bands (700 to 200 cm^{-1}) are recorded [1].

[XeF][Pd₂F₉] ($XeF_2 \cdot 2PdF_4$). This light brown to pink solid forms when [XeF][PdF₅] is heated in vacuo for several hours at 140 to 150°C. The anion is thought to be polymeric with bridging fluorides since triple fluoride bridges are considered unlikely [1, 2]. The Raman spectrum (700 to 100 cm⁻¹) is reproduced and frequencies are tabulated. X-ray powder data are recorded [1]. Thermal decomposition in vacuo at 280°C affords Pd[PdF₆] and very pure XeF₄ [1, 2].

[Xe₂F₃]₂[PdF₆] is thought to form when Pd[PdF₆] is oxidised by liquid XeF₂, or by thermal reaction of PdF₄ with liquid XeF₂ at 140°C [1].

[XeF₅]₂[PdF₆] is prepared as yellow needles from XeF₆ and PdF₄ [3]. The proposed ionic structure has been confirmed by X-ray diffraction methods. The unit cell is orthorhombic, space group either Pca2₁-C_{2v}^5 or Pcam-D_{2h}^{11} with a = 9.346(6), b = 12.786(7), c = 9.397(6) Å, Z = 4; crystallographic density 3.91 g/cm³. The structure contains two crystallographically distinct [XeF₅]⁺ cations and an octahedral [PdF₆]²⁻ anion (Pd–F$_{av}$ = 1.893 Å). The Xe atom of each [XeF₅]⁺ cation makes one short contact with the [PdF₆]²⁻ anion. The Xe···F contacts link [XeF₅]⁺ and [PdF₆]²⁻ ions in rings, see **Fig. 10** [3]. Vibrational spectra (700 to 100 cm⁻¹) have been recorded [1, 4, 7].

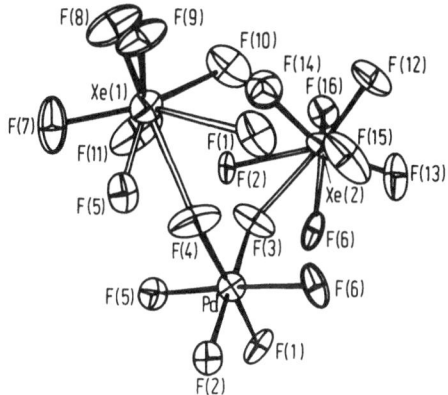

Fig. 10.   The formula unit in [XeF₅⁺]₂[PdF₆²⁻].

[XeF₅][Pd₂F₉]. See 2 PdF₄·XeF₆, p. 53.

[Xe₂F₁₁]₂[PdF₆] is prepared from XeF₆ and PdF₄ [5]. By analogy with the related gold complex [Xe₂F₁₁][AuF₆] the cation is thought to be a fluoride-bridged F₅Xe–F–XeF₅ unit. The conclusion is supported by Raman data (700 to 200 cm⁻¹) which are tabulated [6].

[KrF][PdF₆]. Attempts to isolate this salt from a mixture of KrF₂ and PdF₄ in liquid HF at low temperatures were unsuccessful [7].

References:

[1] Bartlett, N.; Žemva, B.; Graham, L. (J. Fluorine Chem. **7** [1976] 301/20).
[2] Graham, L. (LBL-8088 [1978]; E.R.A. **4** [1979] No. 19893).
[3] Leary, K.; Templeton, D. H.; Zalkin, A.; Bartlett, N. (Inorg. Chem. **12** [1973] 1726/30).
[4] Adams, C. J.; Bartlett, N. (Israel J. Chem. **17** [1978] 114/25).
[5] Leary, K. M. (LBL-3746 [1975] 1/101; C.A. **85** [1976] No. 186104).

[6] Leary, K.; Zalkin, A.; Bartlett, N. (Inorg. Chem. **13** [1974] 775/9).
[7] Sokolov, V. B.; Drobyshevskii, Yu. V.; Prusakov, V. N.; Ryzhkov, A. V.; Khoroshev, S. S. (Dokl. Akad. Nauk SSSR **229** [1976] 641/4; Dokl. Chem. Proc. Acad. Sci. USSR **226/231** [1976] 503/5).

$O_2[PdF_6]$ is obtained as a brown-gold [1] or yellowish brown [2] powder by repeated oxyfluorination of palladium powder at 320°C and 60000 psi for a total of 227 h [1] or by introduction of NaF and O_2 to a mixture of KrF_2 and PdF_4 in liquid HF [2]. It is a powerful oxidising agent, decomposing with loss of O_2 at 40°C and fluorine at 240°C to form $Pd[PdF_6]$ [1]. Vibrational (Raman) spectra have been reported, $v(O_2^+) = 1817$ [2] or 1819 cm^{-1}, anion frequencies $v_1 = 643$, $v_2 = 570$ and $v_5 = 268$ cm^{-1} [1, 3] or 645, 569 and 265 cm^{-1}, respectively [2]. The magnetic susceptibility $\chi = C/(T + \Theta)$ at 290 K is given as 2.55×10^{-3} emu/mol after correction for ferromagnetic impurities. There are indications of a change of C or Θ below 100 K and the compound becomes ferromagnetic at ~12 K [1].

$O_2[Pd_2F_{11}]$ is briefly mentioned as a contaminant in samples of $O_2[PdF_6]$ detected by additional Raman bands at 655, 628, 588, 390 and 287 cm^{-1} [1].

$[NO]_2[PdF_6]$ is prepared from palladium metal and a NOF/F_2 mixture at 300°C over a period of 62 h and has been characterised by its Raman spectrum; $v(NO^+) = 2312$, $v_1 = 573$, $v_2 = 554$, $v_5 = 246$ cm^{-1} [3].

$[BrF_2][PdF_4]$. An "adduct" $PdF_3 \cdot BrF_3$ prepared from $PdBr_2$ or $[SeF_3]_2[PdF_6]$ and excess BrF_3 [5, 6] is probably the salt $[BrF_2][PdF_4]$. It is described as a dark brown paramagnetic solid ($\mu_{eff} = 2.2$ BM at 21°C) which is readily decomposed to "PdF_3" [5].

$[SeF_3]_2[PdF_6]$ is prepared by heating $[BrF_2][PdF_4]$ with a large excess of SeF_4 in a quartz tube, and separates as a yellow solid. It rapidly decomposes to PdF_2 at 155°C, and reacts readily with water to precipitate a brown solid (palladous selenite?). $[SeF_3]_2[PdF_6]$ is diamagnetic ($\chi_g = -0.053 \times 10^{-6}$ cgs units) and isomorphous with $[SeF_3]_2[PtF_6]$ and $[SeF_3]_2[GeF_6]$ [5].

References:

[1] Falconer, W. E.; DiSalvo, F. J.; Edwards, A. J.; Griffiths, J. E.; Sunder, W. A.; Vasile, M. J. (Inorg. Nucl. Chem. H. H. Hyman Mem. Vol. **1976** 59/60).
[2] Sokolov, V. B.; Drobyshevskii, Yu. V.; Prusakov, V. N.; Ryzhkov, A. V.; Khoroshev, S. S. (Dokl. Akad. Nauk SSSR **229** [1976] 641/4; Dokl. Chem. Proc. Acad. Sci. USSR **226/231** [1976] 503/5).
[3] Sunder, W. A.; Wayda, A. L.; Distefano, D.; Falconer, W. E.; Griffiths, J. E. (J. Fluorine Chem. **14** [1979] 299/325).
[4] Woolf, A. A. (Advan. Inorg. Chem. Radiochem. **9** [1966] 217/314, 273).
[5] Bartlett, N.; Quail, J. W. (J. Chem. Soc. **1961** 3728/32).
[6] Sharpe, A. G. (J. Chem. Soc. **1950** 3444/50).

Adducts

$2PdF_4 \cdot XeF_2$. This adduct has been characterised as the salt $[XeF][Pd_2F_9]$ (see p. 51).

$PdF_4 \cdot XeF_2$ is thought to be the salt $XeF[PdF_5]$, see p. 50.

$PdF_4 \cdot nXeF_2$ (n = 2, 3 or 4). These adducts were detected in the time vs. weight-loss curve for the XeF_2/PdF_4 system [1, 2] but all lose XeF_2 at or below 20°C [2].

2 PdF$_4$·XeF$_6$. Raman shifts (700 to 1000 cm^{-1}) have been tabulated for an adduct of this stoichiometry [2]. It is possibly a salt [XeF$_5$][Pd$_2$F$_9$] containing a polymeric fluoride-bridged anion.

PdF$_4$·2XeF$_6$. This "adduct" has been characterised by vibrational spectroscopy and X-ray diffraction as the salt [XeF$_5$]$_2$[PdF$_6$], see p. 51.

PdF$_4$·4XeF$_6$. This "adduct" has been characterised by vibrational spectroscopy and X-ray diffraction as the salt [Xe$_2$F$_{11}$]$_2$[PdF$_6$], see p. 51.

References:

[1] Graham, L. (LBL-8088 [1978]; E.R.A. **4** [1979] No. 19893).
[2] Bartlett, N.; Žemva, B.; Graham, L. (J. Fluorine Chem. **7** [1976] 301/20).

3.7.2 Alkali Metal Fluoropalladates

Lithium Fluoropalladates

Li$_3$[PdF$_6$]. Attempts to prepare this complex from LiF and Pd[PdF$_6$] in sealed tubes at 300°C failed, Li$_2$PdF$_6$ and rutile type PdF$_2$ were obtained instead [1].

Li$_2$[PdF$_6$] is prepared by fluorinating a mixture of Li$_2$CO$_3$ and [NH$_4$]$_2$[PdCl$_6$]. The yellow-brown crystals are monoclinic (Na$_2$SnF$_6$ type of structure), a = 10.09, b = 4.64, c = 4.63 Å, β = 117°, D$_{pyk}$ = 4.00, D$_{X-ray}$ = 4.02 g/cm^3, Z = 2 [2].

References:

[1] Tressaud, A.; Khairoun, S.; Dance, J. M.; Hagenmüller, P. (Z. Anorg. Allgem. Chem. **517** [1984] 43/58).
[2] Henkel, H.; Hoppe, R. (Z. Anorg. Allgem. Chem. **359** [1968] 160/77).

Sodium Fluoropalladates

NaPdF$_3$ is mentioned as a decomposition product of Na$_3$[PdF$_6$] [1].

Na[PdF$_4$] is prepared as a grey hygroscopic powder from NaF and Pd[PdF$_6$] in a sealed platinum tube at ~600°C and >70 kbar pressure. It is stable at room temperature but loses F$_2$ above 60°C. The crystals are monoclinic (KBrF$_4$ type structure), a = 11.84, b = 5.36, c = 5.86 Å, β = 114°. The EPR spectrum at 8 K is reproduced in the paper. The values g$_{\parallel}$ = 2.050 and g$_{\perp}$ = 2.263 are consistent with appreciable axial distortion of the Pd^{3+} coordination polyhedron and the EPR spectrum confirms the presence of low-spin (t$_{2g}^6$e$_g^1$) palladium(III) with significant Jahn-Teller distortion [3].

The presence of Pd^{III} is also confirmed by X-ray photoelectron spectra (reproduced in paper; Pd3d$_{3/2}$ = 343.7; Pd3d$_{5/2}$ = 338.2; F1s = 685.2 and 687.2; Na1s = 1072.5 eV; all ±0.4 eV) [8].

Na$_3$[PdF$_6$] is a light grey powder obtained by heating NaF and Pd[PdF$_6$] in the correct proportions at 600°C under 3 kbar pressure. The crystals are monoclinic (isostructural with cryolite), space group P2$_1$/n-C$_{2h}^5$, a = 5.54 ± 0.01 Å, b = 5.78 ± 0.01 Å, c = 8.22 ± 0.01 Å, β = 91°. The EPR spectrum at 3.7, 200, 230 and 295 K is reproduced in the paper, g$_{\parallel}$ = 2.312, g$_{\perp}$ = 2.025 [1].

The presence of Pd^{III} is confirmed by X-ray photoelectron spectra (reproduced in paper; $Pd3d_{3/2} = 343.5$; $Pd3d_{5/2} = 338$; $F1s = 685.3$ and 687.3; Na ls $= 1072.7$ eV; all ± 0.4 eV) [8].

$Na_5[Pd_3F_{14}]$ is prepared by heating a correctly proportioned mixture of NaF and $Pd[PdF_6]$ in anhydrous HF at 700°C for 2 h [4].

$Na_2[PdF_6]$ is prepared by fluorination of Na_2PdCl_6 using fluorine or ClF_3 [5]. An earlier paper lays claim to the first synthesis but gives no details [6]. The crystals are hexagonal (Na_2SiF_6 type structure) [2]; a = 9.23, c = 5.25 Å [6], a = 9.20, c = 5.16 Å (Debye-Scherrer), a = 9.213, c = 5.170 Å (Guinier), Z = 3, $D_{X-ray} = 3.49$ or 3.51 g/cm^3 [2].

$Na[PdF_6]$ is obtained from KrF_2, PdF_4 and NaF in liquid HF. Raman data $\nu_1 = 642$, $\nu_2 = 566$, $\nu_5 = 267 \text{ cm}^{-1}$. It is reactive, acidifying water and slowly decomposing to Pd^{IV} [7].

References:

[1] Tressaud, A.; Khairoun, S.; Dance, J. M.; Hagenmüller, P. (Z. Anorg. Allgem. Chem. **517** [1984] 43/58).
[2] Henkel, H.; Hoppe, R. (Z. Anorg. Allgem. Chem. **359** [1968] 160/77).
[3] Tressaud, A.; Khairoun, S.; Dance, J. M.; Grannec, J.; Demazeau, G.; Hagenmüller, P. (Compt. Rend. [2] **295** [1982] 183/6).
[4] Knox, K. (U.S. 2945744 [1960]; C.A. **1960** 21549).
[5] Zemskov, S. V.; Nikonorov, Yu. I.; Mit'Kin, V. N.; Lavrova, L. A. (Tezisy. Dokl. 12th Vses. Chugaevskoe Soveshch. Khim. Kompleksn. Soedin., Novosibirsk 1975, Vol. 3, p. 376; C.A. **85** [1976] No. 171050).
[6] Cox, B.; Sharp, D. W. A.; Sharpe, A. G. (J. Chem. Soc. **1956** 1242/4).
[7] Sokolov, V. B.; Drobyshevskii, Yu. V.; Prusakov, V. N.; Ryzhkov, A. V.; Khoroshev, S. S. (Dokl. Akad. Nauk SSSR **229** [1976] 641/4; Dokl. Chem. Proc. Acad. Sci. USSR **226/231** [1976] 503/5).
[8] Tressaud, A.; Khairoun, S.; Touhara, H.; Watanabe, N. (Z. Anorg. Allgem. Chem. **540/541** [1986] 291/9).

Potassium Fluoropalladates

$K[PdF_3]$ is prepared by heating a 1:1 mixture of KF and PdF_2 at 400 to 500°C for 7 to 10 h in a sealed gold tube under argon. The dark brown violet crystals are orthorhombic ($GdFeO_3$ variation of perovskite type), space group $Pbnm-D_{2h}^{16}$ with a = 5.98(6), b = 6.00(1), c = 8.50(3) Å, Z = 4, $D_{X-ray} = 4.36$, $D_{calc} = 4.40 \text{ g/cm}^3$ [1].

$K_2[PdF_4]$ is prepared by heating a 2:1 mixture of KF and PdF_2 at 400 to 500°C for 7 to 10 h in a sealed tube under argon. The complex is diamagnetic, $\chi_{mol} = -107 \times 10^{-6} \text{ cm}^3/\text{mol}$ at room temperature. The crystal structure could not be determined by X-ray analysis [1].

$K[PdF_4]$ is prepared as a brown solid by treating a 1:1 mixture of KCl and $PdCl_2$ with BrF_3, then evaporating to dryness and heating the residual **KPdF$_4$·0.8BrF$_3$** at 200°C [2].

$K_2[PdF_5]$. This pale yellow solid is prepared by treating K_2PdCl_4 with BrF_3, then heating the resultant light brown adduct **K_2PdF_5·0.5 BrF$_3$** at 280°C [2].

$K_3[PdF_6]$ is obtained as a moisture-sensitive light green powder by heating a stoichiometric mixture of KF and $Pd[PdF_6]$ at 300°C for 15 h in a sealed palladium tube. Most of the X-ray diffraction lines of $K_3[PdF_6]$ can be indexed at room temperature with tetragonal symmetry, a = 8.85, c = 8.38 Å. The EPR spectrum (4.2 K, $g_{iso} = 2.206$) is reproduced in the paper [3].

The presence of Pd^{III} is confirmed by X-ray photoelectron spectra (reproduced in paper; $Pd3d_{3/2} = 343.7$; $Pd3d_{5/2} = 338.2$; $F1s = 685.5$, 688.6; $K2p_{3/2, 1/2} = 293.2$, 295.7 eV; all ± 0.4 eV) [17].

$K_2Li[PdF_6]$ is prepared as a light beige, air-sensitive hygroscopic solid by heating homogeneous stoichiometric mixtures of KF, LiF and $Pd[PdF_6]$ in sealed tubes at 400°C and 3 kbar pressure. Below 200 ± 5 K the cubic form of $K_2Li[PdF_6]$ converts to a tetragonal form. Crystal data are for the cubic form, space group $Fm3m-O_h^5$, $a = 8.154$ Å, $Z = 4$, $D_{obs} = 3.71 \pm 0.05$, $D_{calc} = 3.745$ g/cm³; for the tetragonal form, space group $F4/mmm-D_{4h}^{17}$, $a = 8.024 \pm 0.010$, $c = 8.234 \pm 0.010$ Å [3]. Plots of the cell constants [3] and reciprocal magnetic susceptibility [3, 4] against temperature are reproduced in the papers. The EPR spectrum (4.2 K, $g_{iso} = 2.208$) is reproduced in the paper [3], magnetic constants: $C_m(calc) = 0.414$, $C_m(exp) = 0.424$, $\Theta_p(K) = -23$ [3, 4].

$K_2Na[PdF_6]$ is prepared as a light green [3] or beige [5] air-sensitive hygroscopic solid by heating homogeneous, stoichiometric mixtures of KF, NaF and $Pd[PdF_6]$ in sealed tubes at 400°C and 3 kbar pressure [3, 5]. The compound has a tetragonal structure at room temperature but converts to a cubic form on heating above ~ 318 K. Crystal data for the tetragonal form: space group $F4/mmm-D_{4h}^{17}$, $a = 8.30$, $c = 8.72$ Å, $Z = 4$, $D_{obs} = 3.53 \pm 0.05$, $D_{calc} = 3.56$ g/cm³, for the cubic form: space $Fm3m-O_h^5$, $a = 8.40$ Å [3, 5]. In the tetragonal form $[PdF_6]^{3-}$ octahedra are elongated in a ferro-distortive manner: Pd–F, 4 at 1.95, 2 at 2.14 Å. Plots of variation in cell constants and X-ray pattern with temperature are reproduced in the paper [3]. A plot of reciprocal magnetic susceptibility vs. temperature (0 to 300 K) is also reproduced [3, 5]. The EPR spectrum has been recorded (4.2 K, $g_{\parallel} = 2.012$, $g_{\perp} = 2.295$) and is reproduced (4.2 and 295 K) [3, 5]. Magnetic constants: $C_m(calc) = 0.413$, $C_m(exp) = 0.390$, $\Theta_p(K) = -8$ [3].

X-ray photoelectron spectra (reproduced in paper): $Pd3d_{3/2} = 343.7$, satellite 346.3; $Pd3d_{5/2} = 338.2$; $F1s = 684.6$, 688.9; $K2p_{3/2, 1/2} = 293.0$, 295.9; $Na1s = 1077.7$ eV (all ± 0.4 eV) [17].

$K_2[PdF_6]$. This palladium(IV) complex crystallises in hexagonal and trigonal forms. The hexagonal form has been obtained by fluorination of $K_2[PdCl_6]$ at 200 to 300°C [6, 7]. The trigonal form has been obtained by neutralisation of $[SeF_3]_2[PdF_6]$ with $KSeF_5$ in SeF_4 solution [8] and by treatment of $K_2[PdCl_6]$ with BrF_3 [9]. Products of unspecified crystal form are generated by fluorination of $K_2[Pd(CN)_4]$, $K_2[PdCl_4]$ [10] or $K_2[PdCl_6]$ [11, 12]. Differential thermal analysis has been used to optimise conditions for the fluorination reaction [12]. The complex is bright yellow in colour and is sensitive to moisture [6, 9]. Pure material is obtained by refluxing equimolar quantities of KF and $[SeF_3]_2[PdF_6]$ in SeF_4 [8].

The hexagonal form belongs to space group $P\overline{3}m1-D_{3d}^3$ with $a = 5.75$, $c = 9.51$ Å [6, 13], the trigonal form belongs to space group $C\overline{3}m-D_{3d}^3$ with $a = 5.717 \pm 0.003$, $c = 4.667 \pm 0.003$ Å [8, 13]. The complex is diamagnetic, $\chi = -0.23 \times 10^{-6}$ cgs units at 20°C [9]. The ^{19}F NMR of $K_2[PdF_6]$ has been studied in anhydrous HF solution (-60 to $+15$°C, data tabulated in paper) [10] and in the solid state [14], a large temperature-independent diamagnetic ^{19}F shift ($+284$ ppm with respect to doublet signal of $CFCl_2 \cdot CF_2Cl$) is observed in solution [10], solid state spectra show two lines (intensity ratio 1:2) attributed to axial and equatorial fluorides. The first derivative of the ^{19}F spectrum of solid $K_2[PdF_6]$ at -118°C is reproduced in the paper [14]. X-ray and X-ray electron spectroscopy results indicate that replacement of F by Cl in $K_2[PdF_6]$ leads to decrease in ionisation energies of all levels in molecule. The X-ray spectra of Pd and F in $K_2[PdF_6]$ are reproduced in the paper together with a scheme of molecular orbitals and a table of ionisation potentials for $[PdF_6]^{2-}$ [15].

The enthalpy of the process $K_2[PdF_6](c) \rightarrow PdF_2(c) + 2KF(c) + F_2(g)$ has been analysed and the standard enthalpy of formation of $K_2[PdF_6]$ at 25°C has been measured as -2040 kJ/mol [16].

References:

[1] Alter, E.; Hoppe, R. (Z. Anorg. Allgem. Chem. **408** [1974] 115/20).

[2] Sharpe, A. G. (J. Chem. Soc. **1950** 3444/50).

[3] Tressaud, A.; Khairoun, S.; Dance, J. M.; Hagenmüller, P. (Z. Anorg. Allgem. Chem. **517** [1984] 43/58).

[4] Tressaud, A.; Darriet, J.; Lagassié, P.; Grannec, J.; Hagenmüller, P. (Mater. Res. Bull. **19** [1984] 983/8).

[5] Khairoun, S.; Dance, J. M.; Grannec, J.; Demazeau, G.; Tressaud, A. (Rev. Chim. Minerale **20** [1983] 871/6).

[6] Hoppe, R.; Klemm, W. (Z. Anorg. Allgem. Chem. **268** [1952] 364/71).

[7] Klemm, W.; Krause, J.; Wahl, K.; Huss, E.; Hoppe, R.; Weise, E.; Brandt, W. (Forschungsber. Wirtsch.-Verkehrsmin. Nordrhein-Westf. No. 160 [1955]).

[8] Bartlett, N.; Quail, J. W. (J. Chem. Soc. **1961** 3728/32).

[9] Sharpe, A. G. (J. Chem. Soc. **1953** 197/9).

[10] Matwiyoff, N. A.; Asprey, L. B.; Wageman, W. E.; Reisfeld, M. J.; Fukushima, E. (Inorg. Chem. **8** [1969] 750/3).

[11] Zemskov, S. V.; Nikonorov, Yu. I.; Mit'kin, V. N.; Lavrova, L. A. (Tezisy. Dokl. 12th Vses. Chugaevskoe Soveshch. Khim. Kompleksn. Soedin., Novosibirsk 1975, Vol. 3, p. 376; C. A. **85** [1976] No. 171050).

[12] Zemskov, S. V.; Nikonorov, Yu. I.; Pastukhova, E. D.; Mit'kin, V. N. (Izv. Sibirsk. Otd. Akad. Nauk SSSR Ser. Khim. Nauk **1976** 83/8).

[13] Hanic, F. (Chem. Zvesti **20** [1966] 738/51).

[14] Zemskov, S. V.; Gabuda, S. P.; Nikonorov, Yu. I.; Selezneva, V. A. (Zh. Strukt. Khim. **15** [1974] 933/4; J. Struct. Chem. [USSR] **15** [1974] 824/5).

[15] Peshchevitskii, B. I.; Zemskov, S. V.; Sadovskii, A. P.; Kravtsova, E. A.; Mit'kin, V. N. (Koord. Khim. **5** [1979] 1838/45; Soviet J. Coord. Chem. **5** [1979] 1433/9).

[16] Hopkins, K. G. G.; Nelson, P. G. (J. Chem. Soc. Dalton Trans. **1984** 1393/9).

[17] Tressaud, A.; Khairoun, S.; Touhara, H.; Watanabe, N. (Z. Anorg. Allgem. Chem. **540/541** [1986] 291/9).

Rubidium Fluoropalladates

RbPdF$_3$ is prepared by heating RbF/PdF_2 mixtures in a sealed tube under argon at 400 to 500°C for 7 to 10 h. The dark brown crystals are cubic (perovskite type), a = 4.29 Å, D_{cryst} = 5.20, D_{pyk} = 5.16 g/cm^3; X-ray intensity data are tabulated. The Madelung part of the lattice energy has been calculated [1].

X-ray photoelectron spectra (reproduced in paper): $Pd3d_{3/2}$ = 343.2, satellite 346.0; $Pd3d_{5/2}$ = 337.7, satellite 340.0; F1s = 684.9; $Rb4p_{1/2+3/2}$ = 13.8 eV (all ±0.4 eV) [13].

Rb$_3$PdF$_5$ is prepared by heating PdF_2 with excess RbF. The yellow crystals are tetragonal, space group P4/mbm-D_{4h}^5, a = 7.462, c = 6.457 Å, Z = 2, square planar PdF_4 units, Pd–F = 1.926 Å; X-ray data and projections of the structure are reproduced [2].

Rb$_2$Na[PdF$_6$] exists in tetragonal (room temperature) and cubic (high temperature) forms, transition temperature 388 K. The compound is prepared by heating a homogenised stoichiometric mixture of RbF, NaF and $Pd[PdF_6]$ at 400°C and 3 kbar pressure, then quenching to room temperature. It is hygroscopic and darkens in air. The light green tetragonal form, space group F4/mmm-D_{4h}^{17}, has cell constants a = 8.47, c = 8.76 Å, D_{obs} = 4.33 ± 0.05 and D_{calc} = 4.38 g/cm^3, X-ray data are tabulated in paper. For the cubic form (space group Fm3m-O_h^5) a = 8.50 Å at 393 K. The ESR spectrum (4.2 and 295 K) is reproduced in the paper, $g_\perp$ = 2.298,

$g_{\parallel} = 2.013$. Thermal variation plots for cell constants and reciprocal magnetic susceptibility are reproduced in the paper. Magnetic constants: C_m(calc) = 0.414, C_m(exp) = 0.420, Θ_p(K) = −10 [3].

Rb$_2$K[PdF$_6$] exists in tetragonal (room temperature) and cubic (high temperature) forms, transition temperature 493 K. The compound is prepared by heating a homogeneous, stoichiometric mixture of RbF, KF and Pd[PdF$_6$] at 400°C and 3 kbar pressure, then quenching to room temperature. It is hygroscopic and darkens in air. The light beige tetragonal form (space group F4/mmm-D_{4h}^{17}) has cell constants a = 8.74, c = 9.23 Å, D_{obs} = 4.05 ± 0.05 and D_{calc} = 4.06 g/cm^3, X-ray data are tabulated in paper. For the cubic form (space group Fm3m-O_h^5) a = 8.85 Å at 493 K. The ESR spectrum (9 and 295 K) is reproduced in the paper, $g_{\perp}$ = 2.280, $g_{\parallel}$ = 2.052. Thermal variation plots for cell constants and reciprocal magnetic susceptibility are reproduced in the paper. Magnetic constants: C_m(calc) = 0.413, C_m(exp) = 0.417, Θ_p(K) = −2 [3].

Rb$_2$[PdF$_6$] is prepared by fluorination of Rb$_2$[PdCl$_6$] using F$_2$ [4, 7], ClF$_3$ [4, 5] or BrF$_3$ [6]. Optimum fluorination temperature has been established by differential thermal analysis [4]. The yellow compound is air-sensitive and immediately hydrolysed by cold water [4, 7]. Hexagonal and cubic forms have been reported, cell dimensions are a = 5.98, c = 9.70 Å for hexagonal form [8] and a = 8.57 Å for cubic form [7]. The far-infrared spectrum (700 to 400 cm^{-1}) is reproduced in a paper [9]. Solid state [10] and solution [11] ^{19}F NMR data have been reported, two peaks (relative intensity 1:2) observed in the solid state spectrum are assigned to axial and equatorial fluorines, respectively [10]. The X-ray (F K$_\alpha$) spectrum is reproduced in a paper [12].

References:

[1] Alter, E.; Hoppe, R. (Z. Anorg. Allgem. Chem. **408** [1974] 115/20).

[2] Müller, B. G. (Z. Anorg. Allgem. Chem. **491** [1982] 245/52).

[3] Tressaud, A.; Khairoun, S.; Dance, J. M.; Hagenmüller, P. (Z. Anorg. Allgem. Chem. **517** [1984] 43/58).

[4] Zemskov, S. V.; Nikonorov, Yu. I.; Mit'kin, V. N.; Lavrova, L. A. (Tezisy. Dokl. 12th Vses. Chugaevskoe Soveshch. Khim. Kompleksn. Soedin., Novosibirsk 1975, Vol. 3, p. 376; C. A. **85** [1976] No. 171050).

[5] Zemskov, S. V.; Nikonorov, Yu. I.; Pastukhova, E. D.; Mit'kin, V. N. (Izv. Sibirsk. Otd. Akad. Nauk SSSR Ser. Khim. Nauk 3 [1976] 83/8; C.A. **85** [1976] No. 86494).

[6] Sharpe, A. G. (J. Chem. Soc. **1953** 197/9).

[7] Hoppe, R.; Klemm, W. (Z. Anorg. Allgem. Chem. **268** [1952] 364/71).

[8] Cox, B.; Sharpe, A. G. (J. Chem. Soc. **1953** 1783/4).

[9] Zemskov, S. V.; Gabuda, S. P. (Zh. Strukt. Khim. **17** [1976] 904/21; J. Struct. Chem. [USSR] **17** [1976] 772/89).

[10] Zemskov, S. V.; Gabuda, S. P.; Nikonorov, Yu. I.; Selezneva, V. A. (Zh. Strukt. Khim. **15** [1974] 933/4; J. Struct. Chem. [USSR] **15** [1974] 824/5).

[11] Matwiyoff, N. A.; Asprey, L. B.; Wageman, W. E.; Reisfeld, M. J.; Fukushima, E. (Inorg. Chem. **8** [1969] 750/3).

[12] Mazalov, L. N.; Kravtsova, E. A.; Zemskov, S. V.; Nikonorov, Yu., I. (Zh. Strukt. Khim. **18** [1977] 565/72; J. Struct. Chem. [USSR] **18** [1977] 453/9).

[13] Tressaud, A.; Khairoun, S.; Touhara, H.; Watanabe, N. (Z. Anorg. Allgem. Chem. **540/541** [1986] 291/9).

Caesium Fluoropalladates

Cs[PdF$_3$] is left behind as a pinkish brown solid when an equimolar mixture of CsF and PdF$_2$ is heated under reflux in SeF$_4$ for 3 h and the solvent is then removed under reduced pressure [1], or heated under argon in a bomb [2]. The X-ray powder pattern was diffuse and complex but PdF$_2$ lines were absent [1]. There is evidence of a cubic modification (a = 4.13 Å) formed at 400°C and 110 kbar pressure [2]. X-ray intensity data are tabulated [2]. Magnetic data over the temperature range 82 to 286 K are tabulated, μ_{eff} = 1.04 (82.3 K) to 1.60 (286.5 K) [1].

Cs$_3$[PdF$_5$] is prepared from a 1:2 molar mixture of PdF$_2$ and CsF by heating to ~560°C in a sealed gold tube. The yellow crystals are tetragonal, space group P4/mbm-D$_{4h}^5$, Z = 2, a = 7.848, c = 6.688 Å [3].

CsRb$_2$[PdF$_5$] is prepared by heating together CsF, RbF and PdF$_2$ in the correct proportions at ~560°C. The crystals are tetragonal, space group P4/mbm-D$_{4h}^5$, Z = 2, a = 7.579, c = 6.455 Å, square planar PdF$_4$ units Pd–F = 1.940 Å [3].

Cs[Pd$_2$F$_5$] is prepared from CsF and PdF$_2$ at 600°C. The orange-brown crystals are orthorhombic, space group Imma-D$_{2h}^{28}$, a = 6.53$_3$, b = 7.86$_2$, c = 10.79 Å, Z = 4; palladium atoms in octahedral (Pd–F$_{ax}$ = 2.163, Pd–F$_{eq}$ = 2.164 Å) and square-planar (Pd–F = 1.951 Å) coordination [3]. The structure is shown in **Fig. 11**.

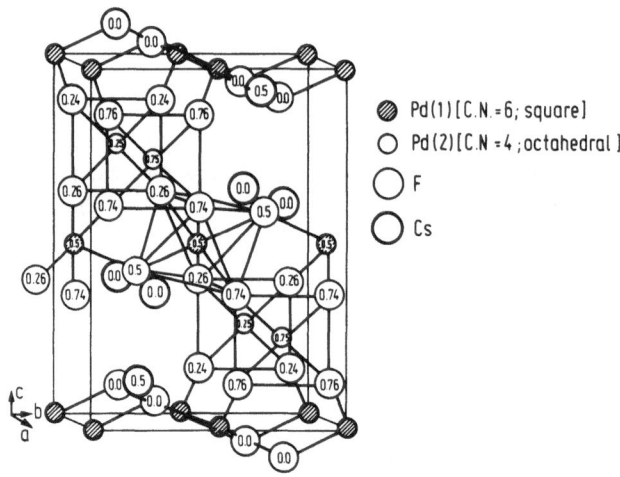

Fig. 11.   Unit cell of CsPd$_2$F$_5$ [3].

Cs$_2$K[PdF$_6$] is prepared by heating a homogeneous stoichiometric mixture of CsF, KF and Pd[PdF$_6$] at 400°C under 3 kbar pressure. It exists in tetragonal (room temperature) and cubic (high temperature) forms, transition temperature 328 ± 5 K. The light beige solid is air-sensitive and hygroscopic. Cell dimensions are a = 9.04, c = 9.32 Å for the tetragonal form (space group F4/mmm-D$_{4h}^{17}$) and a = 9.06 Å at 333 K for the cubic form (space group Fm3m-O$_h^5$). X-ray intensity data are recorded. ESR data are tabulated (orthorhombic symmetry, g$_1$ = 2.297, g$_2$ = 2.197, g$_3$ = 2.012) and the spectrum (8, 154, 212 and 295 K) is reproduced in the paper. Thermal variation plots for cell dimensions and reciprocal magnetic susceptibility are reproduced [4].

Cs$_2$Rb[PdF$_6$] is prepared by heating a homogeneous stoichiometric mixture of CsF, RbF and Pd[PdF$_6$] at 400°C under 3 kbar pressure. It exists in tetragonal (room temperature) and cubic (high temperature) forms, transition temperature 383 ± 5 K. The beige compound is air-sensitive and hygroscopic. Cell dimensions are a = 9.06, c = 9.57 Å for the tetragonal form (space group F4/mmm-D_{4h}^{17}) and a = 9.22 Å at 393 K for the cubic form (space group Fm3m-O_h^5). Observed and calculated densities for the tetragonal form are 4.35 ± 0.05 and 4.80 g/cm³, respectively. X-ray intensity data are recorded. ESR data are tabulated (orthorhombic symmetry, g_1 = 2.287, g_2 = 2.216, g_3 = 2.073) and the spectrum (4.2 and 295 K) is reproduced in the paper. Thermal variation plots for cell dimensions and reciprocal magnetic susceptibility are reproduced [4].

$Cs_2[PdF_6]$ is prepared by fluorinating $Cs_2[PdCl_6]$ using F_2 [5, 6], ClF_3 [5, 7] or BrF_3 [8]. A preparation from $CsSeF_5$ and $[SeF_3]_2[PdF_6]$ in refluxing SeF_4 has also been described [1]. The bright yellow crystals are cubic, space group Fm3m-O_h^5 [9], with a = 9.00 Å [6] or 9.01 Å [8]. The infrared spectrum (700 to 400 cm^{-1}) is reproduced and magnetic shielding constants for fluoride nuclei are tabulated [9]. The electronic spectrum of the solid complex has been reported; $^1A_{1g} \rightarrow {}^3T_{1g}$ = 21000; $^1A_{1g} \rightarrow {}^1T_{1g}$ = 25000; $^1A_{1g} \rightarrow {}^1T_{2g}$ = 30000 cm^{-1}; Δ = 26000 cm^{-1}; B = 400 cm^{-1} [10].

References:

[1] Bartlett, N.; Quail, J. W. (J. Chem. Soc. **1961** 3728/32).
[2] Alter, E.; Hoppe, R. (Z. Anorg. Allgem. Chem. **408** [1974] 115/20).
[3] Müller, B. G. (Z. Anorg. Allgem. Chem. **491** [1982] 245/52).
[4] Tressaud, A.; Khairoun, S.; Dance, J. M.; Hagenmüller, P. (Z. Anorg. Allgem. Chem. **517** [1984] 43/58).
[5] Zemskov, S. V.; Nikonorov, Yu. I.; Mit'kin, V. N.; Lavrova, L. A. (Tezisy. Dokl. 12th Vses. Chugaevskoe Soveshch. Khim. Kompleksn. Soedin., Novosibirsk 1975, Vol. 3, p. 376; C. A. **85** [1976] No. 171050).
[6] Hoppe, R.; Klemm, W. (Z. Anorg. Allgem. Chem. **268** [1952] 364/71).
[7] Zemskov, S. V.; Nikonorov, Yu. I.; Pastukhova, E. D.; Mit'kin, V. N. (Izv. Sibirsk. Otd. Akad. Nauk SSSR Ser. Khim. Nauk **3** [1976] 83/8; C.A. **85** [1976] No. 86494).
[8] Sharpe, A. G. (J. Chem. Soc. **1953** 197/9).
[9] Zemskov, S. V.; Gabuda, S. P. (Zh. Strukt. Khim. **17** [1976] 904/21; J. Struct. Chem. [USSR] **17** [1976] 772/89).
[10] Brown, D. H.; Russell, D. R.; Sharp, D. W. A. (J. Chem. Soc. A **1966** 18/20).

3.7.3 Alkaline Earth Fluoropalladates

Magnesium Fluoropalladates

$Mg[PdF_6]$ is prepared by fluorination of a mixture of $(NH_4)_2PdCl_6$ and basic magnesium carbonate at 100°C. The yellow-brown solid has a hexagonal ($LiSbF_6$ type) structure, cell dimensions are a = 4.98, c = 13.48 Å, D_{X-ray} = 4.21 g/cm³. Intensity data are tabulated in the paper [1].

$MgCs[PdF_5]$ is obtained from MgF_2, CsF and PdF_2 at 650 to 700°C as a yellow solid. The crystals are orthorhombic, space group Imma-D_{2h}^{28}, a = 6.603, b = 7.415, c = 10.548 Å, Z = 4 [2].

References:

[1] Henkel, H.; Hoppe, R. (Z. Anorg. Allgem. Chem. **359** [1968] 160/77).
[2] Müller, B. G. (Z. Anorg. Allgem. Chem. **491** [1982] 245/52).

Calcium Fluoropalladates

Ca[PdF$_4$] is obtained as purple crystals by heating a mixture of CaF$_2$ and PdF$_2$ in a sealed Pd tube at 750 [1] or 820°C over a period of 20 to 30 d [2]; it has also been prepared by heating Ca[PdF$_6$] at 750°C [1, 3]. The tetragonal crystals belong to space group I4/mcm-D$_{4h}^{18}$ (KBrF$_4$ type), a = 5.521, c = 10.570 Å, Z = 4, D$_{X-ray}$ = 4.58, D$_{pyk}$ = 4.59 g/cm^3 [2, 3], square-planar [PdF$_4$]$^{2-}$ ions, Pd–F = 1.96 Å [3] or 1.991 Å [2], Ca–F = 2.40 Å; X-ray intensity data are tabulated [3]. The complex is slightly paramagnetic, χ_{mol}^c = +15 × 10^{-6} cm^3/mol [2, 3]. The Madelung part of the lattice energy (MAPLE) is discussed and data are tabulated [3].

Ca[PdF$_6$] is obtained as yellow-brown solid by direct fluorination of a (NH$_4$)$_2$[PdCl$_6$]/CaCO$_3$ mixture at 100°C. The crystals are hexagonal (LiSbF$_6$ type), a = 5.19, c = 14.59 Å, Z = 3, D$_{X-ray}$ = 3.81, D$_{pyk}$ = 3.80 g/cm^3, Pd–F = 1.89 Å. X-ray intensity data are tabulated [4].

References:

[1] Müller, B.; Hoppe, R. (Naturwissenschaften **58** [1971] 268).
[2] Müller, B. G. (J. Fluorine Chem. **20** [1982] 291/9).
[3] Müller, B.; Hoppe, R. (Mater. Res. Bull. **7** [1972] 1297/306).
[4] Henkel, H.; Hoppe, R. (Z. Anorg. Allgem. Chem. **359** [1968] 160/77).

Strontium Fluoropalladates

Sr[PdF$_4$] is obtained as red crystals by heating a mixture of SrF$_2$ and PdF$_2$ in a closed system at 800°C or by heating Sr[PdF$_6$] under argon at 720°C [1, 2]. The tetragonal crystals are of the K[BrF$_4$] type, a = 5.793, c = 10.747 Å, D$_{X-ray}$ = 5.00, D$_{pyk}$ = 4.97 g/cm^3, square-planar [PdF$_4$]$^{2-}$ ions, Pd–F = 1.96 Å, Sr–F = 2.53 Å. X-ray intensity data are tabulated. The compound is diamagnetic, χ_{mol}^c = −8 × 10^{-6} cm^3/mol. The Madelung part of the lattice energy is discussed and data are tabulated [2].

Sr[PdF$_6$] is obtained, mixed with Sr[PdF$_4$], on heating SrF$_2$ with Pd[PdF$_6$] under argon at 600°C [2]. The crystals are described as trigonal, a = 4.72 Å, α = 98.2° [3].

References:

[1] Müller, B.; Hoppe, R. (Naturwissenschaften **58** [1971] 268).
[2] Müller, B.; Hoppe, R. (Mater. Res. Bull. **7** [1972] 1297/306).
[3] Sharp, D. W. A. (unpublished work) according to Cox, B. (J. Chem. Soc. **1956** 876/8).

Barium Fluoropalladates

Ba[PdF$_4$] is obtained as orange crystals by heating mixtures of BaF$_2$ and PdF$_2$ at 700°C [1] or 760°C for 4 to 14 d [2] in a closed system, or by pyrolysis of Ba[PdF$_6$] under argon at 750°C [1]. The crystals are tetragonal, KBrF$_4$ type, a = 6.120, c = 10.981 Å, D$_{X-ray}$ = 5.16, D$_{pyk}$ = 5.17 g/cm^3, square-planar [PdF$_4$]$^{2-}$ ions, Pd–F = 1.96 Å, Ba–F = 2.68 Å. X-ray intensity data are tabulated. The compound is diamagnetic, χ_{mol}^c = −60 × 10^{-6} cm^3/mol. The Madelung part of the lattice energy is discussed and data are tabulated [2].

Ba[PdF$_6$] is obtained, mixed with Ba[PdF$_4$], on heating BaF$_2$ with Pd[PdF$_6$] at 600°C under argon, on further heating at 750°C it loses F$_2$ to form Ba[PdF$_4$] [2]. The crystals are trigonal, a = 4.88 Å, α = 98.4° [3].

References:

[1] Müller, B.; Hoppe, R. (Naturwissenschaften **58** [1971] 268).
[2] Müller, B.; Hoppe, R. (Mater. Res. Bull. **7** [1972] 1297/306).
[3] Sharp, D. W. A. (unpublished work) according to Cox, B. (J. Chem. Soc. **1956** 876/8).

3.7.4 Fluoropalladates of Zinc, Cadmium and Mercury

Zinc Fluoropalladates

ZnPdF₄ (see $Pd_{1-x}Zn_xF_2$, below)

CsZnPdF₅ is prepared by heating an equimolar mixture of CsF and ZnF_2 with PdF_2 at 650 to 700°C in a sealed tube. The beige crystals are isotypic with $CsPd_2F_5$ (orthorhombic space group Imma-D_{2h}^{28}), a = 6.576(1), b = 7.483(2), c = 10.645(2) Å [1].

Zn[PdF₆] is prepared as a yellow-brown solid by fluorinating a mixture of $(NH_4)_2[PdCl_6]$ and ZnO. The crystals are hexagonal ($LiSbF_6$ type structure), a = 4.95, c = 13.69 Å, $D_{X\text{-ray}}$ = 4.90 g/cm³, Pd–F = 1.85 Å. X-ray intensity data are tabulated in the paper [2].

References:

[1] Müller, B. G. (Z. Anorg. Allgem. Chem. **491** [1982] 245/52).
[2] Henkel, H., Hoppe, R. (Z. Anorg. Allgem. Chem. **359** [1968] 160/77).

Palladium Zinc Fluorides $Pd_{1-x}Zn_xF_2$ (x = 0.1 to 0.8)

These mixed fluorides, prepared by annealing mixtures of PdF_2 and ZnF_2, are paramagnetic [1], magnetic moments increase with increasing zinc content from 1.97 BM (x = 0) to > 2.83 BM (x = 0.8) [2]. A plot of magnetic moment vs. composition is reproduced [1]. The Néel temperature of all samples exhibit a spontaneous magnetisation which indicates a spin ordering with canted spins [2]. All show weak ferromagnetism below the magnetic transition temperature. Plots of reciprocal magnetic moments vs. temperature (0 to 300 K) are reproduced for all values of x. Cell dimensions for different values of x and plots of a and c axes against x are presented [1].

Cadmium Fluoropalladates

Cd[PdF₄] is obtained as blue [3] or dark blue [4] crystals by heating PdF_2 with CdF_2 at 700°C [3] or at 900°C for 20 to 30 d in a sealed platinum tube [4]. The crystals are cubic (CaF_2 variant), space group Pa3-T_h^6 with a = 5.403 Å, Z = 4. The crystal structure is shown in **Fig. 12**, p. 62.

The compound is antiferromagnetic, T_N = 80 K, μ_{eff} = 2.59 BM at 298 K. A plot of reciprocal magnetic moment against temperature (0 to 300 K) is reproduced in the paper [4].

X-ray photoelectron spectra (reproduced in paper): $Pd3d_{3/2}$ = 343.1; $Pd3d_{5/2}$ = 337.7; F1s = 685.5; $Cd3d_{5/2, 3/2}$ = 406.9, 412.7 eV (all ±0.4 eV) [6].

Cd[PdF₆] is obtained as a yellow-brown solid by fluorination of $(NH_4)_2[PdCl_6]/Cd(O_2CCH_3)_2$ mixtures. The crystals are hexagonal ($LiSbF_6$ type structure) with a = 5.08, c = 14.39 Å, $D_{X\text{-ray}}$ = 5.16, D_{pyk} = 5.07 g/cm³. X-ray intensity data are tabulated in the paper [5].

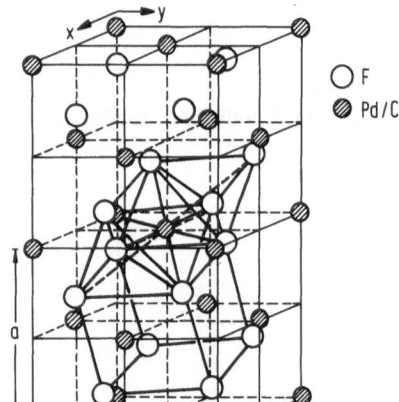

○ F
⬚ Pd/Cd

Fig. 12.   Unit cell of CdPdF$_4$.

References:

[1] Paus, D.; Hoppe, R. (Z. Anorg. Allgem. Chem. **431** [1977] 207/16).
[2] Paus, D.; Müller, B.; Hoppe, R. (Tr. Mezhdunar Konf. Magn., Moscow 1973 [1974], Vol. 3, pp. 32/5; C.A. **84** [1976] No. 37985).
[3] Müller, B., Hoppe, R. (Naturwissenschaften **58** [1971] 268).
[4] Müller, B. G. (J. Fluorine Chem. **20** [1982] 291/9).
[5] Henkel, H.; Hoppe, R. (Z. Anorg. Allgem. Chem. **359** [1968] 160/77).
[6] Tressaud, A.; Khairoun, S.; Tounara, H.; Watanabe, N. (Z. Anorg. Allgem. Chem. **540/541** [1986] 291/9).

Mercury Fluoropalladates

HgPdF$_4$ is obtained as a black crystalline powder by heating HgF$_2$ and PdF$_2$ together in a sealed platinum tube at 700°C under argon for 30 to 45 d. The crystals are cubic, space group Pa3-T$_h^6$ with a = 5.43 Å. The complex is antiferromagnetic [1].

HgPdF$_6$. A light brown-yellow powder of this composition is obtained by fluorination of a Hg(O$_2$CCH$_3$)$_2$/[NH$_4$]$_2$[PdCl$_6$] mixture. The X-ray pattern is similar to that of CdPdF$_6$, cell dimensions are a = 5.0, c = 14.2 Å [2].

References:

[1] Müller, B. G. (J. Fluorine Chem. **20** [1982] 291/9).
[2] Henkel, H.; Hoppe, R. (Z. Anorg. Allgem. Chem. **359** [1968] 160/77).

3.7.5   Miscellaneous Metal Fluoropalladates

CsMPdF$_6$ (M = Al, Ga, In, Sc). These are prepared by heating mixtures of the appropriate metal fluorides in sealed gold tubes at 450 to 700°C for 4 to 10 d [1].

CsAlPdF$_6$ as light violet noncubic crystals [1].

CsGaPdF$_6$ as violet noncubic crystals [1].

CsInPdF$_6$ as brown-violet cubic crystals, a = 10.89 Å, D$_{\text{X-ray}}$ = 5.13, D$_{\text{pyk}}$ = 5.11 g/cm^3 [1].

CsScPdF$_6$ as brown-violet cubic crystals, a = 10.82 Å, D_{X-ray} = 4.18, D_{pyk} = 4.15 g/cm^3 [1].

X-ray intensity data are recorded and the Madelung part of the lattice energy is calculated and discussed for CsInPdF$_6$ and CsScPdF$_6$ [1].

Tl[PdF$_3$] is prepared by heating a 1:1 mixture of TlF and PdF$_2$ at 400 to 700°C in a sealed tube under argon for 7 to 10 h. The dark brown crystals are cubic (perovskite type), a = 4.30 Å, D_{X-ray} = 7.67, D_{pyk} = 7.51 g/cm^3. X-ray intensity data are recorded. The Madelung part of the lattice energy is calculated and discussed [2].

NaPdZr$_2$F$_{11}$ is prepared by heating an equimolar mixture of NaF, PdF$_2$ and ZrF$_4$ in a palladium tube at 700°C for 14 to 20 d. The blue, paramagnetic (μ = 2.88 BM) crystals are triclinic, space group P$\bar{1}$-C$_i^1$ with a = 7.910, b = 5.746, c = 5.745 Å, α = 106.7°, β = 112.2°, γ = 97.9°, Z = 1 [8].

Pd[GeF$_6$] is obtained by adding BrF$_3$ to PdBr$_2$ and GeO$_2$, unit cell parameters are a = 5.53 ± 0.01 Å, α = 54.0 ± 0.02°. The complex is paramagnetic, μ_{eff} = 2.82 BM (temperature-independent), Θ = 31° [3].

Pd[SnF$_6$] is obtained by adding BrF$_3$ to PdBr$_2$ and SnBr$_4$. Unit cell parameters are a = 5.70 ± 0.02 Å, α = 53.13 ± 0.05°. The complex is paramagnetic, μ_{eff} = 2.98 BM (temperature-independent), Θ = 28° [3].

K$_3$[Pd$_3$Sn$_8$F$_{24}$]·10H$_2$O. A product of this formulation has been obtained from a mixture of PdCl$_2$, SnCl$_2$ (and KCl?) in HF, the presence of Pd–Sn bonds is postulated [9].

PbPdF$_4$ is prepared by heating a 1:1 mixture of PbF$_2$ and PdF$_2$ in a sealed tube at 650°C for 2 to 6 d. The red-violet microcrystals are tetragonal (KBrF$_4$ type), a = 5.873, c = 10.833 Å, D_{X-ray} = 6.92, D_{exp} = 6.80 g/cm^3, Pd–F = 1.96 Å, Pb–F = 2.56 Å. The Madelung part of the lattice energy is calculated and discussed [4].

CsMPdF$_6$ (M = Mo, Fe, Rh). These are prepared by heating mixtures of the appropriate metal fluorides in sealed gold tubes at 450 to 700°C for 4 to 10 d.

CsMoPdF$_6$ as red-brown noncubic crystals [1].

CsFePdF$_6$ as red-brown cubic crystals, a = 10.64 Å, D_{X-ray} = 4.51, D_{pyk} = 4.50 g/cm^3 [1].

CsRhPdF$_6$ as red-brown cubic crystals, a = 10.65 Å, D_{X-ray} = 5.02, D_{pyk} = 5.00 g/cm^3 [1].

X-ray intensity data are recorded and the Madelung part of the lattice energy is calculated and discussed for CsFePdF$_6$ and CsRhPdF$_6$ [1].

CsNiPdF$_5$ is prepared by heating CsF, NiF$_2$ and PdF$_2$ in a sealed gold tube at 650 to 700°C. The yellow crystals are orthorhombic, space group Imma-D$_{2h}^{28}$, a = 6.499(1), b = 7.504(2), c = 10.575(3) Å [5].

CsCoPdF$_5$ is prepared by heating CsF, CoF$_2$ and PdF$_2$ in a sealed gold tube at 650 to 700°C. The brown crystals are orthorhombic, space group Imma-D$_{2h}^{28}$, a = 6.527(1), b = 7.553(1), c = 10.659(2) Å [5].

RbCuPdF$_5$ is prepared by heating a mixture of RbF, CuF$_2$ and PdF$_2$ in a Pd-tube for 3 to 5 d at 650°C and forms orange-brown orthorhombic crystals, space group Pnma-D$_{2h}^{16}$, a = 6.269, b = 7.199, c = 10.763 Å, Z = 4 [10].

M$_{0.5}$Pd$_{0.5}$F$_2$ (M = Mn, Co, Cu, Ni). These violet antiferromagnetic mixed crystals have been mentioned in the literature. They have rutile structures with 6-coordinate Pd^{II} [11].

AgPdF$_4$·1.2BrF$_3$, AgPdF$_5$·0.8BrF$_3$. These orange materials are briefly mentioned as products obtained from the reaction of BrF$_3$ with AgCl/PdCl$_2$ mixtures [6].

Ag[PdF$_6$] is prepared by fluorination (F$_2$/Ar) of a (NH$_4$)$_2$[PdCl$_6$]/Ag$_2$SO$_4$ mixture at 550°C for 50 to 60 h. The dark brown Ag^{II}/Pd^{IV} complex is paramagnetic (μ_{eff} = ~1.96 BM) and obeys the Curie-Weiss law (Θ = +1 K), magnetic data over the temperature range 77 to 297 K are recorded. The AgF$_6$ groups are tetragonally distorted by the Jahn-Teller effect [7].

AgPdZr$_2$F$_{11}$ is prepared by heating an equimolar mixture of AgF, PdF$_2$ and ZrF$_4$ in a palladium tube at 700°C for 14 to 20 d. The blue paramagnetic (μ = 2.83 BM) crystals are monoclinic, space group C2/m-C$_{2h}^3$, a = 9.351, b = 6.991, c = 7.801 Å, β = 115.7°, Z = 2 [8].

References:

[1] Jesse, R. R.; Hoppe, R. (Z. Anorg. Allgem. Chem. **428** [1977] 91/6).
[2] Alter, E.; Hoppe, R. (Z. Anorg. Allgem. Chem. **408** [1974] 115/20).
[3] Bartlett, N.; Rao, P. R. (Proc. Chem. Soc. **1964** 393/4).
[4] Müller, B.; Hoppe, R. (Mater. Res. Bull. **7** [1972] 1297/306).
[5] Müller, B. G. (Z. Anorg. Allgem. Chem. **491** [1982] 245/52).
[6] Sharpe, A. G. (J. Chem. Soc. **1950** 3444/50).
[7] Müller, B.; Hoppe, R. (Z. Anorg. Allgem. Chem. **392** [1972] 37/41).
[8] Müller, B. G. (Z. Anorg. Allgem. Chem. **553** [1987] 205/11).
[9] Antonov, P. G.; Kukushkin, Yu. N.; Karymova, R. Kh.; Shtrele, V. G. (Izv. Vyssh. Uchebn. Zaved. Khim. Khim. Tekhnol. **25** [1982] 918/43; C. A. **98** [1983] No. 45770).
[10] Müller, B. G. (Z. Anorg. Allgem. Chem. **556** [1988] 79/84).

[11] Müller, B. G. (J. Fluorine Chem. **20** [1982] 291/9).

Palladium(II) hexafluoropalladate(IV) Pd[PdF$_6$]. This compound was originally formulated as Pd^{III}F$_3$ (see "Palladium" 1942, p. 270) but was subsequently reformulated as the palladium(II) salt of the hexafluoropalladate(IV) anion Pd^{II}[Pd^{IV}F$_6$] [1]. New preparations involve treatment of PdBr$_2$ with BrF$_3$ [1, 2], pyrolysis of XeF[Pd$_2$F$_9$] in a vacuum at 280°C [3], fluorination of PdI$_2$ at ~400°C [4] or fluorination of Pd metal at 500°C [5, 6]. It is described as a black hygroscopic powder [6].

An early X-ray diffraction study described the crystals as rhombohedral, space group R$\overline{3}$c-D$_{3d}^6$ with cell dimensions a = 5.5234 ± 0.0005 Å, α = 53.925 ± 0.005°. Calculated and observed X-ray intensities are tabulated [4]. A more recent neutron diffraction study indicates the rhombohedral space group R$\overline{3}$-C$_{3i}^2$ with a = 5.52(3) Å and α = 53.9°. Both Pd atoms are octahedrally surrounded by fluorines (Pd^{II}-F = 2.17, Pd^{IV}-F = 1.90 Å). Observed and calculated intensity data and a neutron diffraction diagram (4.2 and 25 K) are given in the paper [5].

Early magnetic measurements (97 to 298 K) showed Pd[PdF$_6$] to obey the Curie-Weiss law with Θ = 28° and an effective temperature-independent magnetic moment of 2.88 BM per Pd$_2$F$_6$ unit [1] or 2.05 BM per palladium [7]. More recently a value of 1.75 BM per palladium has been reported [5]. Powdered samples of "PdF$_3$" gave neutron diffraction patterns (4.2 and 298 K) which showed no evidence of magnetic ordering [10]. The complex is ferromagnetic (T$_c$ = 10 K) and the magnetic diagram below this temperature reveals that the moments of the palladium(II) ions are aligned perpendicular to the ternary axis of the rhombohedron [5]. Thermochemical (bomb calorimeter) studies of the reactions between F$_2$ and UPd$_3$ to give Pd[PdF$_6$], and between PF$_3$ and Pd[PdF$_6$] to give Pd and PF$_5$, afford a value of -967.4 ± 7.3 kJ/mol for ΔH$_f^°$(Pd[PdF$_6$]) at 298.15 K [8, 9].

Measurements of activation energy (ΔE) of electrical resistivity (ϱ) at 293 to 500 K indicate that $Pd[PdF_6]$ undergoes an insulator-semiconductor transition starting at ~ 25 kbar pressure corresponding to formation of Pd^{3+} at high pressures by an electronic conproportionation reaction: $Pd^{2+} + Pd^{4+} \rightarrow 2\,Pd^{3+}$. Diagrams of log ϱ vs. $1/T(K)$ and log ϱ vs. pressure are reproduced in the paper. Values of ΔE are 1.34, 0.24 and 0.07 eV for applied pressures of 1, 20 and 60 kbar, respectively. An energy level diagram for $Pd[PdF_6]$ is reproduced in the paper [6].

X-ray photoelectron spectra (reproduced in the paper) gave the data $Pd3d_{3/2} = 342.9$, 344.8; $Pd3d_{5/2} = 337.5$, 339.4; $F1s = 683.8$, 687.1 eV (all ± 0.4 eV) [11].

References:

[1] Bartlett, N.; Rao, P. R. (Proc. Chem. Soc. **1964** 393/4).
[2] Sharpe, A. G. (J. Chem. Soc. **1950** 3444/50).
[3] Bartlett, N.; Žemva, B.; Graham, L. (J. Fluorine Chem. **7** [1976] 301/20).
[4] Hepworth, M. A.; Jack, K. H.; Peacock, R. D.; Westland, G. J. (Acta Cryst. **10** [1957] 63/9).
[5] Tressaud, A.; Wintenberger, M.; Bartlett, N.; Hagenmüller, P. (Compt. Rend. C **282** [1976] 1069/72).
[6] Langlais, F.; Demazeau, G.; Portier, J.; Tressaud, A.; Hagenmüller, P. (Solid State Commun. **29** [1979] 473/6).
[7] Nyholm, R. S.; Sharpe, A. G. (J. Chem. Soc. **1952** 3579/87).
[8] Wijbenga, G. (J. Chem. Thermodyn. **14** [1982] 483/93).
[9] Wijbenga, G.; Johnson, G. K. (J. Chem. Thermodyn. **13** [1981] 471/5).
[10] Wilkinson, M. L.; Wollan, E. O.; Child, H. R.; Cable, J. W. (Phys. Rev. [2] **121** [1961] 74/7).

[11] Tressaud, A.; Khairoun, S.; Touhara, H.; Watanabe, N. (Z. Anorg. Allgem. Chem. **540/541** [1986] 291/9).

66

4 Palladium and Chlorine

4.1 The Palladium-Chlorine System

Fig. 13 shows the ~58 to 68 at% Cl section of the palladium-chlorine system at high temperatures [1]; see also "Thermochemical Data", p. 72/3.

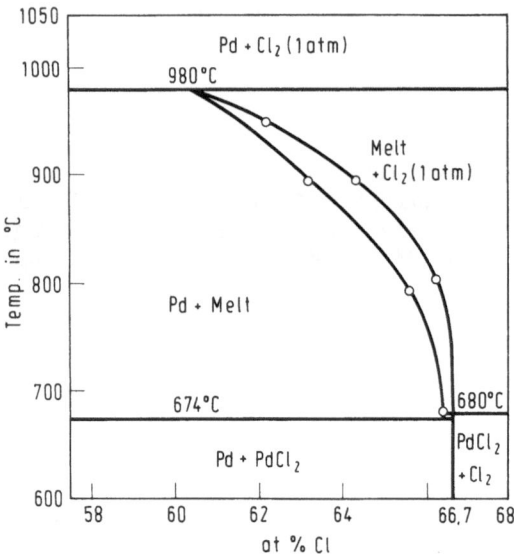

Fig. 13.   Partial Pd–Cl diagram.

Palladium appears to form only one stable chloride, $PdCl_2$. Early reports of a monochloride, PdCl [2], have not been confirmed and there is no evidence for a trichloride. The oxidation state Pd^{IV} is found in the $[PdCl_6]^{2-}$ anion but the simple tetrachloride $PdCl_4$ is unknown, except possibly as an intercalation compound with graphite [1, 2].

References:

[1] Bell, W. E.; Merten, U.; Tagami, M. (J. Phys. Chem. **65** [1961] 510/7).
[2] Gmelin Handbuch "Palladium" 1942, p. 272.

4.2 Palladium Dichloride $PdCl_2$ (see "Palladium" 1942, pp. 272/87)

Palladous chloride exists in three polymorphic forms: α-$PdCl_2$, which is formed at intermediate temperatures and is metastable at room temperature, β-$PdCl_2$, which is formed and is stable at lower temperatures [1, 2], and γ-$PdCl_2$, which appears to be formed at higher temperatures but is also stable at low temperatures. Commercial $PdCl_2$ is usually the γ-form [3]. Transition temperatures have been given as 401 and 504°C [4]. The structures of the three forms are discussed in Section 4.2.2, p. 68.

References:

[1] Maitlis, P. M. (The Organic Chemistry of Palladium, Vol. 1, Academic, New York – London 1971, p. 43).
[2] Henry, P. M. (Palladium Catalysed Oxidation of Hydrocarbons, Reidel, Dordrecht – Boston 1980, p. 6).
[3] Yatsimirski, A.; Ugo, R. (Inorg. Chem. **22** [1983] 1395/7).
[4] Soulen, J. R.; Chappell, W. H. (J. Phys. Chem. **69** [1965] 3669/71).

4.2.1 Preparation

Palladium(II) chloride is the only chloride formed on heating palladium metal in chlorine gas [1, 2], reaction commences at 260°C and is complete at 525°C [2]. Chlorination of palladium powder with aqua regia at $40 \pm 3°C$ followed by evaporation at 60 to 70°C also affords $PdCl_2$ [3]. Solutions of H_2PdCl_4, prepared by chlorination of Pd sponge in methanol solution and subsequent addition of further Pd metal (to reduce Pd^{IV} to Pd^{II}), yield $PdCl_2$ on evaporation [4]. A suspension of palladium in acetic acid reacts with HCl, NO and O_2 to afford $PdCl_2$ [5, 6]. Preparation of $PdCl_2$ by electrochemical dissolution of powdered palladium from graphite anodes in HCl solution has been described [7, 8]. The β-form of $PdCl_2$ has been obtained by treatment of palladium with SO_2Cl_2 at 400°C over a period of 3 d [9], and by adding conc. HCl (2 equivalents per Pd) to $Pd_3(O_2CMe)_6$ in glacial acetic acid [10]. Vacuum sublimation of α- or γ-$PdCl_2$ at 430 to 460°C gives the β-form [11, 12], which converts to the α-form on tempering at 500°C [12]. Whiskers of $PdCl_2$, prepared by treatment of Pd metal with chlorine at >500°C are thought to be a morphological form of one of the high-temperature polymorphs of $PdCl_2$ [13]. The chemical transport of $PdCl_2$ by means of Al_2Cl_6 (gas) leads to formation of oxide-free crystals of α-$PdCl_2$ [14, 15, 16]. Recovery of palladium from spent catalysts as $PdCl_2$ has been achieved by extraction with conc. HCl/HNO_3 solutions or HCl/Cl_2 solutions [17], by heating in aqueous carboxylic acid solution in the presence of air and a chloride salt, MCl_x (M = NH_4 or a group I, II or III metal) [18] or by treatment with HCl, NO and O_2 in anhydrous carboxylic acid solution [5]. Thin layer chromatography using semicrystalline stannic arsenate cation exchanger has been employed to separate $PdCl_2$ from other platinum metal chloro complexes [19]. High speed cooling techniques have been used to trap $PdCl_2$ in a glass-like state [20].

For data on **heats of formation** of $PdCl_2$ see "Thermochemical Data", p. 72.

References:

[1] Bell, W. E.; Merten, U.; Tagami, M. (J. Phys. Chem. **65** [1961] 510/7).
[2] Ivashentsev, Ya. I.; Timonova, R. I. (Zh. Neorgan. Khim. **12** [1967] 592/6; Russ. J. Inorg. Chem. **12** [1967] 308/10).
[3] Ono, T.; Hitachi Chemical Co. Ltd. (Japan. Kokai 74-125, 295 [1974]; C.A. **82** [1975] No. 142234).
[4] Clements, F. S.; Nutt, E. V.; International Nickel Co. (Mond) Ltd. (Brit. 879074 [1961]; C.A. **56** [1962] 8298).
[5] van Helden, R.; Jonkhoff, T.; Shell Oil Co. (U.S. 3210152 [1965]; C.A. **63** [1965] 15895).
[6] Shell Internationale Research Maatschappij N. V. (Belg. 614968 [1962]; C.A. **58** [1963] 8739).
[7] Volfson, A. I.; Ryazanov, A. I.; Chigrinova, G. D. (Zh. Prikl. Khim. **34** [1961] 173/6; J. Appl. Chem. [USSR] **34** [1961] 162/4).
[8] Ryazanov, A. I.; Volfson, A. I.; Chigrinova, G. D. (U.S.S.R. 138922 [1959]; C.A. **56** [1962] 5766).

[9] Dillamore, I. M.; Edwards, D. A. (J. Inorg. Nucl. Chem. **31** [1969] 2427/30).

[10] Yatsimirski, A.; Ugo, R. (Inorg. Chem. **22** [1983] 1395/7).

[11] Schäfer, H.; Wiese, U.; Rabeneck, H. (Z. Anorg. Allgem. Chem. **513** [1984] 157/9).

[12] Schäfer, H.; Wiese, U.; Rinke, K.; Brendel, K. (Angew. Chem. **79** [1967] 244/5; Angew. Chem. Intern. Ed. Engl. **6** [1967] 253/4).

[13] Riebling, E. F. (Science **162** [1968] 467/8).

[14] Schäfer, H.; Nowitzki, J. (J. Less-Common Metals **61** [1978] 47/50).

[15] Schäfer, H.; Trenkel, M. (Z. Anorg. Allgem. Chem. **437** [1977] 10/8).

[16] Schäfer, H.; Binnewies, M.; Domke, W.; Karbinski, J. (Z. Anorg. Allgem. Chem. **403** [1974] 116/26).

[17] Vinarov, I. V.; Orlova, A. I.; Grigor'eva, L. P.; Il'chenko, L. I. (Kompleksn. Ispol'z. Mineral. Syr'ya **1985** No. 1, pp. 39/42; C.A. **103** [1985] No. 25454).

[18] Fernholz, H.; Schmidt, H. J.; Farbwerke Hoechst A.-G. (Ger. 1269114 [1968]; C.A. **69** [1968] No. 28940).

[19] Husain, S. W.; Rasheedzad, S. (Mikrochim. Acta [Wien] **1978** I, pp. 11/8).

[20] Dimitriev, I.; Marinov, M. R. (Dokl. Bolg. Akad. Nauk **22** [1969] 1157/9; C.A. **72** [1970] No. 103291).

4.2.2 Physical Properties

Structure

Palladium chloride exists in at least three polymorphic forms.

α-PdCl$_2$. The crystal structure reported by A. F. Wells ("Palladium" 1942, p. 273) has a [PdCl$_{4/2}$]$_\infty$ chain structure with square-planar PdCl$_4$ units linked by double chloride bridges. The structure has been described in terms of an ideal rhombic prism [1]. Theoretical calculations performed by the method of symmetry of the potential functions indicate that the structure adopted by α-PdCl$_2$ does not give maximum packing density which would require 5-coordinate palladium [2]. Valence state calculations on gaseous PdCl$_2$ point to an angular rather than a linear structure [3]. The energetics of solid α-PdCl$_2$ have been described in terms of small fragments [4].

β-PdCl$_2$. The crystals contain Pd$_6$Cl$_{12}$ molecular units and are isotypic with those of Pt$_6$Cl$_{12}$ [5]. However, chlorine NQR data suggest that the structure of β-Pd$_6$Cl$_{12}$ unlike that of Pt$_6$Cl$_{12}$ is regular [6]. X-ray powder pattern data have been tabulated [7, 8].

γ-PdCl$_2$. Evidence for the existence of this form has been discussed and X-ray powder data reported but no structural details are known [8].

A suggestion that differences in the catalytic activity of PdCl$_2$ samples heated above and below 425°C are due to redox disproportionation [$3\,PdCl_2 \rightleftharpoons PdCl_4 + Pd_2Cl_2$] [9] can now be discounted.

References:

[1] Nyman, H. (J. Solid State Chem. **17** [1976] 75/8).

[2] Zorkii, P. M.; Porai-Koshits, M. A.; Prezman, L. M.; Murav'eva, G. P. (Zh. Strukt. Khim. **10** [1969] 633/7; J. Struct. Chem. [USSR] **10** [1969] 538/42).

[3] Charkin, O. P.; Dyatkina, M. E. (Zh. Strukt. Khim. **6** [1965] 579/90; J. Struct. Chem. [USSR] **6** [1965] 550/60).

[4] Burdett, J. K.; Caneva, D. C. (Inorg. Chem. **24** [1985] 3866/73).

[5] Schäfer, H.; Wiese, U.; Rinke, K.; Brendel, K. (Angew. Chem. **79** [1967] 244/5; Angew. Chem. Intern. Ed. Engl. **6** [1967] 253/4).

[6] van Bronswyk, W.; Nyholm, R. (J. Chem. Soc. A **1968** 2084/6).

[7] Soulen, J. R.; Chappell, W. H. (J. Phys. Chem. **69** [1965] 3669/71).

[8] Yatsimirski, A.; Ugo, R. (Inorg. Chem. **22** [1983] 1395/7).

[9] Moiseev, I. I.; Grigor'ev, A. A. (Dokl. Akad. Nauk SSSR **178** [1968] 1090/3; Dokl. Chem. Proc. Acad. Sci. USSR **178/183** [1968] 132/4).

Spectra

Mass Spectra. The mass spectrum of β-$PdCl_2$ has been reported [1, 2, 3], at low ionisation energies (15 eV) only the parent ion $Pd_6Cl_{12}^+$ is observed [1, 2]. At higher ionisation energies (50 eV) fragmentation occurs [2] and one report lists ions ranging from $Pd_6Cl_{12}^+$ to $Pd_3Cl_3^+$, see table below [3].

Mass spectrum of Pd_6Cl_{12} (β-$PdCl_2$):

m/e	intensity in %	assignment
1062	100	$Pd_6Cl_{12}^+$
1027	46	$Pd_6Cl_{11}^+$
992	62	$Pd_6Cl_{10}^+$
920	15	$Pd_6Cl_8^+$
885	22	$Pd_5Cl_{10}^+$ (or $Pd_6Cl_7^+$)
848	12	$Pd_5Cl_9^+$ (or $Pd_6Cl_6^+$)
814	40	$Pd_5Cl_8^+$ (or $Pd_6Cl_5^+$)
708	30	$Pd_4Cl_8^+$ (or $Pd_5Cl_5^+$ or $Pd_6Cl_2^+$)
672	15	$Pd_4Cl_7^+$ (or $Pd_5Cl_4^+$ or Pd_6Cl^+)
601	7	$Pd_4Cl_5^+$ (or $Pd_5Cl_2^+$)
566	40	$Pd_4Cl_4^+$ (or Pd_5Cl^+)
530	40	$Pd_3Cl_6^+$ (or $Pd_4Cl_3^+$)
495	30	$Pd_3Cl_5^+$ (or $Pd_4Cl_2^+$)
427	75	$Pd_3Cl_3^+$

Numerous peaks below m/e = 400 were observed but could not be assigned [3].

Vibrational Spectra. An early report lists infrared bands for α-$PdCl_2$ at ~348(sh), 340(vs), 297(vw), 187(vw) and ~174 (m-w) cm^{-1} and gives tentative assignment [4]. In a later report absorptions at 340(s, sh), 325(vs, b), 143(s) and 115(w) cm^{-1} are listed [5]. Raman data for $PdCl_2$ in MeCN solution [$Pd_2Cl_4(MeCN)_2$, $\nu(Pd–Cl)_{terminal}$ 380 to 350, $\nu(Pd–Cl)_{bridge}$ 310 to 300 cm^{-1}] have been reported [6]. The infrared spectrum of β-$PdCl_2$ (Pd_6Cl_{12}) has been recorded over the range 450 to 33 cm^{-1} and a normal coordinate analysis has been performed. Absorptions occur at 345(s, sh), 325(vs), 197(s) and 143(s) cm^{-1} [5] or at 340(sh) and 329(s) cm^{-1} [7]. An infrared spectrum (500 to 100 cm^{-1}) of $PdCl_2$ vapour isolated in solid argon is reproduced in a paper and compared with the calculated spectrum of the Pd_6Cl_{12} hexamer [8]. Force constants (mdyn/Å) are $f_r = 1.7$, $f_\alpha = 0.16$ and $f_\gamma = 0.15$ for β-$PdCl_2$ and $f_r = 1.7$, $f_\gamma = 0.13$ for α-$PdCl_2$ [5].

Electronic Spectra. The absorption and reflectance spectra of crystalline α-$PdCl_2$ have been measured over the range 1 to 11 eV at low temperature. The structures found in the spectra and their anisotropies have been related to D_{2h} MO states of a d^8 metal ion in an orthorhombic planar ligand field. The absorption spectrum (2 to 11 eV, 4.2 K) and polarised reflection spectrum (1 to 5 eV, single crystal, 77 K) are reproduced in the paper [9]. Electronic spectra 50 000 to 27 000 cm^{-1} have been measured for the adducts PdCl$_2$(DMF)$_2$ and (PdCl$_2$L)$_2$ (L = MeCN, H$_2$O, CHCl$_3$, EtOH and Me$_2$SO), and explained in terms of charge transfers involving MO's formed with the participation of terminal and bridging chloride ligands [10]. The absorption spectra 48 000 to 24 000 cm^{-1} of PdCl$_2$ in MeCN solutions (5 ×10^{-3}, 5 ×10^{-4} and 5 ×10^{-5}M) are reproduced in a paper together with spectra of similar solutions containing added LiCl (0 to 25 ×10^{-3}M) or [NEt$_4$]Cl. Evidence for formation of the anionic complex anion [Pd$_2$Cl$_6$]$^{2-}$, λ$_{max}$ = 41, 35, 30, 25 and 22 (all ×10^3) cm^{-1}, was observed in the latter spectra [6]. Absorption curves (350 to 625 nm) for 10^{-3}M solutions of PdCl$_2$ in H$_2$O (pH = 5.60) and in DMF are reproduced in [11]. Spectrophotometric measurements (370 to 500 μm) on aqueous Pd(ClO$_4$)$_2$ solutions containing amounts of Cl$^-$ (spectra reproduced in paper) indicate formation of PdCl$^+$, PdCl$_2$, PdCl$_3^-$, PdCl$_4^{2-}$ and possibly PdCl$_5^{3-}$ and PdCl$_6^{4-}$ [12]. The state of palladium ions in aqueous solutions has been studied spectrophotometrically during dilution as a function of the chloride ion concentration at a constant atomic ratio Pd:Cl. Absorption spectra of aqueous PdCl$_2$ solutions, pure and with added HCl or AlCl$_3$ are reproduced in the paper. The concentration limits of existence of the "monomeric" and "dimeric" palladium complexes have been determined [13]. The electronic spectrum (14 000 to 32 000 cm^{-1}) of PdCl$_2$ in fused LiCl/KCl eutectic at 450°C (reproduced in the paper) has been interpreted in terms of an essentially octahedral [PdCl$_6$]$^{4-}$ anion [14].

Nuclear Quadrupole Resonance Spectra. Chlorine-35 NQR spectra have been recorded for α- and β-forms of PdCl$_2$ (spectrum of the β-form reproduced in paper) and used to calculate charge distributions. Data are tabulated below [15]:

compound	freq. (MHz)	η(calc.)	eqQ/h (MHz)	calc. on Cl	charge on Pd
α-PdCl$_2$	18.069	0.157	35.994	− 0.27	+ 0.54
β-PdCl$_2$	20.242	0	40.484	− 0.26	+ 0.52

X-Ray Photoelectron Spectra (ESCA). Data (in eV) for PdCl$_2$ have been given as follows:

Pd 3d$_{5/2}$ = 335.6 [16]; Pd 3d$_{3/2}$ = 343.2 ± 0.4, Pd 3d$_{5/2}$ = 337.9 ± 0.3, shift 2.2, charge on Pd = 0.60 [17]; Pd 3d$_{5/2}$ = 338.0, Cl 2p$_{3/2}$ = 199.1 [18]. Spectra are reproduced in papers [16, 17].

The charges on the palladium atoms in a series of Pd^{II} compounds including PdCl$_2$ have been estimated by the method of self-consistency of the electronegativities. A plot of the Pd 3d$_{5/2}$ binding energy chemical shifts against the charge on the metal atoms, and a table giving data on the dependence of the energy of the one-electron levels of the palladium atom on the charge, are reproduced in the paper [19].

X-Ray Absorption Near Edge Structure (XANES) Spectra. The L-edge absorption spectra of PdCl$_2$, obtained using synchrotron radiation, have been compared with similar data for Pd metal. L$_1$ and L$_{2,3}$ near edge structures of PdCl$_2$ and Pd metal are reproduced in [20]. The palladium Lβ$_{2,15}$ emission, palladium L$_{III}$ absorption, chlorine Kβ emission and chlorine K absorption spectra of α-PdCl$_2$ have been obtained using a 50 cm bent-quartz-crystal vacuum spectrograph, and used to determine electronic band structures. Spectra are reproduced and photon energy values tabulated in [21]. An earlier paper by the same author reports chlorine K absorption curves for PdCl$_2$ (reproduced in paper) and tabulates photon energy values of the maxima [22]. As part of a study of the effect of chemical bonding on the Lα$_{1,2}$ and L$_{β1}$ X-ray lines of palladium, measurements have been reported for PdCl$_2$, PdO and Pd metal [23, 24]. Muonic X-ray spectra of 32 chlorides including PdCl$_2$ have been measured with Ge detectors, Coulomb capture ratios and Lyman X-ray intensities were determined [25].

References:

[1] Schäfer, H.; Wiese, U.; Rinke, K.; Brendel, K. (Angew. Chem. **79** [1967] 244/5; Angew. Chem. Intern. Ed. Engl. **6** [1967] 253/4).

[2] Schäfer, H.; Wiese, U.; Rabeneck, H. (Z. Anorg. Allgem. Chem. **513** [1984] 157/9).

[3] Yatsimirski, A.; Ugo, R. (Inorg. Chem. **22** [1983] 1395/7).

[4] Adams, D. M.; Goldstein, M.; Mooney, E. F. (Trans. Faraday Soc. **59** [1963] 2228/32).

[5] Mattes, R. (Z. Anorg. Allgem. Chem. **364** [1969] 290/6).

[6] Volchenskova, I. I.; Yatsimirskii, K. B. (Zh. Neorgan. Khim. **18** [1973] 1875/82; Russ. J. Inorg. Chem. **18** [1973] 990/5).

[7] Dillamore, I. M.; Edwards, D. A. (J. Inorg. Nucl. Chem. **31** [1969] 2427/30).

[8] Martin, T. P. (J. Chem. Phys. **69** [1978] 2036/42).

[9] Tanino, H.; Kobayashi, K. (J. Phys. Soc. Japan **52** [1983] 3978/84).

[10] Volchenskova, I. I.; Yatsimirskii, K. B. (Teor. Eksperim. Khim. **13** [1977] 197/204; Theor. Exptl. Chem. **13** [1977] 146/51).

[11] Khain, V. S.; Val'kova, V. P. (Zh. Neorgan. Khim. **23** [1978] 3368/70; Russ. J. Inorg. Chem. **23** [1978] 1870/1).

[12] Sundaram, A. K.; Sandell, E. B. (J. Am. Chem. Soc. **77** [1955] 855/7).

[13] Kravchuk, L. S.; Stremok, I. P.; Markevich, S. V. (Zh. Neorgan. Khim. **21** [1976] 728/31; Russ. J. Inorg. Chem. **21** [1976] 391/3).

[14] Dickinson, J. R.; Johnson, K. E. (Can. J. Chem. **45** [1967] 1631/6).

[15] van Bronswyk, W.; Nyholm, R. (J. Chem. Soc. A **1968** 2084/6).

[16] Baldy, A.; Durocher, A.; Rzehak, H.; Limousin-Maire, Y. (Bull. Soc. Chim. France **1981** 67/8).

[17] Kumar, G.; Blackburn, J. R.; Albridge, R. G.; Moddeman, W. E.; Jones, M. M. (Inorg. Chem. **11** [1971] 296/300).

[18] Nefedov, V. I.; Zakharova, I. A.; Moiseev, I. I.; Porai-Koshits, M. A.; Vargaftik, M. N.; Belov, A. P. (Zh. Neorgan. Khim. **18** [1973] 3264/8; Russ. J. Inorg. Chem. **18** [1973] 1738/40).

[19] Gagarin, S. G. (Zh. Strukt. Khim. **19** [1978] 723/4; J. Struct. Chem. [USSR] **19** [1978] 620/1).

[20] Sham, T. K. (Chem. Phys. Letters **101** [1983] 567/72).

[21] Sugiura, C. (J. Chem. Phys. **62** [1975] 1111/5).

[22] Sugiura, C. (J. Chem. Phys. **59** [1973] 4907/10).

[23] Genchev, D. (Dokl. Bolg. Akad. Nauk **30** [1977] 821/3; C.A. **87** [1977] No. 159499).

[24] Genchev, D. (Dokl. Bolg. Akad. Nauk **30** [1977] 997/9; C.A. **87** [1977] No. 191516).

[25] Daniel, H.; Bergmann, R.; Fottner, G.; Hartmann, F. J.; Wilhelm, W. (Z. Physik A **300** [1981] 253/62).

Electronic Configuration

Magnetic data for $PdCl_2$ adsorbed on SiO_2 gel indicate that the Pd^{2+} ions have square-planar dsp^2 electronic configuration.

Reference:

French, C. M.; Howard, J. P. (Trans. Faraday Soc. **52** [1956] 996/8).

4.2.3 Chemical Reactions

Radioactive Decay of $^{103}PdCl_2$

The effect of chemical environment on the rate of ^{103}Pd radioactive decay, λ, has been demonstrated by large changes recorded for $^{103}PdCl_2$ in the solid state and in solution ($\Delta\lambda/\lambda = 0.0126 \pm 0.0012$).

Reference:

Belyavenko, V. S.; Borozenets, G. P.; Vishnevskii, I. N.; Zheltonozhskii, V. A. (Yad. Fiz. **41** [1985] 12/3; C.A. **102** [1985] No. 85363).

Thermochemical Data

New **melting point** determinations for $PdCl_2$ (680 ± 2°C) [1, 2] confirm Puche's original value of 678°C.

$PdCl_2$	T in K	ΔH_T° in kJ/mol	ΔS_T° in $cal \cdot mol^{-1} \cdot K^{-1}$	Ref.
Pd + $Cl_2 \rightleftharpoons PdCl_2$				
solid	—	−189.8		[3]
solid	—	−163		[4]
solid	883 to 953	−138 ± 4	−117 ± 4	[5]
solid	947	−161.3 ± 4	−136 ± 4	[1]
liquid	953	−143 ± 4	−117 ± 4	[1]
gas	1573	+116.6 ± 8	+46.4 ± 8	[1]
liquid	953 to 1030	−90 ± 6	−66.9 ± 4	[5]
$PdCl_2$ (solid) → $PdCl_2$ (vapour)				
solid	883 to 953	+140 ± 6	+113 ± 4	[5]
liquid	953 to 1030	+92 ± 4	+58.5 ± 4	[5]
$PdCl_2$ (solid) → $PdCl_2$ (liquid)		48.0		[5]

Thermochemical data for the reaction of Cl_2 with metallic palladium [14]:

$Pd + Cl_2 \rightleftharpoons PdCl_2 + 184$ kJ/mol; $\Delta G_{298}^\circ = -138$ kJ/mol.

$PdO + Cl_2 \rightleftharpoons PdCl_2 + \frac{1}{2} O_2 + 22$ kJ/mol; $\Delta G_{298}^\circ = -84$ kJ/mol.

$PdCl_2 + \frac{1}{2} O_2 \rightleftharpoons PdO + Cl_2 - 92$ kJ/mol; $\Delta G_{298}^\circ = +77.6$ kJ/mol.

Expressions for the **vapour pressure** (log p_{PdCl_2}) and the **dissociation pressure** (log p_{Cl_2}) of $PdCl_2$ over the temperature range 883 to 1030 K [5]:

log $p_{PdCl_2} = 8.86 - 7452.5/T$ (Torr) at 883 to 953 K; log $p_{PdCl_2} = 6.32 - 5031.8/T$ (Torr) at 953 to 1030 K;

log $p_{Cl_2} = 8.63 - 6728.5/T$ (Torr) at 883 to 953 K; log $p_{Cl_2} = 7.26 - 5422.9/T$ (Torr) at 953 to 1030 K.

Data reproduced in graphical form include plots of log p_{PdCl_2} and log p_{Cl_2} versus reciprocal temperature [5], dissociation pressure of condensed chlorides in palladium-chlorine system, effect of Cl_2 pressure on vapour pressure of $PdCl_2$ (at 850, 900 and 950°C and at 1103, 1304 and

1506°C) and the temperature dependence of $PdCl_2$ pressure at 1 atm Cl_2 pressure. The dissociation pressure of Cl_2 in the $PdCl_2$ system reaches 1 atm at 980°C [1].

A pentamer **Pd_5Cl_{10}** has been claimed as an important species in the vapour phase of the $PdCl_2$ system, and the major one at $<980°C$ and 1 atm Cl_2. Thermochemical data according to [1]:

for $5PdCl_2$ (solid) $\rightleftharpoons Pd_5Cl_{10}$ (gas)
$\Delta H^\circ_{953} = +198 \pm 8$ kJ/mol, $\Delta S^\circ_{953} = +161.3 \pm 8$ $J \cdot mol^{-1} \cdot deg^{-1}$;

for $5PdCl_2$ (liquid) $\rightleftharpoons Pd_5Cl_{10}$ (gas)
$\Delta H^\circ_{953} + 73.5 \pm 4$ kJ/mol, $\Delta S^\circ_{953} + 30.5 \pm 4$ $J \cdot mol^{-1} \cdot deg^{-1}$;

for $5Pd$ (solid) $+ 5Cl_2 \rightleftharpoons Pd_5Cl_{10}$ (gas)
$\Delta H^\circ_{953} = -640 \pm 25$ kJ/mol.

Data presented in graphical form include effect of Cl_2 pressure on Pd_5Cl_{10} pressure at 850, 900 and 950°C, and at 700 and 800°C, and the temperature dependence of the partial pressure of Pd_5Cl_{10} and $PdCl_2$ at 1 atm Cl_2 pressure [1].

A more recent report concludes that "Pd_5Cl_{10}" is in fact the hexamer Pd_6Cl_{12} and gives the following thermochemical data:

$\Delta H^\circ_{953} = 197.8 \pm 8$ kJ/mol, $\Delta S^\circ_{953} = 160$ $J \cdot mol^{-1} \cdot deg^{-1}$,
$\Delta C_p = -41.8$ J/deg, $\Delta H^\circ_{298} = 225.3$ kJ/mol, $\Delta S^\circ_{298} = 208.6$
$J \cdot mol^{-1} \cdot deg^{-1}$ for the reaction

$$6PdCl_2 \rightleftharpoons Pd_6Cl_{12}(gas) \quad [7].$$

Calculated values of $-\Delta H^\circ_{f298}$ (in kJ/mol) for the hydrates $PdCl_2 \cdot nH_2O$ are 480 (n=1), 777.5 (n=2), 1074 (n=3), 1371 (n=4), 1668 (n=5) and 1964 (n=6) [8].

Evaluation of data given in [1] und [2] gave the following bond energies for the Pd–Cl bond [9]:

α-$PdCl_2$ [$(PdCl_{1/2})_\infty$] 200.6 kJ/mol; gaseous β-$PdCl_2$ (Pd_6Cl_{12}) 192.3 kJ/mol; gaseous $PdCl_2$ 259.2 kJ/mol.

The lattice energy of $PdCl_2$ has been given as 2771 [10], 2763 (from thermochemical data) and 2437 kJ/mol (from optical data) [11].

The standard emf (in V) for solid or molten (above 680°C) $PdCl_2$ has been given as 0.373 (25°C in aqueous solution), 0.768 (25°C, solid), 0.714 (100°C), 0.646 (200°C), 0.581 (300°C), 0.549 (350°C), 0.518 (400°C), 0.487 (450°C), 0.457 (500°C), 0.427 (550°C), 0.397 (600°C), 0.331 (800°C) and 737°C (decomposition) [12].

An analysis of the most favourable thermodynamical conditions for performing vapour transport reactions involving $PdCl_2$ has been reported [6].

Thermal Analysis. Differential heating curves of $PdCl_2$ in a stream of O_2, Pd in a stream of HCl, Pd in a stream of Cl_2, Pd in streams of Cl_2 and O_2 (various ratios) [13, 14] and $PdCl_2$ in a stream of Ar [14] have been reported.

References:

[1] Bell, W. E.; Merten, U.; Tagami, M. (J. Phys. Chem. **65** [1961] 510/7).
[2] Oranskaya, M. A.; Mikhailova, N. A. (Zh. Neorgan. Khim. **5** [1960] 12/5; Russ. J. Inorg. Chem. **5** [1960] 5/7).
[3] Long, L. H. (Quart. Rev. [London] **7** [1953] 134/74, 154).
[4] Goldberg, R. N.; Hepler, L. G. (Chem. Rev. **68** [1968] 229/52, 241).

[5] Oranskaya, M. A.; Mikhailova, N. A. (Zh. Neorgan. Khim. **5** [1960] 12/5; Russ. J. Inorg. Chem. **5** [1960] 5/7).

[6] Alcock, C. B.; Jeffes, J. H. E. (Inst. Mining Met. Trans. C **76** [1967] 246/58).

[7] Schäfer, H. (Z. Anorg. Allgem. Chem. **415** [1975] 217/24).

[8] Burylev, B. P. (Zh. Fiz. Khim. **50** [1976] 2689/90; Russ. J. Phys. Chem. **50** [1976] 1608).

[9] Schäfer, H.; Wiese, U.; Rinke, K.; Brendel, K. (Angew. Chem. **79** [1967] 244/5; Angew. Chem. Intern. Ed. Engl. **6** [1967] 253/4).

[10] Karapet'yants, M. Kh. (Zh. Fiz. Khim. **28** [1954] 1136/52, 1151).

[11] Yatsimirskii, K. B. (Zh. Neorgan. Khim. **3** [1958] 2244/52; Russ. J. Inorg. Chem. **3** No. 10 [1958] 26/36).

[12] Hamer, W. J.; Malmberg, M. S.; Rubin, B. (J. Electrochem. Soc. **103** [1956] 8/16).

[13] Ivashentsev, Ya. I.; Timonova, R. I. (Zh. Neorgan. Khim. **12** [1967] 592/5; Russ. J. Inorg. Chem. **12** [1967] 308/10).

[14] Sidorov, N. S.; Karepov, B. G.; Nikolaev, R. K. (Izv. Akad. Nauk SSSR Neorgan. Materialy **24** [1988] 70/2; C.A. **108** [1988] No. 160213).

Solubility

β-PdCl$_2$ dissolves in aromatic solvents; the solubility, which is about 5×10^{-3} M, increases with increasing donor properties of the hydrocarbon [1]. The solubility of "pure" commercial PdCl$_2$ in DMF ranges from 2.663 g/100 ml at $-10°C$ to 5.508 g/100 ml at $+40°C$. A plot of conductivity versus concentration for PdCl$_2$ in DMF solution at 22°C is also reproduced [2].

References:

[1] Yatsimirski, A.; Ugo, R. (Inorg. Chem. **22** [1983] 1395/7).

[2] Khain, V. S.; Val'kova, V. P. (Zh. Neorgan. Khim. **23** [1978] 3368/70; Russ. J. Inorg. Chem. **23** [1978] 1870/1).

Complex Formation

Palladium chloride is rather insoluble in water and almost totally so in noncomplexing organic solvents. Consequently it is frequently passed over as a precursor for the synthesis of palladium complexes in favour of the salt Na$_2$PdCl$_4$ [1] or the adduct PdCl$_2$(PhCN)$_2$ [2]. These are readily soluble in aqueous and nonpolar solvents, respectively, and are both readily accessible from PdCl$_2$. In many cases where PdCl$_2$ is specified as a starting material the reactions are performed in conc. HCl solution and the palladium species present is the anion [PdCl$_4$]$^{2-}$.

However, PdCl$_2$ is capable of reacting directly with a variety of ligands under suitable conditions, and a selection of these reactions is given below.

References:

[1] Brauer, G. (Handbook of Preparative Inorganic Chemistry, Vol. 2, Academic, New York 1965, p. 1584).

[2] Kharasch, M. S.; Seyler, R. C.; Mayo, F. R. (J. Am. Chem. Soc. **60** [1938] 882/4).

Reduction to Metal

The effects of ionogenic surfactants ([cetyl NMe$_3$]Br and Na[dodecyl]SO$_4$) on the reduction of PdCl$_2$ by molecular hydrogen have been investigated; the ammonium salt inhibits, and the sodium salt slightly accelerates the reduction process [1]. Reduction of $PdCl_2$ to palladium black on NaCl plates by CO/H_2 or H_2/He mixtures has been reported [2]. Reduction of $PdCl_2$ by lithium metal is greatly speeded up by the application of ultrasound [3]. The kinetics of reduction of $PdCl_2$ in aqueous HCl solution by carbon monoxide have been attributed to the autocatalytic action of the $[Pd(CO)Cl_3]^-$ anion [4]. Spectrophotometric studies on the kinetics of reduction of Pd^{2+} (in HCl solution of $PdCl_2$) by activated carbon indicate that the Pd^{2+} ions chemisorb on the activated surface and are reduced to metal in a diffusion controlled reaction [5]. The reactions of transition metal chlorides (including $PdCl_2$) with alkali metal tetrahydro-borates have been reviewed [6]. Reduction of $PdCl_2$ to metallic palladium by $NaBH_4$ in aqueous [7] or dimethylformamide [8] solution has been described. The combination of $PdCl_2$ and $NaBH_4$ acts as a powerful reducing agent for aryl ketones, aryl chlorides and benzylic alcohols to the corresponding hydrocarbons [9]. The group IV(14) metal hydrides Et_3MH (M = Si [10], Ge [11] or Sn [12]) reduce $PdCl_2$ to afford metallic palladium, hydrogen and the corresponding chloride Et_3MCl. Reduction of $PdCl_2$ to a dark grey product (Pd metal?) by $(Me_2SiH)_2S$ has also been reported [13]. Reduction of $PdCl_2$ by $Ca(H_2PO_3)_2$ and Na_3PO_3 has been studied at room temperature and at 95 to 98°C [14].

References:

[1] Lisichkin, G. V.; Yuffa, A. Ya.; Khinchagashvili, V. Yu. (Zh. Fiz. Khim. **50** [1976] 2139/40; Russ. J. Phys. Chem. **50** [1976] 1285/6).

[2] Vozdvizhenskii, V. F.; Sokol'skii, D. V. (Dokl. Akad. Nauk SSSR **295** [1987] 121/4; C.A. **107** [1987] No. 103674).

[3] Boudjouk, P.; Thompson, D. P.; Ohrbom, W. H.; Han, B. H. (Organometallics **5** [1986] 1257/60).

[4] Spitsyn, V. I.; Fedoseev, I. V.; Znamenskii, I. V. (Zh. Neorgan. Khim. **25** [1980] 2754/8; Russ. J. Inorg. Chem. **25** [1980] 1518/20).

[5] Kublanovskii, V. S.; Tarasenko, Yu. A.; Danilov, M. O.; Antonov, S. P. (Ukr. Khim. Zh. **51** [1985] 948/50; C.A. **103** [1985] No. 221758).

[6] Mal'tseva, N. N. (Izv. Akad. Nauk SSSR Neorgan. Materialy **14** [1978] 1718/21; Inorg. Materials [USSR] **14** [1978] 1337/9).

[7] Mal'tseva, N. N.; Sterlyadkina, Z. K.; Erusalimchik, I. G.; Mikheeva, V. I. (Zh. Neorgan. Khim. **24** [1979] 822/4; Russ. J. Inorg. Chem. **24** [1979] 459/61).

[8] Khain, V. S.; Val'kova, V. P. (Zh. Neorgan. Khim. **23** [1978] 1117/9; Russ. J. Inorg. Chem. **23** [1978] 617/9).

[9] Satoh, T.; Mitsuo, N.; Nishiki, M.; Nanba, K.; Suzuki, S. (Chem. Letters **1981** 1029/30).

[10] Anderson, H. H. (J. Am. Chem. Soc. **80** [1958] 5083/5).

[11] Anderson, H. H. (J. Am. Chem. Soc. **79** [1957] 326/8).

[12] Anderson, H. H. (J. Am. Chem. Soc. **79** [1957] 4913/5).

[13] Eméleus, H. J.; Smythe, L. E. (J. Chem. Soc. **1958** 609/11).

[14] Sutyagina, A. A.; Gorbunova, K. M. (Zh. Prikl. Khim. **37** [1964] 1676/81; J. Appl. Chem. [USSR] **37** [1964] 1668/72).

Reactions with Oxygen Donors

Dimethylformamide (DMF) reacts with $PdCl_2$ at room temperature to form the complex salt $[Pd(DMF)_4]Cl_2 \cdot DMF$ in which the DMF ligands bond through oxygen [1]. Salts of the general form $[PdL_4][ClO_4]_2$ (L = dimethylformamide, dimethylacetamide, dimethyl sulphoxide) have been obtained by dissolving $PdCl_2$ in the respective neat ligand and adding stoichiometric amounts of anhydrous $AgClO_4$. In the absence of the silver salt adducts of the form $PdCl_2L_2$ are formed [2]; solvates formed by dissolution of $PdCl_2$ in water, ethanol, dimethyl sulphoxide and dimethylformamide have been examined in solution by electronic spectroscopy [3]. Formation of an unstable 1:1 acetone solvate $PdCl_2 \cdot Me_2CO$ (equilibrium constant pK=1.71) in water/acetone solution has been detected spectroscopically [4]. 1-Phenyl-3-methyl-4-benzoyl-pyrazole-5-one (HL) forms a complex, PdL_2, which can be used to extract palladium(II) into organic media [5]. Hydrolysis of $PdCl_2$ in HCl solution by titration with NaOH leads to formation of polymeric products and eventual precipitation of $Pd(OH)_2$ [6]. Chlorine trioxide reacts with $PdCl_2$ to form the covalently bonded perchlorate salts $Pd(ClO_4)_2$ and $ClO_2Pd(ClO_4)_3$ [7]. A palladium oxychloride, Pd_2OCl_2 has been obtained from a mixture of γ-$PdCl_2$ and PdO-nH_2O in a $KNO_3/NaNO_3$ melt at 450°C [8]. The adsorption of $PdCl_2$ onto aluminium oxide has been investigated and evidence for coordinate bond formation presented [9].

References:

[1] Khain, V. S.; Val'kova, V. P. (Zh. Neorgan. Khim. **23** [1978] 3368/70; Russ. J. Inorg. Chem. **23** [1978] 1870/1).

[2] Wayland, B. B.; Schramm, R. F. (Inorg. Chem. **8** [1969] 971/6).

[3] Volchenskova, I. I.; Yatsimirskii, K. B. (Teor. Eksperim. Khim. **13** [1977] 197/204; Theor. Exptl. Chem. [USSR] **13** [1977] 146/51).

[4] Kondratov, P. I.; Lakiza, V. V. (Izv. Vysshikh Uchebn. Zavedenii Khim. Khim. Tekhnol. **21** [1978] 1394/5; C. A. **89** [1978] No. 205047).

[5] Mirza, M. Y.; Bailey, R. T. (J. Inorg. Nucl. Chem. **41** [1979] 772/3).

[6] Wyatt, R. (Chem. Weekblad **62** [1966] 310/4).

[7] Chaabouni, M.; Pascal, J.-L.; Potier, J. (Compt. Rend. C **291** [1980] 125/8).

[8] Dannecker, B.; Thiele, G. (Z. Naturforsch. **41b** [1986] 1363/6).

[9] Kravchuk, L. S.; Koslov, N. S.; Sinyakova, S. V.; Yankova, T. V. (Zh. Fiz. Khim. **57** [1983] 2794/7; Russ. J. Phys. Chem. **57** [1983] 1686/8).

Reactions with Nitrogen Donors

Differential thermal analysis, thermogravimetric analysis and X-ray diffraction data indicate that heated $PdCl_2$ reacts with gaseous ammonia to form $[Pd(NH_3)_4]Cl_2$, $PdCl_2(NH_3)_2$ (>200°C) and palladium metal (>340°C) [1]. Palladium chloride reacts with trimethylamine in acetone solution to form trans-$PdCl_2(NMe_3)_2$ in good yield [2]. Reactions of $PdCl_2$ with various amounts of $NH_2OH \cdot H_2SO_4$ in ethanol at room temperature afford $PdCl_2(NH_2OH)_2$, $PdCl_2(NH_2OH)_2 \cdot EtOH$ and $[Pd(NH_2OH)_4]Cl_2$ [3]. Dissolution of $PdCl_2$ in MeCN leads to formation of the adduct $PdCl_2(MeCN)_2$ [4]. However, spectroscopic studies have provided evidence for formation of a chloro-bridged species $Pd_2Cl_4(MeCN)_2$ [5]. The analogous benzonitrile complex $PdCl_2(PhCN)_2$, a valuable precursor for the synthesis of many other palladium complexes, is readily obtained by warming a suspension of $PdCl_2$ in the neat ligand at 100°C [6, 7]. Solutions of $PdCl_2$ in ethanol [8] or dimethylformamide [9] react with imidazole (and various substituted imidazoles), L, to give adducts of the form $PdCl_2L_2$, use of excess imidazole leads to formation of salts $[PdL_4]Cl_2$. Palladium chloride dissolved in HCl forms adducts $PdCl_2L_2$ with benzotriazole, benzoxazole and benzothiazole [10]. An earlier paper describes formation of 1:1 and 1:2 adducts between $PdCl_2$ and benzotriazole [11]. Pyridine, quinoline and iso-quinoline all react with $PdCl_2$ (formed in situ from Pd-black/HCl/HNO_3) to form adducts $PdCl_2L_2$ [12]. An insoluble polymeric 1:1 adduct has been obtained from $PdCl_2$ and N, N-ethylenedimorpholine in acetone solution [13]. Adducts of the form PdX_2L_2 are obtained when aqueous solutions of $PdCl_2$ react with morpholine, L, in the presence of potassium salts KX (X = Cl, Br, I, NO_2, SCN) [14].

References:

[1] Ryumin, A. I.; Smirnov, I. I.; Blokhina, M. L. (Zh. Neorgan. Khim. **30** [1985] 2849/51; Russ. J. Inorg. Chem. **30** [1985] 1622/3).

[2] Goggin, P. L.; Goodfellow, R. J.; Reed, F. J. S. (J. Chem. Soc. Dalton Trans. **1972** 1298/303).

[3] Shaplygin, I. S.; Lazarev, V. B. (Zh. Neorgan. Khim. **23** [1978] 603/5; Russ. J. Inorg. Chem. **23** [1978] 334/5).

[4] Wayland, B. B.; Schramm, R. F. (Inorg. Chem. **8** [1969] 971/6).

[5] Volchenskova, I. I.; Yatsimirskii, K. B. (Zh. Neorgan. Khim. **18** [1973] 1875/82; Russ. J. Inorg. Chem. **18** [1973] 990/5).

[6] Kharasch, M. S.; Seyler, R. C.; Mayo, F. R. (J. Am. Chem. Soc. **60** [1938] 882/4).

[7] Doyle, J. R.; Slade, P. E.; Jonassen, H. B. (Inorg. Syn. **6** [1960] 216/9).

[8] Hazarika, T. N.; Bora, T. (Transition Metal Chem. [Weinheim] **7** [1982] 210/2).

[9] van Kralingen, C. G.; de Ridder, J. K.; Reedijk, J. (Inorg. Chim. Acta **36** [1979] 69/77).

[10] House, J. E.; Pik-Shing, L. (J. Inorg. Nucl. Chem. **36** [1974] 223/4).

[11] Wilson, R. F.; Wilson, L. E. (J. Am. Chem. Soc. **77** [1955] 6204/5).

[12] Gupte, S. P.; Chaudhari, R. V. (J. Mol. Catal. **24** [1984] 197/210).

[13] Lott, A. L.; Rasmussen, P. G. (J. Am. Chem. Soc. **91** [1969] 6502/3).

[14] Ghosh, S. P.; Prasad, R. K.; Mishra, L. K. (J. Ind. Chem. Soc. **64** [1987] 574/5).

Reactions with Chlorine Donors

Palladium chloride reacts with the hydrochlorides of many organic bases, B·HCl, or with the free base, B, plus hydrochloric acid to yield salts of the general form $[BH]_2[PdCl_4]$. Details of these reactions are given under tetrachloropalladates, pp. 118/25.

For reactions with **chlorides of alkali and alkaline earth metals** see pp. 128/41, with **other metal chlorides** see pp. 141/4.

Reactions with Sulphur Donors

Binuclear chloride-bridged anions $[Pd_2Cl_4(SO_3R)_2]^{2-}$ containing S-bonded SO_3R^- ligands are formed when SO_2 is passed into a solution of $PdCl_2$ in an alcohol, ROH (R = Me, Et, ^nPr), and can be precipitated as tetraphenylphosphonium or -arsonium salts [1, 2]. Adduct formation between $PdCl_2$ and phosphine sulphides $[Ph_3PS]$ or diphosphinedisulphides $[Me_2P(S)P(S)Me_2$ or $Ph_2P(S)CH_2P(S)Ph_2]$ has been described [3, 4]. Addition of disodium-1,2-dicyanoethylene-1,2-dithiolate to an aqueous suspension of $PdCl_2$ affords salts of the complex anion $[Pd\{S_2C_2(CN)_2\}_2]^{2-}$ [5]. Dilute (1%) aqueous solutions of $PdCl_2$ containing a few drops of conc. HCl react with naphthylthiourea (L) in methanol to afford the salt $[PdClL_4]Cl$ [6].

References:

[1] Graziani, M.; Ros, R.; Carturan, G (J. Organometal. Chem. **27** [1971] C 19/C 20).

[2] Ros, R.; Carturan, G.; Graziani, M. (Transition Metal Chem. [Weinheim] **1** [1975/76] 13/7).

[3] King, M. G.; McQuillan, G. P. (J. Chem. Soc. A **1967** 898/901).

[4] Lobana, T. S.; Sharma, K. (Transition Metal Chem. [Weinheim] **7** [1982] 333/5).

[5] Weiher, J. F.; Melby, L. R.; Benson, R. E. (J. Am. Chem. Soc. **86** [1964] 4329/33).

[6] Khan, M. M. (J. Inorg. Nucl. Chem. **36** [1974] 299/302).

Reactions with Selenium Donors

Palladium chloride forms 1:2 adducts with Ph_3PSe [1, 2] and 1:1 adducts with $Ph_2P(Se)(CH_2)_nP(Se)Ph_2$ (n = 1 or 3) [2].

References:

[1] King, M. G.; McQuillan, G. P. (J. Chem. Soc. A **1967** 898/901).
[2] Lobana, T. S.; Sharma, K. (Transition Metal Chem. [Weinheim] **7** [1982] 333/5).

Reactions with Carbon Donors

Intercalation with Graphite

An early paper in the area reports that $PdCl_2$ is not intercalated by graphite but that the normally unstable $PdCl_4$ is so bound [1]. However, laminar compounds of graphite with $PdCl_2$ (up to 58%) have been obtained by heating graphite/$PdCl_2$ in a Cl_2 stream, chlorination of Pd in the presence of graphite and heating graphite/$PdCl_2$ under chlorine in a sealed tube at 450 to 800°C. X-ray analysis gave structural parameters with a repeating thickness of 16.7 Å and a filled layer thickness of 10 Å. Energy values for Pd $3d_{5/2}$ and Cl $2p_{3/2}$ (338 and 199 eV, respectively) corresponding to those in $PdCl_2$ indicate that bonding must be of van der Waals type. Treatment with PhC_6H_4Li in THF gives reduction of Pd^{II} to Pd^0 or Pd^I. The reduced material catalyses acetylene reduction to olefin and double bond isomerisation in olefins [2]. The diffusion coefficient for the two-dimensional intercalation of highly orientated pyrolytic graphite by $PdCl_2$ at 450°C has been given as 0.52×10^{-6} cm²/min [3, 4]. The electronic properties of $PdCl_2$/graphite intercalation compounds have been studied by the Shubnikov-de Haas effect [5 to 7]. The Shubnikov-de Haas spectrum has frequencies of $\sim$27, 80, 380 and 550 T for fields parallel to the c-axis [5]. Current-carrier effective mass ($m^*/m_0 = 0.08$), scattering time (5×10^{-13}s) and mobility (10^4 cm²·V⁻¹·s⁻¹) (all at SdH frequency F_2) have been measured [6]. The Shubnikov-de Haas quantum oscillations of the acceptor compound C_xPdCl_2 have been determined by resistance measurements at 1.5 to 50 K in fields of 22.5 Tesla, and data interpreted in terms of pristine graphite energy bands [7]. The resistivities and carrier mobilities at 300 to 450 K and Shubnikov-de Haas effect and magnetoresistance rotation at liquid helium temperature have been measured for samples of $C_{4n}PdCl_2$ (n = 3) obtained by intercalation of natural single-crystal graphite flakes, highly orientated pyrolytic graphite and pitch-based carbon fibres [8]. Palladium chloride intercalated carbon fibres are reported to have high thermal stability (to $\leqq$300°C), relatively low resistance and small positive temperature coefficients of resistance [9].

References:

[1] Croft, R. C. (Nature **172** [1953] 725/6).
[2] Novikov, Yu. N.; Postnikov, V. A.; Salyn, Ya. V.; Nefedov, V. I.; Vol'pin, M. E. (Izv. Akad. Nauk SSSR Ser. Khim. **1973** 1689; Bull. Acad. Sci. USSR Div. Chem. Sci. **1973** 1653).
[3] Dowell, M. B.; Badorrek, D. S. (Carbon **16** [1978] 241/9).
[4] Dowell, M. B. (Extend. Abstr. Program Biennial Conf. Carbon **13** [1977] 11/2; C. A. **88** [1978] No. 95040).
[5] Woollam, J. A.; Haugland, E.; Dowell, M. B.; Yavrouian, A.; Lozier, A. G.; Matulka, G. (Syn. Metals **2** [1980] 309/20).
[6] Woollam, J. A.; Haugland, E.; Dowell, M. B.; Underhill, C. (Phys. Letters A **82** [1981] 359/61).
[7] Woollam, J. A.; Haugland, E.; Dowell, M. B.; Kambe, N.; Mendez, E.; Hakimi, F.; Dresselhaus, G.; Dresselhaus, M. S. (Extend. Abstr. Program Biennial Conf. Carbon **14** [1979] 320/1; C.A. **91** [1979] No. 116348).

[8] Oshima, H; Woollam, J. A.; Khan, A. A.; Haugland, E. J.; Dowell, M. B.; Brandt, B. (Extend. Abstr. Program Biennial Conf. Carbon **15** [1981] 66/7; C.A. **95** [1981] No. 143140).

[9] Dominguez, D. D.; Murday, J. S. (Extend. Abstr. Program Biennial Conf. Carbon **16** [1983] 276/7; C.A. **99** [1983] No. 214006).

Reactions with Carbon Monoxide or Cyanide

These reactions have been reviewed [1]. Carbonylation of $PdCl_2$ in aqueous dioxan affords a product of approximate composition $[PdCl(CO)]_n$ and is accompanied by CO_2 formation [2]. The complex anion $[PdCl_2H(CO)]^-$ obtained by carbonylation of $PdCl_2$ in acidified 2-methoxy-ethanol [3] has subsequently been reformulated as $[Pd_2Cl_4(CO)_2]^{2-}$ [4]. Carbonylation of $PdCl_2$ in thionyl chloride solution to afford $[Pd_2Cl_4(CO)_2]$ is exothermic ($\Delta H° = -56.8 \pm 0.8$ kJ/mol of dimer, $\Delta S° = -130 \pm 3.3$ J·mol^{-1}·deg^{-1}) [5]. Pressure carbonylation (40 atm) of $PdCl_2$ in acetic anhydride provides a route to $[PdCl(CO)]_n$ in high yield [5]. Finally, carbonylation of $PdCl_2/SnCl_2$ (1:10) mixtures in acidified methanol and subsequent addition of $[NEt_4]Cl$ gives $(NEt_4)[PdCl(SnCl_3)_2CO]$ in low yield [6]. Palladium cyanide is conveniently obtained by treatment of $PdCl_2$ with KCN in aqueous solution, with excess cyanide formation of $K_2[Pd(CN)_4]$ occurs [7].

References:

[1] Maitlis, P. M.; Espinet, P.; Russell, M. J. H. (Comprehensive Organometallic Chemistry, Compounds with Palladium-Carbon σ-Bonds, Vol. 6, Pergamon Press, Oxford 1982, pp. 279/349).

[2] Kushnikov, Yu. A.; Beilina, A. Z.; Vozdvizhenskii, V. F. (Zh. Neorgan. Khim. **16** [1971] 416/19; Russ. J. Inorg. Chem. **16** [1971] 218/20).

[3] Kingston, J. V.; Scollary, G. R. (Chem. Commun. **1969** 455/6).

[4] Goggin, P. L.; Mink, J. (J. Chem. Soc. Dalton Trans. **1974** 534/40).

[5] Dell'amico, D. B.; Calderazzo, F; Zandona, N. (Inorg. Chem. **23** [1984] 137/40).

[6] Kingston, J. V.; Scollary, G. R. (Chem. Commun. **1970** 362).

[7] Bigelow, J. H. (Inorg. Syn. **2** [1946] 245/6).

Reactions with Olefins and Acetylenes

These reactions have been reviewed [1].

Several olefins (cyclohexene, cycloheptene, cyclooctene, 1-methyl-cyclohexene, isobutene and α-methylstyrene) react with $PdCl_2$ in 50% aqueous acetic acid solution at 20°C to form olefin complexes in poor yield [2, 3, 4]. Palladium chloride also reacts with neat olefins to afford olefin complexes but yield and purity are low [5, 6]. The effects of solvent changes on the reaction of $PdCl_2$ with ethylene have been studied, in chloroform solution after 20 h the red solution of $[PdCl_2(C_2H_4)]_2$ gave a yellow precipitate of $PdCl_2(C_2H_4)_2$, stable only under an ethylene atmosphere [7].

Diphenyl acetylene reacts with $PdCl_2$ in ethanol to form a tetraphenyl cyclobutenyl derivative of palladium(II) which on treatment with HCl affords the tetraphenylcyclobutadiene complex $[PdCl_2(C_4Ph_4)]_2$ [8].

For organic reactions involving stoichiometric or catalytic participation of $PdCl_2$ see pp. 80/3.

References:

[1] Maitlis, P. M.; Espinet, P.; Russell, M. J. H. (Comprehensive Organometallic Chemistry, Mono Olefin and Acetylene Complexes of Palladium, Vol. 6, Pergamon, Oxford 1982, pp. 351/62).

[2] Hüttel, R.; Bechter, M. (Angew. Chem. **71** [1959] 456).
[3] Hüttel, R.; Kratzer, J. (Angew. Chem. **71** [1959] 456).
[4] Hüttel, R.; Dietl, H.; Christ, H. (Chem. Ber. **97** [1964] 2037/45).
[5] Pregaglia, G. F.; Donati, M.; Conti, F. (Chem. Ind. [London] **1966** 1923/4).
[6] Pregaglia, G.; Donati, M.; Conti, F. (Chim. Ind. [Milan] **49** [1967] 1277/83).
[7] Ketley, A. D.; Fisher, L. P.; Berlin, A. J.; Morgan, C. R.; Gorman, E. H.; Steadman, T. R. (Inorg. Chem. **6** [1967] 657/63).
[8] Vallarino, L. M.; Santarella, G. (Gazz. Chim. Ital. **94** [1964] 252/86).

Reactions with Phosphorus and Arsenic Donors

Palladium chloride reacts with NOCl in the presence of triphenyl phosphine or -arsine to afford the complexes cis-$PdCl_2(EPh_3)_2$ (E = P or As).

Reference:

Jain, K. C.; Pandey, K. K.; Parashad, R.; Singh, T.; Agarwala, U. C. (Ind. J. Chem. A **19** [1980] 1089/91).

4.2.4 Palladium Dichloride in Organic Synthesis
(see also "Platinum" Suppl. Vol. A 1, 1986, pp. 299/310)

In addition to forming complexes with unsaturated organic molecules, $PdCl_2$, in common with many other palladium compounds, is an active catalyst in a wide range of commercially important organic reactions. It is also used extensively as a catalytic or stoichiometric reagent in laboratory synthesis of fine organic chemicals, although more soluble and/or labile palladium species, notably palladium acetate, are often preferred for this purpose. Recent monographs and reviews covering work in this area are listed below [1 to 6]. Some examples drawn from the huge literature of palladium chloride mediated organic reactions are mentioned below.

References:

[1] Heck, R. F. (Best Synthetic Methods, Palladium Reagents in Organic Synthesis, Academic, London 1985).
[2] Tsuji, J. (Organic Syntheses with Palladium Compounds, Springer, Berlin 1980).
[3] Henry, P. M. (Palladium Catalysed Oxidation of Hydrocarbons, Reidel, New York 1980).
[4] Trost, B. M. (Tetrahedron **33** [1977] 2615/49).
[5] Maitlis, P. M. (The Organic Chemistry of Palladium, Metal Complexes, Vol. I, Academic, New York – London 1971).
[6] Maitlis, P. M. (The Organic Chemistry of Palladium, Catalytic Reactions, Vol. II, Academic, New York – London 1971).

Olefin Oxidation Catalysed by $PdCl_2$

The commercial importance of the Wacker process for olefin oxidation has stimulated much work on the kinetics and mechanisms of the $PdCl_2$ catalysed oxidation of ethylene [1 to 8], higher olefins [4, 8, 9, 10], styrene [11, 12] and allyl alcohol [13, 14].

References:

[1] Cuiec, L.; Beaufils, J. P.; Hellin, M. (Bull. Soc. Chim. France **1977** 214/22).

[2] Henry, P. M. (J. Org. Chem. **38** [1973] 2415/6).

[3] Pestrikov, S. V. (Zh. Fiz. Khim. **39** [1965] 428/9; Russ. J. Phys. Chem. **39** [1965] 218/9).

[4] Pestrikov, S. V.; Moiseev, I. I.; Romanova, T. N. (Zh. Neorgan. Khim. **10** [1965] 2203; Russ. J. Inorg. Chem. **10** [1965] 1199).

[5] Clark, D.; Hayden, P. (Am. Chem. Soc. Div. Petrol. Chem. Preprints **11** [1966] D5/D9; C.A. **66** [1967] No. 37082).

[6] Moiseev, I. I.; Vargaftik, M. N.; Syrkin, Ya. K. (Dokl. Akad. Nauk SSSR **133** [1960] 377/80; Proc. Acad. Sci. USSR Chem. Sect. **130/135** [1960] 801/4).

[7] Lu, G.; Huang, G.; Guo, Y.; Wang, R.; Zhu, R.; Qi, H. (Huagong Xuebao **1986** 368/72; C.A. **106** [1987] No. 35058).

[8] Moiseev, I. I.; Vargaftik, M. N.; Pestrikov, S. V.; Levanda, O. G.; Romanova, T. N.; Syrkin, Ya. K. (Dokl. Akad. Nauk SSSR **171** [1966] 1365/8).

[9] Kaszonyi, A.; Vojtko, J.; Hrusovsky, M. (Petrochemia **19** [1979] 109/14; C.A. **92** [1980] No. 163524).

[10] Belov, A. P.; Pek, G. Yu.; Moiseev, I. I. (Izv. Akad. Nauk SSSR Ser. Khim. **1965** 2204/6; Bull. Acad. Sci. USSR Div. Chem. Sci. **1965** 2170/2).

[11] Zakharova, L. M.; Vargaftik, M. N.; Moiseev, I. I.; Katsman, L. A. (Kinetika Kataliz **10** [1969] 901/3; Kinet. Catal. [USSR] **10** [1969] 736/8).

[12] Uemura, S.; Zushi, K.; Okano, M. (J. Chem. Soc. Chem. Commun. **1972** 234/5).

[13] Zaw, K.; Lautens, M.; Henry, P. M. (Organometallics **2** [1983] 197/9).

[14] Ahmad, I.; Ashraf, C. M. (Intern. J. Chem. Kinet. **11** [1979] 813/9).

Olefin Isomerisation Catalysed by $PdCl_2$

The ability of palladium compounds including $PdCl_2$ to catalyse the migration of olefinic double bonds was first recognised in 1962 [1]. Since then numerous reports of $PdCl_2$ catalysed olefin isomerisation have appeared [2 to 7].

References:

[1] Turner, L.; British Petroleum Co. Ltd. (Brit. 932748 [1963]; C.A. **60** [1964] 405).

[2] Moiseev, I. I.; Pestrikov, S. V. (Dokl. Akad. Nauk SSSR **171** [1966] 722/5).

[3] Moiseev, I. I.; Pestrikov, S. V. (Izv. Akad. Nauk SSSR Ser. Khim. **1965** 1717; Bull. Acad. Sci. USSR Div. Chem. Sci **1965** 1690).

[4] Moiseev, I. I.; Pestrikov, S. V.; Sverzh, L. M. (Izv. Akad. Nauk SSSR Ser. Khim. **1966** 1866/7; Bull. Acad. Sci. USSR Div. Chem. Sci. **1966** 1809).

[5] Moiseev, I. I.; Grigor'ev, A. A.; Pestrikov, S. V. (Zh. Org. Khim. **4** [1968] 354/9; J. Org. Chem. [USSR] **4** [1968] 346/51).

[6] Pestrikov, S. V.; Moiseev, I. I.; Sverzh, L. M. (Kinetika Kataliz **10** [1969] 74/82; Kinet. Catal. [USSR] **10** [1969] 57/64).

[7] Moiseev, I. I.; Grigor'ev, A. A. (Dokl. Akad. Nauk SSSR **1968** 1090/3; Dokl. Chem. Proc. Acad. Sci. USSR **178/183** [1968] 132/4).

Olefin Dimerisation Catalysed by $PdCl_2$

The palladium chloride catalysed dimerisation of ethylene was first reported in 1964 [1]. Subsequent papers have described the $PdCl_2$ catalysed dimerisation of ethylene [2, 3, 4], propylene [3, 4], and a range of higher olefins in a variety of solvents [4].

References:

[1] van Gemert, J. T.; Wilkinson, P. R. (J. Phys. Chem. **68** [1964] 645/7).
[2] Kusunoki, Y.; Katsuno, R.; Hasegawa, N.; Kurematsu, S.; Nagao, Y.; Ishii, K.; Tsutsumi, S. (Bull. Chem. Soc. Japan **39** [1966] 2021/3).
[3] Ketley, A. D.; Fisher, L. P.; Berlin, A. J.; Morgan, C. R.; Gorman, E. H.; Steadman, T. R. (Inorg. Chem. **6** [1967] 657/63).
[4] Barlow, M. G.; Bryant, M. J.; Haszeldine, R. N.; Mackie, A. G. (J. Organometal. Chem. **21** [1970] 215/26).

Oxidative Coupling with PdCl$_2$

Palladium chloride displays important activity in the oxidative coupling of organic substrates.

$$2\,RH + PdCl_2 \rightarrow R - R + 2\,HCl + Pd^0$$

In the presence of a suitable oxidant (O_2/Cu^{II}) the reactions become catalytic. Species coupled in this way include arenes [1, 2] and styrenes [3, 4] which yield biaryls and 1,4-diarylbutadienes, respectively. Coupling of alkenes with arenes [5, 6] and with aryl halides [7 to 10] are likewise catalysed by PdCl$_2$.

Related PdCl$_2$ promoted reactions include the coupling of aryl mercury compounds to form biaryls or, in the presence of CO, ketones and acid chlorides [11, 12, 13] and the coupling of olefins with amines [14] or amides [15]. Palladium chloride also catalyses oxycyanation of olefins [16, 17].

References:

 [1] van Helden, R.; Verberg, G. (Recl. Trav. Chim. **84** [1965] 1263/73).
 [2] Fujiwara, Y.; Moritani, I.; Ikegami, K.; Tanaka, R.; Teranishi, S. (Bull. Chem. Soc. Japan **43** [1970] 863/7).
 [3] Hüttel, R.; Bechter, M. (Angew. Chem. **71** [1959] 456).
 [4] Hüttel, R.; Kratzer, J.; Bechter, M. (Chem. Ber. **94** [1961] 766/80).
 [5] Moritani, I.; Fujiwara, Y. (Tetrahedron Letters **1967** 1119/22).
 [6] Fujiwara, Y.; Moritani, I.; Matsuda, M. (Tetrahedron **24** [1968] 4819/24).
 [7] Plevyak, J. E.; Heck, R. F. (J. Org. Chem. **43** [1978] 2454/6).
 [8] Chalk, A. J.; Magennis, S. A. (J. Org. Chem. **41** [1976] 273/8).
 [9] Mizoroki, T.; Mori, K.; Ozaki, A. (Bull. Chem. Soc. Japan **44** [1971] 581; **46** [1973] 1505/8).
[10] Heck, R. F.; Nolley, J. P. (J. Org. Chem. **37** [1972] 2320/2).

[11] Heck, R. F. (J. Am. Chem. Soc. **90** [1968] 5518/26).
[12] Henry, P. M. (Tetrahedron Letters **1968** 2285/7).
[13] Heck, R. F. (J. Am. Chem. Soc. **90** [1968] 5546/8).
[14] Hegedus, L. S.; Akermark, B.; Zetterberg, K.; Olsson, L. F. (J. Am. Chem. Soc. **106** [1984] 7122/6).
[15] Hirai, H.; Sawai, H. (Bull. Chem. Soc. Japan **43** [1970] 2208/13).
[16] Nakajima, H.; Kimura, T.; Kominami, N.; Miyata, S.; Kobayashi, T. (Kogyo Kagaku Zasshi **74** [1971] 2440/4; C.A. **76** [1972] No. 100093).
[17] Nakajima, H.; Kimura, T.; Kominami, N.; Miyata, S.; Kobayashi, T. (Kogyo Kagaku Zasshi **74** [1971] 2447/50; C.A. **76** [1972] No. 100095).

Carbonylations with PdCl$_2$

Palladium chloride catalyses the carbonylation of many organic substrates including olefins, acetylenes, aromatic hydrocarbons and alkyl halides, a representative selection of examples is given below.

Palladium chloride catalysed oxidative carbonylation of olefins in alcohol solution affords an efficient synthesis of substituted succinate esters [1, 2]. Under similar conditions acetylenes yield substituted maleate esters [2].

Acetylenes and conjugated dienes undergo PdCl$_2$-catalysed carbonylation in alcohols to afford acetylene carboxylic acids [3], and β-, γ-unsaturated esters [4, 5]. In the presence of SnCl$_2$ and PPh$_3$, PdCl$_2$ catalyses carbonylation of alkynols to α-methylene lactones and of alkenols to γ- and δ-lactones [6]. Palladium chloride catalysed carbonylation of primary amines [7, 8] or aromatic nitro compounds [9] affords isocyanates, ArNCO.

References:

[1] Fenton, D. M.; Steinwand, P. J. (J. Org. Chem. **37** [1972] 2034/5).
[2] Heck, R. F. (J. Am. Chem. Soc. **94** [1972] 2712/6).
[3] Tsuji, J.; Takahashi, M.; Takahashi, T. (Tetrahedron Letters **21** [1980] 849/50).
[4] Hosaka, S.; Tsuji, J. (Tetrahedron **27** [1971] 3821/9).
[5] Brewis, S.; Hughes, P. R. (Chem. Commun. **1965** 157/8).
[6] Murray, T. F.; Samsel, E. G.; Varma, V.; Norton, J. R. (J. Am. Chem. Soc. **103** [1981] 7520/8).
[7] Stern, E. W.; Spector, M. L. (J. Org. Chem. **31** [1966] 596/7).
[8] Tsuji, J.; Iwamoto, N. (Chem. Commun. **1966** 828/9).
[9] Weigert, F. J. (J. Org. Chem. **38** [1973] 1316/9).

4.2.5 Applications of PdCl$_2$

These are numerous, only a selection of the more important ones can be included here. The use of PdCl$_2$ in organic synthesis has been covered in Section 4.2.4, pp. 80/3.

In Gas Detectors

Palladium chloride is the active component in detectors for a variety of gases including CO, H$_2$ and hydrocarbons. Visual detection of low concentrations (200 to 500 ppm) of CO has been achieved using PdCl$_2$ mounted on silica gel [1] or activated alumina [2]. Mixtures of silica gel and activated carbon impregnated with 1% PdCl$_2$ serve as efficient CO adsorbants [3]. Stannic oxide based gas sensors doped with PdCl$_2$ and/or an f-block oxide (La$_2$O$_3$, CeO$_2$ or ThO$_2$) show good sensitivity to CO and hydrocarbons (CH$_4$, C$_3$H$_8$) [4, 5, 6]. New ceramic gas sensors, formed by sintering SnO$_2$/TiO$_2$ with small amounts of PdCl$_2$ and Nb$_2$O$_5$, are sensitive to city gas and L.P. gas [7]. Neutron activation analysis of palladium metal formed by CO reduction of PdCl$_2$ has been studied as a method of CO detection (sensitivity 2×10^{-6} ml CO) [8]. Zinc oxide films loaded with PdCl$_2$ from aqueous solution and heat treated ($< 200°C$) function as efficient hydrogen detectors [9, 10]. Hydrogen reduction of PdCl$_2$ to Pd metal catalyses reduction of methylene blue to the leuco derivative and thereby provides a colorimetric method for determination of H$_2$ at low concentrations (0.1 to 10 ppm; 5 to 20 l samples) [11]. The EtSH content of L.P. gases has been determined using PdCl$_2$/silica gel sensors [12].

References:

[1] Fensom, A. (Lab. Pract. **20** [1971] 49, 52).

[2] Harrison, R. M.; Neotronics Ltd. (Ger. 2853430 [1979]; C.A. **91** [1979] No. 111905).

[3] Toppan Printing Co. Ltd. (Japan. 58177149 [1983]; C.A. **100** [1984] No. 90644).

[4] Kanefusa, S.; Nitta, M.; Taketa, Y.; Haradome, M. (Nihon Daigaku Seisankogakubu Hokoku A **10** [1977] 155/9; C.A. **89** [1978] No. 15629).

[5] Park, S.-J.; Lee, J. Y. (Yoop Hakhoechi **20** [1983] 93/8; C.A. **99** [1983] No. 160403).

[6] Coles, G. S. V.; Gallaher, K. J.; Watson, J. (Sens. Actuators **7** [1985] 89/96; C.A. **103** [1985] No. 226465).

[7] Shimizu, H.; Yamamoto, T. (Shizuoka Daigaku Denshi Kogaku Kenkyusho Kenkyu Hokoku **18** [1983] 25/31; C.A. **100** [1984] No. 16824).

[8] Diebolt, J. (Bull. Soc. Chim. France **1964** 1389/92).

[9] Yoneyama, H.; Li, W. B.; Tamura, H. (Anal. Chem. Symp. Ser. **17** [1983] 113/8; C.A. **100** [1984] No. 144170).

[10] Li, W. B.; Yoneyama, H.; Tamura, H. (Mater. Chem. Phys. **10** [1984] 69/81).

[11] Silverman, L.; Bradshaw, W. (Anal. Chim. Acta **15** [1956] 31/42).

[12] Peurifoy, P. V.; O'Neal, M. J.; Shell Oil Co. (U.S. 3208828 [1965]; C.A. **63** [1965] 14626).

In Photosensitive Materials

Palladium chloride is an active component in several photosensitising systems and a very extensive literature on photographic applications exists. Light sensitive photographic films containing $PdCl_2$ with $K_3Fe(C_2O_4)_3$ or $Fe_2(C_2O_4)_3$, and developed with cobalt, nickel or copper salts have been extensively investigated [1 to 4]. Palladium chloride has been used in the formulation of a photosensitiser for optical screens or conducting circuits [5, 6]. The use of $PdCl_2$ (and other noble metal salts) in a direct-positive non-reversal imaging process [7] and in the preparation of lithographic plate material [8] has been described. Photoactivation type physical developers formulated with Pd^{II} salts ($PdCl_2$) are four times faster than similar formulations using Ag^I salts [9]. Palladium chloride has been used to intensify the iodine image ($I_2 \rightarrow PdI_2$) formed by photo-oxidation of 2,3-di(diiodo-2,3-oxypropyl)propylcellulose [10], in the development of a cellulose-anthrone based non-silver light sensitive material [11] and in the formulation of electronically conducting antistatic layers on photographic film [12, 13].

References:

[1] Eroshkin, V. I.; Semeshko, A. V.; Trofimov, A. S. (Avtometriya **1983** 9/13; C.A. **100** [1984] No. 28115).

[2] Gorunov, V. I.; Eroshkin, V. I.; Semeshko, A. V. (Avtometriya **1976** 99/102; C.A. **86** [1977] No. 113648).

[3] Lutskina, T. V.; Rogach, L. P.; Sviridov, V. V.; Boldyrev, V. V. (Zh. Nauchn. Prikl. Fotogr. Kinematogr. **18** [1973] 236; C.A. **79** [1973] No. 85620).

[4] Lutskina, T. V.; Sviridov, V. V. (Vestsi Akad. Navuk Belarusk. SSR Ser. Khim. Navuk **1973** 45/9; C.A. **79** [1973] No. 151594).

[5] Lenoble, J. P.; Albert, B.; Coquard, J.; Feist, J. M.; Norture, R. (Fr. Addn. 94731 [1969]; C.A. **73** [1970] No. 61206).

[6] Lenoble, J. P.; Albert, B.; Bochard, F.; Coquard, J.; Norture R. (Fr. 1548401 [1968]; C.A. **71** [1969] No. 55533).

[7] Miller, R. A.; Minnesota Mining and Manufact. Co. (U.S. 3656952 [1972]; C.A. **77** [1972] No. 54907).

[8] Dainippon Printing Co. Ltd. (Japan. Kokai Tokkyo Koho 57 196 238 [1982]; C.A. **100** [1984] No. 130973).

[9] Tanaka, K.; Kokado, H. (Nippon Kagaku Kaishi **1983** 11/7; C.A. **98** [1983] No. 98741).

[10] Roman, A; Sachetto, J. P.; Wust, M.; Cuccolo, S. (Swiss 604 209 [1978]; C.A. **90** [1979] No. 130689).

[11] Koshevar, V. D.; Ermolenko, I. N. (Zh. Nauchn. Prikl. Fotogr. Kinematogr. **28** [1983] 241/5; C.A. **99** [1983] No. 149499).

[12] Trevoy, D. J.; Eastman Kodak Co. (Brit. 998 642 [1965]; C.A. **63** [1965] 10902).

[13] Kodak Soc. Anon (Belg. 608 321 [1961]; C.A. **56** [1962] 9618).

In Metallised Films and Surfaces

Palladium atoms formed by reduction or ion-beam induced decomposition of $PdCl_2$ promote the electroless deposition of nickel on polyimide [1] or polystyrene [2]. Polyimide films prepared by treating polyamic acids with $PdCl_2$ and heating, show enhanced electrical conductivity [3]. Activation of nylon fibres for chemical metallisation has been achieved using $PdCl_2$/0.01 to 0.4 M HCl solutions [4]. Synthesis of palladium containing catalysts by impregnation of Al_2O_3, SiO_2, MgO and silico-aluminas with $PdCl_2$ in aqueous HCl gives almost total fixing of palladium and significant chlorination by the HCl medium [5]. The adhesive strength of copper coatings on ABS plastics is increased by activating the etched polymer surface with ethanolic rather than aqueous solutions of $PdCl_2$ [6]. Carbon monoxide formed during chromic acid etching of polyethylene and ABS plastics reduces $PdCl_2$ to metallic palladium on the polymer surface [7]. Colloidal dispersions of palladium, prepared by refluxing $MeOH/H_2O$ solutions of $PdCl_2$ in the presence of polyvinyl alcohol as a protective colloid, serve as catalysts for hydrogenation of cyclohexene [8]. Glass fibres [9], anodised aluminium substrates [10], ceramic surfaces [11, 12] and laminated circuit boards [13, 14] have all been activated for surface metallisation using $PdCl_2$ solutions. Bright palladium coatings have been deposited at 70 to 80°C, pH = 6.1 to 6.7, current density 0.6 to 0.8 A/dm^2 from baths containing $PdCl_2$/$(NH_4)_3PO_4$/Na_3PO_4/$PhCH_2OH$ [15].

References:

[1] Eskildsen, S. S.; Soerensen, G. (Nucl. Instrum. Methods Phys. Res. **218** [1983] 485/8).

[2] Mikhailova, N. A.; Kulikovskaya, S. G.; Ermolaev, M. I. (Izv. Vyssikh Uchebn. Zavedenii Khim. Khim. Tekhnol. **21** [1978] 612/3; C.A. **89** [1978] No. 44734).

[3] Carver, V. C.; Furtsch, T. A.; Taylor, L. T.; St. Clair, A. K. (Org. Coat. Plast. Chem. **41** [1979] 150/3; C.A. **94** [1981] No. 175952).

[4] Marcu, G.; Pop, I. (Stud. Univ. Babes-Bolyai Chem. **27** [1982] 28/33; C.A. **98** [1983] No. 171194).

[5] Gomez, R.; Figueras, F. (Rev. Inst. Mex. Petrol. **5** [1973] 47/51; C.A. **79** [1973] No. 10263).

[6] Vaikutyte, A.; Dobreva, E. D.; Encheva, M. A.; Lirkov, A. L.; Petrov, Ch. B.; Salkauskas, M. (Lieturos TSR Mokslu Akad. Darbai B **1983** 32/6; C.A. **99** [1983] No. 159259).

[7] Salkauskas, M.; Kimtiene, D. (J. Appl. Polym. Sci. **26** [1981] 2097/8).

[8] Hirai, H.; Nakao, Y.; Toshima, N. (J. Macromol. Sci. Chem. **13** [1979] 727/50).

[9] Stefan, M.; Hegedus, Z.; Bagi, T.; Dorner, H.; Kanyo, M.; Fuzesi, P. (Fr. Demande 2 486 064 [1982]; C.A. **97** [1982] No. 40388).

[10] Burns, R. S.; Mahmoud, I. S.; International Business Machines Corp. (U.S. 4 431 707 [1984]; C.A. **100** [1984] No. 147507).

[11] Gorker, L. S.; Bertosh, I. G.; Nachinov, G. N.; Negrei, V. P. (U.S.S.R. 990 742 [1983]; C.A. **98** [1983] No. 220731).

[12] Aihara, K.; Uchida, H. (Japan. Kokai Tokkyo Koho 61 199090, 61 199191, 61 199192 [1986]; C.A. **106** [1987] No. 222983, 222984, 222985).
[13] Fiedler, W. E.; Rogass, H.; Jahoda, A. (Ger. [East] 94423 [1972]; C.A. **79** [1973] No. 47101).
[14] Toshiba Corp. (Japan. Kokai Tokkyo Koho 5755240 [1982]; C.A. **98** [1983] No. 216737).
[15] Voronina, M. V.; Makarova, Z. A. (U.S.S.R. 354010 [1972]; C.A. **78** [1973] No. 66291).

Miscellaneous

Thermal decomposition of $PdCl_2$ at 960°C has been employed to form palladium whiskers [1] and to provide a quantitative source of chlorine, particularly ^{36}Cl from $PdCl_2/H^{36}Cl$ exchange [2]. Palladium chloride is one of several transition metal salts used to catalyse aerobic oxidation of organics, notably exchange resins, in the disposal of radioactive waste [3]. The sensitivity of low-level mercury determination by atomic absorption is markedly improved by the presence of $PdCl_2$ in the analytical solution [4]. The use of $PdCl_2$ in fluxes for soldering [5] and brazing [6] has been discussed. Palladium chloride solutions have also been used in the fabrication of electrodes for brine electrolysis [7], ceramic electronic parts [8] and ohmic contacts for group IVa chalcogenides [9]. Small beds of $PdCl_2$ on silicic acid complex and thus remove microgram amounts of unsaturated components from organic mixtures [10].

References:

[1] Riebling, E. F.; Webb, W. W. (Science **126** [1957] 309).
[2] McNeill, I. C. (J. Chem. Soc. **1961** 639/41).
[3] Toyo Engineering Co. Ltd. (Japan. Kokai Tokkyo Koho 58146899 [1983]; C.A. **100** [1984] No. 128621).
[4] Diggs, T. H.; Ledbetter, J. O. (Am. Ind. Hyg. Assoc. J. **44** [1983] 606/8; C.A. **99** [1983] No. 92941).
[5] Okamoto, I.; Omori, A. (Trans. JWRI **4** [1975] 85/90; C.A. **86** [1977] No. 94566).
[6] Okamoto, I.; Omori, A.; Miyake, M. (Trans. JWRI **4** [1975] 223/30; C.A. **86** [1977] No. 125643).
[7] Nakamura, A.; Saito, S.; TDK Electronics Co. Ltd. (Japan. Kokai Tokkyo Koho 79-102290 [1979]; C.A. **91** [1979] No. 219421).
[8] Matsushita Electric Industrial Co. Ltd. (Japan. Kokai Tokkyo Koho 82-79695 [1982]; C.A. **97** [1982] No. 187033).
[9] Hannemann, M.; Herrmann, K.; Sumpf, B. (Ger. [East] 156108 [1982]; C.A. **97** [1982] No. 228572).
[10] Schwartz, D. P. (J. Chromatog. **178** [1979] 105/16).

4.3 Palladium Tetrachloride PdCl₄(?)

Stabilisation of $PdCl_4$ as a graphite intercalation compound has been claimed.

Reference:

Croft, R. C. (Nature **172** [1953] 725/6).

4.4 Palladium Oxochloride Pd₂OCl₂ (see "Palladium" 1942, p. 287)

This oxochloride has been prepared by heating a mixture of γ-$PdCl_2$ and $PdO \cdot nH_2O$ in a $KNO_3/NaNO_3$ melt at 450°C. Single crystals were grown from a TlCl flux. The crystals are tetragonal, space group $I4_1/amd-D_{4h}^{19}$ with a = 6.313(2), c = 9.872(2)Å, Z = 4; D_{calc} = 5.06, $D_{exp} \geq 4.1 \, g/cm^3$.

Reference:

Dannecker, B.; Thiele, G. (Z. Naturforsch. **41b** [1986] 1363/6).

4.5 Chloropalladates (see "Palladium" 1942, pp. 285, 308, 311/3, 322/3, 326/8, 332, 334/40)

General Survey. The $[PdCl_4]^{2-}$ ion is one of the most important species in palladium chemistry, being used as a precursor for many palladium complexes. The strongly oxidising species $[PdCl_6]^{2-}$ is known as is the $\mu\mu'$ dichloro bridged species $[Pd_2Cl_6]^{2-}$, but there is relatively little evidence for the existence of palladium(I) or (III) chloro complexes. There is a short review on chloro complexes of palladium.

Reference:

Prokof'eva, I. V.; Fedorenko, N. V. (Zh. Neorgan. Khim. **13** [1968] 1348/53; Russ. J. Inorg. Chem. **13** [1968] 705/8).

4.5.1 Chloropalladates (I)

None has been isolated in a pure state, but there is ESR evidence for at least two species.

$[PdCl_4]^{3-}$. This is probably formed when crystals of $K_2[PdCl_4]$ or $(NH_4)_2[PdCl_4]$ are γ-irradiated at 77 K [1, 2]; $[PdCl_5]^{2-}$ is also observed [1]. The ion is also observed when crystals of NH_4Cl doped with $PdCl_2$ or $(NH_4)_2[PdCl_4]$ or $K_2[PdCl_4]$ are γ-irradiated [3]. In all cases ESR spectra suggested the presence of the species.

The ion may be present in solutions of $[PdCl_4]^{2-}$ subjected to pulse radiolysis [4].

References:

[1] Krigas, T.; Rogers, M. R. (J. Chem. Phys. **54** [1971] 4769/75).
[2] Fujiwara, S.; Nakamura, M. (J. Chem. Phys. **54** [1971] 3378/80).
[3] Sastry, M. D. (J. Chem. Phys. **64** [1976] 3957/60).
[4] Broszkiewicz, R. (Intern. J. Radiat. Phys. Chem. **6** [1974] 249/58).

4.5.2 Tetrachloropalladates $[PdCl_4]^{2-}$

The normal method of preparation is by reaction of $PdCl_2$ with excess HCl and addition of the appropriate cation.

4.5.2.1 Vibrational Spectra

There are reviews of early data [1, 2]. Raman spectra of $[PdCl_4]^{2-}$ in the presence of concentrated HCl have been measured (100 to 500 cm^{-1}) and also of $(Bu_4^nN)_2[PdCl_4]$ in $CHCl_3$ and acetone solutions (50 to 400 cm^{-1}) [3], and of $[PdCl_4]^{2-}$ in aqueous solution (100 to 400 cm^{-1}) [4]; data are given in the table below. There are also infrared data for $(Bu_4^nN)_2[PdCl_4]$ in $CHCl_3$ (60 to 400 cm^{-1}) [3]. Relative molar intensities of the Raman-active fundamentals and depolarisation ratios of the $\nu_1(A_{1g})$ mode were measured for this and related species [4].

Intensities of Raman bands were investigated as a function of excitation laser frequency in the vicinity of vibronically-allowed ligand-field electronic transitions [5].

The presence of vibrational fine structure as revealed in high-resolution spectra of $[PdCl_4]^{2-}$ in various environments has been reviewed [6].

For more vibrational data see under $K_2[PdCl_4]$, p. 132.

Vibrational spectra of $[PdCl_4]^{2-}$ in solution:

solution	$\nu_1(A_{1g})$	$\nu_2(B_{1g})$	$\nu_3(A_{2u})$	$\nu_4(B_{2g})$	$\nu_6(E_u)$	$\nu_7(E_u)$	Ref.
R a)	303 p	275 dp	—	164 dp	—	—	[3]
R b)	304.3	278.3	—	165.5	—	—	[4]
R c)	300 p	270* dp	—	176 dp	—	—	[3]
IR	—	—	150	—	321	161	[3]
R b)	307	282	—	175	—	—	[5]

a) $K_2[PdCl_4]$ in H_2O containing HCl. – b) $K_2[PdCl_4]$ in H_2O. – c) $(Bu_4^nN)_2[PdCl_4]$ in $CHCl_3$ or acetone. Numbering scheme as in [3]; p = polarised, dp = depolarised.

References:

[1] James, D. W.; Nolan, M. J. (Prog. Inorg. Chem. **9** [1968] 195/275, 236).
[2] Mink, J.; Goggin, P. L. (Kem. Kozlem. **58** [1982] 215/41, 222).
[3] Goggin, P. L.; Mink, J. (J. Chem. Soc. Dalton Trans. **1974** 1479/83).
[4] Bosworth, Y. M.; Clark, R. J. H. (Inorg. Chem. **14** [1975] 170/7).
[5] Stein, P.; Miskowski, V.; Woodruff, W. H.; Griffin, J. P.; Werner, K. G.; Gaber, B. F.; Spiro, T. G. (J. Chem. Phys. **64** [1976] 2159/67).
[6] Ciescak-Golonka, M.; Bartecki, A.; Sinha, S. P. (Coord. Chem. Rev. **31** [1980] 251/88, 278).

Force Constants

Valence force constants (VFF) for $[PdCl_4]^{2-}$ were calculated using solution infrared and Raman results [1]; general valence force field (GVFF) constants [2] using the data of [3]; a modified Urey-Bradley force field [4] using the data of [1]; and GVFF and symmetrised force constants [5] using the data of [1].

Mean amplitudes of vibration for $[PdCl_4]^{2-}$ were calculated [5] using the data of [1].

There are also force constant data for $[PdCl_4]^{2-}$ based on the Raman and infrared spectra of solid $K_2[PdCl_4]$; see p. 132.

The Pd–Cl stretching force constant was compared with those for other metals and ligands, and correlated with ligand-field effects [6]; bond orders in square-planar molecules including $[PdCl_4]^{2-}$ were correlated with stretching force constants [7].

References:

[1] Goggin, P. L.; Mink, J. (J. Chem. Soc. Dalton Trans. **1974** 1479/83).
[2] Pandey, A. N.; Sharma, D. K.; Verma, U. P.; Kumar, V. (J. Raman Spectrosc. **6** [1977] 163/4).
[3] Bosworth, Y. M.; Clark, R. J. H. (Inorg. Chem. **14** [1975] 170/7).
[4] Pandey, A. N.; Verma, U. P. (J. Mol. Struct. **42** [1977] 171/9).
[5] Sanyal, N. K.; Verma, D. N.; Dixit, L. (Indian J. Pure Appl. Phys. **14** [1976] 819/22).
[6] Hancock, R. D.; Evers, A. (J. Inorg. Nucl. Chem. **35** [1973] 2558/61).
[7] Rastogi, V. K.; Kumar, V.; Pandey, A. N. (Indian J. Pure Appl. Phys. **20** [1982] 150/1).

4.5.2.2 Electronic Spectra

For an early review of electronic spectra of square-planar complexes including $[PdCl_4]^{2-}$ see [17].

Aqueous Solutions with Fixed Amounts of Cl^-

The following spectra, over the stated ranges and in the media listed, were reported in the cited papers: in 2M KCl aqueous solution (180 to 750 nm) [1]; of $[Pd(H_2O)_4]^{2+}$ in excess HCl (200 to 700 nm with assignments) [2, 3]; as $PdCl_2$ in excess HCl (200 to 300 nm) [4]; as $[PdCl_4]^{2-}$ in 2M HCl (200 to 700 nm, with assignments) [5]; as $[PdCl_4]^{2-}$ in 2 M HCl (180 to 550 nm, with assignments) [6]; in H_2O, 450 to 550 nm [7]; in aqueous Cl^- (380 to 600 nm) [8]; in 1M HCl (200 to 600 nm) [9]; in 2 M HCl (270 to 450 nm) [10]; as $[PdCl_4]^{2-}$ in 2M HCl (200 to 350 nm, with assignments) [11]; as $[PdCl_4]^{2-}$ in aqueous Cl^- (300 to 500 nm) [12]; as $[PdCl_4]^{2-}$ in 2 M HCl (240 to 560 nm) [13]; as $[PdCl_4]^{2-}$ in 6M HCl (400 to 700 nm) [14]; as $[PdCl_4]^{2-}$ in 1 M $HClO_4$ and in 10M HCl (200 to 600 nm) [15]; as $[PdCl_4]^{2-}$ in excess Cl^- (400 to 560 nm) [16]; as $[PdCl_4]^{2-}$ with excess Cl^- from 200 to 400 nm [17].

Care has to be taken in measuring spectra of $[PdCl_4]^{2-}$ in aqueous solution that sufficient Cl^- (>2 M) is present to suppress aquation [13].

References:

[1] Rush, R. M.; Martin, D. S.; Le Grand, R. G. (Inorg. Chem. **14** [1975] 2543/50).
[2] Elding, L. I.; Olsson, L. F. (J. Phys. Chem. **82** [1978] 69/74).
[3] Elding, L. I. (Inorg. Chim. Acta **6** [1972] 647/51).
[4] Kaszonyi, A.; Vojtko, J.; Hrusovsky, M. (Collection Czech. Chem. Commun. **45** [1980] 179/86, 181).
[5] Nakayama, K.; Komorita, T.; Shimura, Y. (Bull. Chem. Soc. Japan **57** [1984] 972/9).
[6] McCaffery, A. J.; Schatz, P. N.; Stephens, P. J. (J. Am. Chem. Soc. **90** [1968] 5730/5).
[7] Stein, P.; Miskowski, V.; Woodruff, W. H.; Griffin, J. P.; Werner, K. G.; Gaber, B. P.; Spiro, T. G. (J. Chem. Phys. **64** [1976] 2159/67).
[8] Shchukarev, S. A.; Lobaneva, D. A.; Ivanova, M. A.; Kononova, M. A. (Vestn. Leningr. Univ. Fiz. Khim. **16** No. 2 [1961] 152/5; C.A. **1961** 24362).
[9] Cooper, J. C.; Venezky, D. L.; Lorenz, T. (Fundam. Res. Homogeneous Catal. **3** [1979] 847/57).
[10] Mureinik, R. J.; Pross, E. (J. Coord. Chem. **8** [1978] 127/33).

[11] Anex, B. G.; Takeuchi, N. (J. Am. Chem. Soc. **96** [1974] 4411/6).
[12] Kazakova, V. I.; Ptitsyn, B. V. (Zh. Neorgan. Khim. **12** [1967] 620/5; Russ. J. Inorg. Chem. **12** [1967] 323/6).
[13] Sol'shakov, M. A.; Sinitsyn, N. M.; Borbat, V. F.; Selina, L. I.; Rubtsov, M. V. (Zh. Neorgan. Khim. **19** [1974] 122/7; Russ. J. Inorg. Chem. **19** [1974] 66/9).
[14] Bailey, R. A.; McIntyre, J. A. (Inorg. Chem. **5** [1966] 1824/5).
[15] Harris, C. M.; Livingstone, S. E.; Reece, I. H. (J. Chem. Soc. **1959** 1505/11).
[16] Kukushkin, Yu. N.; Simanova, S. A.; Fedorova, G. I. (Zh. Prikl. Khim. **41** [1968] 1604/6; J. Appl. Chem. [USSR] **41** [1968] 1521/3).
[17] Gray, H. B. (Transition Metal Chem. **1** [1965] 239/287, 247).

Aqueous Solutions with Varying Amounts of Cl^-

The spectrum of $[PdCl_4]^{2-}$ in water with varying amounts of Cl^- has been measured as a function of time for a number of systems (spectra reproduced in paper unless otherwise stated). Spectra have been measured for $[PdCl_4]^{2-}$ and Cl^- from 220 to 400 nm [1], $[PdCl_4]^{2-}$ from 200 to 400 nm [2], from 200 to 450 nm (after pulse radiolysis) [3], and from 280 to 550 nm [4]; for $[Pd(H_2O)_4]^{2+}$ and Cl^- from 350 to 520 nm [5] and from 360 to 540 nm [6]; for $PdCl_2$ with varying amounts of Cl^- (200 to 540 nm) [7]; for $PdCl_2$ in varying amounts of HCl (200 to 300 nm) [8]; $[PdCl_4]^{2-}$ with varying amounts of Cl^- (380 to 600 nm) [9]; for $[Pd(H_2O)_4]^{2+}$ and Cl^- (210 to 320 nm) [10]; as $PdCl_2$ in varying amounts of HCl (300 to 580 nm) [11]; as $[PdCl_4]^{2-}$ in KCl solution, (400 to 560 nm) [12]; as Pd^{II} in Cl^- (200 to 310 nm), [13]; $[PdCl_4]^{2-}$ in HCl (350 to 530 nm) [14]; $PdCl_2$ in varying amounts of HCl (200 to 700 nm) [15].

References:

[1] Kazakova, V. I.; Ptitsyn, B. V. (Zh. Neorgan. Khim. **12** [1967] 620/5; Russ. J. Inorg. Chem. **12** [1967] 323/6).
[2] Rittner, W. F.; Gulko, A.; Schmuckler, G. (Talanta **17** [1970] 807/16, 808).
[3] Broszkiewicz, R. K. (Intern. J. Radiat. Phys. Chem. **6** [1974] 249/58, 250).
[4] Grinberg, A. A.; Kiseleva, N. V. (Zh. Neorgan. Khim. **3** [1958] 1804/9; Russ. J. Inorg. Chem. **3** [1958] 119/26).
[5] Biryukov, A. A.; Shlenskaya, V. I. (Zh. Neorgan. Khim. **9** [1964] 813/6; Russ. J. Inorg. Chem. **9** [1964] 450/2).
[6] Sundaram, A. K.; Sandell, E. B. (J. Am. Chem. Soc. **77** [1955] 855/7).
[7] Jackson, E.; Pantony, D. A. (J. Appl. Electrochem. **1** [1971] 283/91).
[8] Kaszonyi, A.; Voitko, J.; Hrusovsky, M. (Collection Czech. Chem. Commun. **45** [1980] 179/86).
[9] Shchukarev, S. A.; Lobaneva, O. A.; Ivanova, M. A.; Kononova, M. A. (Vestn. Leningr. Univ. Fiz. Khim. **16** No. 2 [1961] 152/5; C. A. **1961** 24362).
[10] Shlenskaya, V. I.; Biryukov, A. A. (Vestn. Mosk. Univ. Ser. II Khim. **19** No. 3 [1964] 65/8; C. A. **61** [1964] 9043).

[11] Ananin, V. N.; Trokhimetz, A. I. (React. Kinet. Catal. Letters **27** [1985] 129/32).
[12] Kukushkin, Yu. N.; Simanova, S. A.; Fedorova, G. I. (Zh. Prikl. Khim. **41** [1968] 1604/6; J. Appl. Chem. [USSR] **41** [1968] 1521/3).
[13] de Waal, D. J. A.; Robb, W. (Intern. J. Chem. Kinet. **6** [1974] 309/21, 317).
[14] Grinberg, A. A. (Chem. Zvesti **13** [1959] 201/23, 208).
[15] Kravchuk, L. S.; Stremok, I. P.; Markevich, S. V. (Zh. Neorgan. Khim. **21** [1976] 728/31; Russ. J. Inorg. Chem. **21** [1976] 391/3).

Data for Solutions in Nonaqueous Media

Such spectra were measured (reproduced in papers over the stated ranges) for solutions in acetonitrile as the (Bu_4^nN) salt, 250 to 450 nm [1]; in acetic acid (200 to 350 nm) [2]; for the trioctylammonium salt of $[PdCl_4]^{2-}$ in CH_2Cl_2 and $CHCl_3$ (400 to 500 nm) [3]; for the sodium salt in acetic acid (250 to 400 nm) [4]; as $PdCl_2$ with Cl^- in CH_3COOH (240 to 350 nm) [5]; in acetonitrile with Cl^- ion as $PdCl_2$ (200 to 500 nm) [6, 7]; for the tri- and di-n-octylammonium salts in CCl_4–$CHCl_3$ (240 to 600 nm) [8]; from 300 to 800 nm in $CHCl_3$ as the imipraminium salt [17].

Miscellaneous Data

Data were listed for $[PdCl_4]^{2-}$ as follows: in HCl (300 to 450 nm) [9]; in water (300 to 400 nm) [10]; in water (200 to 600 nm, with assignments) [11]; in water (200 to 500 nm) [12]; in HCl (200 to 500 nm) [13]. No ranges were stated for data listed in [14] and [15].

The spectrum of $Cs_2[PdCl_4]$ in a LiCl–KCl eutectic at 400 and 600°C has also been measured (300 to 700 nm, reproduced in paper) and assigned; data were also listed for the eutectic at 500°C [16]. Spectra (200 to 300 nm) were also measured of $PdCl_2$ in a fused LiCl–KCl eutectic at 450°C [18].

References:

[1] Klotz, P.; Feldberg, S.; Newman, L. (Inorg. Chem. **12** [1973] 164/8).
[2] Alcock, R. M.; Hartley, F. R.; Rogers, D. E.; Wagner, J. L. (J. Chem. Soc. Dalton Trans. **1975** 2189/93).
[3] Vasil'eva, A. A.; Gindin, L. M.; Pelina, G. N.; Mal'chikov, G. D. (Izv. Sibirsk. Otd. Akad. Nauk SSSR Ser. Khim. Nauk **1969** No. 6, pp. 40/6; Sib. J. Chem. **1969** 675/9).
[4] Alcock, R. M.; Hartley, F. R.; Rogers, D. E.; Wagner, J. L. (Coord. Chem. Rev. **16** [1975] 59/66).
[5] Patrick, P. M.; Marks, O. W. (Inorg. Chem. **10** [1971] 373/6).
[6] Volchenskova, I. I.; Yatsimirskii, K. B. (Zh. Neorgan. Khim. **18** [1973] 1875/82; Russ. J. Inorg. Chem. **18** [1973] 990/5).
[7] Volchenskova, I. I. (Zh. Neorgan. Khim. **19** [1974] 2820/5; Russ. J. Inorg. Chem. **19** [1974] 1540/3).
[8] Bol'shakov, K. A.; Sinitsyn, N. M.; Borbat, V. F.; Selina, L. I.; Rubtsov, M. V. (Zh. Neorgan. Khim. **19** [1974] 122/7; Russ. J. Inorg. Chem. **19** [1974] 66/9).
[9] Tomilov, S. B.; Moskvin, L. N.; Krasnoperov, V. M.; Chereshkevich, Yu. L. (Izv. Sibirsk. Otd. Akad. Nauk SSSR Ser. Khim. Nauk **1974** No. 2, pp. 43/6; C.A. **81** [1974] No. 30382).
[10] Babaeva, A. V.; Rudyi, R. L. (Zh. Neorgan. Khim. **1** [1956] 921/9; Russ. J. Inorg. Chem. **1** [1956] 42/51).

[11] Basch, H.; Gray, H. B. (Inorg. Chem. **6** [1967] 365/9).
[12] Victori, L.; Tomás, X.; Malgosa, F. (Afinidad **32** [1975] 867/72).
[13] Broszkiewicz, R. K. (Intern. J. Radiat. Phys. Chem. **6** [1974] 249/58).
[14] Czászár, J.; Balog, J.; Lehotai, L. (Acta Univ. Szegediensis Acta Phys. Chem. [2] **2** [1956] 56/61; C.A. **1957** 16183).
[15] Lobaneva, O. A.; Kononova, M. A. (Probl. Sovrem. Khim. Koord. Soedin. No. 2 [1968] 180/90; C.A. **70** [1969] No. 15682).
[16] Bailey, R. A.; McIntyre, J. A. (Inorg. Chem. **5** [1966] 1824/5).
[17] Dembinski, B. (Chem. Anal. [Warsaw] **28** [1983] 261/5).
[18] Dickinson, J. R.; Johnson, K. E. (Can. J. Chem. **45** [1967] 1631/6).

Magnetic Circular Dichroism (MCD) Spectra

The MCD spectra of $[PdCl_4]^{2-}$ in HCl were measured (200 to 700 nm [1]; 180 to 600 nm [2], reproduced in paper). The method of moments [3] has been applied to interpreting the data of [2].

The MCD spectra of a number of square-planar tetra-halo complexes, including $[PdCl_4]^{2-}$, have been studied theoretically on the basis of MO theory [4].

References:

[1] Nakayama, K.; Komorita, T.; Shimura, Y. (Bull. Chem. Soc. Japan **57** [1984] 972/9).
[2] McCaffery, A. J.; Schatz, P. N.; Stephens, P. J. (J. Am. Chem. Soc. **90** [1968] 5730/5).
[3] Stephens, P. J.; Mowery, R. L.; Schatz, P. N. (J. Chem. Phys. **55** [1971] 224/231, 228).
[4] Kato, H. (Bull. Chem. Soc. Japan **45** [1972] 1281/8).

92

4.5.2.3 Molecular Orbital (MO) Calculations and Studies on $[PdCl_4]^{2-}$

For the most part these have been based on the results of electronic spectral data [1 to 5, 12]. There are CNDO-type MO calculations on $[PdCl_4]^{2-}$ [6], and empirical SCF-MO-LCAO [7] and SCF-X_α-scattered wave MO calculations [8]. Values of H_{ij} matrix elements for MO calculations on $[PdCl_4]^{2-}$ have been computed and a Cotton-Harris MO calculation was carried out [9]. Oscillator strengths of the symmetry-forbidden transitions of $[PdCl_4]^{2-}$ were calculated by evaluation of the molecular orbitals of the distorted anion [10]. The MO ordering in $[PdCl_4]^{2-}$ has been compared with other calculations using CNDO, INDO, X_α and other semi-empirical schemes [11]. Pseudopotential calculations on $[PdCl_4]^{2-}$ and $[PtCl_4]^{2-}$ were carried out and the results compared [13] (see also the section on magnetic circular dichroism spectra, above).

References:

[1] Basch, H.; Gray, H. B. (Inorg. Chem. 6 [1967] 365/9).
[2] Ballhausen, C. J.; Gray, H. B. (ACS Monogr. No. 168 [1971] 3/83, 72).
[3] Gray, H. B.; Ballhausen, C. J. (J. Am. Chem. Soc. 85 [1963] 260/4).
[4] Larsson, S.; Olsson, A. L. F.; Rosen, A. (Intern. J. Quantum Chem. 25 [1984] 201/9).
[5] Baranovskii, V. I.; Davydova, M. K.; Panina, N. S.; Panin, A. I. (Koord. Khim. 2 [1976] 409/15; Soviet J. Coord. Chem. 2 [1976] 308/13).
[6] Sakaki, S.; Kato, H. (Bull. Chem. Soc. Japan 46 [1973] 2227/9).
[7] Tondello, E.; Di Sipio, L.; De Michelis, G.; Oleari, L. (Inorg. Chim. Acta 5 [1971] 305/10).
[8] Messmer, R. P.; Interrante, L. V.; Johnson, K. H. (J. Am. Chem. Soc. 96 [1974] 3847/54).
[9] Harrison, T. G.; Patterson, H. H.; Hsu, M. T. (Inorg. Chem. 15 [1976] 3018/24).
[10] Erny, M.; Moncuit, C. (Theor. Chim. Acta 61 [1982] 29/39).

[11] Ruiz, M. E.; Daudey, J. P.; Novaro, O. (Mol. Phys. 46 [1982] 853/62).
[12] Gray, H. B. (Transition Metal Chem. 1 [1965] 239/87, 247).
[13] Rasch, G.; Bögel, H.; Hrusak, J.; Tobisch, S. (Z. Chem. [Leipzig] 27 [1987] 152/3).

4.5.2.4 Stability Constants

There has been much work on the stability constants of $[PdCl_4]^{2-}$ and a variety of methods used to determine them. The four constants K_1 to K_4 for $[PdCl_4]^{2-}-H_2O$ are given in the following tables.

There is an extensive review of earlier data [1], with an evaluation of the data in that review [2], and a shorter review of early data [3].

From stability constant data distribution curves for the four species $[PdCl_n(H_2O)_{4-n}]^{2-n}$ have been drawn [3 to 6], see **Fig. 14** and **15**.

References:

[1] Victori, L.; Tomás, X.; Malgosa, F. (Afinidad 32 [1975] 867/72).
[2] Kragten, J. (Talanta 27 [1980] 375/7).
[3] Shchukarev, S. A.; Lobaneva, D. A.; Ivanova, M. A.; Kononova, M. A. (Vestn. Leningr. Univ. Fiz. Khim. 16 No. 2 [1961] 152/5; C.A. 1961 24362).
[4] Pearson, R. G.; Hynes, M. J. (Kgl. Tekn. Hoegsk. Handl. No. 285 [1972] 461/9).
[5] Droll, H. A.; Block, B. P.; Fernelius, W. C. (J. Phys. Chem. 61 [1957] 1000/4).
[6] Shlenskaya, V. I.; Biryukov, A. A. (Zh. Neorgan. Khim. 11 [1966] 54/60; Russ. J. Inorg. Chem. 11 [1966] 28/31).

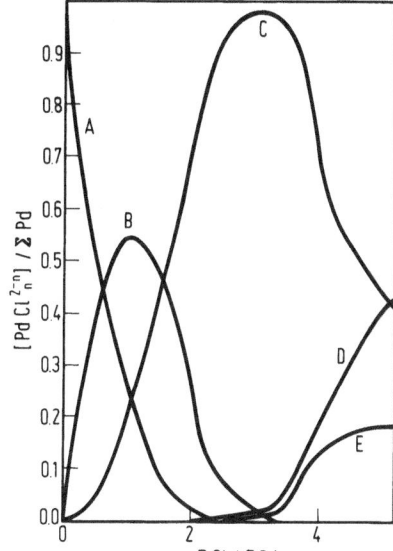

Fig. 14. Distribution of Pd^{II} chloro complexes at 21°C in 0.2 M HClO$_4$, $\mu = 0.44$. A is [Pd(H$_2$O)$_4$]$^{2+}$ + [Pd(OH)$_n$(H$_2$O)$_{3-n}$]$^{2-n}$, B is [PdCl(H$_2$O)$_3$]$^+$, C is PdCl$_2$(H$_2$O)$_2$, D is [PdCl$_3$(H$_2$O)]$^-$, E is [PdCl$_4$]$^{2-}$ [5].

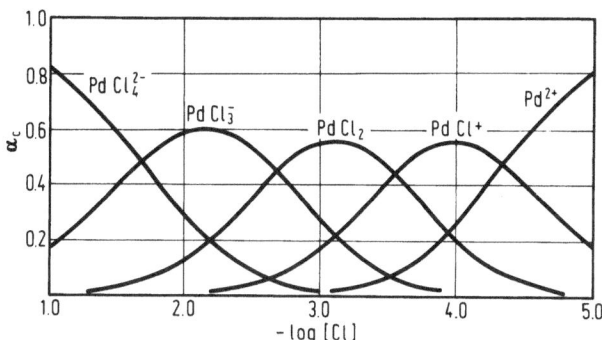

Fig. 15. Distribution of Pd^{II} chloro complexes [4].

Values of K$_1$

Data on K$_1$ are less numerous than for K$_2$ and K$_4$, but there are nevertheless several determinations using a variety of methods and these are listed in the following table. The values have been reviewed [1].

Values of K$_1$ and log K$_1$:

method	t in °C	total ionic strength (M)	K$_1$ (M^{-1})	log K$_1$ (M^{-1})	Ref.
a)	25	0.26 − 1.01	—	4.7 − 3.98	[2]
a)	25	0	—	6.0 ± 0.2	[2]
a)	25	1	—	4.00	[3]
a)	25	1	29.800 ± 600	4.47 ± 0.01	[4]

method	t in °C	total ionic strength (M)	K_1 (M^{-1})	$\log K_1$ (M^{-1})	Ref.
a)	21	0.44	—	6.2 ± 0.1	[5]
a)	20	0.8	—	4.34	[6]
b)	25	1	—	3.88 ± 0.09	[7, 8, 12]
b)	20	0.1	—	5.1	[9]
c)	25	1	—	3.48	[10]
c)	—	—	—	4.70	[11]
b)	20	0.1	—	5.10	[10]
a)	20	—	—	16.7 ± 0.2	[5]

a) Spectrophotometric; b) from AgCl solubility data; c) from competition studies with the Tl^{III}–Tl^{I} system.

References:

[1] Victori, L.; Tomás, X.; Malgosa, F. (Afinidad **32** [1975] 867/72).
[2] Biryukov, A. A.; Shlenskaya, V. I. (Zh. Neorgan. Khim. **9** [1964] 813/6; Russ. J. Inorg. Chem. **9** [1964] 450/2).
[3] Shlenskaya, V. I.; Biryukov, A. A. (Zh. Neorgan. Khim. **11** [1966] 54/60; Russ. J. Inorg. Chem. **11** [1966] 28/31).
[4] Elding, L. I. (Inorg. Chim. Acta **6** [1972] 647/51).
[5] Droll, H. A.; Block, B. F.; Fernelius, W. C. (J. Phys. Chem. **61** [1957] 1000/4).
[6] Shchukarev, S. A.; Lobaneva, D. A.; Ivanova, M. A.; Kononova, M. A. (Vestn. Leningr. Univ. Fiz. Khim. **16** No. 2 [1961] 152/5; C. A. **1961** 24362).
[7] Burger, K.; Dyrssen, D. (Acta Chem. Scand. **17** [1963] 1489/1501, 1496).
[8] Burger, K. (Magyar Kem. Folyoirat **70** [1964] 179/84).
[9] Grinberg, A. A.; Gel'fman, M. I.; Kiseleva, N. V. (Zh. Neorgan. Khim. **12** [1967] 1171/5; Russ. J. Inorg. Chem. **12** [1967] 620/2).
[10] Rittner, W. F.; Gulko, A.; Schmuckler, G. (Talanta **17** [1970] 807/160).

[11] Gel'fman, M. I.; Kiseleva, N. V. (Zh. Neorgan. Khim. **14** [1969] 502/4; Russ. J. Inorg. Chem. **14** [1969] 258/60).
[12] Burger, K. (Acta Chim. Acad. Sci. Hung. **40** [1964] 261/73).

Values of K_2

These are listed in the following table; data are reviewed in [1]. As expected, they are substantially lower than those for K_1 but exceed those for K_3.

Values of K_2, $\log K_2$ and $\log \beta_2$:

method	t in °C	total ionic strength (M)	K_2 (M^{-1})	$\log K_2$ (M^{-1})	$\log \beta_2$ (M^{-1})	Ref.
a), b)	25	1	—	3.06 ± 0.04	—	[2, 11]
a)	25	0	—	4.6 ± 0.1	10.7 ± 0.1	[3]
a)	25	1	—	3.45	—	[4]
a)	25	1	1920 ± 130	3.28	7.76 ± 0.04	[5]
a)	21	0.44	—	4.7 ± 0.1	—	[6]

method	t in °C	total ionic strength (M)	K_2 (M^{-1})	$\log K_2$ (M^{-1})	$\log \beta_2$ (M^{-1})	Ref.
a)	20	0.8	—	3.54	—	[7]
b)	25	1	—	3.03	—	[8]
a)	25	1	—	2.79	—	[9]
c)	—	—	—	3.00	—	[10]
d)	25	1	—	3.06 ± 0.04	9.08	[11]
a)	25	1	—	3.05	—	[2]

a) Spectrophotometric; b) data from AgCl solubility studies; c) from competition studies with the Tl^L–Tl^{III} system; d) potentiometric data.

References:

[1] Victori, L.; Tomàs, X.; Malgosa, F. (Afinidad **32** [1975] 867/72).
[2] Burger, K. (Magyar Kem. Folyoirat **70** [1964] 179/84).
[3] Biryukov, A. A.; Shlenskaya, V. I. (Zh. Neorgan. Khim. **9** [1964] 813/6; Russ. J. Inorg. Chem. **9** [1964] 450/2).
[4] Shlenskaya, V. I.; Biryukov, A. A. (Zh. Neorgan. Khim. **11** [1966] 54/60; Russ. J. Inorg. Chem. **11** [1966] 28/31).
[5] Elding, L. I. (Inorg. Chim. Acta **6** [1972] 647/51).
[6] Droll, H. A.; Block, B. P.; Fernelius, W. C. (J. Phys. Chem. **61** [1957] 1000/4).
[7] Shchukarev, S. A.; Lobaneva, D. A.; Ivanova, M. A.; Kononova, M. A. (Vestn. Leningr. Univ. Fiz. Khim. **16** No. 2 [1961] 152/5; C.A. **1961** 24362).
[8] Burger, K.; Dryssen, D. (Acta Chem. Scand. **17** [1963] 1489/501, 1496).
[9] Rittner, W. F.; Gulko, A.; Schmuckler, G. (Talanta **17** [1970] 807/16).
[10] Gel'fman, M. I.; Kiseleva, N. V. (Zh. Neorgan. Khim. **14** [1969] 502/4; Russ. J. Inorg. Chem. **14** [1969] 259/60).

[11] Burger, K. (Acta Chim. Acad. Sci. Hung. **40** [1964] 261/73).

Values of K_3

These, the least frequently measured of the four stability constants, are listed in the following table and reviewed in [1].

Values of K_3, $\log K_3$ and $\log \beta_3$:

method	t in °C	total ionic strength (M)	K_3 (M^{-1})	$\log K_3$ (M^{-1})	$\log \beta_3$ (M^{-1})	Ref.
b)	25	1	—	2.14 ± 0.05	—	[2, 10]
a)	25	0	—	2.4 ± 0.1	13.1 ± 0.1	[3]
d)	25	9	57	$1.76^{*)}$	—	[4]
a)	25	1	260 ± 20	$2.41^{*)}$	10.17 ± 0.07	[5]
a)	21	0.44	—	2.5 ± 0.1	—	[6]
a)	20	0.8	—	$2.68 - 2.79$	—	[7]
b)	25	1	—	2.18	—	[8]
a)	25	1	—	2.00	—	[9]

method	t in °C	total ionic strength (M)	K_3 (M^{-1})	log K_3 (M^{-1})	log β_3 (M^{-1})	Ref.
c)	25	1	—	2.14 ± 0.05	9.08	[10]
c)	25	1	—	2.17	—	[2]
d)	—	—	—	2.60	—	[11]

*) Calculated from K value; a) spectrophotometric; b) potentiometric; c) potentiometric data and AgCl solubility studies; d) potentiometric; e) calculated [11].

References:

[1] Victori, L.; Tomàs, X.; Malgosa, F. (Afinidad **32** [1975] 867/72).
[2] Burger, K. (Magyar Kem. Folyoirat **70** [1964] 179/84).
[3] Biryukov, A. A.; Shlenskaya, V. I. (Zh. Neorgan. Khim. **9** [1964] 813/6; Russ. J. Inorg. Chem. **9** [1964] 450/2).
[4] Kravtsov, V. I.; Martynova, L. B. (Zh. Neorgan. Khim. **16** [1971] 858/60; Russ. J. Inorg. Chem. **16** [1971] 457/8).
[5] Elding, L. I. (Inorg. Chim. Acta **6** [1972] 647/51).
[6] Droll, H. A.; Block, B. F.; Fernelius, W. C. (J. Phys. Chem. **61** [1957] 1000/4).
[7] Shchukarev, S. A.; Lobaneva, D. A.; Ivanova, M. A.; Kononova, M. A. (Vestn. Leningr. Univ. Fiz. Khim. **16** No. 2 [1961] 152/5; C.A. **1961** 24362).
[8] Burger, K.; Dryssen, D. (Acta Chem. Scand. **17** [1963] 1489/501, 1497).
[9] Rittner, W. F.; Gulko, A.; Schmuckler, G. (Talanta **17** [1970] 807/16).
[10] Burger, K. (Acta Chim. Acad. Sci. Hung. **40** [1964] 261/73).

[11] Gel'fman, M. I.; Kiseleva, N. V. (Zh. Neorgan. Khim. **14** [1969] 502/4; Russ. J. Inorg. Chem. **14** [1969] 258/60).

Values of K_4

There are many of these, collected in the following table, and obtained by a variety of methods. The data up to 1975 has been comprehensively reviewed [1]. The influence of ionic strength [12, 24] and temperature [5, 19, 24] on K_4 has been noted. At unit ionic strength log $K_4 = 5.68 + 1245.90/T$ with T in K [24].

For β_4 values and thermodynamic data see p. 99.

Values of K_4, log K_4 and log β_4:

method	t in °C	total ionic strength (M)	K_4 (M^{-1})	log K_4 (M^{-1})	log β_4 (M^{-1})	Ref.
a)	—	1	18.7 ± 0.13	1.27*)	—	[2]
a)	25	1	—	0.96	—	[3]
b)	25	1	—	1.34 ± 0.02	—	[4]
a)	21	0.44	—	2.6 ± 0.1	—	[5]
a)	20	0.8	—	1.68 to 1.83	12.24	[6]
b)	25	1	—	1.34 ± 0.05	10.42 ± 0.2	[7]
a)	25	0.5	24 ± 1	1.38 ± 0.03	—	[8]
b)	25	1	—	1.34 ± 0.05	—	[23]

method	t in °C	total ionic strength (M)	K_4 (M^{-1})	log K_4 (M^{-1})	log β_4 (M^{-1})	Ref.
a)	25	1	30.67	1.49*)	—	[9]
a)	25	1	—	2.0	—	[10]
a)	25	0	—	2.6 ± 0.1	15.7 ± 0.1	[10]
a)	25	1	—	1.42	—	[11]
a)	25	1 to 4	26.8 to 101	1.43 ± 0.01	—	[12]
a)	25	1	26.8	1.43 ± 0.01	—	[12]
c)	25	1	100	2.0*)	11.14*) (1.4×10^{12})	[13]
c)	25	9	22.6	2.35*)	—	[14]
a)	25	1	23.7 ± 1.0	1.37*)	11.54 ± 0.09	[15]
e)	—	—	—	1.62	11.9	[16]
a)	25	1	6	0.78*)	—	[17]
c)	25	3.4	—	—	11.40	[18]
c)	10 to 60	1	—	—	12.2 to 11.2	[19]
c)	20	1	—	—	12.0	[19]
c)	25	2	—	—	12.15	[13]
c)	25	4.02	—	—	13.22	[20]
c)	20	—	—	—	12.2 ± 0.5	[21]
d)	—	0	—	1.44 ± 0.50	9.38 ± 0.31	[22]
c)	25	1	—	1.34 ± 0.05	10.42	[23]
c)	25	1	—	1.32	—	[7]
a)	25	1	—	1.34	—	[23]
c)	25 to 100	0.1 to 2.0	—	—	9.85 to 11.12	[24]
c)	25	1	—	—	9.85	[24]
a)	30	—	—	—	12.57	[25]

*) Calculated from K_4. – Methods: a) = spectrophotometric; b) = data from AgCl solubility studies; c) = potentiometric; d) = polarographic; e) = from competing Tl^I/Tl^{III} reaction.

References:

[1] Victori, L.; Tomás, X.; Malgosa, F. (Afinidad **32** [1975] 867/72).

[2] Gulko, A.; Schmuckler, G. (J. Inorg. Nucl. Chem. **35** [1973] 603/7).

[3] Rittner, W. F.; Gulko, A.; Schmuckler, G. (Talanta **17** [1970] 807/16).

[4] Burger, K.; Dyrssen, D. (Acta Chem. Scand. **17** [1963] 1489/501, 1497).

[5] Droll, H. A.; Block, B. P.; Fernelius, M. C. (J. Phys. Chem. **61** [1957] 1000/4).

[6] Shchukarev, S. A.; Lobaneva, D. A.; Ivanova, M. A.; Kononova, M. A. (Vestn. Leningr. Univ. Fiz. Khim. **16** No. 2 [1961] 152/5; C.A. **1961** 24362).

[7] Burger, K. (Magyar Kem. Folyoirat **70** [1964] 179/84).

[8] Shlenskaya, V. I.; Biryukov, A. A. (Vestn. Mosk. Univ. Ser. II Khim. **19** No. 3 [1964] 65/8; C.A. **61** [1964] 9043).

[9] Biryukov, A. A.; Shlenskaya, V. I. (Vestn. Mosk. Univ. Ser. II Khim. **19** No. 5 [1964] 81/6; C.A. **62** [1965] 4890).

[10] Biryukov, A. A.; Shlenskaya, V. I. (Zh. Neorgan. Khim. **9** [1964] 813/6; Russ. J. Inorg. Chem. **9** [1964] 450/2).

[11] Shlenskaya, V. I.; Biryukov, A. A. (Zh. Neorgan. Khim. **11** [1966] 54/60; Russ. J. Inorg. Chem. **11** [1966] 28/31).

[12] Levanda, O. G. (Zh. Neorgan. Khim. **13** [1968] 3311/3; Russ. J. Inorg. Chem. **13** [1968] 1707/8).

[13] Kravtsov, V. I.; Simakova, I. V. (Vestn. Leningr. Univ. Fiz. Khim. **1969** No. 4, pp. 124/30; C. A. **72** [1970] No. 125686).

[14] Kravtsov, V. I.; Martynova, L. B. (Zh. Neorgan. Khim. **16** [1971] 858/60; Russ. J. Inorg. Chem. **16** [1971] 457/8).

[15] Elding, L. I. (Inorg. Chim. Acta **6** [1972] 647/51).

[16] Gel'fman, M. I.; Kiseleeva, N. V. (Zh. Neorgan. Khim. **14** [1969] 502/4; Russ. J. Inorg. Chem. **14** [1969] 258/60).

[17] Jørgensen, C. K. (Absorption Spectra and Chemical Bonding in Complexes, Pergamon, Oxford 1962, p. 259).

[18] Levanda, O. G.; Moiseev, I. I.; Vargaftik, M. N. (Izv. Akad. Nauk SSSR Ser. Khim. **1968** 2368/70; Bull. Acad. Sci. USSR Div. Chem. Sci. **1968** 2237/9).

[19] Fasman, A. B.; Kutyukov, G. G.; Sokolovskii, D. V. (Zh. Neorgan. Khim. **10** [1965] 1338/44; Russ. J. Inorg. Chem. **10** [1965] 727/30).

[20] Templeton, D. H.; Watt, G. W.; Garner, C. S. (J. Am. Chem. Soc. **65** [1943] 1608/12).

[21] Grinberg, A. A.; Kiseleva, N. V.; Gel'fman, M. I. (Dokl. Akad. Nauk SSSR **153** [1963] 1327/9; Dokl. Chem. Proc. Acad. Sci. USSR **148/153** [1963] 1025/7).

[22] Jackson, E.; Pantony, D. A. (J. Appl. Electrochem. **1** [1971] 283/91).

[23] Burger, K. (Acta Chim. Acad. Sci. Hung. **40** [1964] 261/73).

[24] Nikolaeva, N. M.; Pogodina, L. P. (Izv. Sibirsk. Otd. Akad. Nauk SSSR Ser. Khim. Nauk **1980** No. 1, pp. 130/4; C. A. **93** [1980] No. 15634).

[25] Shchukarev, S. A.; Lobaneva, D. A.; Ivanova, M. A.; Kononova, M. A. (Vestn. Leningr. Univ. Fiz. Khim. **19** No. 3 [1964] 140/2; C. A. **62** [1964] 3459).

4.5.2.5 Thermodynamic Data on [PdCl$_4$]$^{2-}$

Early data have been reviewed [1, 2]; data for ΔH, ΔG and ΔS have been calculated from stability constant data (see table below) with references cited therein. In addition, the stability constant data of [3] were used to calculate $\Delta H°$, $\Delta G°$ and $\Delta S°$ for [PdCl$_n$(H$_2$O)$_{4-n}$]$^{2-n}$ (25°C, in 1 M ClO$_4^-$) [4]. The values of log K_4 at 10°C (1.5), 25°C (1.42) and 45°C (1.28) in 4 M ClO$_4^-$ obtained spectrophotometrically were used to calculate ΔH and ΔS values [5]; analogous data from 15 to 40°C were similarly used [7].

From the electrochemical data of [8] and stability constant data (see p. 96), $\Delta G°_{298}$ for [PdCl$_4$]$^{2-}$·aq has been calculated as -410 kJ/mol, with $\Delta H°_{298} = -522.5$ kJ/mol and S° $\approx$ 234 J·deg^{-1}·mol [1].

Thermodynamic data for stepwise formation of [PdCl$_n$(H$_2$O)$_{4-n}$]$^{2-n}$ complexes in water:

		t in °C	$\Delta H_f°$ in kJ/mol	$\Delta G_f°$ in kJ/mol	$\Delta S°$ in J·mol^{-1}·K^{-1}	Ref.
log K$_1$	4.47 ± 0.01	25	-12.1 ± 0.04	-25.5 ± 0.04	43.0 ± 0.4	[4]
log K$_1$	6.2 ± 0.1	21	-33 ± 8	-34.7 ± 0.4	4 ± 40	[6]
log β$_2$	7.76 ± 0.04	25	-10.8 ± 0.2	-10.8 ± 0.2	26.8 ± 1.2	[4]
log K$_2$	4.7 ± 0.1	21	-37.6 ± 8	-26.3 ± 0.4	-42 ± 42	[6]
log β$_3$	10.17 ± 0.07	25	-10.7 ± 0.2	-13.7 ± 0.2	10.0 ± 2.4	[4]

	t in °C	ΔH_f° in kJ/mol	ΔG_f° in kJ/mol	ΔS° in $J \cdot mol^{-1} \cdot K^{-1}$	Ref.
log K_3	2.5 ± 0.1	21	-33 ± 8	-14.2 ± 0.4	-58 ± 40	[6]
log β_4	11.54 ± 0.09	25	-14.2 ± 0.7	-7.8 ± 0.6	-21.7 ± 12	[4]
log K_4	2.6 ± 0.1	21	-33.4 ± 8	-14.6 ± 0.4	-62.7 ± 40	[6]

References:

[1] Goldberg, R. N.; Hepler, L. G. (Chem. Rev. **68** [1968] 229/52, 241).
[2] Ashcroft, S. J.; Mortimer, C. T. (Thermochemistry of Transition Metal Complexes, Academic, London 1970, p. 362).
[3] Elding, L. I. (Inorg. Chim. Acta **6** [1972] 647/51).
[4] Ryhl, T. (Acta Chem. Scand. **26** [1972] 2961/2).
[5] Shlenskaya, V. I.; Biryukov, A. A. (Zh. Neorgan. Khim. **11** [1966] 54/60; Russ. J. Inorg. Chem. **11** [1966] 28/31).
[6] Droll, H. A.; Block, B. P.; Fernelius, W. C. (J. Phys. Chem. **61** [1957] 1000/4).
[7] Levanda, O. G. (Zh. Neorgan. Khim. **13** [1968] 3311/3; Russ. J. Inorg. Chem. **13** [1968] 1707/8).
[8] Kravtsov, V. I.; Zelenskii, M. I. (Elektrokhimiya **2** [1966] 1138/43; Soviet Electrochem. **2** [1966] 1042/6).

4.5.2.6 Electrochemical Behaviour

For a review of early data see [1].

Redox Measurements

The potential of the reaction $[PdCl_4]^{2-} + 2e^- \rightarrow Pd$ (solid) is $+0.6124$ V at ionic strength 0.23 M and 0.6119 V at 1.12 M [2]. At 15, 25 and 35°C, $E_0 = 0.623$ V in 0.9952 M HCl at 25°C for $[PdCl_4]^{2-}/Pd$ (solid), and E_0 at 25°C is given by $E_0 = 0.629 - 0.008 \mu$ where μ is the ionic strength from 1 to 4 M [3]. At 50°C, $E_0 = 0.619$ V and at 150°C, $E_0 = 0.531$ V, with $E_0 = 0.619 - 8.8 \times 10^{-4}$ $(T - 323)$ [4]. At 1 M ionic strength, $E_0 = 0.589$ V at 10°C and 0.619 V at 60°C; the corresponding values at zero ionic strength are 0.780 and 0.805 V for the potassium salt, while for the free acid E_0 values at 1 M and zero ionic strength were given as 0.774 and 0.577 V at 10°C and as 0.801 and 0.605 V at 60°C [5]. From polarographic data (see also p. 100) the E_0 value for $[PdCl_4]^{2-}/Pd$ has been found to be 0.64 ± 0.01 V [6], while in 1 M HCl and 3 M H_2SO_4 $E_0 = 0.591 \pm 0.009$ V [7]; at 20°C in 1 M KCl, $E_0 = 0.563$ V. Values for E_0 for $[PdCl_4]^{2-}$ were obtained from a system using automated electrode kinetic measurements [9].

The potential for $Pd + 4 Cl^- \rightarrow [PdCl_4]^{2-} + 2e^-$ is given by $E_0 = 0.62 + 0.118 \, pCl + 0.0295$ log $[PdCl_4]^{2-}$ [10].

Data have also been accumulated for the $[PdCl_6]^{2-}/[PdCl_4]^{2-}$ couple (see also p. 149). In 1 M HCl it is $+1.286$ V and in aqueous M NaCl solution $+1.301$ V [11]; other values given are $+1.0225$ V [12] and $+1.26$ V [1].

At 1000 K the E_0 for Pd^{II}/Pd^0 in molten (1:1 molar) NaCl–KCl is $+0.374 \pm 0.008$ V [13]; the potential in molten $MgCl_2$–KCl (32.5 : 67.5 mol%) at 475°C is $+0.3867 \pm 0.0022$ V [14], while in LiCl–KCl eutectic at 450°C it is $+0.214$ V [15].

References:

[1] Goldberg, R. N.; Hepler, L. N. (Chem. Rev. **68** [1968] 229/52, 240).
[2] Levanda, O. G.; Moiseev, I. I.; Vargaftik, M. N. (Izv. Akad. Nauk SSSR Ser. Khim. **1968** 2368/70; Bull. Acad. Sci. USSR Div. Chem. Sci. **1968** 2237/9).
[3] Templeton, D. H.; Watt, G. W.; Garner, C. S. (J. Am. Chem. Soc. **65** [1943] 1608/12).
[4] Nikolaeva, N. M.; Tsvelodub, L. D.; Erenburg, A. M. (Izv. Sibirsk. Otd. Akad. Nauk SSSR Ser. Khim. Nauk **1978** No. 3, pp. 44/7; C.A. **89** [1978] No. 206412).
[5] Fasman, A. B.; Kutyukov, G. G.; Sokol'skii, D. V. (Zh. Neorgan. Khim. **10** [1965] 1338/44; Russ. J. Inorg. Chem. **10** [1966] 727/30).
[6] Jackson, E.; Pantony, D. A. (J. Appl. Electrochem. **1** [1971] 283/91).
[7] Kravtsov, V. I.; Zelenskii, M. I. (Elektrokhimiya **2** [1966] 1138/43; Soviet Electrochem. **2** [1966] 1042/6).
[8] Grinberg, A. A.; Kiseleva, N. V.; Gel'fman, M. I. (Dokl. Akad. Nauk SSSR **151** [1963] 1327/9; Dokl. Chem. Proc. Acad. Sci. USSR **148/153** [1963] 1025/7).
[9] Harrison, J. A. (Electrochem. Acta **27** [1982] 1123/8).
[10] Kammel, R.; Takei, T.; Winterhager, H.; Yamamoto, H. (Metalloberfläche **22** [1968] 107/11).

[11] Grinberg, A. A.; Shamsiev, A. M. (Zh. Obshch. Khim. **12** [1942] 55/72; C.A. **1943** 1915).
[12] Dubinskii, V. I. (Zh. Neorgan. Khim. **16** [1971] 1145/7; Russ. J. Inorg. Chem. **16** [1971] 607/8).
[13] Combes, R.; Vedel, J.; Trémillon, B. (J. Electroanal. Chem. Interfacial Electrochem. **27** [1970] 174/7).
[14] Gaur, H. C.; Jindal, H. L. (Electrochim. Acta **15** [1970] 1113/26, 1115, 1123).
[15] Laitinen, H. A.; Liu, C. H. (J. Am. Chem. Soc. **80** [1958] 1015/20).

Polarography of $[PdCl_4]^{2-}$

There are early data on the polarographic determination of Pd using $[PdCl_4]^{2-}$ with Pt electrodes [10]. Polarographic reduction of $[Pd(H_2O)_4]^{2+}$ with varying Cl^- concentrations at a boron carbide electrode was measured; in 0.3 M Cl^-, $E_{1/2} = -0.229 \pm 0.021$ V (vs. standard calomel electrode). $E_{1/2} = 0.468 \pm 0.014 + (0.178 \pm 0.007) \log Cl^-$ [1].

Polarographic reduction of $[PdCl_4]^{2-}$ at various concentrations in 1 M HCl was studied at a dropping mercury electrode; the diffusion coefficient for $[PdCl_4]^{2-}$ at 10^{-3} M concentration in 1 M HCl was found to be 1.28×10^{-6} cm²/s [2]. A diffusion coefficient of 1.39×10^{-6} cm²/s was obtained by a chronopotentiometric method for $[PdCl_4]^{2-}$ [3]. Polarography of $[PdCl_4]^{2-}$ in aqueous 1 M KCl solution at the dropping mercury electrode showed that the reduction began without any applied voltage: i.e., the dissolution wave of mercury is followed by the reduction wave of $[PdCl_4]^{2-}$ [4]. For voltammetric studies of $[PdCl_4]^{2-}$ in 0.2 M HCl in the presence of $[IrCl_6]^{2-}$ see [5]. The dependence of diffusion current on time for the polarographic solution of $[PdCl_4]^{2-}$ at the dropping mercury electrode has been plotted [6].

Polarography of $PdCl_2$ dissolved in a molten KCl–LiCl eutectic at 450°C has been studied using intermittently polarised Pt indicator electrodes and an approximate value of 1.39×10^{-5} cm²/s deduced for the diffusion coefficient of the metal cation (presumably $[PdCl_4]^{2-}$ in this case) [7].

The mechanism of the polarographic reduction of $[PdCl_4]^{2-}$ at a mercury dropping electrode has been discussed: the first step is likely to be the dissociation of two chloro ligands [8, 9, 13]. Galvanostatic study of the kinetic currents during reduction of $[PdCl_4]^{2-}$ in 3 M H_2SO_4 at a Pd electrode gave evidence for the reactions $[PdCl_4]^{2-} \rightarrow [PdCl_3]^- + Cl^-$ (where $[PdCl_3]^-$ is presumably $[PdCl_3(H_2O)]^-$), followed by $[PdCl_3(H_2O)]^- \rightarrow PdCl_2$ (the latter presumably as $PdCl_2(H_2O)_2$) [11].

Electrical conductances of $[PdCl_4]^{2-}$ in methanol, ethanol or mixtures of these with acetone were used to calculate dissociation constants and limiting conductance values [12]. Chronopotentiometric data were recorded for $PdCl_2$ dissolved in a LiCl–KCl eutectic at 450°C; a diffusion coefficient of $(1.68 \pm 0.05) \times 10^{-5}$ cm^2/s was calculated for the solute, which is presumably $[PdCl_4]^{2-}$ [14].

References:

[1] Jackson, E.; Pantony, D. A. (J. Appl. Electrochem. **1** [1971] 283/91).

[2] Kravtsov, V. I.; Shereshevskaya, I. I. (Elektrokhimiya **5** [1969] 985/8; Soviet Electrochem. **5** [1969] 926/8).

[3] Yur'ev, B. P.; Shkuryakova, S. P.; Borisoglebskii, Yu. V. (Zh. Prikl. Khim. **56** [1983] 1892/4; J. Appl. Chem. [USSR] **56** [1983] 1767/9).

[4] Hirota, M.; Umezawa, Y.; Nakamura, M.; Fujiwara, S. (J. Inorg. Nucl. Chem. **33** [1971] 2617/21).

[5] Harrar, J. E.; Shain, I. (Anal. Chem. **38** [1966] 1148/58, 1156).

[6] Kravtsov, V. I.; Shereshevskaya, I. I. (Vestn. Leningr. Univ. Fiz. Khim. **1974** No. 10, pp. 147/9; C.A. **81** [1974] No. 98 578).

[7] Schmidt, E.; Pfander, H.; Siegenthaler, H. (Electrochim. Acta **10** [1965] 429/43, 436).

[8] Kravtsov, V. I. (Elektrokhimiya **6** [1970] 275/7; Soviet Electrochem. **6** [1970] 265/7).

[9] Kravtsov, V. I.; Shereshevskaya, I. I. (Elektrokhimiya **7** [1971] 99/102; Soviet Electrochem. **7** [1971] 91/4).

[10] Bardin, M. B.; Lyalikov, Yu. S.; Temyanko, V. S. (Anal. Blagorodn. Metal. **1959** 80/7; C.A. **1960** 13962).

[11] Zelenskii, M. I.; Kravtsov, V. I. (Elektrokhimiya **6** [1970] 793/8; Soviet Electrochem. **6** [1970] 768/72).

[12] Kreshkov, A. P.; Yarovenko, A. N.; Bartikova, O. D. (Zh. Obshch. Khim. **43** [1973] 714/8; J. Gen. Chem. [USSR] **43** [1973] 714/8).

[13] Kravtsov, V. I.; Shereshevskava, I. I. (Elektrokhimiya **7** [1971] 618/25; Soviet Electrochem. **7** [1971] 597/603).

[14] Scrosati, B. (Anal. Chem. **38** [1966] 1588/9).

Electrodeposition Studies

The mechanisms of the anodic dissolution of Pd and electrodeposition of the metal in solutions containing Cl^- and $[PdCl_4]^{2-}$ in 3 M H_2SO_4 have been investigated. In electrodeposition "$[PdCl_3]^-$" (presumably $[PdCl_3(H_2O)]^-$) and "$PdCl_2$" (presumably $PdCl_2(H_2O)_2$) are implicated as well as $[PdCl_4]^{2-}$ [1]. The electrodeposition of Pd from $[PdCl_4]^{2-}$ in HCl together with $[PtCl_6]^{2-}$ has been studied [2], as has the electrochemical dissolution of Pd in 10 M HCl at 25 to 60°C [3]. The discharge mechanism of $[PdCl_4]^{2-}$ from electrodeposition studies of Pd from NH_4Cl baths has been shown to be a two-step process [4], and the mechanism of the anodic dissolution of Pd in 6 M and 10 M HCl has been discussed [5]. The electrodeposition of Pd from $[PdCl_4]^{2-}$ on a carbon electrode and on a Pd-covered gold electrode has been studied [6]. Potentiodynamic methods have been used to study the anodic dissolution of Pd films from an insulating layer with aqueous solutions of KCl using Ti as an intermediate layer [7]. Electrode kinetics of Pd deposition-dissolution reactions were investigated on the assumption that $[PdCl_n(H_2O)_{4-n}]^{2-n}$ species were involved [8].

For electroreflectance determinations carried out on surface energy levels created by the electrodeposition of Pd from $[PdCl_4]^{2-}$ on TiO_2 films see [9].

102

References:

[1] Kravtsov, V. I.; Zelenskii, M. I. (Elektrokhimiya **2** [1966] 1138/43; Soviet Electrochem. **2** [1966] 1042/6).

[2] Panova, I. B.; Ramm, K. S.; Krasikov, B. S. (Zashchitn. Metal. Oksidnye Pokrytiya Korroz. Metal. Issled. Obl. Elektrokhim. **1965** 82/6; C.A. **65** [1966] 3338).

[3] Vol'fson, A. I.; Ryazanov, A. I.; Chigrinova, G. D. (Zh. Vses. Khim. Obshchestva **5** [1960] 712; C.A. **1961** 10138).

[4] Makarova, N. G.; Lainer, V. I. (Zashch. Metal. **11** [1975] 453/7; C.A. **83** [1975] No. 154546).

[5] Vol'fson, A. I.; Ryazanov, A. I.; Chigrinova, G. D. (Zh. Prikl. Khim. **34** [1961] 173/6; J. Appl. Chem. [USSR] **34** [1961] 162/4).

[6] Bell, M. F.; Harrison, J. A. (J. Electroanal. Chem. Interfacial Electrochem. **41** [1973] 15/25).

[7] Sorokin, I. N.; Alekhin, A. P.; Lavrishchev, V. P. (Zh. Fiz. Khim. **48** [1974] 3067/70; Russ. J. Phys. Chem. **48** [1974] 1793/5).

[8] Crosby, J. N.; Harrison, J. A.; Whitfield, T. A. (Electrochim. Acta **26** [1981] 1647/51).

[9] Strel'tsov, E. A.; Pakhomov, V. P.; Lazorenko-Manevich, R. M.; Kulak, A. I. (Elektrokhimiya **19** [1983] 365/8; Soviet Electrochem. **19** [1983] 323/5).

Electrophoretic and Photoelectrochemical Data

The high-voltage electrophoretic separation of chloro complexes of the platinum-group metals, including $[PdCl_4]^{2-}$, in a variety of solvents has been investigated [1].

Principles of the photoreduction of metal-containing ions, including $[PdCl_4]^{2-}$, on a ZnO surface were studied by a photoelectrochemical investigation [2].

References:

[1] Meier, H.; Zimmerhackl, E.; Albrecht, W.; Bösche, D.; Hecker, W.; Menge, P.; Unger, E.; Zeitler, G. (Mikrochim. Acta **1971** 811/20).

[2] Fateev, V. N.; Kondrat'ev, V. A.; Pakhomov, V. P. (Zh. Nauchn. Prikl. Fotogr. Kinematogr. **27** [1982] 284/6).

4.5.2.7 Extraction of $[PdCl_4]^{2-}$

Extraction of $[PdCl_4]^{2-}$ and "$[HPdCl_4]^{-}$" from HCl solutions into trilaurylamine-xylene has been studied [1]; extraction by tetraoctylammonium chloride $(C_8H_{17})_4NCl$ gave $[PdCl_4]^{2-}$ and $[Pd_2Cl_6]^{2-}$ species in $C_2Cl_2H_4$ and $CHCl_3$ [2, 3]. The application of pyrazole and methylpyrazoles to extraction of $[PdCl_4]^{2-}$ from HCl solutions has been studied [4], and likewise the extraction of $[PdCl_4]^-$ with tetraheptylammonium and trilaurylammonium salts has been accomplished [5]. The influence of counter-ion absorption on ion-exchange kinetics by alkylammonium monolayers has been investigated using $[Au(CN)_2]^- - [PdCl_4]^{2-} - RNHCl$ $(R = (n\text{-}C_8H_{17})_3N)$; ion-exchange processes are first order in $[Au(CN)_2]^-$ and $[PdCl_4]^{2-}$ [6].

Both $[PdCl_4]^{2-}$ and $[PdCl_3(H_2O)]^-$ species can be extracted from solutions of $PdCl_2$ in HCl with tri-isobutylphosphine sulphide, and extraction rates and kinetics were measured [7]. Extraction of Pd from solutions of "$H_2[PdCl_4]$" by RMe_3NCl, RMe_2PhCH_2NCl and R_2Me_2NCl $(R = C_{10}$ to C_{16} or C_{17} to $C_{20})$ was compared with that for $H_2[PtCl_6]$, $H_3[RhCl_6]$, $H_3[IrCl_6]$ and $H_2[RuCl_6]$ with a variety of solvents [8]. Absorption parameters for $(Bu_3NH)Cl$ or $(octyl_3NH)_2$-$[PdCl_4]$ from the organic phase at a water-toluene interface were measured and the character of surface pressure isotherms discussed [9], and infrared spectroscopy was used to show that extraction of $[PdCl_4]^{2-}$ and $[PtCl_6]^{2-}$ proceeds via a cation-exchange mechanism [4].

Paper and thin-layer extraction chromatography of chloro complexes of the platinum-group metals, including $[PdCl_4]^{2-}$ and gold have been studied [10]. Flotation of $[PdCl_4]^{2-}$ and $[PtCl_6]^{2-}$ from dilute aqueous solution with organic ionic collectors $(R_4N)^+$ and $(R_3NH)^+$ was studied and a mechanism of flotation suggested [11]. Extraction of chloro complexes of platinum-group metals, including $[PdCl_4]^{2-}$, with $[(C_8H_{17})_4N]ClO_4$ (tetraoctylammonium perchlorate) was studied [12], as was extraction of $[PdCl_4]^{2-}$ by tri-octylammonium salts, $[(C_8H_{17})_3NH]Cl$ [13]; by tetraethyldiamide heptylphosphate [14]; by tetraoctylammonium and other quaternary ammonium bases [15]; and by aliphatic amines [16].

The kinetics of extraction of $[PdCl_4]^{2-}$ by trioctylamine chloride in toluene show the process to be diffusion-controlled, with an effective activation energy of extraction of 16 kJ/mol. The passage of an electrical current through the extraction mixture produces very little change in the extraction rate [17]. Kinetics of Cl^- exchange between $[PdCl_4]^{2-}$ and the chloride form of a strong-base amberlite ion-exchange resin were measured [18]. Crown ethers such as dibenzo-18-crown-6 have been used to extract $[PdCl_4]^{2-}$ from $KCl-PdCl_2$ in $CHCl_3-CH_3CN$ [19].

References:

[1] Zveguintzoff, D.; Gourisse, D. (Bull. Soc. Chim. France **1979** I 179/84).

[2] Vasil'eva, A. A.; Gindin, L. M.; Pelina, G. N.; Mal'chikov, G. D. (Izv. Sibirsk. Otd. Akad. Nauk SSSR Ser. Khim. Nauk **1969** No. 6, pp. 40/6; Sib. Chem. J. **1969** 675/9).

[3] Ivanova, S. N.; Gindin, L. M.; Chernyaeva, A. P.; Chernobrov, A. S.; Shaidurova, N. N. (Izv. Sibirsk. Otd. Akad. Nauk SSSR Ser. Khim. Nauk **1978** No. 6, pp. 68/74; C.A. **90** [1979] No. 110710).

[4] Pronin, V. A.; Usol'tseva, M. V.; Shastina, E. N.; Volkov, A. N.; Seraya, V. I. (Zh. Neorgan. Khim. **19** [1974] 800/2; Russ. J. Inorg. Chem. **19** [1974] 434/5).

[5] Zveguintseff, D.; Gourisse, D.; Kikindai, T. (Bull. Soc. Chim. France **1983** I 14/20).

[6] Popov, A. N.; Timofeeva, S. K.; Kulikova, L. D. (Latvijas PSR Zinatnu Akad. Vestis Khim. Ser. **1982** No. 3, pp. 374/5; C.A. **97** [1982] No. 61877).

[7] Baba, Y.; Ohshima, M.; Inove, K. (Bull. Chem. Soc. Japan **59** [1986] 3829/33).

[8] Borbat, V. F.; Kouba, E. F. (Protsessy Zhidk. Ekstr. Tr. 2nd Vses. Nauchn. Tekhn. Soveshch., Leningrad 1964 [1966], pp. 299/306; C.A. **65** [1964] 11423).

[9] Serga, V.; Kulikova, L. D.; Popov, A. N. (Latvijas PSR Zinatnu Akad. Vestis Kim. Ser. **1986** No. 2, pp. 207/13; C.A. **105** [1986] No. 30560).

[10] Przeszlakowski, S.; Flieger, A. (Chem. Anal. [Warsaw] **22** [1977] 431/44).

[11] Walkowiak, W. (Pr. Nauk Inst. Chem. Nieorg. Met. Pierwiastkow Rzadkich Politech. Wroclaw No. 17 [1973] 433/42; C.A. **81** [1974] No. 28035).

[12] Ivanova, S. N.; Gindin, L. M.; Toloknova, V. N. (Izv. Sibirsk. Otd. Akad. Nauk SSSR Ser. Khim. Nauk **1969** No. 6, pp. 46/50; Sib. Chem. J. **1969** 680/3).

[13] Ivanova, S. N.; Gindin, L. M.; Chernyaeva, A. P.; Chernobrov, A. S.; Shaidurova, N. N. (Izv. Sibirsk. Otd. Akad. Nauk SSSR Ser. Khim. Nauk **1978** No. 6, pp. 68/74; C.A. **90** [1979] No. 110710).

[14] Fadeeva, V. I.; Nasonovskii, I. S.; Minochkina, L. N.; Volynskii, A. B.; Zorov, N. B. (Vestn. Mosk. Univ. Ser. II Khim. 21 No. 1 [1980] 98; C.A. **92** [1980] No. 170088).

[15] Gindin, L. M.; Ivanova, S. N.; Mazurova, A. A.; Vasil'eva, A. A.; Mironova, L. Y.; Sokolov, A. P.; Smirnov, P. P. (Izv. Sibirsk. Otd. Akad. Nauk SSSR Ser. Khim. Nauk **1967** No. 1, pp. 89/96; Sib. Chem. J. **1967** 66/71).

[16] Raber, H.; Soraya, B. (Mikrochim. Acta **1975** II 37/44).

[17] Sadyrbaeva, T. Z.; Kulikova, L. D.; Pichugin, A. A. (Latvijas PSR Zinatnu Akad. Vestis Kim. Ser. **1986** No. 6, pp. 682/8; C.A. **106** [1987] No. 73778).

[18] Nativ, M.; Goldstein, S.; Schmuckler, G. (J. Inorg. Nucl. Chem. **37** [1975] 1951/6).

[19] Poladyan, V. E.; Burtnenko, L. M.; Avlasovich, L. M.; Andrianov, A. M. (Zh. Neorgan. Khim. **32** [1987] 737/40; Russ. J. Inorg. Chem. **32** [1987] 413/5).

Ion Exchange and Adsorption of [PdCl$_4$]$^{2-}$

A general discussion of exchange of halogeno ions including [PdCl$_4$]$^{2-}$ on anion exchangers has been given [1]. Gel-permeation chromatography of a number of complex ions, including [PdCl$_4$]$^{2-}$, on biogels has been investigated [2]. The adsorption from solutions of [PdCl$_4$]$^{2-}$ in HCl on hydrated ZrO$_2$ has been studied [3, 4]. In treatment of Al$_2$O$_3$ by "H$_2$[PdCl$_4$]" solutions, a uniform Pd adlayer is formed on some of the Al$_2$O$_3$ particles, and this Pd coating brings about a decrease in electron-positron annihilation [5]. There are also other studies on [PdCl$_4$]$^{2-}$–Al$_2$O$_3$ sorption [6,7].

Anion-exchange sorption of [PdCl$_4$]$^{2-}$ and [PtCl$_6$]$^{2-}$ have been compared as a function of the nature of the solvent, the resin and the acid [8]. Conditions for efficient adsorption of Pd and Pt as chloro complexes from industrial solutions on the ion-exchange resin EDE-IOP were investigated [9], and the behaviour of platinum-group chloro complexes, including [PdCl$_4$]$^{2-}$, towards amberlite [10, 11] and TEAE-cellulose [12] resins discussed. The sorption of [PdCl$_4$]$^{2-}$ on modified S-containing polyvinylalcohol fibres was investigated [13], as was sorption of chloro complexes of the platinum-group metals (including [PdCl$_4$]$^{2-}$) from 2M HCl on ion-exchange resins containing pyridine or ammonium ionogenic groups [14]. The anion-exchange resins EDE-IOP and AN-ZF for the separation of Pd and Pt have been assessed [15]. Sorption of a variety of chloro complexes of the platinum-group metals including [PdCl$_4$]$^{2-}$ on strongly basic anion exchangers was investigated [16]. The use of [N(CH$_3$)$_3$]$^+$Cl$^-$ groups in Dowex and Amberlite resins to support [PdCl$_4$]$^{2-}$ anions was studied, and infrared spectra of polymer-supported [PdCl$_4$]$^{2-}$ (as polymer [N(CH$_3$)$_3$]$_2$[PdCl$_4$]) measured (230 to 430 cm^{-1}, reproduced in paper [17]). Adsorption of [PdCl$_4$]$^{2-}$ from "H$_2$[PdCl$_4$]" on Al$_2$O$_3$ has been measured in competition with other adsorbates [18]. Adsorption of unspecified Pd chloro complexes on Al$_2$O$_3$ has been measured as a function of time of contact [19].

References:

[1] Knothe, M. (Z. Anorg. Allgem. Chem. **463** [1980] 204/12).

[2] Limoni, B.; Schmuckler, G. (J. Chromatog. **135** [1977] 173/82).

[3] Abovskaya, N. V.; Boichinova, E. S.; Simonova, S. A.; Pak, V. N. (Zh. Prikl. Khim. **59** [1986] 278/83; J. Appl. Chem. [USSR] **59** [1986] 252/6).

[4] Boichinova, E. S.; Kryuchkova, N. Ya.; Brynzova, E. D.; Simanova, S. A.; Abovskaya, N. V. (Zh. Prikl. Khim. **57** [1984] 1217/2; J. Appl. Chem. [USSR] **57** [1984] 1127/30).

[5] Dekhtar, I. Ya.; Sagov, Yu. M.; Fedchenko, R. G. (Khim. Fiz. **1983** 1573/6; C. A. **100** [1984] No. 13164).

[6] Zhou, Z.; Mizuno, K.; Suzuki, M. (Ranliao Huaxue Xuebao 11 [1983] 69/74; C. A. **100** [1984] No. 57360).

[7] Ananin, Y. N.; Trokhimetz, A. I. (React. Kinet. Catal. Letters **27** [1985] 129/32).

[8] Popov, A. N.; Kononov, Yu. S.; Gorbachev, V. M. (Izv. Sibirsk. Otd. Akad. Nauk SSSR Ser. Khim. Nauk **1965** No. 3, pp. 141/4; C. A. **64** [1965] 16683).

[9] Vasilev, K.; Petkova, E.; Mekhandzhiev, M.; Chimbulev, M.; Neikova, E. (God. Khim. Tekhnol. Inst. 11 [1964] 199/205; C. A. **65** [1966] 13229).

[10] MacNevin, W. M.; Crummett, W. B. (Anal. Chem. **25** [1953] 1628/30).

[11] Berman, S. S.; McBryde, W. A. E. (Can. J. Chem. **36** [1958] 845/52).

[12] Kuroda, R.; Takahashi, T.; Oguma, K. (Bull. Chem. Soc. Japan **49** [1976] 815/6).

[13] Simanova, S. A.; Bobritskaya, L. S.; Kukushkin, Yu. N.; Kolontarov, I. Ya.; Kuznetsova, G. V. (Ionnyi Obmen Khromatogr. **1984** 43/5; C.A. **103** [1985] No. 11976).

[14] Kukushkin, Yu. N.; Paramonova, V. I.; Simanova, S. A.; Kopalkin, Yu. A.; Popik, V. P.; Mel'gunova, L. G.; Gur'yanova, G. P. (Zh. Prikl. Khim. **47** [1974] 554/9; J. Appl. Chem. [USSR] **47** [1974] 556/60).

[15] Korobkin, A. A.; Plaksin, I. N. (Nauchn. Dokl. Vysshei Shkoly Met. **1959** No. 1, pp. 14/8; C.A. **1960** 16988).

[16] Blasius, E.; Wachtel, U. (Z. Anal. Chem. **142** [1954] 341/56).

[17] Acocdedji, E.; Leito, J.; Aune, J. P. (J. Mol. Catal. **22** [1983] 73/8).

[18] Guo, Z.; Zhang, Z. (Ying-Yong Huaxue **4** [1982] 31/5; C.A. **107** [1987] No. 65355).

[19] Kravchuk, L. S.; Kozlov, N. S.; Sinvakova, S. V.; Yankova, T. V. (Zh. Fiz. Khim. **57** [1983] 2794/7; Russ. J. Phys. Chem. **57** [1983] 1686/8).

4.5.2.8 Chemical Reactions

As mentioned earlier, $[PdCl_4]^{2-}$ is a useful starting material for many other palladium complexes, but it also has considerable catalytic uses; these include olefin oxidation and oligomerisation, oligomerisation of dienes and acetylenes, vinylation, acetoxylation, halogenation, arene coupling and carbonylation. Here we give a selection only of its reactions with a variety of ligands.

For general reviews including much of the reaction chemistry of $[PdCl_4]^{2-}$ see [1 to 4], and for the vast amount of catalytic work see [1 to 6]; see also "Platinum" Suppl. Vol. A 1, 1986, pp. 299/308.

References:

[1] Maitlis, P. M. (The Organic Chemistry of Palladium, Metal Complexes, Vol. I, Academic, London 1971).

[2] Maitlis, P. M. (The Organic Chemistry of Palladium, Catalytic Reactions, Vol. II, Academic, London 1971).

[3] Heck, R. F. (Palladium Reagents in Organic Syntheses, Academic, London 1985).

[4] Maitlis, P. M.; Espinet, P.; Russell, M. J. H. (in: Wilkinson, G.; Stone, F. G. A.; Abel, E. W., Comprehensive Organometallic Chemistry, Palladium Introduction and General Principles, Vol. 6, Pergamon, Oxford 1982, pp. 233/42).

[5] Tsuji, J. (Organic Synthesis by Means of Transition Metal Complexes, Springer, Berlin 1975).

[6] Colqohoun, H. M. (Chem. Ind. [London] **1987** 612/7).

Reactions with Hydrogen

Reduction kinetics of $[PdCl_4]^{2-}$ and other Pd[II] complexes by H_2 were studied as a function of acidity, concentration and temperature; an unstable palladium(II) hydride, $[PdHCl_3]^-$, may be involved [1]. Isotope effects (using D_2O) on the catalytic activation of molecular H_2 by metal ions including $[PdCl_4]^{2-}$ were evaluated [2]. Calculations of kinetic parameters based on experimental curves have been made using the $[PdCl_4]^{2-}-H_2$ reaction as an example [3].

References:

[1] Fasman, A. B.; Golodov, V. A.; Pustyl'nikov, L. M.; Luk'yanov, A. T.; Baranovskii, B. P.; Darinskii, Y. V. (Izv. Sibirsk. Otd. Akad. Nauk SSSR Ser. Khim. Nauk **1968** No. 3, pp. 144/8; Sib. Chem. J. **1968** 352/5).

[2] Harrod, J. F.; Halpern, J. (J. Phys. Chem. **65** [1961] 563/4).

[3] Pustyl'nikov, L. M.; Golodov, V. A.; Luk'yanov, A. T.; Fasman, A. B.; Baranovskii, B. P. (Zh. Fiz. Khim. **44** [1970] 2392/4; Russ. J. Phys. Chem. **44** [1970] 1354/6).

Reactions with Nitrogen Donors

Monodentate Ligands. In buffered solution $[PdCl_4]^{2-}$ reacts with ammonia to give mixtures of $Pd(NH_3)_2Cl_2$, $[Pd(NH_3)_3Cl]^+$ and $[Pd(NH_3)_4]^{2+}$; kinetics of these reactions were studied [1].

With acetamide (L), paramagnetic complexes $Pd_4(C_2H_4ON)_7(OH)_2$ are formed from $[PdCl_4]^{2-}$ [2], while 2-octylaminopyridine (L') gives $PdCl_2L'_2$ [3] and α-hydroxylamino oximes (L'') give $PdL''Cl_4$ [4]. Reaction of 4-methylpyrimidine (L) with $[PdCl_4]^{2-}$ gives the N-bonded complex trans-$PdCl_2L_2$ as well as a σ-metallated species $Pd_2L(L-H)Cl_3$ [5].

With theophylline, cytidine and 5-azacytosine (L), $PdCl_2L_2$ is formed [6], and the nucleoside complexes $PdCl_2L'_2$ (L' = inosine, guanosine) obtained from "$H_2[PdCl_4]$" in HCl with the nucleoside [7].

Polydentate Ligands. Reaction of $[PdCl_4]^{2-}$ with 1,10-phenanthroline (phen) gives $PdCl_2phen$; the kinetics of this reaction were measured [8, 9], as were the kinetics of the $[PdCl_4]^{2-}$-ethylenediamine (en) reaction to yield $PdCl_2(en)$ and $[Pden_2]^{2+}$ [10, 11]. Production of the latter from the $[PdCl_4]^{2-}$–en reaction together with formation of $[Pden_2][PdCl_4]$ was noted, and reaction of 2,2'-bipyridyl (bipy) gave $[Pdbipy_2][PdCl_4]$ and $PdbipyCl_2$ [12]. With the dipyrromethane ligands 4,4'-bis(ethoxycarbonyl)-3,3',5,5'-tetramethyldipyrromethene and 3,4'-bis(ethoxycarbonyl)-5-chloro-3',4,5'-trimethyldipyrromethene (L), complexes of the form $Pd_2Cl_2L_2$ are formed [13]. With amines of the form $ArCH(R^2)NHR^1$ (L), $[PdCl_4]^{2-}$ gives trans-$PdCl_2L_2$ [14], while kinetics of the reaction between $[PdCl_4]^{2-}$ and N- and N,N'-methylethylenediamines (L) have been investigated in aqueous acid; the product is $PdCl_2L$ [15].

Kinetics of the reaction of $[PdCl_4]^{2-}$ and $[PtCl_4]^{2-}$ with 2,2'-bipyridyl (bipy) to give MCl_2L (M = Pd, Pt) have been studied [16]. The N-methyl-2,2'-bipyridinium ion $(bipyMe)^+$ reacts with $[PdCl_4]^{2-}$ to give $Pd(bipyMe)Cl_3$ [17], while orotic, 3-methylorotic, 5-nitroorotic acids and orotic acid methyl ester (L) all give $K_4[PdL_2]$ with $[PdCl_4]^{2-}$ in base [18]. Histidine (HL) behaves as a bidentate ligand in $Pd(HL)Cl_2$, which is made from $[Pd(HL)_2]Cl_2$ and $[PdCl_4]^{2-}$ [19]. Reaction of 8-aminoquinoline, 5-nitro-1,10-phenanthroline, o-phenylenediamine and 1,8-naphthalenediamine (L) with $[PdCl_4]^{2-}$ gives $PdCl_2L$ [20]. The bidentate 3,3'-dicarbomethoxy-2,2'-bipyridyl ligand L' gives $PdCl_2L'$ with $[PdCl_4]^{2-}$ [21].

The N,N,N',N'-tetramethyl-o-phenylenediamine ligand o-$(Me_2N)_2C_6H_4$ (L) gives the very stable $PdLCl_2$ with $[PdCl_4]^{2-}$ [22].

Alkylated polyamines such as $N(CH_2CH_2NMe_2)_3$ (L) and $Me_2NC_2H_4NMeC_2H_4NMeC_2H_4NMe_2$ (L') react with "$H_2[PdCl_4]$", followed by treatment with base to give $[PdClL]Cl$ and $[PdL']Cl_2$, respectively [23].

References:

[1] Reinhardt, R. A.; Monk, W. W. (Inorg. Chem. **9** [1970] 2026/30).

[2] Durand, S.; Jugie, G.; Laurent, J. P. (Transition Metal Chem. **7** [1982] 310/2).

[3] Borshch, N. A.; Petrukhin, O. M.; Zolotov, Yu. A.; Zhumadilov, E. G.; Nefedov, V. I.; Sokolov, A. B.; Marov, L. N. (Koord. Khim. **7** [1981] 1242/9; Soviet J. Coord. Chem. **7** [1981] 616/23).

[4] Naumova, K. P.; Stetsenko, A. I.; Volodarskii, L. B. (Zh. Obshch. Khim. **49** [1979] 2564/9; J. Gen. Chem. [USSR] **49** [1979] 2268/72).

[5] Krylova, L. F.; Luk'yanova, I. G.; Dikanskaya, L. D.; Podoplelov, A. V.; Dubovemko, Z. D. (Koord. Khim. **12** [1986] 1117/21; Soviet J. Coord. Chem. **12** [1986] 645/9).
[6] Umapathy, P.; Harnesswala, R. A.; Dorai, C. S. (Polyhedron **4** [1985] 1595/602).
[7] Hadjilladis, N. (Inorg. Syn. **23** [1985] 51/3).
[8] Rund, J. V. (Inorg. Chem. **13** [1974] 738/40).
[9] Rund, J. V. (Inorg. Chem. **9** [1970] 1211/5).
[10] de Waal, D. J. A.; Robb, W. (Intern. J. Chem. Kinet. **6** [1974] 309/21).

[11] Roulet, R.; Ernst, R. (Helv. Chim. Acta **54** [1971] 2357/62).
[12] McCormick, B. J.; Jayes, E. N.; Kaplan, R. I. (Inorg. Syn. **13** [1972] 216/8).
[13] March, F. C.; Fergusson, J. E.; Robinson, W. T. (J. Chem. Soc. Dalton Trans. **1972** 2069/76).
[14] Dunina, V. V.; Zalevskaya, O. A.; Smolyakova, I. P.; Potapov, V. M.; Kuz'mina, L. G.; Struchkov, Yu. T.; Reshetova, L. N. (Zh. Obshch. Khim. **56** [1986] 1164/75; J. Gen. Chem. [USSR] **56** [1986] 1023/33).
[15] de Waal, D. J. A.; Robb, W. (Intern. J. Chem. Kinet. **6** [1974] 323/36).
[16] Schwab, D. E.; Rund, J. V. (Inorg. Chem. **11** [1972] 499/503).
[17] Dholakia, S.; Gillard, R. D.; Wimmer, F. L. (Inorg. Chim. Acta **69** [1983] 179/81).
[18] Arrizabalaga, P.; Castan, P.; Laurent, J.-P. (J. Inorg. Biochem. **20** [1984] 215/24).
[19] Chernova, N. N.; Konovalov, L. V. (Zh. Neorgan. Khim. **32** [1987] 722/7; Russ. J. Inorg. Chem. **32** [1987] 404/7).
[20] Umapathy, P.; Harnesswala, R. A.; Dorai, C. S. (Polyhedron **4** [1985] 1595/602).

[21] Dholakia, S.; Gillard, R. D.; Wimmer, F. L. (Polyhedron **4** [1985] 791/5).
[22] Stewart, F. H. C. (Chem. Ind. [London] **1958** 264).
[23] Walker, F. S.; Bhattacharya, S. N.; Senoff, C. V. (Inorg. Syn. **21** [1982] 129/30).

Reactions with Oxygen Donors

Reactions with O-Donors. Benzoxazole and 2-methylbenzoxazole (L) with $[PdCl_4]^{2-}$ yield $PdCl_2L_2$ [1], while isoxazole and 3,5-dimethylisoxazole also give $PdCl_2L_2$ [2]. With K_2H_2L (H_4L = trans-cyclohexanediaminetetracetic acid) and caesium ions, $Pd(H_2L) \cdot H_2O$, and $Cs[Pd(HL)] \cdot H_2O$ can be obtained [3]. The catalytic decomposition of H_2O_2 has been studied with a variety of Pd complexes including $[PdCl_4]^{2-}$ and activation energies determined for the reactions [4]. Reaction of $[PdCl_4]^{2-}$ with caffeine (L) yields $K[PdLCl_3]$ [5]. Kinetics of formation of $[PdLCl_3]^-$ from $[PdCl_4]^{2-}$ and a series of amines L (L = 3- and 4-chloropyridines, 2-, 3- and 4-methylpyridines, pyridine, morpholine, diethylamine, 2,4- and 2,6-dimethylpyridine, 2,3,6- and 2,4,6-trimethylpyridine were measured and compared [6].

With the N–O donor carbohydrazide (L), $[PdCl_4]^{2-}$ yields $PdLCl_2$ and $Pd(LH)_2$ [7], while N-arylglycines (2,3- and 2,4-dimethylphenyl-glycine, 2,6-dimethylphenyl(carboxymethyl)-glycine and 4-ethylphenyl(carboxymethyl)glycine (LH)) with $[PdCl_4]^{2-}$ yield PdL_2 complexes [8].

Hydrolysis and Aquation. The hydrolysis of $[PdCl_4]^{2-}$ by alkali has been followed spectrophotometrically (spectra 200 to 500 nm reproduced in paper) and the hydrolysis constant of $[PdCl_4]^{2-}$ evaluated as approximately 5×10^5 [9]. Reaction of $[PdCl_4]^{2-}$ with 2M OH^- yields $[Pd(OH)_4]^{2-}$ [10]. Rates of hydrolysis of $[PdCl_4]^{2-}$, $[PdBr_4]^{2-}$ and the corresponding platinum complexes have been compared; the change in pH of a solution of $[PdCl_4]^{2-}$ was measured against time [11] as were spectra and pH of a solution of $[PdCl_4]^{2-}$ [12]. Effects of Cl^-, pH and ageing on the precipitation of "$Pd(OH)_2$" from $[PdCl_4]^{2-}$ have been studied [13]. With aqueous mercuric perchlorate, $[PdCl_4]^{2-}$ gives $[Pd(H_2O)_4]^{2+}$ [14]. Aquation and hydrolysis of a number of

Pd^{II} species including $[PdCl_4]^{2-}$ were investigated by electrophoretic methods [15]. With aqueous mercuric perchlorate, $[PdCl_4]^{2-}$ gives $[Pd(H_2O)_4]^{2+}$ [19].

Kinetics of the aquation reaction $[PdCl_4]^{2-} \rightleftharpoons [PdCl_3(H_2O)]^- + Cl^-$ and the reverse anation reaction in 2M $HClO_4$ in total ionic strength 2.1M at 15°C have been studied by stopped-flow methods [16]; at 18 to 25°C using a temperature-jump method, thermodynamic data for the reaction were obtained [17]; from 13 to 35°C by stopped-flow methods [18]; by stopped-flow and temperature-jump methods [19]. Data from [16, 19] were critically assessed [20, 21].

Thermodynamic data for the aquation reaction were calculated [17, 18]; see p. 172.

References:

[1] Massacesi, M.; Pinna, R.; Biddau, M.; Ponticelli, G.; Zakharova, I. A. (Inorg. Chim. Acta **80** [1983] 151/5).

[2] Pinna, R.; Ponticelli, G.; Preti, C. (J. Inorg. Nucl. Chem. **37** [1975] 1681/4).

[3] Ezerskaya, N. A.; Solovykh, T. P.; Shubochkin, L. H. (Koord. Khim. **6** [1980] 536/41; Soviet J. Coord. Chem. **6** [1980] 536/41).

[4] Grinberg, A. A.; Kukushkin, Yu. N.; Vlasova, R. A. (Zh. Neorgan. Khim. **13** [1968] 2177/83; Russ. J. Inorg. Chem. **13** [1968] 1126/9).

[5] Pneumatikakis, G. (Inorg. Chim. Acta **93** [1984] 5/11).

[6] Cattalini, L.; Orio, A.; Martelli, M. (Chim. Ind. [Milan] **49** [1967] 625/8).

[7] Ivanova, M. G.; Kalinichenko, L. I. (Zh. Neorgan. Khim. **29** [1984] 1237/40; Russ. J. Inorg. Chem. **29** [1984] 708/10).

[8] Rodriguez, E. C.; Peregrin, J. M. S.; Varela, J. S.; Roson, J. C. A. (Syn. React. Inorg. Metal-Org. Chem. **15** [1985] 681/98, 685).

[9] Kazakova, V. I.; Psitsyn, B. V. (Zh. Neorgan. Khim. **12** [1967] 620/5; Russ. J. Inorg. Chem. **12** [1967] 323/6).

[10] Bardin, M. B.; Ketrush, P. M. (Teor. Prakt. Polyarogr. Metodov. Anal. **1973** 62/8; C.A. **83** [1975] No. 21648).

[11] Grinberg, A. A.; Nikol'skaya, L. E.; Kiseleva, N. V. (Zh. Neorgan. Khim. **1** [1956] 220/4; Russ. J. Inorg. Chem. **1** No. 2 [1956] 31/5).

[12] Grinberg, A. A.; Kiseleva, N. V. (Zh. Neorgan. Khim. **3** [1958] 1804/9; Russ. J. Inorg. Chem. **3** No. 8 [1958] 119/26).

[13] Chueh, C.-H.; Yung, T.-L. (Huaxue Tongbao **11** [1963] 49/54; C.A. **60** [1964] 12652).

[14] Jørgensen, C. K. (Absorption Spectra and Chemical Bonding in Complexes, Pergamon, Oxford 1962, p. 259).

[15] Davydova, I. Yu.; Ermakov, A. N.; Korchemnaya, E. K. (Tezisy Dokl. 10th Vses. Soveshch. Khim. Anal. Tekhnol. Blagorodn. Met., Novosibirsk 1976, Vol. 1, p. 24; C.A. **90** [1979] No. 62019).

[16] Bekker, P. van Z.; Robb, W. (J. Inorg. Nucl. Chem. **37** [1975] 829/34).

[17] Vargaftik, M. N.; Igoshin, V. A.; Syrkin, Ya. K. (Izv. Akad. Nauk SSSR Ser. Khim. **1972** 1426/8; Bull. Acad. Sci. USSR Div. Chem. Sci. **1972** 1380/2).

[18] Elding, L. I. (Inorg. Chim. Acta **6** [1972] 683/8).

[19] Pearson, R. G.; Haynes, M. J. (Kgl. Tek. Hoegsk. Handl. No. 285 [1972] 1461/9).

[20] Elding, L. I. (Inorg. Chim. Acta **7** [1973] 581/8).

[21] Elding, L. I. (Inorg. Chim. Acta **15** [1973] L9/L11).

Reactions with Halogen Donors

Reaction of $[PdCl_4]^{2-}$ with Cl_2 gives $[PdCl_6]^{2-}$; the mechanism was studied [1] and kinetics of the Cl^-–$[PdCl_4]^{2-}$ exchange interaction on a strong-base anion exchange resin investigated [2].

With Br^-, $[PdCl_4]^{2-}$ gives $[PdCl_nBr_{4-n}]^{2-}$; formation constants were determined [3, 4]; see also p. 203 for reactions of the $[PdCl_4]^{2-}$–Br^- system.

Formation constants for the intermediate species $[PdCl_nI_{4-n}]^{2-}$ were determined [5]; see also p. 215 for the $[PdCl_4]^{2-}$–I system.

References:

[1] Mureinik, R. J.; Pross, E. (J. Coord. Chem. **8** [1978] 127/33).
[2] Nativ, M.; Goldstein, S.; Schmuckler, G. (J. Inorg. Nucl. Chem. **37** [1975] 1951/6).
[3] Srivastava, S. C.; Newman, L. (Inorg. Chem. **5** [1966] 1506/10).
[4] Klotz, P.; Feldburg, S.; Newman, L. (Inorg. Chem. **12** [1973] 164/8).
[5] Srivastava, S. C.; Newman, L. (Inorg. Chem. **11** [1972] 2855/9).

Reactions with Sulphur Donors

With thioproline (L), $[PdCl_4]^{2-}$ gives $PdCl_2L_2$ [1]. Quantitative precipitation of Pd from other elements can be achieved by using controlled amounts of S^{2-} on chloro complexes of Au, Pd, Ir, Rh and Pt, the Pd being present as $[PdCl_4]^{2-}$ (misprinted as $[PdCl_4]^-$ in abstract) [2]. Reaction of $[PdCl_4]^{2-}$ in tetrahydrofuran with $S(OPr^i)_2$ (L) gives PdL_2Cl_2 [3]. Rates and activation energy parameters for reaction of $[PdCl_4]^{2-}$ with thiourea [4] and N-alkyl substituted thioureas (L) to give $PdCl_2L_2$ and $[PdL_4]Cl_2$ have been measured [5]; with thiourea (L) itself, mixed complexes $[PdCl_nL_{4-n}]^{2-n}$ are formed [6]. With hydrogenated thiamines (L), $[PdCl_4]^{2-}$ yields $PdCl_2L_2$ [7]. Oxythiamine gives a 1:1 complex with $[PdCl_4]^{2-}$ [8]. Sulphaguanidine (L) gives $PdCl_2L_2$ according to spectrophotometric measurements; formation constants were determined [9]. With diphenyl disulphide the polymeric $[Pd(SC_6H_5)Cl]_n$ is formed [10], while ethyl cysteinate gave $Pd(S_2C_{10}H_{20}O_4N_2)_2$ [11]. Glutathione (H_3G) reacts with $[PdCl_4]^{2-}$ to give $[Pd(H_2G)Cl_{4-n}]^{3-n}$; equilibrium constant data were obtained for this [12].

Reaction of $[PdCl_4]^{2-}$ with o-methyl mercaptobenzoic acid (LH) gives PdL_2 and $PdCl_2(LH)_2$ [13]. The complexes PdL_2 and $Pd_3L_4(LH)Cl_2 \cdot 4H_2O$ were obtained by reaction of $[PdCl_4]^{2-}$ with 6-mercaptopurine (L) and 2-amino-6-mercaptopurine (thioguanine) (L'), respectively [14]. Dimethyl, tetramethylene- and phenylmethylsulphoxides RR^1SO react with $[PdCl_4]^{2-}$; the kinetics of these reactions were measured, and for Me_2SO the existence of the species $[Pd(Me_2SO)Cl_3]^-$ and trans-$Pd(Me_2SO)_2Cl_2$ postulated [15]. With arenesulphinic acids, $[PdCl_4]^{2-}$ is reduced to Pd, SO_2 evolved and the diaryls formed [16].

The potentially tetradentate ligand cis-1,2-bis(o-methylthiophenylthio)ethane (L) gave Pd_2LCl_4 with $[PdCl_4]^{2-}$ [17], while the dithioether 1,12-bis(phenylthio)dodecane (dpd) reacts with $[PdCl_4]^{2-}$ in C_2H_5OH–CH_2Cl_2 to give trans-$PdCl_2(dpd)$ [18]. The sulphur-nitrogen donor 3-(mercaptomethyl)piperidine (HL) gives $Pd(HL)Cl_2$ with $[PdCl_4]^{2-}$ in which the zwitterionic ligand is probably bound via S and N donor centres [19]. The thiocyanate ion gives $[Pd(SCN)_4]^{2-}$ [20].

References:

[1] Craciunescu, D. G.; Doadrio, A.; Furlani, A.; Scarcia, V. (Inorg. Chim. Acta **67** [1982] L11/L13).
[2] Cheng, J.; Yang, Z.-F. (Ziran Zazhi **3** [1980] 558/9; C.A. **93** [1980] No. 210919).

110

[3] Wenschuh, E.; Schleif, P.; Kolbe, A. (Z. Chem. [Leipzig] **19** [1979] 414).
[4] Bekker, P. van Z.; Robb, W. (Intern. J. Chem. Kinet. **7** [1975] 87/98).
[5] Bekker, P. van Z. (Inorg. Chim. Acta **31** [1978] 109/15).
[6] Shlenskaya, V. I.; Biryukov, A. A.; Moskovkina, E. M. (Zh. Neorgan. Khim. **11** [1966] 600/5; Russ. J. Inorg. Chem. **11** [1966] 325/8).
[7] Hadjilladis, N.; Markopoulos, J. (J. Chem. Soc. Dalton Trans. **1981** 1635/44).
[8] Adeyemo, A.; Shamim, A.; Turner, A. (Inorg. Chim. Acta **78** [1983] L23/L24).
[9] Rittner, W. F.; Gulko, A.; Schmuckler, G. (Talanta **17** [1970] 807/16).
[10] Boschi, T.; Crociani, B.; Toniolo, L.; Bellucco, U. (Inorg. Chem. **9** [1970] 532/7).

[11] Hay, R. W.; Banerjee, P. (Inorg. Chim. Acta **44** [1980] L205/L207).
[12] Zegzhda, G. D.; Zegzhda, T. V. (Zh. Neorgan. Khim. **23** [1978] 3293/6; Russ. J. Inorg. Chem. **23** [1978] 1826/8).
[13] Livingstone, S. E.; Plowman, R. A.; Sorensen, S. (J. Proc. Roy. Soc. New South Wales **84** [1950/51] 28/9).
[14] Das, M.; Livingstone, S. E. (Brit. J. Cancer **38** [1978] 325/8).
[15] Canovese, L.; Cattalini, L.; Marangoni, G.; Michelon, G.; Nicolini, M.; Tobe, M. L. (J. Coord. Chem. **12** [1982] 63/9).
[16] Garves, K. (J. Org. Chem. **35** [1970] 3273/5).
[17] McAuliffe, C. A.; Murray, S. G. (Inorg. Nucl. Chem. Letters **12** [1976] 897/8).
[18] McAuliffe, C. A.; Soutter, H. E.; Levason, W.; Hartley, F. R.; Murray, S. G. (J. Organometal. Chem. **159** [1978] C25/C28).
[19] Sola, J.; Yanez, R. (J. Chem. Soc. Dalton Trans. **1986** 2021/4).
[20] Yamada, S.; Tsuchida, R. (Bull. Chem. Soc. Japan **26** [1953] 489/93).

Reactions with Selenium and Tellurium Donors

Activation energy parameters for reaction of $[PdCl_4]^{2-}$ in HCl with selenourea (L) to give $PdCl_2L_2$ were measured and the mechanism of the reaction studied [1]. The tetradentate selenoether 1,3-bis(methylselenoethylseleno)propane (L) reacts with $[PdCl_4]^{2-}$ to give $Pd_2L_4Cl_2$ [2]. Reaction between $[PdCl_4]^{2-}$ and both 2,5-diselena-3,3,4,4-tetrafluorohexane and 2,5-diselena-1,1,1,6,6,6-hexafluorohexane gives 1:1 complexes [3].

Reaction of $TeEt_2$ with $[PdCl_4]^{2-}$ yields $PdCl_2(TeEt_2)_2$ [4].

References:

[1] Bekker, P. van Z.; Robb, W. (Intern. J. Chem. Kinet. **7** [1975] 87/98).
[2] Levason, W.; McAuliffe, C. A.; Murray, S. G. (J. Chem. Soc. Dalton Trans. **1976** 269/71).
[3] Bhasin, K. K.; Cross, R. J.; Rycroft, D. S.; Sharp, D. W. A. (J. Fluorine Chem. **14** [1979] 171/6).
[4] Chatt, J.; Venanzi, L. M. (J. Chem. Soc. **1957** 2351/6).

Reactions with Carbon Donors

With CO, CN and Isonitriles. Such reactions have been reviewed [1]. It is likely that palladium(II) mono- and dicarbonyl species are intermediates in the reduction of CO by "$H_2[PdCl_4]$" in water-dioxan from 5 to 60°C; kinetic measurements were carried out [2]. The effect of addition of Rh^{III}, Pt^{IV}, Os^{IV}, Ir^{IV} and Ru^{IV} complexes on the induction period of this reduction of CO by $[PdCl_4]^{2-}$ has been studied [3]. Activation of CO and its effect on the kinetics of reduction of p-benzoquinone by CO has been studied [4, 5]. The effect of O_2 on the reduction of CO by $[PdCl_4]^{2-}$ has been investigated [6]. In the catalytic reduction of Fe^{III} salts by

CO in the presence of aqueous $[PdCl_4]^{2-}$ a palladium(II) carbonyl chloro complex is thought to be involved [7].

Reaction of $[PdCl_4]^{2-}$ in aqueous solution with KCN yields $Pd(CN)_2$, redissolving in excess gives $[Pd(CN)_4]^{2-}$ [8], while hydroxocobalamin can be made from cyanocobalamin and $[PdCl_4]^{2-}$ in solution [9].

References:

[1] Maitlis, P. M.; Espinet, P.; Russell, M. J. H. (in: Wilkinson, G.; Stone, F. G. A.; Abel, E. W., Comprehensive Organometallic Chemistry, Compounds with Palladium-Carbon σ-Bonds, Vol. 6, Pergamon, Oxford 1982, pp. 279/349).

[2] Golodov, V. A.; Kutyukov, G. G.; Fasman, A. B.; Sokol'skii, D. V. (Zh. Neorgan. Khim. **9** [1964] 2319/24; Russ. J. Inorg. Chem. **9** [1964] 1257/9).

[3] Spitzyn, V. I.; Fedoseev, I. V.; Znamenskii, I. V. (Dokl. Akad. Nauk SSSR **238** [1978] 1402/3; Dokl. Phys. Chem. Proc. Acad. Sci. USSR **238/243** [1970] 192/3).

[4] Markov, V. D.; Golodov, V. A.; Kutyukov, G. G.; Fasman, A. B. (Zh. Fiz. Khim. **40** [1966] 1527/33; Russ. J. Phys. Chem. **40** [1966] 827/30).

[5] Fasman, A. B.; Golodov, V. A. (Kinetika Kataliz **6** [1965] 956/7; Kinet. Catal. [USSR] **6** [1965] 868).

[6] Golodov, V.; Kuksenko, E. L. (C$_1$ Mol. Chem. **1** [1984] 109/13).

[7] Golodov, V. A.; Glubokovskikh, N. G.; Taneeva, G. V. (React. Kinet. Catal. Letters **22** [1983] 101/5).

[8] Bigelow, J. H. (Inorg. Syn. **2** [1946] 245/6).

[9] Bieganowski, R.; Kiar, G. (Chem. Zeitung **106** [1982] 235).

With Unsaturated Organic Substrates. The subject has been reviewed for complexes with mono-olefins [1], dienes [2], π-allylic complexes [3], acetylenes [1, 4], arenes and cyclopenta-dienes [5].

The reaction of $[PdCl_4]^{2-}$ with olefins to give aldehydes is of great industrial importance; for example the Wacker process is a catalytic method for the oxidation of ethylene to acetalde-hyde using air and copper(II) chloride [6]. Thus the study of $[PdCl_4]^{2-}$ multiple bond reactions is of considerable importance, and only a very brief selection is offered here.

The kinetics of oxidation of olefins by $[PdCl_4]^{2-}$ have been measured, and formation of intermediate Pd^{II} olefin complexes postulated [7]. The effect of temperature on the equilibrium of π-complex formation between $[PdCl_4]^{2-}$ and ethylene, propylene and 1-butene in water was determined [8, 9]. The interaction of $[PdCl_4]^{2-}$ with ethylene is believed to proceed via $[PdCl_3(C_2H_4)]^-$ and $PdCl_2(C_2H_4)(H_2O)$; by using a gas-liquid reactor of high efficiency the conditions under which acetaldehyde was produced by the C_2H_4-$[PdCl_4]^{2-}$ reaction were identified [10]. Kinetics of oxidation of ethylene and propylene by $[PdX_4]^{2-}$ (X = Cl, Br, I) were also studied [11]. Formation of olefin complexes from $[PdCl_4]^{2-}$ and 1-pentene, 1-hexene, 1-heptene, 1-octene and 1-nonene in acetic acid was investigated [12].

The mechanism of hydroxypalladation by $[PdCl_4]^{2-}$ and $[PdCl_3(H_2O)]^-$ has been discussed: cyclic dienes react much less quickly than do mono-olefins or acyclic dienes [13].

Reaction of "$H_2[PdCl_4]$" with 1,5-cyclo-octadiene (COD) gives $Pd(COD)Cl_2$ [14], while with 2-methyl-2-pentene (L) the dimeric π-allylic complex $L_2Pd_2Cl_2$ is produced [15]. With allyl chloride in water-methanol, $(\pi$-$C_3H_5)_2Pd_2Cl_2$ is formed [16]. Butenylethers have been made by reaction of $[PdCl_4]^{2-}$ with 1,3-butadiene, isoprene, or 2,3-dimethyl-1,3-butadiene [17].

With vinylic and allylic epoxides, $[PdCl_4]^{2-}$ yields functionally substituted allylic alcohols via a regio- and stereoselective catalytic route in the presence of RHgCl and copper(II) chloride [18].

112

The kinetics of the oxidation of xylenes to diarylmethanes by $[PdCl_4]^{2-}$ in acidic media have been investigated using o- and p-xylenes [19].

References:

[1] Maitlis, P. M.; Espinet, P.; Russell, M. J. H. (in: Wilkinson, G.; Stone, F. G. A.; Abel, E. W., Comprehensive Organometallic Chemistry, Mono-olefin and Acetylene Complexes of Palladium, Vol. 6, Pergamon, Oxford 1982, pp. 351/62).

[2] Maitlis, P. M.; Espinet, P.; Russell, M. J. H. (in: Wilkinson, G.; Stone, F. G. A.; Abel, E. W., Comprehensive Organometallic Chemistry, Diene Complexes of Palladium, Vol. 6, Pergamon, Oxford 1982, pp. 363/84).

[3] Maitlis, P. M.; Espinet, P.; Russell, M. J. H. (in: Wilkinson, G.; Stone, F. G. A.; Abel, E. W., Comprehensive Organometallic Chemistry, Allylic Complexes of Palladium(II), Vol. 6, Pergamon, Oxford 1982, pp. 385/446).

[4] Maitlis, P. M.; Espinet, P.; Russell, M. J. H. (in: Wilkinson, G.; Stone, F. G. A.; Abel, E. W., Comprehensive Organometallic Chemistry, Palladium Complexes derived from Reaction with Acetylenes, Vol. 6, Pergamon, Oxford 1982, pp. 455/69).

[5] Maitlis, P. M.; Espinet, P.; Russell, M. J. H. (in: Wilkinson, G.; Stone, F. G. A.; Abel, E. W., Comprehensive Organometallic Chemistry, Cyclopentadienyl and Aryl Complexes of Palladium(II), Vol. 6, Pergamon, Oxford 1982, pp. 447/54).

[6] Bäckvall, J. E.; Åkermark, B.; Ljunggren, S. O. (J. Am. Chem. Soc. **101** [1979] 2411/6).

[7] Moiseev, I. I.; Levanda, O. G.; Vargaftik, M. N. (J. Am. Chem. Soc. **96** [1974] 1003/7).

[8] Pestrikov, S. V.; Moiseev, I. I.; Romanova, T. N. (Zh. Neorgan. Khim. **10** [1965] 2203; Russ. J. Inorg. Chem. **10** [1965] 1199).

[9] Ashcroft, S. J.; Mortimer, C. T. (Thermochemistry of Transition-Metal Complexes, Academic, London 1970, p. 363).

[10] Pandey, R. N.; Henry, P. M. (Can. J. Chem. **57** [1979] 982/9).

[11] Osipov, A. M.; Likholobov, V. A.; Eremeeva, N. M. (Izv. Vysshikh Uchebn. Zavedenii Khim. Khim. Tekhnol. **15** [1972] 1485/8; C.A. **78** [1973] No. 83590).

[12] Alcock, R. M.; Hartley, F. R.; Rogers, D. E.; Wagner, J. L. (Coord. Chem. Rev. **16** [1975] 59/66).

[13] Asfour, H. M.; Green, M. (J. Organometal. Chem. **292** [1985] C25/C27).

[14] Drew, D.; Doyle, J. R. (Inorg. Syn. **13** [1972] 47/55, 52).

[15] Lukas, J. (Inorg. Syn. **15** [1974] 75/81).

[16] Tatsuno, Y.; Yoshida, T. (Inorg. Syn. **19** [1979] 220/3).

[17] Lawrence, R. V.; Ruff, J. K.; Taylor, R. C. (J. Chem. Soc. Chem. Commun. **1976** 9/10).

[18] Larock, R. C.; Ilkka, S. J. (Tetrahedron Letters **27** [1986] 2211/4).

[19] Kozhevnikov, I. V.; Kim, V. I.; Talzi, E. P. (Izv. Akad. Nauk SSSR Ser. Khim. **1985** 2167/74; Bull. Acad. Sci. USSR Div. Chem. Sci. **1985** 2001/7).

Reaction of 1,3-pentadiene with $[PdCl_4]^{2-}$ in aqueous solution gave hydroxylated π-allylic complexes [1]. Lithium tetrachloropalladate reacts with optically active allylsilanes to give π-allylic complexes with retention of chirality [2]. The $[PdX_4]^{2-}$ (X = Cl, Br) ions will react with 2-methylenebicyclo-(2.2.1)heptane to give $[RPdX]_2$, where R = π-allylnorcamphane [3].

Rates of isomerisation of allyl-1,1-d_2 alcohol to allyl-3,3-d_2 alcohol in the presence of $[PdCl_4]^{2-}$ have been studied, and are found to be identical with the rates of exchange of the nondeuteriated alcohol with $H_2^{18}O$ [4]. Reaction of 1,4- and 1,5-dienes with $[PdCl_4]^{2-}$ to give π-allylic complexes has been noted [5], while $[PdCl_4]^{2-}$ and allyl alcohol give β-hydroxy-propanal, hydroxyacetone, acrolein, propanol and propene [6]. In the presence of $Li_2[PdCl_4]$, β-malonyl allyl sulphides and amines cyclise to give fused bicyclic palladocycles both regio-specifically and stereospecifically [7]. Allylmercuric chloride will react with $[PdCl_4]^{2-}$ to give $[(\pi-C_3H_5)PdCl]_2$ [8].

References:

[1] Andari, M. K.; Krylov, A. V.; Belov, A. P. (Zh. Strukt. Khim. **25** [1984] 167/9; J. Struct. Chem. [USSR] **25** [1984] 322/4).
[2] Hayashi, T.; Konishi, M.; Kumada, M. (J. Chem. Soc. Chem. Commun. **1983** 736/7).
[3] Castanet, Y.; Petit, F. (J. Chem. Res. Synop. **1982** 238/9).
[4] Gregor, N.; Zaw, K.; Henry, P. M. (Organometallics **3** [1984] 1251/6).
[5] Carock, R. C.; Takagi, K. (Tetrahedron Letters **24** [1983] 3457/60).
[6] Zaw, K.; Lautens, M.; Henry, P. M. (Organometallics **4** [1985] 1286/91).
[7] Holton, R. A.; Zoeller, J. R. (J. Am. Chem. Soc. **107** [1985] 2124/31).
[8] Nesmeyanov, A. N.; Rubezhov, A. Z. (J. Organometal. Chem. **164** [1978] 259/75, 260).

σ-Bonded Complexes. This subject has been reviewed by [1]. Formation of Pd–C σ-bonds is called "palladation".

Reaction of N-substituted α-methyl benzylamines gives σ-palladated dimeric complexes bound also to the N-donor sites and with bridging chloro ligands [2]. Reaction of 4-chroma-none oxime with $[PdCl_4]^{2-}$ yields a σ-palladated binuclear chloro-bridged complex also bound at the N atom [3]. Methylcobalamin and other organocobalt species in the presence of $[PdCl_4]^{2-}$ will alkylate mono- and disubstituted olefins [4], and $[PdCl_4]^{2-}$ will react with organocobalamins RCoBz to give $[RCoBz]^+[PdCl_3]^-$ [5]. The mechanism of methyl transfer from methylcobalamin by $[PdCl_4]^{2-}$ to give Pd, CH_3Cl and aqua- and chloro-cobalamin has been investigated [6].

Cyclometallated complexes involving Pd–C σ-bonds were made from $[PdCl_4]^{2-}$, 1,10-phenanthroline or 2,2′-bipyridyl with carbomethoxyethanes [7]. Reaction of N,N′-di(2-pyridyl)-1,3-diamino-5-chlorobenzene (LH) with $[PdCl_4]^{2-}$ gives PdClL, in which there is one σ-metallated carbon and two nitrogen donor centres on the L ligand [8]. The cyclopalladated species $[Pd(Me_2NCH(COOEt)C_6H_4Cl]_2$ is formed from $[PdCl_4]^{2-}$ and N,N-dimethyl-C-phenylgly-cine ethyl ester [9]. A reactive species, the apparently two coordinated $ClPdC_6H_4COOH$, is made from $[PdCl_4]^{2-}$ and o-chloromercuriobenzoic acid [10]. There is σ-metallation in the reaction product $(Pd_2Cl_2L_2)$ of $[PdCl_4]^{2-}$ and N,N-dimethylbenzylamine; the species is dimeric with chloro bridges [11].

References:

[1] Maitlis, P. M.; Espinet, P.; Russell, M. J. H. (in: Wilkinson, G.; Stone, F. G. A.; Abel, E. W., Comprehensive Organometallic Chemistry, Vol. 6, Pergamon, Oxford 1982, pp. 351/84).
[2] Zalevskaya, O. A.; Dunina, V. V.; Potapov, V. M.; Kuz'mina, L. G.; Struchkov, Yu. T. (Zh. Obshch. Khim. **55** [1985] 1332/41; J. Gen. Chem. [USSR] **55** [1985] 1191/8).
[3] Izumi, T.; Katou, T.; Kasahara, A.; Hanaya, K. (Bull. Chem. Soc. Japan **51** [1978] 3407/8).
[4] Vol'pin, M. E.; Volkova, L. G.; Levitin, I. Ya.; Boronina, N. N.; Yurkevich, A. M. (J. Chem. Soc. D **1971** 849/50).
[5] Chauser, E. G.; Rudakova, I. P.; Yurkevich, A. M. (Zh. Obshch. Khim. **45** [1975] 1197/8; J. Gen. Chem. [USSR] **45** [1975] 1181/2).
[6] Scovell, W. M. (J. Am. Chem. Soc. **96** [1974] 3451/6).
[7] Newkome, G. R.; Kiefer, G. E.; Frere, Y. A.; Onishi, M.; Gupta, V. K.; Fronczek, F. R. (Organometallics **5** [1986] 348/55).

114

[8] Nonoyama, M. (Polyhedron **4** [1985] 765/8).
[9] Ryabov, A. D.; Polyakov, V. A.; Yatsimirsky, A. K. (Inorg. Chim. Acta **91** [1984] 59/65).
[10] Horino, H.; Inoue, N. (Heterocycles **11** [1978] 281/6).

[11] Cope, A. C.; Friedrich, E. C. (J. Am. Chem. Soc. **90** [1968] 909/13).

Reactions with Phosphorus, Arsenic and Antimony Donors

With $P(OEt)_3$ in CH_3OH and Et_2NH, $[PdCl_4]^{2-}$ yields $Pd[P(OEt_3)_4]$ [1], while $[PdCl_4]^{2-}$ and PBu_3^n in water-ethanol gives $PdCl_2(PBu_3^n)_2$ [2]. Reaction of diphenylphosphine Ph_2PH with $[PdCl_4]^{2-}$ in ethanol yielded $[PdCl(PPh_2)(PHPh_2)]_2$ and $[Pd(PPh_2)_2(PHPh_2)_2]$, in which the PPh_2 moiety has a bridging role [3]. With $Pd(PR_3)_2Cl_2$, $[PdCl_4]^{2-}$ gives $Pd_2Cl_4(PR_3)_2$ (R = Ph, Et) [4].

Treatment of $[PdCl_4]^{2-}$ with tri-2-thienylphosphine (L) gives $PdCl_2L_2$ [5], while 1,8-bis-(diphenylphosphine)-3,6-dioxooctane (L') gives cis and trans $PdL'Cl_2$ [6]. With ethyldithienyl- and phenyldithienylphosphines (L), $[PdCl_4]^{2-}$ yields $PdLCl_2$ in which L apparently functions as a P, S donor [7].

Reaction of 1,3-bis(dimethylarsino)propane with $[PdCl_4]^{2-}$ and I^- in ethanol gives $PdI_2(Me_2As(CH_2)_3AsMe_2)$ [8], while $[PdCl_4]^{2-}$ and $SbEt_3$ gives $Pd_2Cl_4(SbEt_3)_2$ [4].

References:

[1] Meier, M.; Basolo, F. (Inorg. Syn. **13** [1972] 112/7).
[2] Saito, T.; Munakata, H.; Imoto, H. (Inorg. Syn. **17** [1977] 83/91, 87).
[3] Hayter, R. G. (J. Am. Chem. Soc. **84** [1962] 3046/53).
[4] Chatt, J.; Venanzi, L. M. (J. Chem. Soc. **1957** 2351/6).
[5] Gol'dfarb, Ya. L.; Dudinov, A. A.; Litvinov, V. P.; Yufit, D. S.; Struchkov, Yu. T. (Khim. Geterosikl. Soedin. **1982** 1326/32; C.A. **98** [1983] No. 89625).
[6] Hill, W. E.; Taylor, J. G.; Falshaw, C. P.; King, T. J.; Beagley, B.; Tonge, D. M.; Pritchard, R. G.; McAuliffe, C. A. (J. Chem. Soc. Dalton Trans. **1986** 2289/95).
[7] Polovnyak, V. K.; Sharafieva, E. S.; Slobodina, V. S.; Krasil'nikova, E. A. (Zh. Obshch. Khim. **53** [1983] 2148/9; J. Gen. Chem. [USSR] **53** [1983] 1938).
[8] Gray, L. R.; Gulliver, D. J.; Levason, W.; Webster, M. (Inorg. Chem. **22** [1983] 2362/6).

Reactions with $SnCl_3^-$

With $[PdCl_4]^{2-}$, $SnCl_2$ in HCl gives $[PdCl_3(SnCl_3)]^{2-}$, $[PdCl_2(SnCl_3)_2]^{2-}$ and $[Pd(SnCl_3)_5]^{3-}$ as well as polymeric species.

Reference:

Elizarova, G. L.; Matvienko, L. G.; Yurchenko, E. N.; Stukova, R. N. (Koord. Khim. **6** [1980] 1731/6; Soviet J. Coord. Chem. **6** [1980] 867/72).

Miscellaneous Reactions

The effect of ligands on the reactivity of central metal atoms towards hydrated electrons has been investigated for a wide variety of complexes; rate constants have been determined and for Pd^{II} the two species chosen were $[PdCl_4]^{2-}$ and $[Pd(CN)_4]^{2-}$ [1]. Reaction between gelatin and $[PdCl_4]^{2-}$ showed that a stable complex was formed below pH = 6 [2]. The effects on viscosity, density and electrical conductivity of water-dioxan mixtures consequent on addition of $[PdCl_4]^{2-}$ have been noted [3]. The effect of adding $[PdCl_4]^{2-}$ to a solution of L-aspa-ragine containing asparaginase was observed to moderate the enzymic hydrolysis; its effects were compared with those of other added ions [4]. Radiolysis of aqueous solutions of $[PdCl_4]^{2-}$ gives Pd metal, though the metal does inhibit further reduction [5].

Gelatin reacts with $[PdCl_4]^{2-}$ in photographic emulsions and so Pd is a less effective photographic sensitiser than some other metal species [6]. Reaction of 2-allyl-3,6-diamino-5-methyl-1,4-benzoquinone with $[PdCl_4]^{2-}$ gives a cyclisation reaction yielding quinolinoquinone [7]. The effect of addition of organic solvents to aqueous solutions of chloropalladates, including $[PdCl_4]^{2-}$ and their composition and reactivity towards organic solvents, has been studied [8].

References:

[1] Anbar, M.; Hart, E. J. (J. Phys. Chem. **69** [1965] 973/7).
[2] Tanaka, K. (Nippon Shashin Gakkaishi **37** [1974] 133/7; C.A. **82** [1975] No. 132105).
[3] Golodov, V. A.; Fasman, A. B.; Roganov, V. V.; Enker, K. P. (Zh. Neorgan. Khim. **15** [1970] 236/9; Russ. J. Inorg. Chem. **15** [1970] 121/3).
[4] Charlson, A. J.; Coman, A. J.; Karossi, T. A.; Stephens, F. S.; Vagg, R. S.; Watton, E. C. (Inorg. Chim. Acta **28** [1978] 217/22).
[5] Balandin, A. A.; Spitsyn, V. I.; Duzhenkov, V. I.; Barsova, L. I. (Tr. Tashk. Konf. Mirnomu Ispol'z. At. Energ., Tashkent, USSR, 1959 [1960/61], Vol. 1, pp. 289/95).
[6] Narath, A.; Tiilikka, A. (J. Phot. Sci. **9** [1961] 303/11).
[7] Weider, P. R.; Hegedus, L. S.; Asada, H.; D'Andreq, S. V. (J. Org. Chem. **50** [1985] 4276/81).
[8] Kaszonyi, A.; Hrusorsky, M.; Ilavsky, J. (Petrochemia **26** [1986] 62/8; C.A. **107** [1987] No. 88386).

4.5.2.9 Biological Effects of $[PdCl_4]^{2-}$

The $[PdCl_4]^{2-}$ ion has a greater antiphage activity than has $[PdCl_6]^{2-}$ although it is still inferior in this respect to $[Pd(NO_2)_4]^{2-}$ [1]. The anaphylactic and anaphylactoid properties of $[PdCl_4]^{2-}$ towards guinea pigs and cats have been measured [2]. The antiphage properties of a wide range of Pd and Pt complexes including $[PdCl_4]^{2-}$, were compared using extracellular bacteriophage T4 [3, 4]. The $[PdCl_4]^{2-}$ ion causes, when injected intravenously into unanaesthetised rats, cardiac arrythmias [5]. The $[PdCl_4]^{2-}$ ion acts as an inhibitor of mitochondrial electron transport in rat livers [6], but the ion is not effective in promoting antileukemic activity in mice [7]. Pretreatment effects on the LD_{50} values for a number of heavy metal complexes, including $[PdCl_4]^{2-}$, have been measured [8]. The effect of $[PdCl_4]^{2-}$ on the in vitro activity of urease in rats has been compared with that of other ions [9]. Anti-viral properties have been found for $[PdCl_4]^{2-}$ in in vivo and in vitro studies [10].

The mechanism of the effect of biologically active materials, including $[PdCl_4]^{2-}$, on membrane-bound mitochondrial monoamine oxidase [11] and calcium-magnesium-dependent ATP-ase in the sarcoplasmic reticulum [11 to 13] has been assessed, and antitumour activity of $[PdCl_4]^{2-}$ was also examined [14]. No substantial inhibition of lateral root formation in corn by $[PdCl_4]^{2-}$ or $[PdBr_4]^{2-}$ was noted [15].

References:

[1] Zakharova, I. A.; Moshkovskii, Yu. S.; Suraikina, T. I.; Fonshtein, L. M. (Dokl. Akad. Nauk SSSR **222** [1975] 1229/31; Proc. Acad. Sci. USSR Biochem. Sect. **220/225** [1975] 252/3).
[2] Tomilets, V. A.; Zakharova, I. A. (Farmakol. Toksikol. [Moscow] **42** [1979] 170/3; C.A. **91** [1979] No. 13476).
[3] Zakharova, I. A.; Moshkovskii, Yu. S.; Suraikina, T. I.; Fonshtein, L. M. (Koord. Khim. **2** [1976] 1642/5; Soviet J. Coord. Chem. **2** [1976] 1263/6).

116

[4] Fonshtein, L. M.; Suraikina, T. I.; Tal, E. K.; Moshkovskii, Yu. S. (Genetika [Moscow] 11 [1975] 128/34; C.A. 83 [1975] No. 158451).
[5] Wiester, M. J. (EHP Environ. Health Perspect. 12 [1975] 41/4).
[6] Biagini, R. E.; Moorman, W. J.; Winston, G. W. (Toxicol. Letters 12 [1982] 165/70).
[7] Banner, R. J.; Charlson, A. J.; Gale, R. P.; McArdle, N. T.; Trainor, K. E.; Watton, E. C. (Cancer Treat. Rept. 61 [1977] 469/70).
[8] Weaver, A. D. (Toxicol. Appl. Pharmacol. 49 [1979] 41/4).
[9] Olson, D. L.; Christensen, G. M. (Bull. Environ. Contam. Toxicol. 28 [1982] 439/45).
[10] Graham, R. D.; Williams, D. R. (J. Inorg. Nucl. Chem. 41 [1979] 1245/9).

[11] Tat'yanenko, L. V.; Sokolova, N. V.; Moshkovskii, Yu. S. (Vopr. Med. Khim. 28 [1982] 126/31; C.A. 98 [1983] No. 30410).
[12] Letuchii, Ya. A.; Tat'yanenko, L. V.; Moshkovskii, Yu. S.; Khidekel', M. L. (Koord. Khim. 9 [1983] 238/42; Soviet J. Coord. Chem. 9 [1983] 151/5).
[13] Tat'yanenko, L. V.; Raikhman, L. M.; Toshcheva, T. A.; Zakharova, I. A.; Moshkovskii, Yu. S. (Biokhimiya [Moscow] 41 [1976] 1516/21; C.A. 85 [1976] No. 138745).
[14] Gill, D. (Develop. Oncol. 17 [1984] 267/8; C.A. 100 [1984] No. 150637).
[15] Bystrova, E. I.; Zakharova, I. A.; Ivanov, V. B.; Ivanova, N. A. (Izv. Akad. Nauk SSSR Ser. Biol. 1986 No. 6, pp. 884/8; C.A. 106 [1987] No. 113216).

4.5.2.10 Applications of $[PdCl_4]^{2-}$

These are mainly dealt with under the appropriate salts, though in most if not all cases it is likely to be the ion which is the active species.

Solutions of $[PdCl_4]^{2-}$ produce colours on silver bromide photographic emulsions, and sensitisation was also observed; the colour is removed by HCl [1]. The attachment of palladium to silica catalysts via amine groups has been accomplished by using $[PdCl_4]^{2-}$ and treating the silica with such reagents as $(EtO)_3Si(CH_2)_3NH_2$ [2], and treatment of resin surfaces with $[PdCl_4]^{2-}$ can be used to obtain supported catalysts for the decomposition of cyclohexyl hydroperoxide [3]. As "$H_2[PdCl_4]$" the ion will, in organic solvents, activate electroless plating [4], as will $[PdCl_4]^{2-}$ on plastics in aqueous HCl [5]. The use of the ion as a pretreatment reagent for development nuclei in reception layers imparts greater stability and improved images in silver salt photographic emulsions [6]. The use of $[PdCl_4]^{2-}$ in tetrahydrofuran solutions to amplify developed electrographic images has been studied [7].

Layered zirconium phosphonate as the anthranilic acid has been treated with $[PdCl_4]^{2-}$ to give supported Pd catalysts which will catalyse the hydrogenation of organic materials [8]. A method for determination of CO with a $[PdCl_4]^{2-}$-cacotheline reagent has been developed; such a solution is stable and responsive to CO in the 0 to 600 ppm range [9].

Bipolar membranes of low electrical resistance can be produced by the use of, amongst other reagents, $[PdCl_4]^{2-}$ by joining together separate anion exchange and cation exchange films [10]. Salts of $[PdCl_4]^{2-}$ have been used to catalyse the hydrogenation of nitro complexes when on ion-exchange resin supports [11]. The species has been used in the electroless plating of circuit board substrates [12].

References:

[1] Faelens, P. (Sci. Ind. Phot. 28 [1957] 191/3; C.A. 1957 10279).
[2] Sharf, V. Z.; Gurovets, A. S.; Finn, L. P.; Slinyakova, I. B.; Krutii, V. N.; Freidlin, L. K. (Izv. Akad. Nauk SSSR Ser. Khim. 1979 104/8; Bull. Acad. Sci. USSR Div. Chem. Sci. 1979 93/6).

[3] Acodedji, E.; Lieto, J.; Aune, J. P. (J. Mol. Catal. **22** [1983] 73/8).

[4] Sirinyan, K.; Wolf, G. D.; v. Gizycki, U.; Merten, R.; Bayer A.-G. (Eur. 177862 [1986]; C.A. **105** [1986] No. 89803).

[5] Sakai, N.; Kijima, M.; Kaneko, M.; Iwasawa, T.; Aoki, K. (Japan. 76-17599 [1976]; C.A. **85** [1976] No. 125347).

[6] Kodak, N.V. (Neth. Appl. 66-07192 [1966]; C.A. **66** [1967] No. 80807).

[7] Lelental, M.; Kaukeinen, J. K. (Res. Discl. No. 162 [1977] 15/7).

[8] Lane, R. H.; Callahan, K. P.; Cooksey, R.; Giacomo, P. M. D.; Dines, M. B.; Griffith, P. C. (Prepr. Am. Chem. Soc. Div. Petrol. Chem. **27** [1982] 624/31; C.A. **101** [1984] No. 130760).

[9] Lambert, J. L.; Chiang, Y. C. (Anal. Chem. **55** [1983] 1829/30).

[10] (Brit. Appl. 2122543 [1984]; C.A. **100** [1984] No. 141278).

[11] Zhou, Z.; Zhang, M.; Chen, C.; Sun, Y. (Cuihua Xuebao **8** [1987] 69/75; C.A. **107** [1987] No. 9292).

[12] Sirinyan, K.; Wolf, G. D.; v. Gizycki, U.; Merten, R.; Bayer A.-G. (Eur. Appl. 177862 [1986]; C.A. **105** [1986] No. 89803).

4.5.2.11 Miscellaneous Data for [PdCl$_4$]$^{2-}$

The ionicity of the Pd–Cl bond in [PdCl$_4$]$^{2-}$ has been calculated by use of the electronegativity method [1]. The degree of covalency of an orbital doublet in [PdCl$_4$]$^{2-}$ has been obtained by measurement of the spin-orbit splitting when Jahn-Teller coupling in the K$_2$[PdCl$_4$]–Cu^{2+} system is quenched [2]. Energies of relaxation of the valence levels in a number of square-planar Pd^{II} complexes, including [PdCl$_4$]$^{2-}$, were calculated by using second-order perturbation theory [3]. The quenching rate constants of the triplet state of various organic molecules by [PdCl$_4$]$^{2-}$ and other square-planar complexes was studied [4].

Flotation of [PdCl$_4$]$^{2-}$ by diethylcetylammonium chloride and cetylpyridinium chloride was traced using radioactive ^{103}Pd [5].

References:

[1] Batsanov, S. S. (Zh. Neorgan. Khim. **9** [1964] 1323/7; Russ. J. Inorg. Chem. **9** [1964] 722/5).

[2] Moreno, M. (J. Phys. C **10** [1977] L183/L186).

[3] Voityuk, A. A.; Mazalov, L. N. (Zh. Strukt. Khim. **23** [1982] 61/6; J. Struct. Chem. [USSR] **23** [1982] 50/4).

[4] Marshall, K. C.; Wilkinson, F. (Z. Physik. Chem. [N.F.] **101** [1976] 67/78).

[5] Walkowiak, W.; Bartecki, A. (Nukleonika **18** [1973] 133/41; C.A. **79** [1973] No. 138284).

4.5.2.12 Tetrachloropalladates of Nonmetals

There are many of these; some have found application in the extraction of Pd as the [PdCl$_4$]$^{2-}$ anion.

H$_2$[PdCl$_4$]. This is made in solution by oxidation of Pd metal with aqua regia [1, 2, 19], and from Pd metal sponge suspended in methanol with Cl$_2$ [3]. It has also apparently been obtained as a solid by reaction of Pd with aqua regia [2]. It gives no luminescence with SnCl$_4$ in a reducing flame, unlike H[AuCl$_4$] which gives a green luminescence [4].

118

The material has been used to impart anti-static and low-fogging characteristics to monochrome silver halide photographic emulsions [5]. It has also been used in a catalytic process for the heterogeneous Wacker gas-phase oxidation of alkenes and cyclo-alkenes by impregnation of a γ-Al_2O_3 surface with NH_4VO_3 and "$H_2[PdCl_4]$" solution [6], and for ethene chlorination and oxidative coupling over SiO_2, Al_2O_3 and TiO_2 supports impregnated with a variety of "chloro acids" including $H_2[PdCl_4]$ [7]. Adsorption of the material on Al_2O_3 has been studied [20].

For "$[HPdCl_4]^-$" or "$H[PdCl_4]^-$" see under Hydrido-chloro Complexes, p. 171.

With $K_2S_2O_8$, $K_2[PdCl_6]$ is formed [21].

$[C(NH_2)_3]_2[PdCl_4]$. This is made by reaction of $PdCl_2$ and guanidinium hydrochloride [8]. It is brown [9].

The X-ray crystal structure shows the salt to be tetragonal, space group $P4_2/n$-C_{4h}^4, $Z = 4$; $a = 10.755(1)$, $c = 11.006(1)$ Å; measured density 1.93 g/cm^3, calculated density 1.94 g/cm^3. There is planar coordination in the anion, two Pd–Cl at 2.287(2), two at 2.294(2) Å. The guanidinium cations are planar, see **Fig. 16**.

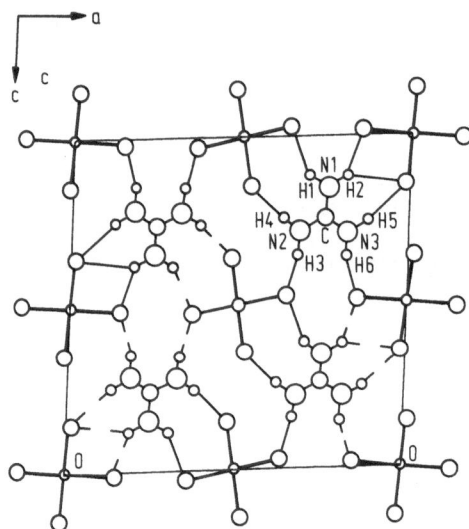

Fig. 16.   Crystal structure of $[C(NH_2)_3]_2[PdCl_4]$ [9].

The nuclear quadrupole resonance (NQR) spectrum of the salt was measured from 77 K to room temperature [8, 10].

^{1}H NMR studies were made on the motion of the guanidinium ions in the salt and dipolar-quadrupolar cross relaxations determined; there was an unusual temperature dependence behaviour of ^{1}H T_1 over the range 100 to 400 K [11].

$[(CH_3)_3NH]_2[PdCl_4] \cdot (CH_3)_3NHCl$. This is made from $PdCl_2$ and trimethylamine in excess HCl [12].

It forms brown prismatic crystals, of low stability in air. They are orthorhombic, space group Pnma-D_{2h}^{16}, $Z = 4$; $a = 9.90(1)$, $b = 12.09(1)$, $c = 17.21(2)$ Å; experimental density 1.45, calculated density 1.49 g/cm^3. The unit cell is shown in **Fig. 17** and contains square-planar $[PdCl_4]^{2-}$

anions (mean Pd–Cl 2.30(3) Å) with two cations cis to the $[PdCl_4]^{2-}$ plane, while a third $(CH_3)_3NHCl$ molecule is inserted in the lattice as an ion pair [12].

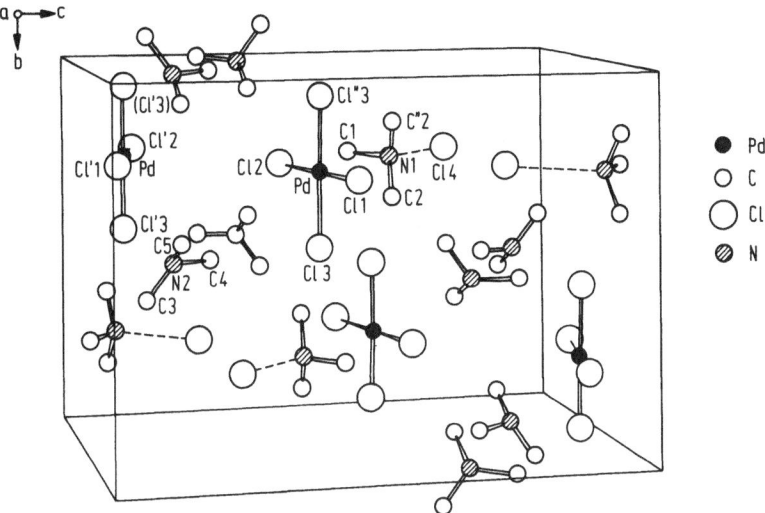

Fig. 17.   Unit cell of $[(CH_3)_3NH]_2[PdCl_4] \cdot (CH_3)_3NHCl$ [12].

[Me₂NH₂]₂[PdCl₄]. This can be made by reaction of Pd in HCl with $(Me_2NH_2)Cl$ in the presence of O_2. It melts at 116 to 118°C, and the infrared spectrum in the 1400 to 4000 cm^{-1} region was measured [13].

(Me₃NH)₂[PdCl₄] is made by reaction of Pd in HCl with $(Me_3NH)Cl$ in the presence of O_2. It melts at 162 to 163°C, and the infrared spectrum in the 1400 to 4000 cm^{-1} region was measured [13].

(Me₄N)₂[PdCl₄]. This is made from $PdCl_2$ in conc. HCl with $(Me_4N)Cl$. Its thermal decomposition begins at 205°C [14].

The far-infrared spectrum of the solid has been measured, $\nu_{PdCl} = 324$ cm^{-1}, $\delta_{PdCl} = 192$ and 150 cm^{-1} [15].

(C₂H₅NH₃)₂[PdCl₄]. This is made from $PdCl_2$ and $(Et_4N)Cl$ in HCl [16] by the method of [17].

The crystals are orthorhombic, a = 7.44, b = 20.51, c = 8.02 Å from powder diffraction data [18]. Infrared (4000 to 600 cm^{-1}) and near infrared (1000 to 2500 nm) spectra were measured of single crystals of the salt at room temperature (reproduced in paper) and assigned, and also Raman spectra at low temperatures [16].

[(C₂H₅)₄N]₂[PdCl₄]. The far-infrared spectrum has been reproduced (230 to 430 cm^{-1}) and ν_{PdCl} listed as 319 cm^{-1} with other modes at 192, 150 and 93 cm^{-1} [15].

(C₃H₂N₂H)₂[PdCl₄]. The use of 3(5)-methylpyrazole to study the extraction of $[PdCl_4]^{2-}$ from HCl solution was investigated, and the infrared spectrum of an unspecified solid complex, presumably $(C_3H_3NH_2)_2[PdCl_4]$, was measured from 2600 to 4000 cm^{-1} (reproduced in paper) [18].

120

References:

[1] Rush, R. M.; Martin, D. S.; LeGrand, R. G. (Inorg. Chem. **14** [1975] 2543/50, 2544).

[2] Kauffman, G. B.; Tsai, J. H. (Inorg. Syn. **8** [1966] 234/8).

[3] Clements, F. S.; Nutt, E. V.; International Nickel Co. (Mond) Ltd. (Brit. 879074 [1961]; C.A. **56** [1962] 8298).

[4] Renaud, P. (Compt. Rend. **242** [1956] 1477/9).

[5] Fuji Photo Film Co. Ltd. (Japan. Kokai Tokkyo Koho 6080847 [1985]; C.A. **103** [1985] No. 113267).

[6] Scholten, J. J. S.; van der Steen, P. J. (Eur. 210705 [1987]; C.A. **106** [1987] No. 104235).

[7] Rollins, K.; Sermon, P. A. (J. Chem. Soc. Chem. Commun. **1986** 1171/3).

[8] Gima, S.; Furukawa, Y.; Ikeda, R.; Nakamura, D. (J. Mol. Struct. **111** [1983] 189/94).

[9] Kiriyama, H.; Matsushita, N.; Yamagata, Y. (Acta Cryst. C **42** [1986] 277/80).

[10] Furukawa, Y.; Nakamura, D. (Z. Naturforsch. **41a** [1986] 416/20).

[11] Furukawa, Y.; Gima, S.; Nakamura, D. (Ber. Bunsenges. Physik. Chem. **89** [1985] 863/75).

[12] Zveguintzoff, D.; Bois, C.; Dao, N. Q. (J. Inorg. Nucl. Chem. **43** [1981] 3183/5).

[13] Letuchii, Ya. A.; Lavrent'ev, I. P.; Khidekel', M. L. (Izv. Akad. Nauk SSSR Ser. Khim. **1979** 718/24; Bull. Acad. Sci. USSR Div. Chem. Sci. **1979** 669/74).

[14] Kukushkin, Yu. N.; Sedova, G. N.; Vlasova, R. A. (Zh. Neorgan. Khim. **23** [1978] 737/40; Russ. J. Inorg. Chem. **23** [1978] 406/8).

[15] Acodedji, E.; Lieto, J.; Aune, J. P. (J. Mol. Catal. **22** [1983] 73/8).

[16] Hagemann, H.; Bill, H. (Chem. Phys. Letters **90** [1982] 282/6).

[17] Arend, H.; Huber, W.; Mischgofsky, F. H.; Richter van-Leeuwen, G. K. (J. Cryst. Growth **43** [1978] 213/23).

[18] Pronin, V. A.; Vsol'tseva, M. V.; Shastina, E. N.; Volkov, A. N.; Seraya, V. I. (Zh. Neorgan. Khim. **19** [1974] 800/2; Russ. J. Inorg. Chem. **19** [1974] 434/5).

[19] Livingstone, S. E. (Syn. Inorg. Metal-Org. Chem. **1** [1971] 1/7).

[20] Guo, Z.; Zhan, Z. (Yingyong Huaxue **4** [1987] 31/5; C.A. **107** [1987] No. 65355).

[21] Bardin, M. B.; Ketrush, P. M.; Goncharenko, V. P. (Metallkhelaty IKH. Svoistva **1985** 63/7).

(enH$_2$)[PdCl$_4$]. This is made from PdCl$_2$en (en = ethylenediamine or 1,2-diaminoethane) and conc. HCl at 80°C, and forms dark red crystals [1]; see also [21, 22]. It decomposes at 183°C [21, 22].

The X-ray crystal structure shows the crystals to be monoclinic, space group P2$_1$/a-C$_{2h}^5$, Z = 2; a = 7.34(2), b = 7.66(2), c = 7.91(2) Å, β = 91.90(1)°; calculated density 2.318, density by flotation 2.33 g/cm^3 [2]. Earlier values were a = 7.92(2), b = 7.73(4), c = 7.39(4) Å, β = 91.3(4)° [1]. The [PdCl$_4$]$^{2-}$ anions are planar with Pd–Cl distances of 2.31(1) Å. The (enH$_2$)$^{2+}$ cation has a trans conformation, and the structure consists of stacks of [PdCl$_4$]$^{2-}$ ions with interleaving (enH$_2$)$^{2+}$ cations [1, 2], see **Fig. 18**.

Infrared spectra have been measured on powdered and single-crystal samples of the salt (30 to 4000 cm^{-1}, reproduced in paper) at room and liquid nitrogen temperatures, and the Raman spectrum measured at room temperature (20 to 450 cm^{-1}, reproduced in paper). Assignments of cation and anion vibrations were proposed [1] (see table on p. 132).

A salicylidene-ethylenediaminium salt of [PdCl$_4$]$^{2-}$ has been made from PdCl$_2$ and HCl with the base: it decomposes at 175°C [22].

(en-EtOH$_3$)[PdCl$_4$]. This is made by reaction of N-(2-aminoethyl)-2-amino-1-ethanol in HCl with PdCl$_2$, and forms wine-brown crystals [3].

The infrared spectrum was measured (100 to 450 cm^{-1}, reproduced in paper) [3].

(C$_4$H$_6$N$_3$O)$_2$[PdCl$_4$]. Brick-red crystals of this cytosinium salt are made by slow evaporation of a solution of cytosine acidified to pH = 1 with HCl and K$_2$[PdCl$_4$] [4].

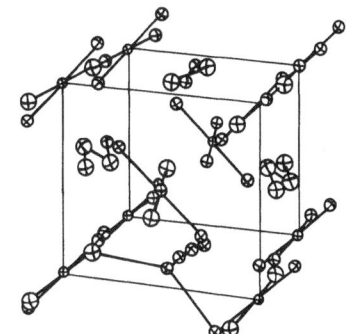

Fig. 18.   Unit cell of (enH)$_2$[PdCl$_4$] [2].

The crystals are monoclinic, space group P2$_1$/c-C$_{2h}^5$, a = 8.437(2), b = 13.776(4), c = 7.191(2) Å, β = 111.07(1)°, measured density 2.01, calculated density 2.00 g/cm^3. There are discrete [PdCl$_4$]$^{2-}$ anions (mean Pd–Cl = 2.296(1) Å) with a complex network of hydrogen bonding, see **Fig. 19**, the protonated cytosine rings being planar [4].

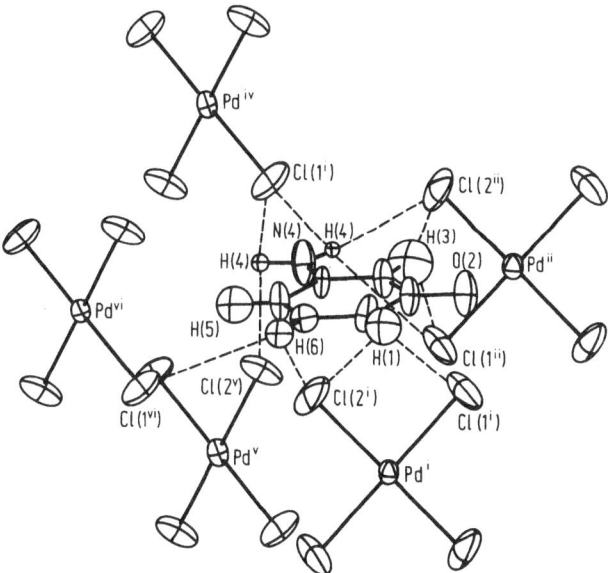

Fig. 19.   Crystal structure of (cytosine H)$_2$[PdCl$_4$] [4].

(C$_3$H$_7$NH$_3$)$_2$[PdCl$_4$]. This, the bis(propylammonium) salt of [PdCl$_4$]$^{2-}$, is made by refluxing an equimolar mixture of (C$_3$H$_7$NH$_3$)Cl and PdCl$_2$ in acetone for several days. It forms red-brown crystals [5].

It forms monoclinic crystals, space group $P2_1/c\text{-}C_{2h}^5$, a = 12.42(1), b = 8.038(4), c = 7.404(3) Å, β = 109.20(2)°. The structure consists of discrete planar $[PdCl_4]^{2-}$ ions (Pd–Cl 2.308(8) and 2.317(7) Å, mean 2.313 Å). There is hydrogen bonding such that the cations form a two-dimensional sandwich around each Pd–Cl layer, see **Fig. 20** [5].

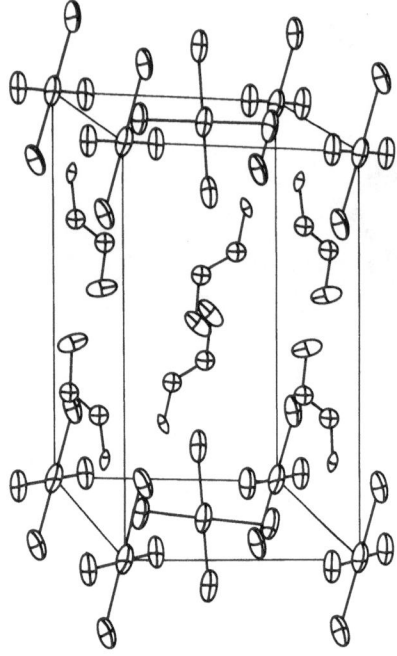

Fig. 20.   Crystal structure of $(C_3H_7NH_3)_2[PdCl_4]$ [5].

$(Bu_4^nN)_2[PdCl_4]$. Attempts to make this from $[PdCl_4]^{2-}$ and $(Bu_4^nN)Cl$ gave mixtures of $(Bu_4^nN)Cl$ with $(Bu_4^nN)_2[Pd_2Cl_6]$ [6], but the salt was made from $(Bu_4^nN)_2[Pd_2Cl_6]$ and $(Bu_4^nN)Cl$ in acetone [7] and also from $PdCl_2$ and $(Bu_4^nN)Cl$ in acetone [8]. It is red-brown, melting at 145°C [7].

The Raman and infrared spectra of the solution were measured in chloroform (see table on p. 132), and also the infrared spectrum of the solid (40 to 400 cm^{-1}) [7].

Values of ν_{PdCl} (in cm^{-1}) from infrared spectra for $(RH_2)[PdCl_4]$ salts [11].

R	ν_{PdCl}	R	ν_{PdCl}	R	ν_{PdCl}
pyridine	327	2,5-lutidine	325	α-picoline	325
2,4-lutidine	325	3,5-lutidine	330	β-picoline	328
2,6-lutidine	325	2,4,6-collidine	325	γ-picoline	325

The electronic spectrum of the salt in acetonitrile was recorded (250 to 450 nm) [8].

$(C_5H_5N_2H)_2[PdCl_4]$. The X-ray photoelectronic spectrum of this 1-vinylazo salt of $[PdCl_4]^{2-}$ shows the binding energies (in eV) to be 338.4 for Pd 3d$_{5/2}$, 198.5 for Cl 2p$_{3/2}$ and 401.7 for N(1s) [9].

$(C_5H_5NH)_2[PdCl_4]$. This pale yellow pyridinium salt, soluble in CH_2Cl_2, is made from C_5H_5N and "$H_2[PdCl_4]$" [10] or from (pyH)Cl and $PdCl_2$ in conc. HCl [11]. It decomposes at 140°C [21, 22]. It can also be made from Pd metal and O_2 with pyridine and HCl; it melts between 248 and 250°C [13].

The infrared spectrum was measured in the 2800 cm^{-1} region [10, 13]; see also table above [11].

(C$_5$H$_{10}$NH)$_2$[PdCl$_4$]. This piperidinium salt has been briefly mentioned as having been prepared from (C$_5$H$_{10}$NH$_2$)Cl and PdCl$_2$ in HCl–C$_2$H$_5$OH at pH = 1 to 2 [12]; see also [21]. It decomposes at 202°C [21].

(C$_6$H$_5$NH$_3$)$_2$[PdCl$_4$]. The anilinium salt has been made by Pd metal in HCl with aniline in the presence of O$_2$ [3] or from PdCl$_2$ and aniline in conc. HCl [21]. It melts at 248 to 250°C; the infrared spectrum from 1400 to 4000 cm^{-1} was reported [13]. It decomposes at 134°C [21].

R$_2$[PdCl$_4$] (R = substituted aniline : aniline itself, 2-fluoro-5-amino toluene, 3-fluoro-5-amino toluene, 3-fluoro-6-amino toluene, 2-amino-4-fluoro toluene, p-fluoro aniline, p-chloro-aniline, p-toluidine, o-toluidine, m-amino-benzotrifluoride, p-amino-benzotrifluoride). All these were made at pH = 1 to 2 from (RH)Cl in HCl–C$_2$H$_5$OH with PdCl$_2$. The salts are readily hydrolysed by water, soluble in alcohols and also in alcohols mixed with HCl but insoluble in other organic solvents; their infrared spectra were measured from 200 to 4000 cm^{-1}. For all of them the ν_{PdCl} stretch was assigned to a band at 340 cm^{-1} [12].

(C$_6$H$_6$N$_2$O$_2$H)$_2$[PdCl$_4$]. This m-nitroanilinium salt is made from PdCl$_2$ in conc. HCl and m-nitroaniline. It decomposes at 152°C [21].

(C$_6$H$_7$NH)$_2$[PdCl$_4$]. The α-, β- and γ-picolinium (i.e., the 2-, 3- and 4-methylpyridinium salts) are made by reaction of PdCl$_2$ in 6 M HCl with the pyridine and the minimum quantity of ethanol. The mixture is heated to give the solid salts, which are soluble in HCl and pyridine but insoluble in benzene, toluene, carbon disulphide and chloroform [11]. The γ-picolinium salt decomposes at 150°C [21].

Infrared spectra were measured (200 to 2000 cm^{-1}); the ν_{PdCl} bands are listed in the table on p. 122 [11].

(C$_6$H$_{11}$NH$_3$)$_2$[PdCl$_4$]. This dark red cyclohexylammonium salt is made from [PdCl$_4$]$^{2-}$ with cyclohexylamine in C$_2$H$_5$OH. It is soluble in alcohols and acetone, with or without mineral acids, but is apparently insoluble in other organic solvents [14].

The infrared spectrum was measured (1400 to 4000 cm^{-1}); ν_{PdCl} is at 320 cm^{-1} [14].

A dicyclohexylammonium salt of [PdCl$_4$]$^{2-}$ has been briefly mentioned, made from the organic cation and PdCl$_2$ in HCl–C$_2$H$_5$OH at pH = 1 to 2 [12].

(C$_7$H$_5$N$_2$H$_2$)$_2$[PdCl$_4$]. The thermal gravimetric analysis of this benzimidazolium salt shows that (C$_7$H$_5$N$_2$H)$_2$[PdCl$_4$] gives PdCl$_2$(C$_7$H$_5$N$_2$)$_2$ [15].

(C$_7$H$_9$NH)$_2$[PdCl$_4$]. This p-toluidinium salt is made from p-toluidine and PdCl$_2$ in conc. HCl. It decomposes at 150°C [21].

(C$_7$H$_9$NH)$_2$[PdCl$_4$]. These 2,4-, 2,5-, 2,6-, and 3,5-lutidinium (i.e., dimethylpyridinium) salts were made by reaction of PdCl$_2$ in 6 M HCl with the appropriate lutidine and the minimum quantity of ethanol. The salts are soluble in HCl but insoluble in benzene, toluene, carbon disulphide and chloroform. Infrared spectra (200 to 2000 cm^{-1}) were measured and the ν_{PdCl} modes assigned (see table on p. 122) [11].

(C$_8$H$_{11}$NH)$_2$[PdCl$_4$]. This collidinium salt (i.e., that of 4-ethyl-2-methylpyridinium) is made from PdCl$_2$ in 6 M HCl with collidine and the minimum quantity of ethanol. The salt is soluble in HCl but insoluble in C$_6$H$_6$, toluene, CS$_2$ and CHCl$_3$. The infrared spectrum (200 to 2000 cm^{-1}) was measured [11]; for ν_{PdCl} see table on p. 122.

[(C$_8$H$_{17}$)$_3$NH]$_2$[PdCl$_4$]. This, the tri-n-octylammonium salt, was made from (C$_8$H$_{17}$)$_3$N in CCl$_4$ with "H$_2$[PdCl$_4$]". It is dark red and easily soluble in CH$_3$OH, C$_2$H$_5$OH, CHCl$_3$ and C$_6$H$_6$, but virtually insoluble in water [10].

The infrared spectrum was measured from 700 to 4000 cm^{-1} (reproduced in paper), and the electronic spectrum in CCl$_4$ and CHCl$_3$ (240 to 600 nm, reproduced in paper).

The molar conductances in CH$_3$OH, C$_2$H$_5$OH, CH$_2$Cl$_2$ and C$_6$H$_6$ are 102, 44.8, 0.24 and 0.1 cm$^2 \cdot \Omega^{-1}$, respectively [10].

[(C$_8$H$_{17}$)$_2$NH$_2$]$_2$[PdCl$_4$]. This di-n-octylammonium salt is a peach-coloured material, soluble in CH$_3$OH, C$_2$H$_5$OH and CHCl$_3$, but only sparingly soluble in C$_6$H$_6$ and CH$_2$Cl$_2$. It is made from (C$_8$H$_{17}$)$_2$NH and "H$_2$[PdCl$_4$]" in hexane. The infrared spectrum was measured (700 to 4000 cm^{-1}, reproduced in papers), as was the electronic spectrum (240 to 560 nm, reproduced in paper) in CCl$_4$ and CHCl$_3$ [10].

(C$_9$H$_7$NH)$_2$[PdCl$_4$]. The quinolinium salt is made from "H$_2$[PdCl$_4$]" and C$_9$H$_7$N in ether. The dark yellow salt is almost insoluble in organic solvents. The infrared spectrum was measured (2000 to 3000 cm^{-1}) [10], or from PdCl$_2$ in concentrated HCl with quinoline. It decomposes at 160°C [21].

(C$_9$H$_7$ONH)$_2$[PdCl$_4$]. This 8-hydroxyquinolinium salt is made from PdCl$_2$, concentrated HCl and 8-hydroxyquinoline. It decomposes at 160°C [21, 22].

(AsPh$_3$Me)$_2$[PdCl$_4$]. Reaction of (AsPh$_3$Me)$_2$[Pd$_2$Cl$_8$] in hot acetone with (AsMe$_3$Ph)Cl yields the compound as pink crystals [16].

The molar conductance in nitrobenzene is 51.0 cm$^2 \cdot \Omega^{-1}$ [16].

The electronic absorption spectrum of the solid was measured (200 to 700 nm, spectrum reproduced in paper) and also the spectra (200 to 600 nm, maxima listed in paper) in nitromethane and acetonitrile [17].

(PPh$_4$)$_2$[PdCl$_4$]. This is made from [PdCl$_4$]$^{2-}$ and PPh$_4$Cl [18].

The X-ray photoelectronic spectra gave a binding energy of 198.6 ± 0.2 eV for Cl 2p$_{3/2}$ [18], Pd 3d$_{5/2}$ 338.0 eV and P 2p 133.0 eV. For [PdX$_4$]$^{2-}$ the Pd 3d$_{5/2}$ binding energies decrease in the sequence X = NCO$^-$ > Cl$^-$ > SCN$^-$ [19].

(C$_{10}$N$_4$H$_{14}$)[PdCl$_4$]. This, the salt of the dication meso-3,7-diazonia tricyclo[4,2,2,2^{2,5}]-dodeca-3,7,9,11-tetraen-4,8-diamine, is made from trans-PdCl$_2$L$_2$ (L = 2-aminopyridine) in dimethylformamide and HCl.

The X-ray crystal structure shows the crystals to belong to the P2$_1$/a-C$_{2h}^5$ space group; Z = 2; a = 12.422(2), b = 8.810(1), c = 6.687(1) Å, β = 101.94(1)°, density 2.03 g/cm^3. The dication is formed from two aminopyridine rings, and the anion is square-planar, Pd–Cl = 2.303(2) Å [20]; see **Fig. 21**.

(C$_{12}$H$_8$N$_2$H$_2$)[PdCl$_4$]. This 1,10-phenanthrolinium salt is made from PdCl$_2$ in conc. HCl with 1,10-phenanthroline. It decomposes at 70°C [21].

(C$_{12}$H$_{25}$NH$_3$)$_2$[PdCl$_4$]. This n-dodecylammonium salt is made as a silvery brown material from "H$_2$[PdCl$_4$]" with n-(C$_{12}$H$_{25}$)NH$_2$ in CHCl$_3$. It is very slightly soluble in CH$_3$OH, C$_2$H$_3$OH and CHCl$_3$, and almost insoluble in other organic solvents [10].

The infrared spectrum was measured (700 to 4000 cm^{-1}; reproduced in paper) [10].

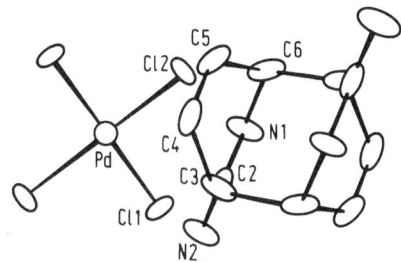

Fig. 21. Crystal structure of $(C_{10}N_4H_{14})[PdCl_4]$ showing one di-cation and one anion; hydrogen atoms omitted for clarity [20].

References:

[1] Berg, R. W. (Spectrochim. Acta A **32** [1976] 1747/57).

[2] Berg, R. W.; Søtofte, I. (Acta Chem. Scand. A **30** [1976] 843/4).

[3] Radova, Z.; Kalalova, E.; Kalal, J.; Kukushkin, Yu. N.; Simanova, S. A.; Konovalov, L. V.; Pak, V. N. (Angew. Makromol. Chem. **81** [1979] 55/62).

[4] Kindberg, B. L.; Amma, E. L. (Acta Cryst. B **31** [1975] 1492/4).

[5] Willett, R. D.; Willett, J. J. (Acta Cryst. B **33** [1977] 1639/41).

[6] Mason, W. R.; Gray, H. B. (J. Am. Chem. Soc. **90** [1968] 5721/9, 5722).

[7] Goggin, P. L.; Mink, J. (J. Chem. Soc. Dalton Trans. **1974** 1479/83).

[8] Klotz, P.; Feldberg, S.; Newman, L. (Inorg. Chem. **12** [1973] 164/8).

[9] Skvortsova, G. G.; Domnina, E. S.; Voropaev, V. N.; Teterin, Yu. A. (Koord. Khim. **12** [1986] 1370/2; C.A. **106** [1987] No. 40787).

[10] Bol'shakov, K. A.; Sinitsyn, N. M.; Borbat, V. F.; Selina, L. I.; Rubtov, M. V. (Zh. Neorgan. Khim. **19** [1974] 172/7; Russ. J. Inorg. Chem. **19** [1974] 66/9).

[11] Doadrio, A.; Craciunescu, D.; Ghirvu, C. (Rev. Chim. Minerale **12** [1975] 129/33).

[12] Craciunescu, D.; Doadrio, A. (Rev. Chim. Minerale **10** [1973] 615/9).

[13] Letuchii, Ya. A.; Lavrent'ev, I. P.; Khidekel', M. L. (Izv. Akad. Nauk SSSR Ser. Khim. **1979** 718/24; Bull. Acad. Sci. USSR Div. Chem. Sci. **1979** 669/74).

[14] Doadrio, A.; Craciunescu, D. (Anales Quim. **70** [1974] 85/7).

[15] Fedyanin, N. P.; Kukushkin, Yu. N.; Khamnuev, G. K.; Pyzhova, L. Ya.; Vatolina, N. L. (Zh. Neorgan. Khim. **26** [1981] 2200/3; Russ. J. Inorg. Chem. **26** [1981] 1183/5).

[16] Harris, C. M.; Livingstone, S. E.; Stephenson, N. C. (J. Chem. Soc. **1958** 3697/702).

[17] Harris, C. M.; Livingstone, S. E.; Reece, I. H. (J. Chem. Soc. **1959** 1505/11).

[18] Folkesson, B.; Carssen, R. (Chem. Scr. **10** [1976] 105/9).

[19] Nefedov, V. I.; Koehler, H.; Fiedler, R. (Koord. Khim. **9** [1983] 1561/2; C.A. **100** [1984] No. 42324).

[20] Navarro-Ranninger, M. C.; Martinez-Carrera, S.; Garcia-Blanco, S. (Polyhedron **4** [1985] 1379/81).

[21] Kukushkin, Yu. N.; Sedova, G. N.; Vlasova, R. A. (Zh. Neorgan. Khim. **23** [1976] 737/40; Russ. J. Inorg. Chem. **23** [1976] 406/8).

[22] Kukushkin, Yu. N.; Belyaev, A. N.; Vlasova, R. A.; Lobadyuk, V. I. (Zh. Neorgan. Khim. **27** [1982] 1500/3; Russ. J. Inorg. Chem. **27** [1982] 844/6).

(C₁₂N₂H₁₄)[PdCl₄]. This, the $[PdCl_4]^{2-}$ salt of the N,N'-dimethyl-4,4'-dipyridylium (paraquat, pqt) di-anion, is made from $PdCl_2$ and $(pqt)Cl_2$ [1].

The crystals are orthorhombic, space group Ibam-D_{2h}^{26}, Z = 4; a = 8.43(2), b = 13.50(2), c = 13.41(2) Å; density by flotation 1.872 g/cm³. The $[PdCl_4]^{2-}$ has planar coordination and the di-cation is twisted. The Pd–Cl distance is 2.32(1) Å. The structure is shown in **Fig. 22** and **Fig. 23** [2].

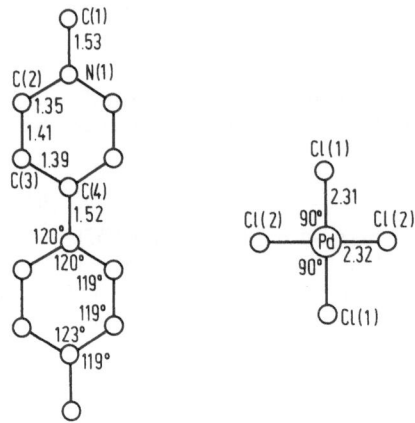

Fig. 22. The anion and cation in (C₁₂N₂H₁₄)[PdCl₄] [2].

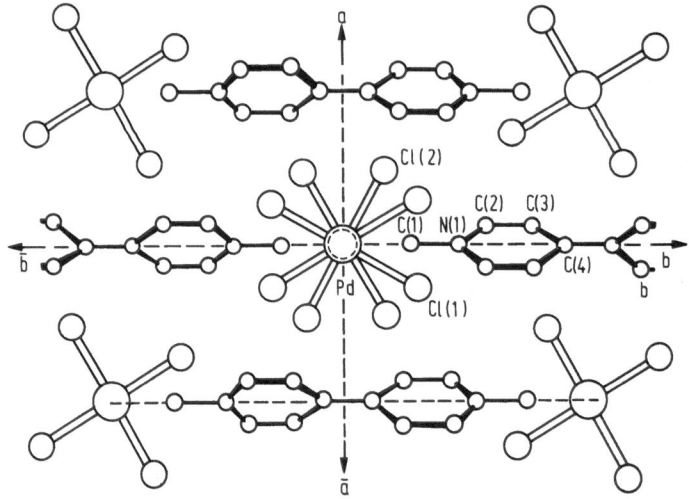

Fig. 23. The structure of (C₁₂N₂H₁₄)[PdCl₄] projected down the c-axis [2].

The electronic absorption spectrum of the solid was measured (200 to 500 nm, data listed and partially assigned in paper) [1].

(C₁₃H₁₈NO)₂[PdCl₄]. This, the iso-indolinium salt, is made by reaction of 3-buten-2-one and the o-palladated N,N-dimethylbenzylamine palladium complex [(C₈H₁₀N)PdCl]₂ [3].

The X-ray crystal structure shows the salt to be triclinic, space group $P\bar{1}$-C_i^1, $Z = 1$; $a = 7.7867(6)$, $b = 8.9834(6)$, $c = 11.7100(8)$ Å, $\alpha = 97.984(5)°$, $\beta = 104.106(6)°$, $\gamma = 110.184(5)°$. The cation is shown in **Fig. 24**; the anion is square-planar (Pd–Cl 2.303(1) Å) [3].

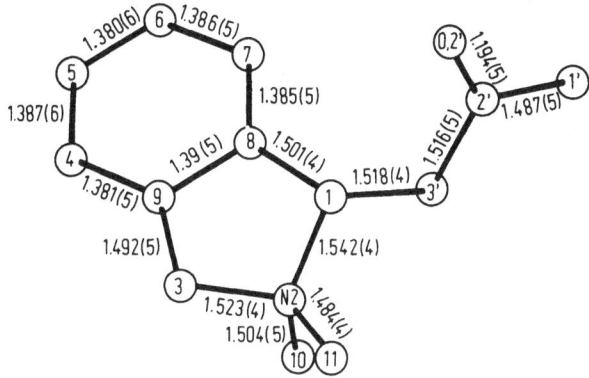

Fig. 24.   X-ray crystal structure of the cation in $(C_{13}H_{18}NO)_2[PdCl_4]$ [3].

$(C_{16}H_{33}NH_3)_2[PdCl_4]$. This cetylammonium salt is made as a dark brown material, sparingly soluble in CH_3OH and C_2H_5OH only, from $C_{16}H_{33}NH_2$ and "$H_2[PdCl_4]$" in $CHCl_3$ [11].

The infrared spectrum was measured from 3000 to 4000 cm^{-1} [11].

$(C_{19}H_{24}N_2H)_2[PdCl_4]$. This, the imipramine [5(3-dimethyl-aminopropyl)-10,11-dihydro-5-dibenzyl azepine] salt, has been made and its electronic spectrum in $CHCl_3$ measured (300 to 800 nm, reproduced in paper) [4].

$Na(C_{19}H_{42}N)[PdCl_4]$ or $H(C_{19}H_{42}N)[PdCl_4]$. This, a cetyltrimethylammonium salt, is made from $Na_2[PdCl_4]$ and cetyltrimethylammonium bromide. The salt is used amongst others for potentiometric titration of a wide variety of anions; its stoichiometry is uncertain [5].

Miscellaneous $R_2[PdCl_4]$ Salts

$(diAmH_2)[PdCl_4]$. This is made by reaction of $PdCl_2$ in HCl with the macroporous copolymer prepared from glycidylmethacrylate and ethylenedimethylacrylate with ethylenediamine. The infrared spectrum was measured (reproduced in paper, 100 to 420 cm^{-1}) [6].

$(PQ^{2+})[PdCl_4]$ is probably formed when $[PdCl_4]^{2-}$ in solution exchanges with an N,N′-dialkyl-4,4′-bipyridinium (PQ^{2+}) derivative; its use in a chemically derivatised electrode for the electrochemical reduction of HCO_3^- to HCO_2^- has been described [7, 8].

$(KL_3)_2[PdCl_4]$ (L_3 = the macrocyclic polyether dibenzo-18-crown-6). This is made from $K_2[PdCl_4]$ and the ether in methanol [9].

$(SO_2)_2[PdCl_4]$. This salt has been briefly mentioned [10].

References:

[1] MacFarlane, A. J.; Williams, R. J. P. (J. Chem. Soc. A **1969** 1517/20).
[2] Prout, C. K.; Murray-Rust, P. (J. Chem. Soc. A **1969** 1520/3).
[3] Ryabov, A. D.; Sakodinskaya, I. K.; Dvoryantsev, S. N.; Eliseev, A. V.; Yatsimirsky, A. K.; Kuz'mina, L. G.; Struchkov, Yu. T. (Tetrahedron Letters **27** [1986] 2169/72).

[4] Dembinski, B. (Chem. Anal. [Warsaw] **28** [1983] 161/5).

[5] Selig, W. (Mikrochim. Acta II **1979** 373/81).

[6] Radova, Z.; Katalova, E.; Kalal, J.; Kukushkin, Yu. N.; Simanova, S. A.; Konovalov, L. V.; Pak, V. N. (Angew. Makromol. Chem. **81** [1979] 55/62).

[7] Stalder, C. J.; Chao, S.; Wrighton, M. S. (J. Am. Chem. Soc. **106** [1986] 3673/5).

[8] Bruce, J. A.; Murahashi, T.; Wrighton, M. S. (J. Phys. Chem. **86** [1982] 1552/63).

[9] Poladyan, V. E.; Ermakov, A. A.; Sokolova, N. P.; Avlasovich, L. M.; Pakhol'chuk, S. F.; Mukelo, K. A. (Zh. Neorgan. Khim. **29** [1984] 1192/6; Russ. J. Inorg. Chem. **29** [1984] 683/5).

[10] Letuchii, Ya. A. (Fiz. Khim. Protsessy Gazov. Kondens. Fazakh. **1979** 41; C. A. **93** [1980] No. 60170).

[11] Bol'shakov, K. A.; Sinitsyn, N. M.; Borbat, V. F.; Selina, L. I.; Rubtsov, M. V. (Zh. Neorgan. Khim. **19** [1974] 122/7; Russ. J. Inorg. Chem. **19** [1974] 66/9).

4.5.2.13 Tetrachloropalladates of Alkali Metals and Ammonium

The LiCl–PdCl$_2$ System. The LiCl–PdCl$_2$ phase diagram is given in **Fig. 25**. Phase equilibria were studied by DTA and X-ray diffraction [1]. A polymorphic transformation of PdCl$_2$ was found at 392°C [1, 3]. In the solid state, at 351°C, Li$_2$[PdCl$_4$] is formed from a mixture of solid solutions based on the low-temperature form of PdCl$_2$(α) and LiCl(β). The thermal effect at 405°C are due to a peritectoid reaction. A eutectic with 67 mol% LiCl forms at 457°C [1].

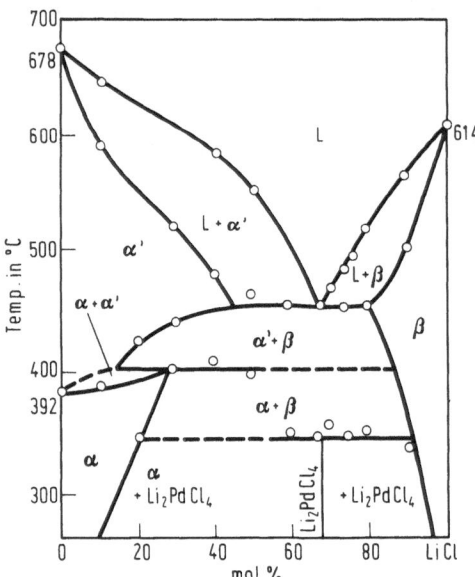

Fig. 25. Equilibrium diagram of the LiCl–PdCl$_2$ system; L = liquid.

Li$_2$[PdCl$_4$] (see also preceding paragraph). This is formed when PdCl$_2$ and LiCl are heated together [1].

The densities of polyimide films into which various species have been incorporated, including Li$_2$[PdCl$_4$], were measured [2].

References:

[1] Safonov, V. V.; Mireev, V. A.; Dubina, S. K. (Zh. Neorgan. Khim. **29** [1984] 1900/1; Russ. J. Inorg. Chem. **29** [1984] 1089/90).
[2] Clar, A. K. S.; Taylor, L. T. (J. Appl. Polym. Sci. **28** [1983] 2393/400).
[3] Blachnik, R.; Alberts, J. E. (Z. Anorg. Allgem. Chem. **489** [1982] 161/72, 163).

The NaCl–PdCl₂ System. The pseudo-binary phase diagram according to DTA results is shown in **Fig. 26.** $Na_2[PdCl_4]$ melts congruently at 430°C. Eutectics containing 20 and 40 mol% $PdCl_2$ melt at 410 and 385°C, respectively [7].

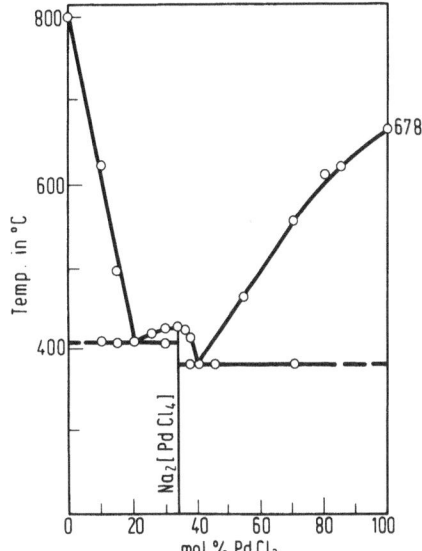

Fig. 26.   The NaCl–PdCl₂ phase diagram.

$Na_2[PdCl_4]$ (see also preceding paragraph). This is made by chlorination of a Pd–NaCl mixture at 650 to 700°C [1]; in solution from $PdCl_2$ and HCl [2]; or, as a brown deliquescent material soluble both in water and in alcohol, by reaction of $PdCl_2$ with an aqueous solution of a stoichiometric quantity of NaCl [3, 4]. It is also made from Pd sponge in water with NaCl and Cl_2, giving a solution of $[PdCl_6]^{2-}$, which is then reduced by addition of more Pd sponge [5], and in solution from "$H_2[PdCl_4]$" and NaCl [6].

The infrared spectrum of $Na_2[PdCl_4]$ has been recorded (40 to 400 cm⁻¹, reproduced in paper) [7] (see table on p. 132).

The Auger binding energies and parameters for a wide variety of complexes including $Na_2[PdCl_4]$ were determined [9, 10].

Uses. The salt has been used in organic solvents to activate substrates for electroless plating [11] and, as a dry powder additive in polycarbonate, as an ignition depressant [8].

References:

[1] Filippov, A. A.; Bazhenov, V. P.; Chumakov, V. G. (Povedenie Blagorod. Metal. Nekot. Met. Protsess **1969** No. 1, pp. 36/47; C. A. **76** [1972] No. 28 203).
[2] Sharpe, A. G. (J. Chem. Soc. **1953** 4177/9).

[3] Brauer, G. (Handbook of Preparative Inorganic Chemistry, Vol. 2, Academic, New York 1965, p. 1584).

[4] de Waal, D. J. A.; Robb, W. (Intern. J. Chem. Kinet. **6** [1974] 309/21, 310).

[5] Clements, F. S.; Nutt, E. V.; International Nickel Co. (Mond), Ltd. (Brit. 922838 [1963]; C. A. **59** [1963] 1312).

[6] Kauffman, G. B.; Tsai, J. H. (Inorg. Syn. **8** [1966] 234/8).

[7] Safonov, V. V.; Chaban, N. G. (Zh. Neorgan. Khim. **24** [1979] 265/8; Russ. J. Inorg. Chem. **24** [1979] 149/50).

[8] Thomas, L. S.; Petrella, R. V.; Dow Chemical Co. (U.S. 4366283 [1982]; C. A. **98** [1983] No. 73426).

[9] Wagner, C. D. (Faraday Discussions Chem. Soc. **60** [1975] 291/300).

[10] Wagner, C. D. (J. Electron Spectrosc. Relat. Phenom. **10** [1977] 305/15).

[11] Sirinyan, K.; Merten, R.; Wolf, G. D. (Eur. Appl. 166360 [1986]; C. A. **105** [1986] No. 80163).

The KCl–PdCl$_2$ System

According to DTA results, K$_2$[PdCl$_4$] is formed which melts congruently at 534°C; see **Fig. 27**. The formation of a new phase was concluded from X-ray diffraction and IR spectroscopic measurements (absorption band at 334 cm^{-1}). A eutectic between KCl and K$_2$[PdCl$_4$] is formed at 511°C, 16 mol% PdCl$_2$, and another one between K$_2$[PdCl$_4$] and PdCl$_2$ containing 47 mol% PdCl$_2$ melts at 379°C.

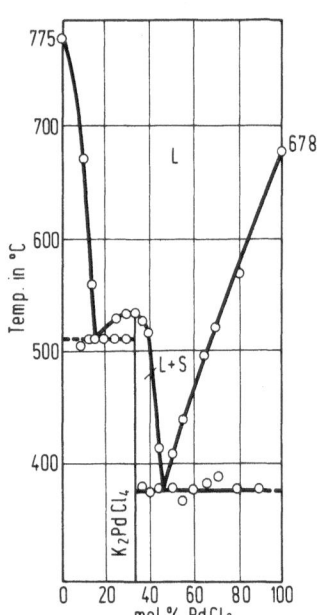

Fig. 27. The KCl–PdCl$_2$ phase diagram.

Reference:

Safonov, V. V.; Mireev, V. A. (Zh. Neorgan. Khim. **28** [1983] 1779/82; Russ. J. Inorg. Chem. **28** [1983] 1004/6).

Potassium Tetrachloropalladate K$_2$[PdCl$_4$]

This is made from a solution of "H$_2$[PdCl$_4$]" and KOH brought to pH = 2, followed by recrystallisation from water-ethanol [1]; from "H$_2$[PdCl$_4$]" and KCl [2]; from Pd in aqua regia followed by treatment with HCl and addition of KCl [3, 12], from PdCl$_2$ and HCl with aqueous KCl solution [4, 5, 13]; from Pd sponge in methanol with Cl$_2$ and KCl [2]; from K$_2$[PdCl$_6$] in water with Pd sponge [6]. Crystals of K$_2$[PdCl$_4$] have been obtained by gel growth techniques [7], and the salt is also the final product of the reaction of K$_2$[Pd(NO$_2$)$_4$] and NOCl [11].

It forms brown [7], orange-brown [8] or olive [13] crystals, moderately soluble in cold water and readily soluble in hot water [6]. It is insoluble in ethanol and acetone [3], and precipitates as golden yellow lamellar crystals by addition of alcohol to its hot aqueous solution [4]. The question of standards of colour in chemicals has been discussed with K$_2$[PdCl$_4$], described as light olive, as an example [17].

Solutions of K$_2$[PdCl$_4$] in water are unstable above 100°C [18].

The X-ray crystal structure shows the salt to be tetragonal, space group P4/mmm-D$_{4h}^1$, Z = 1; a = b = 7.075(2), c = 4.112(3) Å [7, 9]; density by flotation 2.67, calculated density 2.67 g/cm^3 [9]. The coordination about palladium is planar, Pd–Cl = 2.313(2) Å [9].

The heat of formation of crystalline K$_2$[PdCl$_4$], ΔH_{298}, has been calculated as −1093 kJ/mol [14]. The mean Pd–Cl bond energy in K$_2$[PdCl$_4$], using the value of ΔH_{298} = −1082 kJ/mol [15], has been calculated as 351 kJ/mol (lattice energy for the salt 1567 kJ/mol) using the Born-Mayer method, or as 339 kJ/mol (lattice energy 1622 kJ/mol) using the Kapustinskii method [16].

The enthalpies of complexation by K$_2$[PdCl$_4$] and pseudo-lattice energies for the salt have been calculated [19].

The molar magnetic susceptibility of K$_2$[PdCl$_4$] is −132.3 × 10^{-6} cgsu [10].

References:

[1] Rush, R. M.; Martin, D. S.; LeGrand, R. G. (Inorg. Chem. **14** [1975] 2543/50, 2544).

[2] Clements, F. S.; Nutt, E. V.; International Nickel Co. (Mond), Ltd. (Brit. 879074 [1961]; C. A. **56** [1962] 8298).

[3] Livingstone, S. E. (Syn. Inorg. Metal-Org. Chem. **1** [1971] 1/7).

[4] Brauer, G. (Handbook of Preparative Inorganic Chemistry, Vol. 2, Academic, New York 1965, p. 1584).

[5] Perry, C. H.; Athans, D. P.; Young, E. F.; Durig, J. R.; Mitchell, B. R. (Spectrochim. Acta A **23** [1967] 1137/47).

[6] Clements, F. S.; Nutt, E. V.; International Nickel Co. (Mond), Ltd. (Brit. 922838 [1963]; C. A. **59** [1963] 1312).

[7] Marchee, W. G. J.; van Rosmalen, G. M. (J. Cryst. Growth **39** [1977] 358/61).

[8] Yamada, S. (J. Am. Chem. Soc. **73** [1951] 1182/4).

[9] Mais, R. H. B.; Owston, P. G.; Wood, A. M. (Acta Cryst. B **28** [1972] 393/9).

[10] Belova, V. I. (Izv. Sekt. Platiny Drugikh Blagorodn. Metal. Inst. Obshch. Neorgan. Khim. Akad. Nauk SSSR No. 31 [1955] 39/43; C. A. **1956** 15148).

[11] Eskenazi, R.; Raskovan, J.; Levitus, R. (Anales Asoc. Quim. Arg. **51** [1963] 306/14).

[12] Harris, C. M.; Livingstone, S. E.; Reese, I. H. (J. Chem. Soc. **1959** 1505/11).

[13] Kazakova, V. I.; Ptitsyn, B. V. (Zh. Neorgan. Khim. **12** [1967] 620/5; Russ. J. Inorg. Chem. **12** [1967] 323/6).

[14] Shidlovskii, A. A. (Zh. Fiz. Khim. **36** [1962] 1773/6; Russ. J. Phys. Chem. **36** [1962] 957/9).

[15] Goldberg, R. N.; Hepler, L. G. (Chem. Rev. **68** [1968] 229/52, 241).

[16] Hartley, F. R. (Nature Phys. Sci. **236** [1972] 75/7).
[17] Wimmer, F. L.; Poncini, L. (Acta Chim. [Budapest] **120** [1985] 235/7).
[18] Katsurai, T.; Sone, K. (Bull. Chem. Soc. Japan **41** [1968] 519/20).
[19] de Jonge, R. M. (J. Inorg. Nucl. Chem. **38** [1976] 1821/6).

Vibrational Spectra

The data for solid $M_2[PdCl_4]$ salts are listed in the table below; most of them concern the potassium salt. There is an early review [1].

Vibrational spectra of solid $M_2^I[PdCl_4]$ salts:

cation		$\nu_1(A_{1g})$	$\nu_2(B_{1g})$	$\nu_3(A_{2u})$	$\nu_4(B_{2g})$	$\nu_5(A_{2u})$	$\nu_6(E_u)$	$\nu_7(E_u)$	Ref.
(enH$_2$)	IR			183(179)*			336(341)*	(194)*	[15]
(enH$_2$)	R	313	294, 278						[15]
(Bu$_4^n$N)	IR			150			317	167	[11]
Na	IR			160			330	210	[12]
K	R	303.6	269.9					195.8	[2]
K	R	307	274					195	[5]
K	R, IR	310	275		194		336	193	[3]
K	R	310	275						[4]
K	IR						336	193	[3]
K	IR			172			335(242)**	191	[13]
K	IR			170?			325	190	[6]
K	IR			—			336	193	[7]
K	IR			168			332	190.5	[4, 14]
(NH$_4$)	IR	304(305)**		173(188)**			332(338)**	210(219)**	[13]
(NH$_4$)	IR			175?			327	205	[7]
(ND$_4$)	IR			165			333(337)**	200	[13]
Rb	IR			166?			331	188	[7]
Cs	IR			160?			328	183	[7]

Frequencies numbered as in [11]; frequencies in parentheses at * 100 K or ** 20 K; others at ambient temperatures.

The Raman spectrum of solid $K_2[PdCl_4]$ has been measured (20 to 400 cm^{-1}, reproduced in [3]) [2, 3, 4]. There is a brief mention of a Raman study on a single crystal of $K_2[PdCl_4]$ covering ν_1, ν_2 and ν_3 [5]. There are a number of measurements of the infrared spectra (50 to 400 cm^{-1}). The infrared spectrum from 40 to 400 cm^{-1} has been measured at room and at liquid nitrogen temperatures with the data reproduced in the paper [7]. A spectrum reproduced in the paper from 230 to 430 cm^{-1} gives the ν_6 (E_u) band at 335 cm^{-1} [8], and values of 335 cm^{-1} [9] and 334 cm^{-1} [10] have also been given.

References:

[1] Jones, D. W.; Nolan, N. J. (Progr. Inorg. Chem. **9** [1968] 195/275, 236).
[2] Bosworth, Y. M.; Clark, R. J. H. (Inorg. Chem. **14** [1975] 170/7).
[3] Hendra, P. J. (J. Chem. Soc. A **1967** 1298/301).

[4] Hendra, P. J. (Nature **212** [1961] 179).
[5] Beattie, I. R.; Gilson, T. R. (Proc. Roy. Soc. [London] A **307** [1968] 407/29, 412).
[6] Hiraishi, J.; Shimanouchi, T. (Spectrochim. Acta **22** [1966] 1483/91).
[7] Perry, C. H.; Athans, D. P.; Young, E. F.; Durig, J. R.; Mitchell, B. R. (Spectrochim. Acta A **23** [1967] 1137/47).
[8] Acocedji, E.; Lieto, J.; Aune, J. P. (J. Mol. Catal. **22** [1983] 73/8).
[9] Durig, J. R.; Layton, R.; Sink, D. W.; Mitchell, B. R. (Spectrochim. Acta **21** [1965] 1367/78).
[10] Hendra, P. J.; Sadasivan, N. (Spectrochim. Acta **21** [1965] 1271/5).

[11] Goggin, P. L.; Mink, J. (J. Chem. Soc. Dalton Trans. **1974** 1479/83).
[12] Safonov, V. V.; Chaban, N. G. (Zh. Neorgan. Khim. **24** [1979] 265/8; Russ. J. Inorg. Chem. **24** [1979] 149/50).
[13] Adams, D. M.; Berg, R. W. (J. Chem. Soc. Dalton Trans. **1976** 52/8).
[14] Adams, D. M.; Morris, D. M. (Nature **208** [1965] 283).
[15] Berg, R. W. (Spectrochim. Acta A **32** [1976] 1747/57).

Force Constants

Using the Raman and infrared data of [1], Urey-Bradley (UBFF) and generalised valence force field (GVFF) constants were calculated for the anion in solid $K_2[PdCl_4]$ [1]. A modified Urey-Bradley force field (MUBFF) was applied to the compound using far-infrared data [2], a valence force field [3] using the data of [2]; a general valence force field (GVFF) (for "$PdCl_4$", presumably $[PdCl_4]^{2-}$) [4] using the data of [1]; general valence force field [5] using the data of [6]. Force constants for "$[PdCl_4]^-$" [7] (again presumably $[PdCl_4]^{2-}$) and $[PdCl_4]^{2-}$ [8] were evaluated without data sources being listed.

Coriolis coupling constants and mean amplitudes of vibration for $[PdCl_4]^{2-}$ in $K_2[PdCl_4]$ using the data of [1] were calculated [9]. Mean amplitudes of vibration of $[PdCl_4]^{2-}$ using the data of [6] were calculated [5], and also those of "$PdCl_4$" (presumably $[PdCl_4]^{2-}$) were calculated [10] using the data of [1]. Potential energy data for the vibrations of the ion were also calculated [5] using the data of [6].

References:

[1] Hendra, P. J. (J. Chem. Soc. A **1967** 1298/301).
[2] Hiraishi, J.; Shimanouchi, T. (Spectrochim. Acta **22** [1966] 1483/91).
[3] Tranquille, M.; Furel, M. T. (J. Chim. Phys. **68** [1971] 471/4).
[4] Thirugnanasambandam, P.; Mohan, S. (Indian J. Pure Appl. Phys. **12** [1974] 206/9).
[5] Durig, J. R.; Nagarajan, G. (Monatsh. Chem. **100** [1969] 1960/72, 1966).
[6] Perry, C. H.; Athans, D. P.; Young, E. F.; Durig, J. R.; Mitchell, B. R. (Spectrochim. Acta A **23** [1967] 1137/47).
[7] Thyagarajan, G.; Subhedar, M. K. (Indian J. Pure Appl. Phys. **12** [1974] 309/11).
[8] Sanyal, N. K.; Dixit, L.; Pandey, A. N.; Singh, H. S.; Singh, B. P. (Indian J. Pure Appl. Phys. **10** [1972] 493/4).
[9] Lyvin, S. J.; Cyvin, B. N.; Müller, A.; Krebs, B. (Z. Naturforsch. **23a** [1968] 479/81).
[10] Thirugnanasambandam, P.; Mohan, S. (Indian J. Pure Appl. Phys. **13** [1975] 398/401).

Electronic Spectra of Crystalline $K_2[PdCl_4]$

There is a short review on the subject covering the earlier references [1].

The polarised transmission spectra of single crystals of the salt were measured (300 to 700 nm, reproduced in paper) and assigned [2]; also polarised single-crystal reflection [3, 4] and absorption spectra (200 to 700 nm, reproduced in paper with assignments) [4].

Single-crystal absorption spectra of the salt were measured at temperatures from 10 and 295 K (400 to 700 nm, listed in paper and assigned); 10 Dq was given as 23400 cm^{-1} [5]. Spectra of crystalline $K_2[PdCl_4]$ were measured and assigned at 12 and 300 K from 250 to 1000 nm (reproduced in paper) [6]. There are also earlier single-crystal data [7] and dichroic electronic spectra on the solid (200 to1000 nm [8, 9], reproduced in paper [8]). The reflectance spectrum of the powder has also been reported (200 to 800 nm, reproduced in paper) [10], and 300 to 700 nm, reproduced in paper [11]. Assignments of the bands observed for solid $K_2[PdCl_4]$ in [2, 3, 4] have been proposed [12].

For M.O. treatments of the electronic spectra of $[PdCl_4]^{2-}$ see p. 92.

References:

[1] Hush, N. S.; Hobbs, R. J. M. (Progr. Inorg. Chem. **10** [1968] 259/486, 431).
[2] Day, P.; Orchard, A. F.; Thomson, A. J.; Williams, R. J. P. (J. Chem. Phys. **42** [1965] 1973/81).
[3] Anex, B. G.; Takeuchi, N. (J. Am. Chem. Soc. **96** [1974] 4411/6).
[4] Rush, R. M.; Martin, D. S.; LeGrand, R. G. (Inorg. Chem. **14** [1975] 2543/50).
[5] Tuszynski, W.; Gliemann, G. (Z. Naturforsch. **34a** [1979] 211/9).
[6] Francke, E.; Moncuit, C. (Compt. Rend. B **271** [1970] 741/4).
[7] Lobaneva, O. A.; Kononova, M. A. (Probl. Sovrem. Khim. Koord. Soedin. Leningr. Gos. Univ. **1968** No. 2, pp. 180/90; C.A. **70** [1969] No. 15682).
[8] Yamada, S. (J. Am. Chem. Soc. **73** [1951] 1182/4).
[9] Yamada, S.; Tsuchida, R. (Bull. Chem. Soc. Japan **26** [1953] 489/93).
[10] Harris, C. M.; Livingstone, S. E.; Reece, I. H. (J. Chem. Soc. **1959** 1505/11).

[11] Yamada, S. (Nippon Kagaku Zasshi **86** [1965] 753/67; C.A. **64** [1965] 10740).
[12] Elding, L. I.; Olsson, L. F. (J. Phys. Chem. **82** [1978] 69/74).

X-Ray Photoelectronic Spectroscopy (ESCA) Data

ESCA measurements on $K_2[PdCl_4]$ gave binding energies of 338.4 (Pd $3d_{5/2}$), 199.0 (Cl $2p_{3/2}$) and 293.2 eV (K $2p_{3/2}$) [1]; also 343.7 (Pd $3d_{3/2}$) and 338.4 eV (Pd $3d_{5/2}$) [2] and 350.0 (Pd $3d_{3/2}$) and 344.7 eV (Pd $3d_{5/2}$) [3].

A quantum-chemical analysis of ESCA data on hexa- and tetra-halo complexes, including $[PdCl_4]^{2-}$, has been made [4].

References:

[1] Nefedov, V. I.; Zakharova, I. A.; Moiseev, I. I.; Porai-Koshits, M. A.; Vargaftik, M. N.; Belov, A. P. (Zh. Neorgan. Khim. **18** [1973] 3264/8; Russ. J. Inorg. Chem. **18** [1973] 1738/40).
[2] Kumar, G.; Blackburn, J. R.; Aldridge, R. G.; Moddeman, W. E.; Jones, M. M. (Inorg. Chem. **11** [1972] 296/300).
[3] Jørgensen, C. K.; Berthou, H. (Kgl. Danske Videnskab. Selskab Mat. Fys. Medd. **38** [1972] 1/93, 73; C.A. **79** [1973] No. 11782).
[4] Zasukha, V. A.; Volkov, S. V. (Ukr. Khim. Zh. **52** [1986] 227/32; Soviet Progr. Chem. **52** [1986] 227/32).

Miscellaneous Data

The position of the $K\beta_1$ line cf Cl in $K_2[PdCl_4]$ amongst other chloride-containing complexes was used in analysing the electronic structures of such species [1]; inter-ligand interaction in $K_2[PdCl_4]$ was assessed using such data [2]. The experimentally determined profiles of the $L\beta_2$ lines of Pd in $K_2[PdCl_4]$ were resolved into components from the valence molecular orbitals to the L_{III} level of Pd and K level of Cl [3]. Results of $K\beta_1$ and $L\beta_2$ spectra for $K_2[PdCl_4]$ have also been given [4]. Electronic structures of the anions in $K_2[PdCl_4]$ and $K_2[PdCl_6]$ were correlated with the experimental profile of the $K\beta_1$ lines of Cl [5], the Cl K absorption-edge spectra (XANES) of $K_2[PdCl_4]$ and $(NH_4)_2[PdCl_4]$ [6, 7] and L_{III} X-ray absorption edge of these salts were recorded [8]. The fine structure near the ionisation threshold in X-ray L-absorption spectra of $K_2[PdCl_4]$ was calculated using an SCF X_α SW method [5].

$K_2[PdCl_4]$, when γ-irradiated, gives a species which may contain $[PdCl_4]^{3-}$ or $[PdCl_5]^{4-}$ [9].

Empirical numerical effective charges on the atoms in $K_2[PdCl_4]$ were calculated [10]. The molar refraction of $K_2[PdCl_4]$ is 35.62 [11].

References:

[1] Narbutt, K. I.; Nefedov, V. I.; Porai-Koshits, M. A.; Kochetkova, A. P. (Zh. Strukt. Khim. **13** [1972] 451/7; J. Struct. Chem. [USSR] **13** [1972] 422/7).

[2] Mazalov, L. N.; Voityuk, A. A.; Kravtsova, E. A. (Zh. Strukt. Khim. **22** [1981] 169/72; J. Struct. Chem. [USSR] **22** [1981] 282/5).

[3] Nefedov, V. I.; Sadovskii, A. P.; Mazalov, L. N.; Belyaev, A. V.; Gluskin, E. S. (Zh. Strukt. Khim. **12** [1971] 681/8; J. Struct. Chem. [USSR] **12** [1971] 617/22).

[4] Kravtsova, E. A.; Zakharova, I. A.; Mazalov, L. N.; Sadovskii, A. P. (Izv. Sibirsk. Otd. Akad. Nauk SSSR Ser. Khim. Nauk **1973** No. 2, pp. 138/40).

[5] Nefedov, V. I.; Narbutt, K. N. (Zh. Strukt. Khim. **12** [1971] 1019/25; J. Struct. Chem. [USSR] **12** [1971] 937/42).

[6] Sugiura, C.; Muramutsu, S. (J. Phys. Chem. Solids **46** [1985] 1215/9).

[7] Sugiura, C.; Ohashi, M. (J. Chem. Phys. **78** [1983] 88/90).

[8] Sugiura, C.; Muramatsu, S. (J. Chem. Phys. **82** [1985] 2191/5).

[9] Krigas, T.; Nakamura, M. (J. Chem. Phys. **54** [1971] 3378/80).

[10] Kaganyuk, D. S.; Kyskin, V. L.; Kazin, I. V. (Koord. Khim. **10** [1984] 773/5; C.A. **101** [1984] No. 98683).

[11] Shamsiev, A. S.; Obel'chenko, P. F. (Trudy Sredneaziat. Gos. Univ. [2] **33** No. 4 [1952] 57/62; C.A. **1956** 3830).

Reactions of $K_2[PdCl_4]$

Thermal gravimetry was used to study the thermal decomposition under H_2 of single crystals and of powdered $K_2[PdCl_4]$ and the rate of decomposition. The activation energy is 63.5 ± 2 kJ/mol for single crystals and 56.4 ± 1.6 kJ/mol for powders [1]. Orientation studies on the decomposition products (KCl, HCl and Pd) after heating $K_2[PdCl_4]$ in H_2 at 120 to 200°C were made; these orientations arise from parallelism between definite crystallographic directions in the single crystals used. Comparisons between $K_2[PdCl_4]$ in this respect and $K_2[PtCl_4]$ were made [2].

References:

[1] Marchée, W. G. J.; van Rosmalen, G. M.; Hakvoort, G. (Thermochim. Acta **25** [1978] 91/9).

[2] van Rosmalen, G. M.; Burgers, W. G. (Proc. Koninkl. Ned. Akad. Wetenschap. B **77** [1974] 376/97; C.A. **83** [1975] No. 19149).

Ammonium Tetrachloropalladate $(NH_4)_2[PdCl_4]$

This is made by reaction of $PdCl_2$ with aqueous NH_4Cl solutions [1]; from "$H_2[PdCl_4]$" and NH_4Cl [2]; from $PdCl_2$ in HCl with NH_4Cl [3, 4]; from Pd metal and NH_4Cl with Fe_2O_3 at 300°C [5]; from HCl and Pd catalytically oxidised by O_2 [6]; from Pd sponge in methanol with Cl_2 and NH_4Cl [7]; by heating $(NH_4)_2[PdCl_6]$ to 300°C [9]; by heating together NH_4Cl and $Pd(NH_3)_2Cl_2$ [10]

and from oxidation of Pd metal in CH_3CN–HCl mixtures. The deuteriate $(ND_4)_2[PdCl_4]$ is made by repeated recrystallisation of the normal salt from D_2O [12].

Crystals can be grown by slow evaporation of the aqueous solution [5].

The salt forms brown tetragonal needles [4] or long olive-coloured prisms [1]. Thermal decomposition starts at 216°C [3].

The tetragonal crystals, space group $P4/mmm$-D^1_{4h} have unit cell dimensions $a = 7.205(6)$, $c = 4.26(2)$ Å; density by flotation 2.1, calculated density 2.14 g/cm³. The Pd–Cl distance is 2.299(4) Å [7]. The standard X-ray diffraction powder pattern has been recorded for the salt [8]. The unit cell dimensions have also been given as $a = 7.212(4)$, $b = 7.212(4)$, $c = 4.245(2)$ Å, density 2.150 g/cm³ [10].

A neutron diffraction study has been made of $(ND_4)_2[PdCl_4]$ at 295 and 115 K. The crystals are tetragonal, space group $P4/mmm$-D^1_{4h}; $Z = 1$. At 295 K, $a = 7.22(1)$, $c = 4.24(1)$ Å; at 125 K $a = 7.20(1)$, $c = 4.21(1)$ Å. The density is 2.196 g/cm³ at 295 K. The Pd–Cl distance in the square-planar anion is 2.315(7) Å at 295 K and 2.313(5) Å at 125 K; the N–D distances are 1.023(9) and 1.044(4) Å at 295 and at 125 K, respectively [12].

The heat of formation of crystalline $(NH_4)_2[PdCl_4]$ was calculated as -856.1 kJ/mol [13, 14]. Enthalpies of complexation and pseudo-lattice energy calculations have been performed for $(NH_4)_2[PdCl_4]$ [15].

References:

[1] Brauer, G. (Handbook of Preparative Inorganic Chemistry, Vol. 2, Academic, New York 1965, p. 1584).

[2] Clements, F. S.; Nutt, E. V.; International Nickel Co. (Mond), Ltd. (Brit. 879074 [1961]; C.A. **56** [1962] 8298).

[3] Kukushkin, Yu. N.; Sedova, G. N.; Vlasova, R. A. (Zh. Neorgan. Khim. **23** [1978] 737/40; Russ. J. Inorg. Chem. **23** [1978] 406/8).

[4] Perry, C. H.; Athans, D. P.; Young, E. F.; Mitchell, B. R. (Spectrochim. Acta A **23** [1967] 1137/47).

[5] Zvyagintsev, O. E.; Filimonova, V. N. (Izv. Sekt. Platiny Drugikh Blagorodn. Metal. Inst. Obshch. Neorgan. Khim. Akad. Nauk SSSR No. 26 [1951] 69/77; C.A. **1954** 13512).

[6] Letuchii, Ya. A. (Fiz. Khim. Protsessy Gazov. Kondens. Fazakh. **1979** 41; C.A. **93** [1980] No. 60170).

[7] Bell, J. D.; Hall, D.; Waters, T. N. (Acta Cryst. **21** [1966] 440/2).

[8] Swanson, H. E.; Gilfrich, N. T.; Cook, M. I. (NBS-C-539 [1956]; C.A. **1957** 1686).

[9] Sidorenko, Yu. A.; Filippov, A. A.; Lukashenko, E. E.; Chumakov, V. G. (Zh. Neorgan. Khim. **11** [1972] 2234/8; Russ. J. Inorg. Chem. **17** [1972] 1165/7).

[10] Smirnov, I. I.; Ryumin, A. I.; Biokhina, M. L. (Zh. Neorgan. Khim. **30** [1985] 3139/42; Russ. J. Inorg. Chem. **30** [1985] 1783/8).

[11] Makitova, D. D.; Krasochka, O. N.; Atovmyan, L. O.; Lavrent'ev, I. P.; Shul'ga, Y. M.; Revenko, L. V.; Khidekel', M. L. (Koord. Khim. **13** [1987] 383/7).

[12] Larsen, F. K.; Berg, R. W. (Acta Chem. Scand. A **31** [1977] 375/8).

[13] Shidlovskii, A. A. (Zh. Fiz. Khim. **36** [1962] 1773/6; Russ. J. Phys. Chem. **36** [1962] 957/61).

[14] Shidlovskii, A. A. (Teor. Vzryvchatykh Veshchestv **1963** 543/6; C.A. **59** [1963] 12245).

[15] de Jonge, R. M. (J. Inorg. Nucl. Chem. **38** [1976] 1821/6).

Vibrational and Electronic Spectra

The far-infrared spectrum (20 to 400 cm^{-1}, data listed in the table on p. 132) has been measured at room temperature [1, 2] and at 80 K [2]. The infrared spectrum of partially deuteriated $(NH_4)_2[PdCl_4]$ at 80 K has been recorded (2600 to 1200 cm^{-1}, reproduced in paper) and the results interpreted in terms of the symmetry of the $(NH_4)^+$ ion in the crystals [3]; these data were compared with those for other ammonium salts [4].

The electronic absorption spectrum of solid $(NH_4)_2[PdCl_4]$ from 300 to 400 nm was measured [5, 6]; the spectrum was mentioned but no data given [7].

References:

[1] Perry, C. H.; Athans, D. P.; Young, E. F.; Mitchell, B. R. (Spectrochim. Acta A **23** [1967] 1137/47).
[2] Adams, D. M.; Berg, R. W. (J. Chem. Soc. Dalton Trans. **1976** 52/8).
[3] Oxton, I. A.; Knop, O.; Falk, M. (J. Phys. Chem. **80** [1976] 1212/7).
[4] Knop, O.; Oxton, I. A.; Falk, M. (Can. J. Chem. **57** [1979] 404/23, 414).
[5] Babaeva, A. V.; Rudyl, R. I. (Zh. Neorgan. Khim. **1** [1956] 921/9; Russ. J. Inorg. Chem. **1** No. 5 [1956] 42/51).
[6] Babaeva, A. V.; Rudyl, R. I. (Izv. Akad. Nauk SSSR Ser. Fiz. **18** [1954] 729/30; Bull. Acad. Sci. USSR Phys. Ser. **18** [1954] 410/11; C.A. **1956** 7588).
[7] Babaeva, A. V.; Rudyl, R. I. (Zh. Neorgan. Khim. **2** [1957] 552/4; Russ. J. Inorg. Chem. **2** No. 3 [1957] 122/7).

Miscellaneous Properties

A phase transition of $(NH_4)_2[PdCl_4]$ was confirmed by proton T_1 data; the transition is likely to be accompanied by an order-disorder change in an ammonium ion [1]. The thermal decomposition of the salt in N_2, NH_3 and HCl was measured (DTA and TGA curves reproduced, 200 to 500°C for the N_2 and NH_3 atmospheres). It decomposes above 300°C to $PdCl_2$ [2], giving $PdCl_2(NH_3)_2$ above 216°C [11]. It decomposes at 25°C when mixed with Na_2CO_3, the intermediate product being $Na_2[PdCl_4]$ [3]. $(NH_4)_2[PdCl_4]$, when γ-irradiated either alone [4] or doped in NH_4Cl [5], gives ESR signals which may arise from $[PdCl_4]^{3-}$ [4, 5], $[PdCl_5]^{2-}$ [5] or $[PdCl_6]^{3-}$ [6].

The K absorption edge X-ray spectrum of the Cl ligand in $(NH_4)_2[PdCl_4]$ [7] and the L_{III} X-ray absorption edge structure for these salts have been measured [8]. Changes in the dielectric loss of photographic emulsions with frequency in the presence of $(NH_4)_2[PdCl_4]$ as well as of other reagents have been noted [9].

The diamagnetic molar magnetic susceptibility is -126.7×10^{-6} cgsu [10].

References:

[1] Asai, T.; Kiriyama, H. (Chem. Letters **1979** 397/8).
[2] Ryumin, A. I.; Smirnov, I. I.; Biokhina, M. L. (Zh. Neorgan. Khim. **30** [1985] 2582/4; Russ. J. Inorg. Chem. **30** [1984] 1470/1).
[3] Ryumin, A.; Smirnov, I. I.; Bizyukina, N. V.; Krapivko, A. A. (Izv. Vysshikh Uchebn. Zavedenii Tsvetn. Met. **1986** No. 4, pp. 62/5; C.A. **106** [1987] No. 88094).
[4] Krigas, T.; Rogers, M. T. (J. Chem. Phys. **54** [1971] 4769/75).
[5] Fujiwara, S.; Nakamura, M. (J. Chem. Phys. **54** [1971] 3378/80).
[6] Sastry, M. D. (J. Chem. Phys. **64** [1976] 3957/60).
[7] Sugiura, C.; Muramutsu, S. (J. Phys. Chem. Solids **46** [1985] 1215/9).
[8] Sugiura, C.; Muramutsu, S. (J. Chem. Phys. **82** [1985] 2191/5).

138

[9] Peng, B.; Peng, Y.; Li, Z.; Wang, R.; Gao, X.; Wu, X.; Fan, S.; Chen, L.; Wang, W. (Wuhan Daxue Xuebao Ziran Kexueban **1983** 95/111; C.A. **101** [1984] No. 161109).

[10] Belova, V. I. (Izv. Sekt. Platiny Drugikh Blagorodn. Metal. Inst. Obshch. Neorgan. Khim. Akad. Nauk SSSR No. 31 [1955] 39/43; C.A. **1956** 15148).

[11] Kukushkin, Yu. N.; Sedova, G. N.; Vlasova, R. A. (Zh. Neorgan. Khim. **23** [1978] 737/40; Russ. J. Inorg. Chem. **23** [1978] 406/8).

Applications of (NH$_4$)$_2$[PdCl$_4$]

The ammonium salt has been used to sensitise the latent image in AgBr photographic emulsions; such sensitised emulsions were shown to display a much greater oxidisability of the sensitivity centres [1]. The salt imparts higher sensitivity and lowers fog characteristics of silver halide photographic emulsions [2].

References:

[1] Faelens, P.; van Veelen, G. F. (Phot. Korr. **99** [1963] 185/7; C.A. **60** [1964] 2476).
[2] Hoshino, H.; Matsuzaka, M.; Takiguchi, H.; Konishiroku Photo Industry Co. Ltd. (Japan. Kokai Tokkyo Koho 6167845 [1986]; C.A. **105** [1986] No. 124181).

Rubidium Tetrachloropalladates

The PdCl$_2$–RbCl System. The phase diagram of this pseudo-binary system is shown in **Fig. 28**; it was obtained by DTA, TGA and X-ray diffraction. The compound Rb$_2$[PdCl$_4$] melts congruently at 526°C; thermal decomposition begins at 655°C. The compound RbCl·3PdCl$_2$ decomposes in the subsolidus region above 340°C. There is a polymorphic transformation of PdCl$_2$ at 392°C. The eutectic of Rb$_2$[PdCl$_4$] with low-temperature PdCl$_2$ melts at 368°C, it contains 46 mol% RbCl. The Rb$_2$[PdCl$_4$]–RbCl eutectic with 72 mol% RbCl melts at 489°C [1].

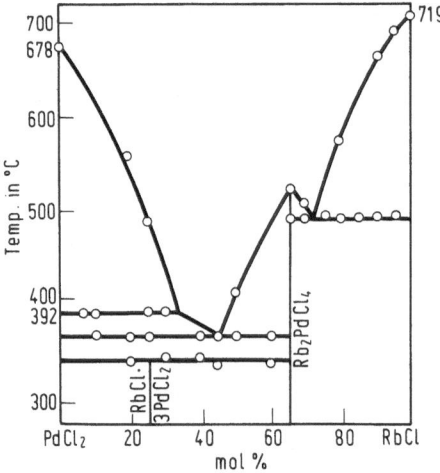

Fig. 28. The PdCl$_2$–RbCl phase diagram.

Rb$_2$[PdCl$_4$]. The compound is made from saturated aqueous K$_2$[PdCl$_4$] and RbCl [2]; from PdCl$_2$ and RbCl in HCl [3]. Crystals of Rb$_2$[PdCl$_4$] were obtained from RbCl, PdCl$_2$ and HCl at pH 1 using a gel-growth technique. It is brown [4].

Crystals of the salt, which is very soluble in water, can be made by adding acetone to the aqueous solution [5].

X-ray diffraction data for $Rb_2[PdCl_4]$ have been given; the salt belongs to the tetragonal $P4/mmm-D_{4h}^1$ space group, $Z=1$; $a=7.27$, $c=4.78$ Å; measured density 2.77, calculated density 2.80 g/cm³ [1]. The values $a=7.228$ and $c=4.340$ Å have also been determined [5]. The Pd–Cl distance in $Rb_2[PdCl_4]$ has been given as 2.309(4) Å [7]. Decomposition begins at 655°C and is appreciable at 850°C for $Rb_2[PdCl_4]$; it melts at 526°C [1].

The far-infrared spectrum (20 to 400 cm⁻¹, reproduced in paper for normal and for liquid nitrogen temperatures) has been recorded (see table on p. 132) [3].

The electronic absorption spectrum of $Rb_2[PdCl_4]$ doped in $Rb_2[HfCl_6]$ has been measured at 2 K (200 to 300 nm) [2].

Thermal decomposition of $Rb_2[PdCl_4]$ under H_2 has been studied [6].

RbCl·3PdCl₂. This species, decomposing at 340°C, was detected in the $RbCl-PdCl_2$ system [1]; see Fig. 26 on p. 138.

References:

[1] Safonov, V. V.; Mireev, V. A. (Zh. Neorgan. Khim. **29** [1984] 1832/5; Russ. J. Inorg. Chem. **29** [1984] 1050/2).
[2] Harrison, T. G.; Patterson, H. H.; Godfrey, J. J. (Inorg. Chem. **15** [1976] 1291/8).
[3] Perry, C. H.; Athans, D. P.; Young, E. F.; Durig, J. R.; Mitchell, B. R. (Spectrochim. Acta A **23** [1967] 1137/47).
[4] Marchée, W. G. J.; van Rosmalen, G. M. (J. Cryst. Growth **39** [1977] 358/61).
[5] Brouwer, G.; van Rosmalen, G. M.; Bennema, P. (J. Cryst. Growth **23** [1974] 228/32).
[6] Marchée, W. G. J.; van Rosmalen, G. M.; Hakvoort, G. (Thermochim. Acta **25** [1978] 91/9).
[7] Keller, H.-L.; Schröder, L. (J. Less-Common Metals **133** [1987] 287/96).

Caesium Tetrachloropalladates

The PdCl₂–CsCl System. The phase diagram constructed after DTA, TGA and X-ray diffraction measurements is shown in **Fig. 29**, p. 140. The compound $Cs_2[PdCl_4]$ melts congruently at 486°C, it begins to decompose at 655°C. The compound $CsCl·3PdCl_2$ melts incongruently at 361°C; perhaps it is partly dissociated below this temperature. The $Cs_2[PdCl_4]$–CsCl ·3PdCl₂ eutectic contains 52 mol% CsCl and melts at 331°C. The $Cs_2[PdCl_4]$–CsCl eutectic containing 68 mol% CsCl melts at 482°C [1].

Cs₂[PdCl₄]. The salt can also be made from $K_2[PdCl_4]$ with CsCl solution [2] or from CsCl, HCl and $PdCl_2$ [3, 4]. Crystals were obtained from CsCl and $PdCl_2$ in HCl at pH=1 using a gel-growth technique [4]. The brown salt is moderately soluble in water [5].

From X-ray data the salt was shown to belong to the tetragonal $P4/mmm-D_{4h}^1$ space group, $Z=1$, with $a=7.53$ and $c=4.65$ Å. The measured density is 3.39 and the calculated density 3.43 g/cm³ [1]. Values of $a=7.438$ and $c=4.656$ Å have also been reported [5].

The far-infrared spectrum (40 to 4000 cm⁻¹, reproduced in paper for normal and liquid nitrogen temperatures) has been recorded (see table on p. 132) [3].

Electronic absorption spectra of $Cs_2[PdCl_4]$ doped in $Cs_2[ZrCl_6]$ at 2 K have been measured (200 to 500 nm, reproduced in paper) and bands assigned [2]. The spectrum of the salt in a LiCl–KCl eutectic at 400°C (300 to 700 nm, reproduced in paper) has been measured and assigned; data were listed also for the eutectic at 500 and 600°C [4].

140

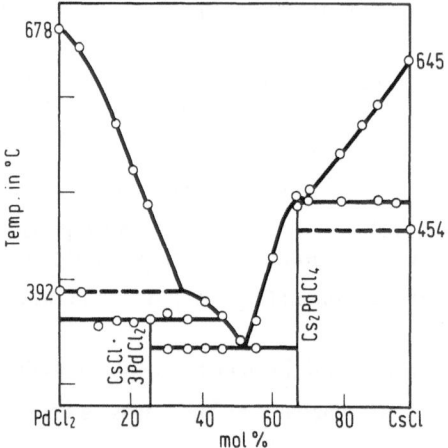

Fig. 29.   The PdCl$_2$–CsCl phase diagram.

Nuclear quadrupole interaction studies using γ-ray angular correlations have been made on a number of salts, including Cs$_2$[PdCl$_4$] as a host lattice [6].

CsCl·3PdCl$_2$. This material, detected in the CsCl–PdCl$_2$ system, decomposes at 361°C [1].

For the infrared spectrum see table on p. 132.

References:

[1] Safonov, V. V.; Mireev, V. A. (Zh. Neorgan. Khim. **29** [1984] 1832/5; Russ. J. Inorg. Chem. **29** [1984] 1050/2).
[2] Harrison, T. G.; Patterson, H. H.; Godfrey, J. J. (Inorg. Chem. **15** [1976] 1291/8).
[3] Perry, C. H.; Athans, D. P.; Young, E. F.; Durig, J. R.; Mitchell, B. R. (Spectrochim. Acta A **23** [1967] 1137/47).
[4] Bailey, R. A.; McIntyre, J. A. (Inorg. Chem. **5** [1966] 1824/5).
[5] Marchée, W. G. J.; van Rosmalen, G. M. (J. Cryst. Growth **39** [1977] 358/61).
[6] Haas, H.; Shirley, D. A. (J. Chem. Phys. **58** [1973] 3339/55, 3350).

4.5.2.14 Alkaline-Earth and Heavy-Metal Chloropalladates

Ca[PdCl$_4$]. This has been used in organic solution as an activating substrate for electroless plating [1].

The PdCl$_2$–SrCl$_2$ System. The phase diagram, see **Fig. 30**, shows that Sr[PdCl$_4$] melts congruently at 585°C and that two eutectics occur, at 567±5°C and at 576±5°C, at SrCl$_2$ concentrations of 47 and 60 mol%, respectively [18].

Sr[PdCl$_4$]. This is made by heating a 1:1 SrCl$_2$:PdCl$_2$ mixture in a sealed and evacuated quartz ampoule to 500°C for 24 h.

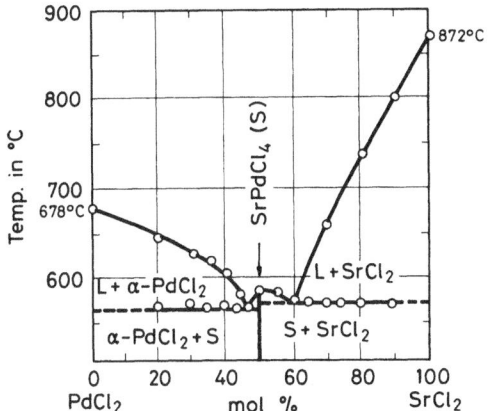

Fig. 30. Phase diagram $PdCl_2$–$SrCl_2$ [18].

Its thermal decomposition was studied from 20 to 1000°C (differential thermal analysis and thermal gravimetric analysis curves reproduced in paper) [15].

The infrared spectrum of the complex was measured (200 to 500 cm^{-1}, reproduced in paper); ν_{PdCl} lies at 328 cm^{-1} [18].

Ba[PdCl$_4$]. This is made by heating a 1:1 $SrCl_2$: $PdCl_2$ mixture in a sealed and evacuated quartz ampoule to 500°C for 24 h.

Its thermal decomposition was studied from 20 to 1000°C (differential thermal analysis and thermal gravimetric analysis curves reproduced in paper) [15].

The PdCl$_2$–TlCl System. This system has been studied by DTA, X-ray diffraction and IR spectroscopy; the phase diagram is shown in **Fig. 31**, p. 142. The two compounds formed melt incongruently, $Tl_2[PdCl_4]$ at 359°C, $TlCl \cdot 3PdCl_2$ at 352°C. The eutectic contains 52 mol% TlCl and melts at 301°C. $PdCl_2$ and $Tl_2[PdCl_4]$ undergo polymorphic transformations at 396 and 321°C, respectively [2].

Tl$_2$[PdCl$_4$]. This has been made from $PdCl_2$–TlCl melts.

X-ray powder diffraction data have been given for the salt. There is a low-temperature modification (existing below 321°C) which belongs to the tetragonal 84/mmm-D_{4h}^1 space group, $Z = 1$; $a = 7.56$, $c = 4.75$ Å; observed density 3.98 g/cm^3, calculated density 4.01 g/cm^3. It has the $K_2[PtCl_4]$ structure. The infrared spectrum was measured (200 to 500 cm^{-1}, reproduced in paper); the Pd–Cl stretch appears at 335 cm^{-1} [2].

TlCl·3PdCl$_2$. This species, identified from the TlCl–PdCl$_2$ system, may contain $[Pd_2Cl_6]^{2-}$ anions. The infrared spectrum was measured (200 to 500 cm^{-1}, reproduced in paper) [2].

Tl$_3$PdCl$_5$. This is made by fusion of solid TlCl with $PdCl_2$ in a 3:1 ratio under vacuum at 400°C followed by slow cooling. It melts at 315 ± 5°C.

It is orthorhombic, space group Pbca-D_{2h}^{15}, $Z = 8$; $a = 7.475(4)$, $b = 16.240(7)$, $c = 16.666(5)$ Å. The arrangement of $[PdCl_4]^{2-}$ units in the unit cell is shown in **Fig. 32**, p. 142; the palladium has approximately square-planar coordination with Pd–Cl distances of 2.27(1), 2.30(2) and 2.31(2) Å [14].

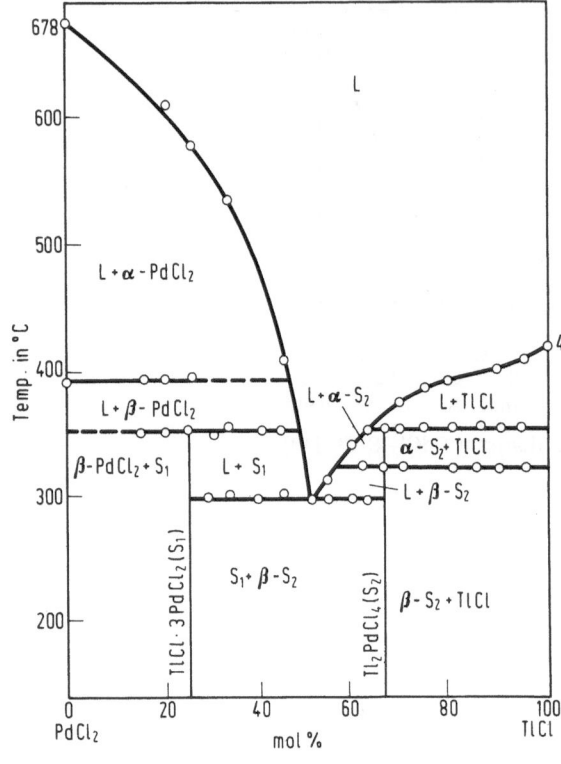

Fig. 31. Equilibrium diagram of the PdCl$_2$–TlCl system [2].

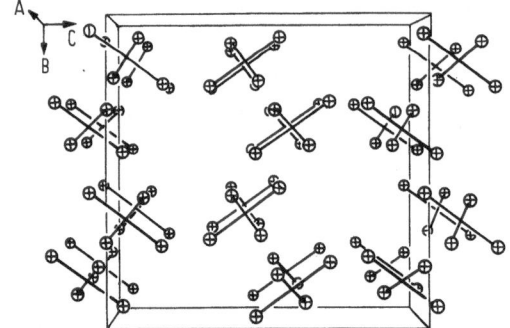

Fig. 32. Arrangement of [PdCl$_4$]$^{2-}$ groups in the unit cell of Tl$_3$PdCl$_5$ [14].

Germaniumchloropalladate(?)

Heating PdCl$_2$ with [NEt$_4$][GeCl$_3$] at 120°C affords a red dispersion which is catalytically active in olefin isomerisation, hydrogenation and carbonylation reactions [33].

The PdCl$_2$–SnCl$_2$ System

Reactions between PdCl$_2$·2H$_2$O, HCl and SnCl$_2$·2H$_2$O (or Sn) at temperatures in the range 300 to 400°C afford alloys of composition Pd$_3$Sn, Pd$_2$Sn, PdSn, PdSn$_2$ and Pd$_3$Sn$_2$ as well as solid solutions of tin in palladium [19]. The addition of ~0.1 to 3% $PdCl_2$ to molten $[NEt_4][SnCl_3]$ at 113°C affords a deep red dispersion capable of catalysing isomerisation, hydrogenation and carbonylation of olefins [20]. Methanolic HCl solutions of palladium ($PdCl_2$?) and $SnCl_2$ are reported to yield $[AsPh_4]_4[(Cl_3Sn)_2PdCl_2Pd(SnCl_3)_2]$ [21] and $[NBu_4]_2[Pd(SnCl_3)_2I_2]$ [22] when treated with $[AsPh_4]Cl$ and $[NBu_4]$ I, respectively. Numerous papers, only a selection of which are listed here [23 to 27], report on the use of $PdCl_2/SnCl_2$ activating solutions ("combined activating solutions") for electroless metal coating of nonconducting surfaces. X-ray diffraction data indicate that the active species is $SnPd_7Cl_{16}$ [5, 9]. Spectrophotometric studies on solutions of $PdCl_2/SnCl_2$ in aqueous HCl have provided evidence of formation of a palladium hydrosol and yellow, green or red-brown complexes of unspecified stoichiometry. Spectra for solutions of varying concentration and pH (300 to 800 nm) are reproduced in the paper [28].

The $PdCl_2$–$PbCl_2$ System

The phase (fusion) diagram of this system (reproduced in paper) is of the eutectic type, the eutectic mixture melts at 430°C and contains 50 mol% $PdCl_2$ [29].

The $PdCl_2$–$NbCl_5$ System. The equilibrium diagram is of the eutectic type. The position of the liquidus line could be established only in the concentration range 75 to 100 mol% $NbCl_5$ and up to the critical temperature of $NbCl_5$ (371°C). Thermal effects were observed at 244 to 270°C and attributed to the boiling of $NbCl_5$. The eutectic melting at 197°C is claimed to contain 96.5 mol% $NbCl_5$ [16].

Systems of $PdCl_2$ with $FeCl_2$ and $FeCl_3$

$PdCl_2$–$FeCl_2$. The phase (fusion) diagram of this system, reproduced in the paper, is of the eutectic type. The eutectic mixture melts at 528 ± 5°C and contains ~56% $FeCl_2$ [18].

$PdCl_2$–$FeCl_3$. The phase (fusion) diagram of this system (reproduced in paper) is of the eutectic type, the eutectic mixture melts at 288°C and contains ~5% $PdCl_2$ [29].

Reactions with Copper(II) Chloride

Separation of $PdCl_2$ and $CuCl_2$ from $PdCl_2/CuCl_2$ catalyst solutions by precipitation of copper oxalate [30] and by selective crystallisation [31] has been reported. The solubilities of $CuCl_2$ and $PdCl_2$, and their mixture in ethanol, acetal and ethanol/acetal have been investigated, calculated apparent heats of solution for $PdCl_2$ are 50 kJ/mol (EtOH) and 2.5 kJ/mol (acetal) [32].

The $AgCl$–$PdCl_2$ System. This is shown in **Fig. 33**, p. 144, as constructed according to DTA and X-ray measurements. The α-β transition of $PdCl_2$ was found at 669 K. Two compounds formed peritectically were claimed: $Ag_2[PdCl_4]$ at 582 K and another one at 610 K containing ~75 mol% $PdCl_2$, tentatively designated $AgPd_3Cl_7$ [3]. A eutectic with 50% $PdCl_2$ melting at 302°C was claimed by [18]. For the AgCl–NaCl–$PdCl_2$ system see p. 144.

$AgPd_3Cl_7$. This species was tentatively identified from the phase diagram shown in Fig. 31 [3].

$Ag_2[PdCl_4]$. The existence of this was established from differential thermal analysis and X-ray methods and a phase diagram obtained (Fig. 31) for the AgCl–$PdCl_2$ system. In addition to $Ag_2[PdCl_4]$ two other unidentified phases were observed [3].

$[Pd(NH_3)_4][PdCl_4]$. This is made by mixing equimolar proportions of $[Pd(NH_3)_4]^{2+}$ and $[PdCl_4]^{2-}$ and cooling to 0°C [4 to 7] or from $PdCl_2$ and a slight excess of NH_4OH [8]. It is formed during the dissolution of Pd metal in a solution of NH_4Cl by an alternating electric current [8].

It is pink, decomposing above 180°C to trans-$PdCl_2(NH_3)_2$ [9]. The salt is tetragonal, a = 8.96, c = 6.49 Å [7] or a = 8.98, c = 6.49 Å [9].

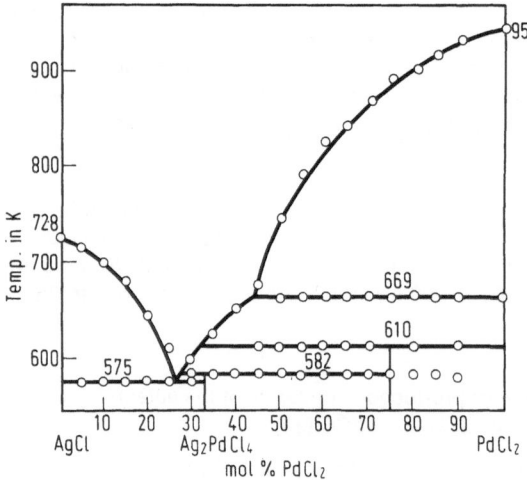

Fig. 33. The AgCl–PdCl₂ phase diagram [3].

The polarised single-crystal electronic absorption spectrum (200 to 600 nm, reproduced in paper [4] and at 6 and 300 K from 250 to 800 nm, reproduced in paper [7]) gave evidence for there being a delocalised excited state as shown by an intense ultraviolet transition [4]. Thermal decomposition studies were carried out; it decomposes at 227°C to give $PdCl_2(NH_3)_2$: ΔH is -26.7 kJ/mol for this reaction [11]; ΔH and the energy of activation for the process were given as -22.6 and 353 kJ/mol [11].

[Pden₂][PdCl₄]. This is made from $K_2[PdCl_4]$ and $[Pden_2]Cl_2$ in water [1, 7].

The trigonal crystals have cell dimensions a = 12.294, b = 8.754, c = 6.870 Å; $\alpha = 110.6°$, $\beta = 82.9°$, $\gamma = 109.9°$. The polarised electronic spectrum for a single crystal was measured from 300 to 700 nm at 6 and 300 K (reproduced in paper) [7].

It decomposes at 139°C to $PdCl_2$en, and the ΔH of decomposition is -3.3 kJ/mol [11].

[Pdbipy₂][PdCl₄]. This pink material is made from 2,2′-bipyridyl (bipy) and $[PdCl_4]^{2-}$ in water [10].

[Pd(NH₂CH₃)₄][PdCl₄]. The decomposition temperature of this is 139°C, and ΔH for the process is -14.2 kJ/mol. The product of decomposition is $PdCl_2(NH_2CH_3)_2$ [11]; $\Delta H = -20.9$ kJ/mol; the energy of activation of thermal decomposition was found to be 163 kJ/mol [17].

[Pd(NH₂C₂H₅)₄][PdCl₄]. The decomposition temperature of this is 120°C and the associated ΔH is -6.7 kJ/mol [11]; $\Delta H = -14.6$ kJ/mol; energy of activation of the process: 149.2 kJ/mol [17].

[Pd(C₄H₈O₂NS)₂Cl₂][PdCl₄] is made from an aqueous solution of $Na_2[PdCl_4]$ and S-methyl-L-cysteine, and is yellow [12].

[Pd(CH₃CH(NH₂)CH₂NH₂)₂][PdCl₄]. This decomposes at 254°C to $Pd(CH_2CH(NH_2)CH_2NH_2)_2Cl_2$ [11]. The thermal decomposition was followed from 20 to 300°C [17].

References:

[1] Sirinyan, K.; Wolf, G. D.; Bayer A.-G. (Eur. Pat. Appl. EP 166360 [1986]; C.A. **105** [1986] No. 80163).

[2] Mireev, V. A.; Safonov, V. V. (Zh. Neorgan. Khim. **30** [1985] 777/9; Russ. J. Inorg. Chem. **30** [1985] 435/6).

[3] Blachnik, R.; Alberts, J. E. (Z. Anorg. Allgem. Chem. **489** [1982] 161/72).

[4] Anex, B. G.; Foster, S. I.; Fucaloro, A. F. (Chem. Phys. Letters **18** [1973] 126/8).

[5] Miller, J. R. (J. Chem. Soc. **1961** 4452/7).

[6] Kauffman, G. B.; Tsai, J. H. (Inorg. Syn. **8** [1966] 234/8).

[7] Rodgers, M. L.; Martin, D. S. (Polyhedron **6** [1987] 225/54, 226, 232, 247).

[8] Fokin, M. N.; Makarova, N. G.; Mazurina, I. I. (Zashch. Metal. **10** [1974] 82/3; C.A. **80** [1974] No. 127325).

[9] Miller, J. R. (Proc. Chem. Soc. **1960** 3/8).

[10] McCormick, B. J.; Jaynes, E. N.; Kaplan, R. I. (Inorg. Syn. **13** [1972] 216/8).

[11] Kukushkin, Yu. N.; Sedova, G. N.; Budanova, V. F. (Zh. Neorgan. Khim. **25** [1980] 200/6; Russ. J. Inorg. Chem. **25** [1980] 108/12).

[12] McAuliffe, C. A. (Inorg. Chem. **12** [1973] 1699/701).

[13] Brauer, G. (Handbook of Preparative Inorganic Chemistry, Vol. 2, Academic, New York 1965, p. 1585).

[14] Keller, H.-L.; Schröder, L. (J. Less-Common Metals **133** [1987] 287/96).

[15] Mireev, V. A.; Podorzhyni, A. M.; Safonov, V. V. (Zh. Neorgan. Khim. **32** [1987] 1767/9; Russ. J. Inorg. Chem. **32** [1987] 1048/9).

[16] Safonov, V. V.; Mireev, V. A.; Dubina, S. K. (Zh. Neorgan. Khim. **29** [1984] 1900/1; Russ. J. Inorg. Chem. **29** [1984] 1089/90).

[17] Kukushkin, Yu. N.; Bakhireva, S. I. (Zh. Neorgan. Khim. **21** [1976] 2721/3; Russ. J. Inorg. Chem. **21** [1976] 1497/9).

[18] Safonov, V. V.; Mireev, V. A. (Zh. Neorgan. Khim. **33** [1988] 1076/7; Russ. J. Inorg. Chem. **33** [1988] 610/1).

[19] Evstigneeva, T. L.; Nekrasov, I. Ya. (Dokl. Akad. Nauk. SSSR **238** [1978] 229/32; C.A. **88** [1978] No. 95569).

[20] Parshall, G. W.; Du Pont de Nemours, E. I. & Co. (U.S. 3565823 [1971]; C.A. **74** [1971] No. 146809).

[21] Khattak, M. A.; Magee, R. J. (Chem. Commun. **1965** 400).

[22] Patron, L.; Contescu, A.; Rolea, G.; Gheorghiu, C. (Rev. Roumaine Chim. **31** [1986] 341/4).

[23] Wang, Y. H.; Wan, C. C. (Plat. Surf. Finish. **69** [1982] 59/61; C.A. **98** [1983] No. 7188).

[24] Cohen, R. L.; West, K. W. (J. Electrochem. Soc. **120** [1973] 502/8).

[25] de Minjer, C. H.; v. d. Boom, P. F. J. (J. Electrochem. Soc. **120** [1973] 1644/50).

[26] Osaka, T.; Takematsu, H.; Nihei, K. (J. Electrochem. Soc. **127** [1980] 1021/9).

[27] Rantell, A.; Holtzman, A. (Plating **61** [1974] 326; C.A. **81** [1974] No. 98824).

[28] Zayats, A. I.; Psareva, T. S.; Shabanov, V. F. (Zh. Neorgan. Khim. **21** [1976] 732/7; Russ. J. Inorg. Chem. **21** [1976] 393/6).

[29] Safonov, V. V.; Mireev, V. A. (Zh. Neorgan. Khim. **24** [1979] 2577/9; Russ. J. Inorg. Chem. **24** [1979] 1433/4).

[30] Igarishi, Y.; Fukuhara, Y.; Harada, T.; Morishige, K.; Muratani, T.; Murakami, M.; Japan Gas Chem. Ind. Co. Ltd. (Japan. 6802210 1968/1963; C.A. **69** [1968] No. 68661).

[31] Masaki, S.; Fujiwara, Y.; Japan Oil Co. Ltd. (Japan. 13354 [1963]; C.A. **60** [1964] 6521).

[32] Gurevich, V. R.; Prudnikova, G. S.; Tret'yakov, V. F. (Khim. Prom. [Moscow] **44** [1968] 389/90; C.A. **69** [1968] No. 30664).

[33] Parshall, G. W.; Du Pont de Nemours, E. I. & Co. (U.S. 3565823 [1971]; C.A. **74** [1971] No. 146809).

4.5.3 Chloropalladates with Coordination Numbers >4

Absorption spectral data on $[PdCl_4]^{2-}$ in HCl have been interpreted as indicating that five- or six-coordinate $[PdCl_5]^{3-}$ or $[PdCl_6]^{4-}$ species are formed in aqueous solution [1, 2, 3], the stability of the chloro species being less than those of the possible corresponding bromo complexes [1].

However, it has been suggested that the spectral shifts could equally well be ascribed to changes brought about by aquation or hydrolysis [4, 5, 6]. Electronic spectra of $PdCl_2$ in excess HCl [7, 8] or in a LiCl–KCl eutectic at 450°C were not inconsistent, however, with the presence of octahedral or distorted octahedral chloropalladates [7].

For **chloro-bromo** complexes $[PdCl_nBr_{4-n}]$ see Section 5.2.9, p. 203; for **chloro-iodo** complexes $[PdCl_nI_{4-n}]^{2-}$ see Section 6.3, p. 215.

References:

[1] Grinberg, A. A.; Kiseleva, N. V. (Zh. Neorgan. Khim. **3** [1958] 1804/9; Russ. J. Inorg. Chem. **3** [1958] 119/26).
[2] Harris, C. M.; Livingstone, S. E.; Reece, I. H. (J. Chem. Soc. **1959** 1505/11).
[3] Sundaram, A. K.; Sandell, E. B. (J. Am. Chem. Soc. **77** [1955] 855/7).
[4] Jørgensen, C. K. (Absorption Spectra and Chemical Bonding in Complexes, Pergamon, New York 1962, p. 259).
[5] Shchukarev, S. A.; Lobaneva, D. A.; Ivanova, M. A.; Kononova, M. A. (Vestn. Leningr. Univ. Fiz. Khim. **16** No. 2 [1961] 152/5; C.A. **1961** 24362).
[6] Kukushkin, Yu. N.; Simanova, S. A.; Fedorova, G. I. (Zh. Prikl. Khim. **41** [1968] 1604/6; J. Appl. Chem. [USSR] **41** [1968] 1521/3).
[7] Dickinson, J. R.; Johnson, K. R. (Can. J. Chem. **45** [1962] 1631/6).
[8] Kravchuk, L. S.; Stremok, I. P.; Markevich, S. V. (Zh. Neorgan. Khim. **21** [1976] 228/31; Russ. J. Inorg. Chem. **21** [1976] 391/3).

4.5.4 Hexachloropalladates(II) $[PdCl_6]^{4-}$

There is no direct evidence for the existence of such a species, though $[PdCl_6]^{4-}$ is a possible intermediate in the reaction of $[PdCl_4]^{2-}$ with OH radicals [1]. An electronic spectrum has been briefly mentioned for $[PdCl_6]^{4-}$ [2]; shifts in the electronic spectrum of $[PdCl_4]^{2-}$ in 0.1M aqueous $HClO_4$ consequent on addition of Cl^- provide some evidence for its existence [3], and this was also the conclusion reached from spectrophotometric data on $[PdCl_4]^{2-}$ in water with added Cl^- [4]. The ion is probably involved in the oxidation by Cl_2 of $[PdCl_4]^{2-}$ [5].

References:

[1] Broszkiewicz, R. K. (Intern. J. Radiat. Phys. Chem. **6** [1974] 249/58).
[2] Kiss, A. (Z. Anorg. Allgem. Chem. **282** [1955] 141/8, 147).
[3] Sundaram, A. K.; Sandell, E. B. (J. Am. Chem. Soc. **77** [1955] 855/7).
[4] Grinberg, A. A.; Kiseleva, N. V. (Zh. Neorgan. Khim. **3** [1958] 1804/9; Russ. J. Inorg. Chem **3** No. 8 [1958] 119/26).
[5] Mureinik, R. J.; Pross, E. (J. Coord. Chem. **8** [1978] 127/33).

4.5.5 Chloropalladates(III)

Evidence for these relies mainly on ESR evidence [1, 2].

[PdCl$_4$]$^-$. The possible existence of this species, generated by reaction of OH radicals with [PdCl$_4$]$^{2-}$, has been mentioned [3].

[PdCl$_5$]$^{2-}$. This may be the species produced by γ-irradiation of K$_2$[PdCl$_4$] or (NH$_4$)$_2$[PdCl$_4$] at 77 K; ESR data were obtained [1].

[PdCl$_6$]$^{3-}$. It is possible that this could be present in one of the species generated by γ-irradiation of NH$_4$Cl crystals doped with PdCl$_2$ or (NH$_4$)$_2$[PdCl$_4$] at room temperature [2]; also by γ-irradiation of K$_2$[PdCl$_4$] or (NH$_4$)$_2$[PdCl$_4$] at 77 K, ESR data were obtained [1].

References:

[1] Krigas, T.; Rogers, M. T. (J. Chem. Phys. **54** [1971] 4769/75).
[2] Sastry, M. D. (J. Chem. Phys. **64** [1976] 3957/60).
[3] Broszkiewicz, R. (Intern. J. Radiat. Phys. Chem. **6** [1974] 249/58).

4.5.6 Hexachloropalladates(IV) [PdCl$_6$]$^{2-}$

See "Palladium" 1942, pp. 285, 308, 314/5, 323/4, 328/9, 332, 334/8, 340.

Formation. Preparation

These strongly oxidising species are normally made by oxidation of [PdCl$_4$]$^{2-}$ with Cl$_2$ or ClO$^-$ in HCl.

The mechanism of formation of [PdCl$_6$]$^{2-}$ from oxidation of [PdCl$_4$]$^{2-}$ by Cl$_2$ in aqueous solution probably involves the intermediacy of [PdCl$_6$]$^{4-}$ [1]. The [PdCl$_6$]$^{2-}$ ion is formed by anodic dissolution of Pd in HCl [2].

References:

[1] Mureinik, R. J.; Pross, E. (J. Coord. Chem. **8** [1978] 127/33).
[2] Kammel, R.; Takei, T.; Winterhager, H.; Yamamoto, H. (Metalloberfläche **22** [1968] 107/11).

Vibrational Spectra

There is an early review on the vibrational spectra of hexahalo ions, including [PdCl$_6$]$^{2-}$ [1].

The Raman spectrum of [PdCl$_6$]$^{2-}$ in aqueous solution has been measured and fundamental modes assigned [2, 3], see table below.

Raman spectra of $[PdCl_6]^{2-}$:

$\nu_1(A_{1g})$	$\nu_2(E_g)$	$\nu_3(F_{1u})$	$\nu_4(F_{1u})$	$\nu_5(F_{2g})$	$\nu_6(F_{2g})$	Ref.
317.0	293.4			154.3		[2]
317	292			164		[3]

Intensities of the $\nu_1(A_{1g})$ fundamentals were used to obtain values for the palladium-chloride bond-polarisability derivatives at zero exciting frequency and the data compared with those obtained for a number of other halogeno complexes [2]. The polarised resonance Raman spectrum was measured for $[PdCl_6]^{2-}$ in 2M HCl containing dissolved Cl_2 (spectrum 100 to 800 cm^{-1} reproduced in paper) [4].

Most force constant data for $[PdCl_6]^{2-}$ are given under $K_2[PdCl_6]$ on p. 152; mean amplitudes of vibration, however, for the anion using the data of [3] have been calculated [5, 6], as have some force constants [6].

References:

[1] Clark, R. J. H. (Halogen Chem. **3** [1967] 85/121, 95).
[2] Bosworth, Y. M.; Clark, R. J. H. (J. Chem. Soc. Dalton Trans. **1974** 1749/61, 1750).
[3] Woodward, L. A.; Creighton, J. A. (Spectrochim. Acta **17** [1961] 594/9).
[4] Hamaguchi, H. (J. Chem. Phys. **69** [1978] 569/78, 572).
[5] Nagarajan, G. (Current Sci. [India] **32** [1963] 448/9).
[6] Sundaram, S. (Proc. Phys. Soc. [London] **91** [1967] 764/7).

Electronic Spectra

For a review of early data see [1,2]. Electronic spectra of $[PdCl_6]^{2-}$ were measured over the stated range and reproduced in the papers: 200 to 750 nm, in 2 M HCl containing dissolved Cl_2 [3]; in 2 M HCl containing dissolved Cl_2, 200 to 600 nm [4]; from 200 to 500 nm in HCl with Cl_2 [5]; in HCl containing dissolved Cl_2, 210 to 600 nm [6].

The necessity of using strong oxidising agents with $[PdCl_6]^{2-}$ for recording the electronic spectrum of the ion has been emphasised; HCl must also be present, for aqueous solutions, at concentrations in excess of 1 M [7].

Data were listed (200 to 400 nm) for $[PdCl_6]^{2-}$ in HCl + ClO^- [7]. Assignments of bands were proposed [3, 5].

The magnetic circular dichroism (MCD) spectrum was recorded (200 to 750 nm, reproduced in paper) and assigned [3].

A modified INDO-MO study of the electronic spectra of a number of halogeno-ions, including $[PdCl_6]^{2-}$, using the data of [2], has been reported [8]; the same data of [2] were used for CNDO-MO calculations [9].

References:

[1] Jørgensen, C. K. (Progr. Inorg. Chem. **12** [1970] 101/58, 109).
[2] Jørgensen, C. K. (Advan. Chem. Phys. **5** [1963] 33/146, 110).
[3] Henning, G. N.; Dobosh, P. A.; McCaffery, A. J.; Schatz, P. N. (J. Am. Chem. Soc. **92** [1970] 5377/82).
[4] Hamaguchi, H. (J. Chem. Phys. **69** [1978] 569/78, 572).
[5] Jørgensen, C. K. (Mol. Phys. **2** [1959] 309/32, 313).
[6] Cohen, A. J.; Davidson, N. (J. Am. Chem. Soc. **73** [1951] 1955/8).

149

[7] Broszkiewicz, R. K. (Intern. J. Radiat. Phys. Chem. **6** [1974] 249/58).
[8] Sakaki, S.; Nishikawa, M.; Tsuru, N.; Yamashita, T.; Ohyoshi, A. (J. Inorg. Nucl. Chem. **41** [1979] 673/80).
[9] Sakaki, S.; Kato, H. (Bull. Chem. Soc. Japan **46** [1973] 2227/9).

Electrode Potentials. Stability Constants. Thermodynamic Data

The half-wave electrode potential for the reversible $[PdCl_6]^-/[PdCl_6]^{2-}$ couple (as Bu_4^nN salts in CH_2Cl_2, versus the standard calomel electrode) is $+2.92$ V. The value was compared with that for a number of other hexahalo salts [1]. The E_0 value for $[PdCl_6]^{2-}/[PdCl_4]^{2-}$ is given as $+1.0225$ V [2], and as 1.26 V [3], with 0.92 V for $[PdCl_6]^{2-}/Pd$ [3]. In 1 M HCl the $[PdCl_6]^{2-}/[PdCl_4]^{2-}$ potential is $+1.286$ V, and in 1 M NaCl solution it is $+1.301$ V [6]. The E_0 for $[PdCl_6]^{2-} + 2e^- \rightarrow [PdCl_4]^{2-} + 2Cl^-$ is given by $E_0 = 1.288 + 0.295 \log([PdCl_6]^{2-}/[PdCl_4]^{2-}) + 0.59$ pCl [7].

A redox potentiometric method gave $k_5 \times k_6 = (6.02 \pm 0.45) \times 10^{-5}$ (instability constants), so $k_5 \times k_6$ (stability constants) would be 1.67×10^4 [2].

The standard molar entropy of $[PdCl_6]^{2-}$ has been estimated as 412.2 J·mol^{-1}·K^{-1} [4] using the vibrational data of [5]. Data from 100 to 1000 K: $C_p = 91.3$ J·deg^{-1}·mol^{-1} at 100 K, 156.71 J·deg^{-1}·mol^{-1} at 1000 K; $C_p/C_v = 1.101$ at 100 K, 1.056 at 1000 K; S (in J·deg^{-1}·mol^{-1}) = 278.3 at 100 K and 547.2 at 1000 K, respectively; values of $H-H_0$ and $-(G-H_0)/T$ were also listed [8]. The conventional and absolute standard molar enthalpies of hydration of $[PdCl_6]^{2-}$ at 298.15 K were calculated as -2936 and -730 kJ/mol, respectively [9].

References:

[1] Heath, G. A.; Moock, K. A.; Sharp, D. W. A.; Yellowlees, L. J. (J. Chem. Soc. Chem. Commun. **1985** 1503/4).
[2] Dubinskii, V. I. (Zh. Neorgan. Khim. **16** [1971] 1145/7; Russ. J. Inorg. Chem. **16** [1971] 607/8).
[3] Goldberg, R. N.; Hepler, L. G. (Chem. Rev. **68** [1968] 229/52, 242).
[4] Loewenschuss, A.; Marcus, Y. (Chem. Rev. **84** [1984] 89/115, 107).
[5] Debeau, M.; Poulet, H. (Spectrochim. Acta A **25** [1969] 1553/62).
[6] Grinberg, A. A.; Shamsiev, A. M. (Zh. Obsh. Khim. **12** [1942] 55/72; C.A. **1943** 1915).
[7] Kammel, R.; Takei, T.; Winterhager, H.; Xamamoro, H. (Metalloberfläche **22** [1968] 107/11).
[8] Loewenschuss, A.; Marcus, Y. (J. Phys. Chem. Ref. Data **16** [1987] 61/89, 83).
[9] Marcus, Y. (J. Chem. Soc. Faraday Trans. I **83** [1987] 339/49, 347).

Reactions of $[PdCl_6]^{2-}$

Reduction of $[PdCl_6]^{2-}$ by Cu powder gave Pd metal quantitatively [1]. With dithiomalonamide (HL), $[PdCl_6]^{2-}$ yields $[Pd(HL)_2]Cl_2$ and $Pd(HL)Cl_3$ [2]. With IO_4^- and $S_2O_8^{2-}$, $K_2[PdCl_6]$ in aqueous solution with KOH yields $K_6[Pd(IO_6)_2]\cdot KOH\cdot 12H_2O$ [3]. Pulse radiolysis of $[PdCl_6]^{2-}$ yields $[PdCl_4]^{2-}$, probably via $[PdCl_6]^{3-}$ [4]. With OH$^-$, $[Pd(OH)_6]^{2-}$ is formed [5].

References:

[1] Aoyama, S.; Watanabe, K. (Nippon Kagaku Zasshi **75** [1954] 20/3; C.A. **1954** 8118).
[2] Bret, J. M.; Castan, P.; Laurent, J.-P. (J. Chem. Soc. Dalton Trans. **1984** 1975/80).
[3] Siebert, H.; Mader, W. (Z. Anorg. Allgem. Chem. **351** [1967] 146/51).
[4] Broszkiewicz, R. K. (Intern. J. Radiat. Phys. Chem. **6** [1974] 249/58).
[5] Venskovski, N. V.; Ivanov-Emin, B. N.; Linko, I. V. (Izv. Vysshikh. Uchebn. Zavedenii Khim. Khim. Tekhnol. **24** [1981] 531/4; C.A. **95** [1981] No. 125328).

150

Miscellaneous Data on [PdCl$_6$]$^{2-}$

The thermochemical radius of the [PdCl$_6$]$^{2-}$ ion has been estimated as 3.19 ± 0.12 Å [1]; the Pd–Cl distance has been calculated as 2.32 Å in [PdCl$_6$]$^{2-}$ [2]. The ^{105}Pd resonance in [PdCl$_6$]$^{2-}$ has been observed [3].

The effect of [PdCl$_6$]$^{2-}$ and of other oxidising hexachloro ions on the delayed emission at 600 nm at liquid N$_2$ temperatures of cubic AgBr grains in photographic emulsions has been studied, as was the effect on photographic sensitivity [4] and on gradation characteristics in photographic materials [7]. Complexes containing [PdCl$_6$]$^{2-}$ have been used as hydrosilylation and hydrogenation catalysts [5].

Cellular resistance to the toxic effects of a number of Pd and Pt compounds including [PdCl$_6$]$^{2-}$ in Chinese hamster ovary cells has been assessed [6]. The anti-cancer properties of [PdCl$_6$]$^{2-}$ cells have been studied in vivo on mice [7], and it has been shown that [PdCl$_4$]$^{2-}$ has a greater antiphage activity than [PdCl$_6$]$^{2-}$ when tested against extracellular bacteriophage T$_4$ [8]. Adsorption of [PdCl$_6$]$^{2-}$ and other hexa- and tetrahalo ions on anion exchange columns was investigated [10].

References:

[1] Jenkins, H. D. B.; Thakur, K. P. (J. Chem. Educ. **56** [1979] 576/7).
[2] Loewenschuss, A.; Marcus, Y. (Chem. Rev. **84** [1984] 89/115, 107).
[3] Tani, T.; Saito, M. (Oyo Butsuri **42** [1973] 581/6; C.A. **80** [1974] No. 21362).
[4] Fedotov, M. A.; Likhobolov, V. A. (Izv. Akad. Nauk SSSR Ser. Khim. **1984** 1917/8; Bull. Acad. Sci. USSR Div. Chem. Sci. **1984** 1751).
[5] Panster, P.; Englisch, M.; Kleinschmit, K. (Eur. Appl. 151991 [1985]; C.A. **104** [1986] No. 34461).
[6] Smith, B. L.; Hanna, M. L.; Taylor, R. T. (J. Environ. Sci. Health A **19** [1984] 267/98, 283).
[7] Graham, R. D.; Williams, D. R. (J. Inorg. Nucl. Chem. **41** [1979] 1245/9).
[8] Zakharova, I. A.; Moshkovskii, Y. S.; Suraikina, T. J.; Fonshtein, L. M. (Dokl. Akad. Nauk SSSR **222** [1975] 1229/31; Dokl. Biochem. Sect. Proc. Acad. Sci. USSR **222** [1975] 252/3).
[9] Menjo, H.; Watanabe, Y.; Sakamoto, N. (Eur. Appl. 91788 [1983]; C.A. **100** [1984] No. 15274).
[10] Knothe, M. (Z. Anorg. Allgem. Chem. **463** [1980] 204/12).

4.5.6.1 Salts of [PdCl$_6$]$^{2-}$ with Nonmetallic Cations

H$_2$[PdCl$_6$]

This can be made in solution from Pd metal with 3:1 HCl:HNO$_3$ containing dissolved Cl$_2$ [1]; from Cl$_2$ bubbled into a suspension of Pd sponge in methanol [2]; from Pd metal in HCl-Cl$_2$ at 20 to 80°C [10]; from PdCl$_2$ in HCl saturated with Cl$_2$ [3, 4]. Dissolution of Pd in aqua regia yields [PdCl$_6$]$^{2-}$ [5]. The roles of "H$_2$[PdCl$_6$]" and (NH$_4$)$_2$[PdCl$_6$] in isolating pure Pd from mixtures of Pd, Pt, Rh and Au have been discussed [5].

The ^{105}Pd NMR resonance of [PdCl$_6$]$^{2-}$ was observed in a solution of "H$_2$[PdCl$_6$]" in 3:1 HCl [1]. Reaction of K$_2$[PdCl$_6$] with K$_3$[RhCl$_6$] [6], or Na$_3$[RhCl$_6$] with H$_2$[PdCl$_4$] [7], or RhCl$_3$ with KCl [8] gives K$_2$[Pd, RhCl$_6$]. Kinetics of formation of [PdCl$_6$]$^{2-}$ from [PdCl$_4$]$^{2-}$, Cl$^-$ and Cl$_2$ have been studied [9].

References:

[1] Fedotov, M. A.; Likholobov, V. A. (Izv. Akad. Nauk SSSR Ser. Khim. **1984** 1917/8; Bull. Acad. Sci. USSR Div. Chem. Sci. **1984** 1751).

[2] International Nickel Co. Ltd. (Brit. 879074 [1961]; C.A. **56** [1962] 8298).

[3] Brouwer, G.; van Rosmalen, G. M.; Bennema, P. (J. Cryst. Growth **23** [1974] 228/32).

[4] Cohen, A. J.; Davidson, N. (J. Am. Chem. Soc. **73** [1951] 1955/8).

[5] Rubinshtein, A. M.; Sokol, S. K. (Zh. Prikl. Khim. **32** [1959] 930/1; J. Appl. Chem. [USSR] **32** [1959] 947/8).

[6] Kiriyama, R.; Ogawa, K.; Azumi, M. (Nippon Kagaku Zasshi **82** [1961] 328/30; C.A. **56** [1962] 12395).

[7] Watanabe, K. (Nippon Kagaku Zasshi **78** [1957] 1204/7; C.A. **1958** 13511).

[8] Watanabe, K. (Nippon Kagaku Zasshi **78** [1957] 246/52; C.A. **1958** 7000).

[9] Mureinik, R. J.; Pross, E. (J. Coord. Chem. **8** [1978] 127/33).

[10] Kovylaev, A. D.; Ivanovskii, M. D. (Izv. Vysshikh. Uchebn. Zavedenii Tsvetn. Met. **12** [1969] 58/61; C.A. **71** [1969] No. 108395).

Reactions of $H_2[PdCl_6]$ and Derivatives

$H_2[PdCl_6]$. With $[Pt(NH_3)_4]Cl_2$, $H_2[PdCl_6]$ yields $[PtCl_2(NH_3)_4][PdCl_4]$ [1]; the dissolution of gold in $H_2[PdCl_6]$ solution was studied by a rotating disk method, and is first order with respect to Pd^{IV} in the solution [2].

$(CH_3NH_3)_2[PdCl_6]$. Nuclear quadrupole resonance (NQR) of this was used to study phase transitions in the material; the temperature dependence of the ^{35}Cl NQR from 20 to 300 K was plotted (reproduced in paper) [3].

$(C_{21}H_{38}N)_2[PdCl_6]$. This salt, of cetylpyridinium chloride, is apparently reduced to the $[PdCl_4]^{2-}$ salt on precipitation from $[PdCl_6]^{2-}$ [4].

References:

[1] Smirnov, I. I.; Volkov, V. E.; Volkova, G. V.; Chumakov, V. G. (Zh. Neorgan. Khim. **26** [1981] 2553/4; Russ. J. Inorg. Chem. **26** [1981] 1372/5).

[2] Filippov, A. A.; Volkova, G. V. (Povedenie Blagorod. Metal. Nekot. Met. Protsess. No. 1 [1969] 25/36; C.A. **76** [1972] No. 28167).

[3] Kume, Y.; Ikeda, R.; Nakamura, D. (J. Magn. Resonance **33** [1979] 331/44; 336).

[4] Selig, W. S. (Z. Anal. Chem. **320** [1985] 562/5).

4.5.6.2 Salts of $[PdCl_6]^{2-}$ with Metallic Cations and Ammonium

$Na_2[PdCl_6]$ is made in solution from PdO dissolved in HCl and then treated with $NaClO_3$ [1].

$K_2[PdCl_6]$. This is made by addition of KCl to a solution of PdO in HCl containing $NaClO_3$ [1]; from aqueous $K_2[PdCl_4]$ and Cl_2 [2]; by adding KCl to a solution of $[PdCl_4]^{2-}$ saturated with Cl_2 [3, 4]; from $PdCl_2en$ and Cl_2 with KCl [8].

The growth of single crystals of the salt by using silica gel media has been studied [5].

It is sparingly soluble in water [1], even less soluble in KCl solutions [4], forming brick-red [2] or bright red [4] crystals.

The salt has the $K_2[PtCl_6]$ structure with $a = 9.74$ Å [1]; a later determination gave $a = 9.7066(3)$ Å [6]. The density is 2.81 g/cm³ [7].

References:

[1] Sharpe, A. G. (J. Chem. Soc. **1953** 4177/9).
[2] Ito, K.; Nakamura, D.; Kurita, Y.; Ito, K.; Kubo, M. (J. Am. Chem. Soc. **83** [1961] 4526/8).
[3] Cohen, A. J.; Davidson, N. (J. Am. Chem. Soc. **73** [1951] 1955/8).
[4] Brauer, G. (Handbook of Preparative Inorganic Chemistry, Vol. 2, Academic, New York 1965, p. 1584).
[5] Brouwer, G.; van Rosmalen, G. M.; Bennema, P. (J. Cryst. Growth **23** [1974] 228/32).
[6] Adams, J. M. (Acta Cryst. B **30** [1974] 555).
[7] Benko, J. (Acta Chim. Acad. Sci. [Budapest] **35** [1963] 447/63, 459).
[8] Kukushkin, Y. N. (Zh. Neorgan. Khim. **8** [1963] 817/22; Russ. J. Inorg. Chem. **8** [1963] 417/20).

Thermodynamics. The topic has been briefly reviewed [1]. For $K_2[PdCl_6]$ ΔH°_{298} is given as approximately -948 kJ/mol [1], ΔG°_{298} as $+1028$ kJ/mol [1] and S°_{298} as approximately 347 $J \cdot deg^{-1} \cdot mol^{-1}$ [1]. The standard molar entropy of $[PdCl_6]^{2-}$ in $K_2[PdCl_6]$ has been calculated as 412.2 $kJ \cdot K^{-1} \cdot mol^{-1}$ [2], using the vibrational data of [3].

References:

[1] Goldberg, R. N.; Hepler, L. G. (Chem. Rev. **68** [1968] 229/52, 240).
[2] Loewenschuss, A.; Marcus, Y. (Chem. Rev. **84** [1984] 89/115, 107).
[3] Debeau, M.; Poulet, H. (Spectrochim. Acta A **25** [1969] 1553/62).

Vibrational Spectra. There have been a number of studies on this topic with accompanying force constant calculations. Data on fundamental frequencies from Raman and infrared spectra are collected in the table below. From the Raman and infrared data on the crystalline solid [1] stretching force constants were calculated and compared with those for other hexahalo species [1]. From the infrared spectra in [2] and the Raman data of [3], molecular force field calculations on $[PdCl_6]^{2-}$ were performed [2]. Raman and infrared data were obtained and force constants calculated on the basis of a general valence force field (GVFF) [4]. There are a number of values given for the $\nu_3(F_{1u})$ mode, viz. 360 cm^{-1} [5], 357 cm^{-1} [6], 356 cm^{-1} [7], 355 cm^{-1} [8] and 357.6 cm^{-1} [9]. Raman data on solid $K_2[PdCl_6]$ have been given for ν_1, ν_2 and ν_4 [23]. Values of ν_3 and ν_4 for $K_2[^{104}PdCl_6]$ have been given [10]. The effect of externally applied pressures on the ν_3 and ν_4 modes and on an F_{1u} lattice mode at 94 cm^{-1} [5] and at 95 cm^{-1} [6] have been studied [5,6]. Raman spectra of single crystals of $K_2[PdCl_6]$ have been measured [24].

Vibrational spectra of $[PdCl_6]^{2-}$.

	$\nu_1(A_{1g})$	$\nu_2(E_g)$	$\nu_3(F_{1u})$	$\nu_4(F_{1u})$	$\nu_5(F_{2g})$	$\nu_6(F_{1g})$	Ref.
$K_2[PdCl_6]$	324	293	356	174	170	(110)[1]	[1][2]
			340	175			[2]
	317	292	358	175	164		[4]
	324	294		172			[23][3]
	325	294		172			[24][3]
$(NH_4)_2[PdCl_6]$	318	289	346	200	178	(83)	[1][2]
$Rb_2[PdCl_6]$	314	287	351	161	175	(135)	[1][2]
$Cs_2[PdCl_6]$	310	270	345	156	173	(136)	[1][2]

[1] Calculated; [2] Raman and IR data; [3] Raman data.

In addition to the force constant calculations mentioned above there have also been studies, using the data of [4], on the Urey-Bradley (UBFF), modified Urey-Bradley (MUBFF), orbital valence (OVFF), modified orbital valence (MOVFF) and general valence (GVFF) force fields [11], and from the shifts observed in v_3 and v_4 for $K_2[^{104}PdCl_6]$, for $K_2[^{110}PdCl_6]$ and also for ^{35}Cl and ^{37}Cl, accurate force constants for the anion were computed [10]. There are UBFF, OVFF and GVFF calculations using the data of [4] with Coriolis coupling constants and potential energy distribution data [12]. Urey-Bradley (UBFF) constants using the data of [2] for $[PdCl_6]^{2-}$ and for other hexahalo complexes were calculated by [13], while general valence force field (GVFF) constants using the data of [4] were calculated by [14]. The data of [1] were used to compute, using the method of progressive rigidity and the Gordy formula, valence force constants for $[PdCl_6]^{2-}$ [15]. The method of kinetic force restraint was used to evaluate symmetric force constants for $[PdCl_6]^{2-}$, using the data of [1], and other hexahalo molecules, and the influence of atomic mass on the Coriolis coupling constant considered [16]. Kinetic and potential energy constants, vibrational mean amplitudes, Coriolis coupling constants and rotational distortion constants at 298 K, using the data of [4], were also calculated [14]. Using the data of [2], mean amplitudes of vibration in $[PdCl_6]^{2-}$ were calculated [17]. With the data of [3] some force constants were obtained [18], but no source data were quoted for other force constant computations [19]. Methods for the solutions of secular equations for molecules including $[PdCl_6]^{2-}$ have been given [20] and comparisons drawn between force constants of $[PdCl_6]^{2-}$ in $K_2[PdCl_6]$ and in $[PdCl_4]^{2-}$ [21]. Force and compliance constants for $[PdCl_6]^{2-}$ in $K_2[PdCl_6]$, apparently using the data of [4], have been calculated [22], and GVFF constants using unreferenced data [25, 26].

References:

[1] Debeau, M.; Poulet, H. (Spectrochim. Acta A **25** [1969] 1553/62).

[2] Hiraishi, J.; Nakagawa, I.; Shimanouchi, T. (Spectrochim. Acta **20** [1964] 819/28).

[3] Woodward, L. A.; Creighton, J. A. (Spectrochim. Acta **17** [1961] 594/9).

[4] Hendra, P. J.; Park, P. J. D. (Spectrochim. Acta A **23** [1967] 1635/40).

[5] Ferraro, J. R. (J. Chem. Phys. **53** [1970] 117/9).

[6] Adams, D. M.; Payne, S. J. (J. Chem. Soc. Dalton Trans. **1974** 407/10).

[7] Adams, D. M.; Gebbie, H. A. (Spectrochim. Acta **19** [1963] 925/30).

[8] Brown, D. H.; Dixon, K. A.; Livingston, C. M.; Nuttall, R. H.; Sharp, D. W. A. (J. Chem. Soc. A **1967** 100/4).

[9] Adams, D. M. (Proc. Chem. Soc. **1961** 335).

[10] Müller, A.; Mohan, N.; Königer, F.; Chakravorti, M. C. (Spectrochim. Acta A **31** [1975] 107/16).

[11] Labonville, P.; Ferraro, J. R.; Wall, M. C.; Basile, L. J. (Coord. Chem. Rev. **7** [1972] 257/87, 274).

[12] Rai, S. N.; Thakur, S. N.; Rai, D. K. (Proc. Indian Acad. Sci. A **74** [1971] 243/54).

[13] Venkateswarlu, K.; Devi, V. N. (Current Sci. [India] **37** [1968] 370/1).

[14] Mohan, S.; Mukunthan, A. (Indian J. Phys. B **56** [1982] 91/5).

[15] Phan, D. Q.; Kovrikov, A. B. (Zh. Prikl. Spektrosk. **15** [1971] 81/5; J. Appl. Spectrosc. [USSR] **15** [1971] 899/902).

[16] Elumalai, K.; Radhakrishnan, M. (Czech. J. Phys. B **28** [1978] 761/72).

[17] Nagarajan, G.; Hariharan, T. A. (Acta Phys. Austriaca **21** [1966] 366/82, 369).

[18] Sundaram, S. (Proc. Phys. Soc. **91** [1967] 764/7).

[19] Gopinath, C. R.; Rao, K. S. R.; Vishwanath, J. (Current Sci. [India] **53** [1984] 839/41).

[20] Thyagarajan, G.; Subhedar, M. K. (Indian J. Pure Appl. Phys. **12** [1974] 309/11).

[21] Sayal, K. N.; Verma, D. N.; Dixit, L. (Indian J. Pure Appl. Phys. **14** [1976] 819/22).
[22] Raja, S. V. K.; Savariraj, G. A. (Act. Ciencia Indica **2** [1975] 180/4).
[23] Hendra, P. J. (Spectrochim. Acta A **23** [1967] 2871/4).
[24] Beattie, I. R.; Gilson, T. R. (Proc. Roy. Soc. [London] A **307** [1968] 407/29, 413).
[25] Wendling, E.; Makmudi, S. (Opt. Spektroskopiya **32** [1972] 492/500; Opt. Spectrosc. [USSR] **32** [1972] 257/61).
[26] Geldard, J. F.; McDowell, H. K. (Spectrochim. Acta A **43** [1987] 439/43).

Electronic Spectra. The effect of applied pressure on the 27 200 cm^{-1} charge-transfer band of $K_2[PdCl_6]$ (misprinted in paper as $K_2[PbCl_6]$) has been determined from 0 to 70 kbar.

Reference:

Balchan, A. S.; Drickamer, H. G. (J. Chem. Phys. **35** [1961] 356/8).

Nuclear Quadrupole Resonance (NQR). For a general review see [13]. Pure NQR frequencies of ^{35}Cl in $K_2[PdCl_6]$ were recorded at liquid N_2 temperatures, at 200 K and at 300 K [1], and at 100 to 340 K [15]. A comparative study of the NQR spectra of $K_2[PdCl_6]$ and $K_2[PtCl_6]$ reveals that the temperature-dependence of both the NQR resonance frequencies and the spin-lattice relaxation times of the ^{35}Cl nuclei are dominated by the F_{1g} rotatory mode of frequency ~ 41 cm^{-1} [2]. The Townes-Dailey procedure was used to calculate the ionic character of M-X bonds in a number of species, including $K_2[PdCl_6]$, on the basis of the NQR spectra from -70 to 0°C [3]. Incipient phase transitions in $K_2[PdCl_6]$ have been observed from the temperature-dependence of the ^{35}Cl NQR resonance [20 to 500 K] [4]. NQR data on $K_2[PdCl_6]$ and other species have been correlated with electronic spectral data [14].

X-Ray Photoelectronic Spectroscopy (ESCA). Binding energies in $K_2[PdCl_6]$ from the ESCA spectrum (in eV): Pd $3d_{3/2} = 340.5$; Cl $2p_{1/2} = 198.9$; K $2p_{3/2} = 293.2$ [5]; Pd$_{3/2} = 345.6$, $3d_{5/2} = 340.3$ [12]. Assignments of level ordering in the ESCA spectrum of $K_2[PdCl_6]$, $K_2[PtCl_4]$ and $Na_2[PtCl_4]$ have been discussed [6]. ESCA data on a number of hexahalo complexes, including $K_2[PdCl_6]$, have been compared [11].

Comparisons of the electronic structures of $[PdX_6]^{2-}$ (X = F, Cl, Br) and $[RhX_6]^{3-}$ (X = F, Cl) have been made on the basis of X-ray and X-ray electronic spectra [7]. The chlorine K absorption spectra of $K_2[MCl_6]$ and $K_2[MCl_4]$ (M = Pd, Pt) have been measured; an intense "white line" was observed at the absorption threshold, the chemical shifts being closely related to the ionisation potentials of the metals, corresponding to transitions for the Cl 1s level to the lowest unoccupied antibonding orbitals originating from Pd 4d (or Pt 5d) and Cl 3p orbitals [8]. The Pd L_{III} absorption spectra of $K_2[PdCl_6]$ also show a "white line" at the absorption threshold [9], and the position of the Kβ_1 line a number of compounds including $K_2[PdCl_6]$ measured [10]. The fine structure of the X-ray K-absorption spectrum of $K_2[PdCl_6]$ was measured and reproduced in [16]. The Cl Kβ, line of $K_2[PdCl_6]$ was correlated with the electronic structure of the anion [17].

References:

[1] Ito, K.; Nakamura, D.; Kurita, Y.; Ito, K.; Kubo, M (J. Am. Chem. Soc. **83** [1961] 4526/8).
[2] Armstrong, R. L.; Cooke, D. F. (Can. J. Phys. **47** [1969] 2165/9).
[3] Kubo, M.; Nakamura, D. (Colloq. Intern. Centre Natl. Rech. Sci. [Paris] No. 191 [1970] 117/23; C.A. **77** [1972] No 81770).
[4] Armstrong, R. L.; van Driel, H. M. (Can. J. Phys. **50** [1972] 2048/53).
[5] Nefedov, V. I.; Zakharova, I. A.; Moiseev, I. I.; Porai-Koshits, M. A.; Varaftik, M. N.; Belov, A. P. (Zh. Neorgan. Khim. **18** [1973] 3264/8; Russ. J. Inorg. Chem. **18** [1973] 1738/40).

[6] Bilden, P.; Prins, R. (Chem. Phys. Letters **16** [1972] 611/3).

[7] Peshchevitskii, B. I.; Zemskov, S. V.; Sadovskii, A. P.; Kravtsova, E. A.; Mit'kin, V. N. (Koord. Khim. **5** [1979] 1838/45; Soviet J. Coord. Chem. **5** [1979] 1433/9).

[8] Sugiura, C.; Ohashi, M. (J. Chem. Phys. **78** [1983] 88/90).

[9] Sugiura, C.; Muramatsu, S. (J. Chem. Phys. **82** [1985] 2191/5).

[10] Narbutt, K. I.; Nefedov, V. I.; Porai-Koshits, M. A.; Kochetkova, A. P. (Zh. Strukt. Khim. **13** [1972] 451/7; J. Struct. Chem. [USSR] **13** [1972] 422/7).

[11] Zasukha, V. A.; Volkov, S. V. (Ukr. Khim. Zh. **52** [1986] 227/32; Soviet Progr. Chem. **52** No. 3 [1986] 1/6).

[12] Kumar, G.; Blackburn, J. R.; Aldridge, R. G.; Moddeman, W. E.; Jones, M. M. (Inorg. Chem. **11** [1972] 296/300).

[13] van Bronswyk, W. (Struct. Bonding [Berlin] **7** [1970] 87/113, 97).

[14] Machmer, P. (Z. Naturforsch. **24 b** [1969] 193/9).

[15] Armstrong, R. L.; Baker, G. L.; Jeffrey, K. R. (Phys. Rev. [3] B **1** [1970] 2847/51).

[16] Bhat, N. V. (Spectrochim. Acta B **28** [1973] 257/61).

[17] Nefedov, V. I.; Narbutt, K. N. (Zh. Strukt. Khim. **12** [1971] 1014/25; J. Struct. Chem. [USSR] **12** [1971] 937/42).

Miscellaneous Data. The empirically numerical effective charge method has been used to evaluate enthalpies of formation of $K_2[PdCl_6]$ [1]. The density of $K_2[PdCl_6]$ (2.81 g/cm^3) and the densities of a large number of other complexes have been related to the masses and atomic numbers of the constituent atoms [2].

Reactions. With F_2 or ClF_3, $K_2[PdCl_6]$ yields $K_2[PdF_6]$ [3]. On heating $K_2[PdBr_4]$ is formed [4]. Boiling an aqueous solution of $K_2[PdCl_6]$ gives $[Pd(OH)Cl]_n$, via $K_2[PdCl_4]$ [5].

References:

[1] Kaganyuk, D. S.; Kyskin, V. I.; Kazin, I. V. (Koord. Khim. **10** [1984] 773/5; C.A. **101** [1984] No. 98683).

[2] Benko, J. (Acta Chim. [Budapest] **35** [1963] 447/63, 439).

[3] Zemskov, S. V.; Nikonorov, Yu. I.; Mit'kin, V. N.; Lavrova, L. A. (Tezisy Dokl. 12th Vses. Chugaevskoe Soveshch. Khim. Kompleksn. Soedin.; Novosibirsk 1976 Vol. 3, p. 376; C.A. **85** [1976] No. 171050).

[4] Shubochkin, L. K.; Sorokina, L. D.; Shubochkina, E. F. (Zh. Neorgan. Khim. **21** [1976] 2367/9; Russ. J. Inorg. Chem. **21** [1976] 1413/4).

[5] Watanabe, K. (Nippon Kagaku Zasshi **77** [1956] 1675/81; C.A. **1958** 2636).

$(NH_4)_2[PdCl_6]$

This is made by addition of NH_4Cl to a solution in HCl of PdO in the presence of $NaClO_3$ [1], from NH_4Cl and "$H_2[PdCl_6]$" [2] or from $PdCl_2$ and Cl_2 in aqueous NH_4Cl solution [3].

It forms wine-red octahedra [4] or bright red crystals [4], sparingly soluble in water [1]. The salt has the $K_2[PtCl_6]$ structure, space group Fm3m-O_h^5, a = 9.84(1) Å, Z = 4. The Pd–Cl distance is 2.300(7) Å [5]. The cell constant has also been determined as 9.84 Å [1] and as 9.8222(4) Å [6].

Vibrational and X-Ray Photoelectronic Spectra. Raman and infrared spectra of solid $(NH_4)_2[PdCl_6]$ were measured (see table on p. 152) and assigned; the Pd–Cl stretching force constant was correlated with Pd–Cl covalent bond character for this and other hexahalo complexes [7] (see also miscellaneous data below). The frequencies of the $(NH_4)^+$ ion in $(NH_4)_2[PdCl_6]$ have been measured in the infrared [8].

The X-ray photoelectronic (ESCA) spectrum of $(NH_4)_2[PdCl_6]$ shows binding energies of 345.3 (Pd $3d_{5/2}$) and 349.5 eV (Pd $3d_{3/2}$) [9].

References:

[1] Sharpe, A. G. (J. Chem. Soc. **1953** 4177/9).
[2] Rubinshtein, A. M.; Sokol, S. K. (Zh. Prikl. Khim. **32** [1959] 930/1; J. Appl. Chem. [USSR] **32** [1959] 947/8).
[3] Brauer, G. (Handbook of Preparative Inorganic Chemistry, Vol. 2, Academic, New York 1965, p. 1584).
[4] Sidorenko, Yu. A.; Filippov, A. A.; Lukashenko, E. E.; Chumakov, V. G. (Zh. Neorgan. Khim. **17** [1972] 2234/8; Russ. J. Inorg. Chem. **17** [1972] 1165/7).
[5] Bell, J. D.; Hall, D.; Waters, T. N. (Acta Cryst. **21** [1966] 440/2).
[6] Adams, J. M. (Acta Cryst. B **30** [1974] 555).
[7] Debeau, M.; Poulet, H. (Spectrochim. Acta A **25** [1969] 1553/62).
[8] Dunsmuir, J. T. R.; Lane, A. P. (Spectrochim. Acta A **28** [1972] 45/50).
[9] Jørgensen, C. K.; Berthou, H. (Kgl. Danske Videnskab. Selskab Mat. Fys. Medd. **38** No. 15 [1972] 1/93, 73; C.A. **79** [1973] No. 11782).

Miscellaneous Data. The pressure dependence of the ground state tunnel splitting and the energy transition to the first excited librational state of the $(NH_4)^+$ ion in $(NH_4)_2[PdCl_6]$ has been measured by inelastic neutron scattering [1]. The rotational potential functions of the $(NH_4)^+$ ion in $(NH_4)_2[PdCl_6]$ have been derived from atom-atom potentials of the Buckingham (exp-6) type, by matching computed and observed librational and tunnelling frequencies in the compound [2]. Activation energies for threefold and fourfold reorientation of the $(NH_4)^+$ ion in the cubic phase of the salt have been derived from a potential function, and activation energies calculated [3].

The temperature-dependence (from 60 to 300 K) of the proton spin-lattice relaxation times for polycrystalline $(NH_4)_2[PdCl_6]$ shows that reorientation of the NH_4^+ ion is the dominant relaxation mechanism [4], Proton relaxation and tunnelling have been measured in a number of species $(NH_4)_2[MCl_6]$ (M = Pd, Pt, Te, Ru, Ti, Ir, Re) [5] and low-temperature NMR data have been used to study proton tunnelling in the salt [6].

Inelastic neutron scattering has been used to obtain phonon frequency spectra for powdered cubic $(NH_4)_2[PdCl_6]$ and other hexachloro complexes [7]. Thermal decomposition of $(NH_4)_2[PdCl_6]$ with Na_2CO_3 starts at 175°C; the intermediate is $Na_2[PdCl_4]$ [8].

References:

[1] Prager, M.; Press, W.; Heidemann, A.; Vettier, C. (J. Chem. Phys. **80** [1984] 2777/81).
[2] Smith, D. (J. Chem. Phys. **82** [1985] 5133/9).
[3] Smith, D. (J. Chem. Phys. **74** [1981] 6480/6).
[4] Bonori, M.; Terenzi, M. (Chem. Phys. Letters **27** [1974] 281/4).
[5] Svare, I.; Raaen, A. M.; Thorkildsen, G. (J. Phys. C **11** [1978] 4069/76).
[6] Svare, I.; Raaen, A. M.; Fimiand, B. O. (Physica B+C **128** [1985] 144/60).
[7] Otnes, K.; Svare, I. (J. Phys. C **12** [1979] 3899/905).
[8] Ryuman, A.; Smirnov, I. I.; Bizyukina, N. V.; Krapivko, A. A. (Izv. Vyssh. Uchebn. Zavedenii Tsvetn. Met. **4** [1986] 62/5; C.A. **106** [1987] No. 88094).

Reactions. The thermal decomposition of solid $(NH_4)_2[PdCl_6]$ has been studied over the range 20 to 450°C (thermograms reproduced in paper). At 300°C it forms the yellow-brown $(NH_4)_2[PdCl_4]$ and then the bright yellow trans-$Pd(NH_3)_2Cl_2$ mixed with Pd; NH_4Cl, Cl_2 and HCl are the gaseous decomposition products [1]. Reaction of $(NH_4)_2[PdCl_6]$ with Ag_2SO_4 and F_2 at 550°C yields $Ag[PdF_6]$ [2].

References:

[1] Sidorenko, Yu. A.; Filippov, A. A.; Lukashenko, E. E.; Chumakov, V. G. (Zh. Neorgan. Khim. **17** [1972] 2234/8; Russ. J. Inorg. Chem. **17** [1972] 1165/7).

[2] Müller, B.; Hoppe, R. (Z. Anorg. Allgem. Chem. **392** [1972] 37/41).

Applications. The salt has been used to separate Pd from Ir by virtue of differences in the solubility and thermal stability of $(NH_4)_2[PdCl_6]$ and $(NH_4)_2[IrCl_6]$. The metals are produced after separation by thermal decomposition in vacuum [1]. The polycyclic aromatic hydrocarbon content of smoke from tobacco treated with $(NH_4)_2[PdCl_6]$ is much reduced from that found in palladium-free tobacco [2]. The use of $(NH_4)_2[PdCl_6]$ in effecting separations of Pd and Pt from other platinum metals and in the isolation of pure Pd has been discussed [3].

References:

[1] Ceng, H. (Huaxue Shijie **25** [1984] 388/90; C.A. **102** [1985] No. 29191).

[2] Bryant, H. G.; Norman, V.; Williams, T. B.; Liggett Group Inc. (U.S. 4177822 [1979]; C.A. **92** [1980] No. 107620).

[3] Rubinshtein, A. M.; Sokol, S. K. (Zh. Prikl. Khim. **32** [1959] 930/1; J. Appl. Chem. [USSR] **32** [1959] 947/8).

$Rb_2[PdCl_6]$ and $Cs_2[PdCl_6]$

$Rb_2[PdCl_6]$ is made by addition of RbCl to a solution in HCl of PdO with $NaClO_3$ [1], or from RbCl and $K_2[PdCl_6]$ in aqueous solution [2]. The growth of crystals of the salt by using silica gel has been demonstrated [3].

The structure is of the $K_2[PtCl_6]$ type with a = 9.87 Å [1]. It is sparingly soluble in water [1].

Vibrational Spectra. The Raman and infrared spectra of solid $Rb_2[PdCl_6]$ were measured (40 to 700 cm^{-1}, reproduced in paper); the stretching force constant was correlated with metal-halide covalency in this and other hexahalo complexes [4] (see table on p. 152). The $v_3(F_{1u})$ frequency for the solid is at 352 cm^{-1} [5].

Reactions. On heating, $Rb_2[PdCl_6]$ decomposes to $Rb_2[PdCl_4]$ [2].

$Cs_2[PdCl_6]$ is made by reaction of CsCl with a solution of PdO in HCl with $NaClO_3$ [1], or from $K_2[PdCl_6]$ and CsCl in aqueous solution [2]. The growth of single crystals of the salt by using silica gel media has been reported [3].

It has the $K_2[PtCl_6]$ structure with a = 10.18 Å [1]. It is sparingly soluble in water [1].

Vibrational Spectra. Raman and infrared spectra of solid $Cs_2[PdCl_6]$ have been measured (see table on p. 152) and assigned; the degree of metal-halide covalency in this and other hexahalo complexes was correlated with the Pd–Cl stretching force constant [4]. The $v_3(F_{1u})$ stretch lies at 344 cm^{-1} [5].

Reactions. On heating, $Cs_2[PdCl_6]$ decomposes to $Cs_2[PdCl_4]$ [2].

References:

[1] Sharpe, A. G. (J. Chem. Soc. **1953** 4177/9).

[2] Shubochkin, L. K.; Sorokina, L. D.; Shubochkina, E. F. (Zh. Neorgan. Khim. **21** [1976] 2567/9; Russ. J. Inorg. Chem. **21** [1976] 1413/4).

[3] Brouwer, G.; van Rosmalen, G. M.; Bennema, P. (J. Cryst. Growth **23** [1974] 228/32).
[4] Debeau, M.; Poulet, H. (Spectrochim. Acta A **25** [1969] 1553/62).
[5] Brown, D. H.; Dixon, K. R.; Livingston, C. M.; Nuttall, R. H.; Sharp, D. W. A. (J. Chem. Soc. A **1967** 100/4).

[Ni(H₂O)₆][PdCl₆]

The heat capacity and low-field magnetic susceptibility of this and a number of other isostructural complexes were measured down to 0.05 K.

Reference:

Friedberg, S. A.; Karnezos, M.; Meier, D. (AD-A 010880 [1975] 1/7; Got. Rept. Announce. Index [U.S.] **75** [1975] 53; C. A. **83** [1975] No. 171774).

K₂[Pd, RhCl₆]

Mixed crystals of this material, isomorphous with $K_2[PtCl_6]$, were obtained from $K_2[PdCl_6]$ and $K_3[RhCl_6]$ under strongly oxidising conditions. It was suggested that Rh^{IV} centres are distributed in a random fashion in the $K_2[PdCl_6]$ lattice [1]. It can also be made from $H_2[PdCl_4]$ and $Na_3[RhCl_6]$ [2] or $H_2[PdCl_6]$ and $Na_3[RhCl_6]$ [3] or with addition of KCl; magnetic evidence suggests that the crystals contain both rhodium(III) and rhodium(IV) [2,3].

References:

[1] Kiriyama, R.; Ogawa, K.; Azumi, M. (Nippon Kagaku Zasshi **82** [1961] 328/30; C. A. **56** [1962] 12395).
[2] Watanabe, K. (Nippon Kagaku Zasshi **78** [1957] 1204/7; C. A. **1958** 13511).
[3] Watanabe, K. (Nippon Kagaku Zasshi **78** [1957] 246/52; C. A. **1958** 7000).

Chloro-aquo Complexes

$[PdCl_5(H_2O)]^-$ has been mentioned in connection with aquation studies on $[PdCl_6]^{2-}$.

Reference:

Dubinskii, V. I. (Zh. Neorgan. Khim. **16** [1971] 1145/7; Russ. J. Inorg. Chem. **16** [1971] 607/8).

4.5.7 Hexachloropalladate(V) [PdCl₆]⁻

The half-wave reversible electrode potential of the $[PdCl_6]^-/[PdCl_6]^{2-}$ couple, as the $(Bu_4^n N)$ salts, has been determined as $+2.92$ V (in CH_2Cl_2 versus the standard calomel electrode).

Reference:

Heath, G. A.; Moock, K. A.; Sharp, D. W. A.; Yellowlees, L. J. (J. Chem. Soc. Chem. Commun. **1985** 1503/4).

4.5.8 Hexachloropalladate(VI) PdCl₆(?)

Although force constant data for "PdCl₆" have been given [1] there is no evidence for the existence of this species. The report in [1] appears to be based on a misreading of [2].

References:

[1] Elumalai, K.; Radhakrishnan, M. (Czech. J. Phys. B **28** [1978] 761/72, 765).
[2] Hiraishi, J.; Nakagawa, I.; Shimanouchi, T. (Spectrochim. Acta **20** [1964] 819/28).

4.5.9 Binuclear Chloropalladates

4.5.9.1 The $[Pd_2Cl_4]^{2-}$ Ion

It has been briefly mentioned that the E_0 for $[AlCl_4]^{2-}/[Pd_2Cl_4]^{2-}$ is < 0.5 V and for $[PdCl_4]^{2-}/Pd^0 > 0.74$ V [1], and may be involved in the oxidation by $[PdCl_4]^{2-}$ of alkenes [2].

References:

[1] Temkin, O. N.; Bruk, L. G. (Uspekhii Khim. **52** [1983] 206/43; Russ. Chem. Revs. **52** [1983] 117/37, 124).
[2] Moiseev, I. I.; Levanda, O. G.; Vargaftik, M. N. (J. Am. Chem. Soc. **96** [1974] 1003/7).

4.5.9.2 Salts of $[Pd_2Cl_6]^{2-}$

As with the bromo- and iodo-palladates(II), there are also dimeric chloro-bridged species, stabilised in the solid state by organic counterions.

Vibrational Spectra of $[Pd_2Cl_6]^{2-}$

The data are collected in the table below; references to the solid state spectra of individual salts are also given in the appropriate sections.

Raman spectra in CH_2Cl_2 and infrared spectra in $CHCl_3$ and in CH_2Br_2 of $(Pn_4^nN)_2[Pd_2Cl_6]$ have been given [1] and the stretching force constants of terminal and bridging Pd–Cl bonds calculated for the in-plane vibrations of the anion [2]. There are similar in-plane force constant data for $[Pd_2Cl_6]^{2-}$ in $R_2[Pd_2Cl_6]$ ($R = Bu_4^n$, Ph_4As, n-cetyltrimethylammonium) based on infrared spectra (100 to 400 cm^{-1}) of these solid salts [3]. Infrared spectra (80 to 400 cm^{-1}, reproduced in paper) have also been presented for $R_2[Pd_2Cl_6]$ ($R = Ph_4As$, $(Me_2N)_3C_3$, $(n-C_3H_7)_3C_3$), and Raman data for the (Bu_4^nN) and $[(CH_3)_2N]_3C_3^+$ salts in CH_3CN show ν_1 at 348 and 346 cm^{-1}, respectively, and ν_6 at 302 and 301 cm^{-1} [4]. Brief reference has been made to Raman spectra of $PdCl_2$ in CH_3CN which suggest that a bridging Pd_2Cl_2 system could be present in such solutions [5, 6].

Raman and infrared spectra of salts of $R_2[Pd_2Cl_6]$:

R	Raman[1)	(Pn_4^nN) IR[1)	IR[2)	(Bu_4^nN) IR[2)	(Me-cetylN) IR[2)	(Ph_4As)
$\nu_1(A_g)\,\nu_{PdCl}$	346[a)					
$\nu_2(A_g)\,\nu_{PdCl}$	302[a)					
$\nu_3(A_g)\,\delta_{PdCl}$	169[a)					
$\nu_4(A_g)\,\delta_{PdCl}$	119[a)					
$\nu_6(B_{1g})\,\nu_{PdCl}$	328[a)					
$\nu_7(B_{1g})\,\nu_{PdCl}$	265[b)					
$\nu_9(B_{1u})\,\delta_{PdCl}$		156[a)	152	155	146	156

(table continued)

R	Raman[1]	(Pn₄ⁿN) IR[1]	IR[2]	(Bu₄ⁿN) IR[2]	(Me-cetylN) IR[2]	(Ph₄As)
$\nu_{12}(B_{2u})\nu_{PdCl}$		335[b]	336	340	336	353
$\nu_{13}(B_{2u})\nu_{PdCl}$		262[b]		302	305	305
$\nu_{16}(B_{3u})\nu_{PdCl}$		343[b]	347	349	339	353
$\nu_{17}(B_{3u})\nu_{PdCl}$		297[b]	301	264	263	264
$\nu_{18}(B_{3u})\delta_{PdCl}$				161	160	186
Ref.	[1]	[1]	[1]	[3]	[3]	[3]

[1] Spectra for solutions; [a] in $CHCl_3$, [b] in CH_2Br_2; [2] spectra for solids (mulls).

Mode numbering from [3] adjusted to that of [1].

Structures. Factors affecting the adoption of a planar or angular conformation in $[Pd_2Cl_6]^{2-}$ and $[Pd_2Br_6]^{2-}$ have been discussed.

References:

[1] Goggin, P. L. (J. Chem. Soc. Dalton Trans. **1974** 1483/6).
[2] Goggin, P. L.; Mink, J. (Inorg. Chim. Acta **26** [1978] 119/24).
[3] Adams, D. M.; Chandler, P. J. (J. Chem. Soc. A **1967** 1272/4).
[4] Harris, D. C.; Gray, H. B. (Inorg. Chem. **13** [1974] 2250/5).
[5] Volchenskova, I. I.; Yatsimirskii, K. B. (Zh. Neorgan. Khim. **18** [1973] 1875/82; Russ. J. Inorg. Chem. **18** [1973] 990/5).
[6] Shestakov, A. F.; Makitova, D. D.; Atovmyar, L. O.; Krasochka, O. N. (Koord. Khim. **13** [1987] 949/52; C.A. **107** [1987] No. 183846).

Electronic Spectra

Spectra of the following salts in the specified solvents have been measured: $(Bu_4^nN)_2[Pd_2Cl_6]$ from 180 to 600 nm in a mixture of methanol with 2-methyltetrahydrofuran; bands assigned [1]; in methanol (200 to 400 nm) [2], in methanol and acetone (200 to 800 nm) [3], and in methanol, ethanol and $ClCH_2 \cdot CH_2Cl$ (200 to 400 nm) [4]; in acetonitrile, 200 to 350 nm, spectrum reproduced in paper and with assignments [5].

Spectra of $(AsPh_3Me)_2[Pd_2Cl_6]$ (200 to 600 nm) in nitrobenzene, acetic anhydride, nitromethane, acetone, methanol, and in acetonitrile have been listed [6]. The electronic spectrum of $R_2[Pd_2Cl_6]$ (R = tetraoctylammonium) reproduced from 200 to 300 nm [7] shows that $[Pd_2Cl_6]^{2-}$ is present in Bu_3PO_4 solutions of palladium(II) chloro complexes [8].

Assignment of electronic absorption bands arising from bridging and terminal modes in $[Pd_2Cl_6]^{2-}$ has been attempted [9] and the electronic structure and charge-transfer spectra of $[Pd_2Cl_6]^{2-}$ calculated [10].

References:

[1] Mason, W. R.; Gray, H. B. (J. Am. Chem. Soc. **90** [1968] 5721/9).
[2] Lobaneva, O. A.; Kononova, M. A.; Kunaeva, N. T.; Davydova, M. K. (Zh. Neorgan. Khim. **17** [1972] 3011/4; Russ. J. Inorg. Chem. **17** [1972] 1583/5).

[3] Lobaneva, O. A.; Kononova, M. A.; Davydova, M. K. (Dokl. Akad. Nauk SSSR **194** [1970] 133/5; Dokl. Phys. Chem. Proc. Acad. Sci. USSR **190/195** [1970] 669/70).

[4] Lobaneva, O. A.; Kononova, M. A.; Davydova, M. K. (Vestn. Leningr. Univ. Fiz. Khim. **1971** No. 4, pp. 149/50; C. A. **76** [1972] No. 92298).

[5] Batiste, J. L. H.; Rumfeldt, R. (Can. J. Chem. **52** [1974] 174/81).

[6] Harris, C. M.; Livingstone, S. E.; Reece, I. H. (J. Chem. Soc. **1959** 1505/11).

[7] Selezneva, I. A.; Ivanova, S. N.; Gindin, L. M. (Izv. Sibirsk. Otd. Akad. Nauk SSSR Ser. Khim. Nauk **1982** No. 6, pp. 107/12; C. A. **98** [1983] No. 60702).

[8] Tomilov, S. B.; Moskvin, L. N.; Krasnoperov, V. M.; Chereshkevich, Yu. L. (Izv. Sibirsk. Otd. Akad. Nauk SSSR Ser. Khim. Nauk **1974** No. 2, pp. 43/6; C. A. **81** [1974] No. 30382).

[9] Lobaneva, O. A.; Kononova, M. A.; Davydova, M. K. (Zh. Neorgan. Khim. **20** [1975] 2648/50; Russ. J. Inorg. Chem. **20** [1975] 1577/8).

[10] Baranovskii, V. I.; Davydova, M. K.; Panina, N. S.; Panin, A. I. (Koord. Khim. **2** [1976] 409/15; Soviet J. Coord. Chem. **2** [1976] 308/13).

Existence of $[Pd_2Cl_6]^{2-}$ in Solutions

Spectrophotometric studies on $[PdCl_4]^{2-}$ in acetic acid suggest that $[Pd_2Cl_6]^{2-}$ may be one of the species present therein [1]; similar conclusions have been drawn from measurements of electronic and Raman spectra of solutions of $PdCl_2$ in acetonitrile [2]. The existence of $[Pd_2Cl_6]^{2-}$ from solutions of LiCl and $PdCl_2$ in acetic acid and also from LiCl and $PdCl_2$ in the same solvent has been deduced from molecular weight and spectrophotometric data and equilibrium constants obtained (see below) [3]; spectroscopic data also suggest the presence of $[Pd_2Cl_6]^{2-}$ in solutions of NaCl and $PdCl_2$ in glacial acetic acid, and equilibrium constant data were likewise obtained (below) [1, 4].

Evidence from mean molar masses of $PdCl_2$ in aqueous solutions of $PdCl_2$ at different concentrations suggests that the quantities of $[Pd_2Cl_6]^{2-}$ in such solutions are small, especially at lower concentrations [5].

References:

[1] Alcock, R. M.; Hartley, F. R.; Rogers, D. E.; Wagner, J. L. (Coord. Chem. Rev. **16** [1975] 59/66).

[2] Volchenskova, I. I.; Yatsimirskii, K. B. (Zh. Neorgan. Khim. **18** [1973] 1875/82; Russ. J. Inorg. Chem. **18** [1973] 990/5).

[3] Henry, P. M.; Marks, O. W. (Inorg. Chem. **10** [1971] 373/6).

[4] Alcock, R. M.; Hartley, F. R.; Rogers, D. E.; Wagner, J. L. (J. Chem. Soc. Dalton Trans. **1975** 2189/93).

[5] Kaszonyi, A.; Vojtko, J.; Hrusovsky, M. (Collection Czech. Chem. Commun. **43** [1978] 3002/6).

Equilibrium Constants

From spectrophotometric data the equilibrium constant K_1 for $2[PdCl_4]^{2-} \rightleftharpoons [Pd_2Cl_6]^{2-} + 2Cl^-$ has been calculated as $0.56 \pm 0.02\,M^{-1}$ in glacial acetic acid at 25°C using NaCl and $PdCl_2$ [1, 2]; for the corresponding $LiCl–PdCl_2$ system in the same solvent, K_1 was found to be $0.1\,M^{-1}$ at 25°C [3].

References:

[1] Alcock, R. M.; Hartley, F. R.; Rogers, D. E.; Wagner, J. L. (Coord. Chem. Rev. **16** [1975] 59/66, 62).

[2] Alcock, R. M.; Hartley, F. R.; Rogers, D. E.; Wagner, J. L. (J. Chem. Soc. Dalton Trans. **1975** 2189/93).

[3] Henry, P. M.; Marks, O. W. (Inorg. Chem. **10** [1971] 373/6).

Extraction of $[Pd_2Cl_6]^{2-}$

The exchange interphase reaction between various platinum metals in organic and aqueous phases has been investigated, e.g., $6[PdCl_4]^{2-} + 2[Rh_2Cl_9]^{2-} + 4H_2O \rightarrow 3[Pd_2Cl_6]^{2-} + 4[Rh(H_2O)Cl_5]^{2-} + 4Cl^-$ where R = dimethyloctylbenzylammonium chloride, tetraoctylammonium chloride and other quaternary bases [1]. Extraction constants for $[Pd_2Cl_6]^{2-}$ from solutions containing Pd with tetraoctylammonium chloride have been measured [2,3]. The ion has also been detected in $PdCl_2$–HCl solutions extracted into xylene-trilaurylamine mixtures [4] and in solutions of Pd-chloro complexes in $C_2H_4Cl_2$ or $CHCl_3$ in the presence of tetraoctylammonium chloride [5].

References:

[1] Gindin, L. M.; Ivanova, S. N.; Mazurova, A. A.; Vasil'eva, A. A.; Mironova, L. Y.; Sokolov, A. P. (Sint. Ochistka Anal. Neorgan. Mater. Tr. Konf. Nauka-Proizvod., Novosibirsk 1965 [1971], pp. 41/51; C.A. **77** [1972] No. 51318).

[2] Vasil'eva, A. A.; Mal'chikov, G. D.; Gindin, L. M. (Izv. Sibirsk. Otd. Akad. Nauk SSSR Ser. Khim. Nauk **1971** No. 3, pp. 67/71; C.A. **76** [1972] No. 145503).

[3] Selezneva, I. A.; Ivanova, S. N.; Gindin, L. M. (Izv. Sibirsk. Otd. Akad. Nauk SSSR Ser. Khim. Nauk **1982** No. 6, pp. 107/12; C.A. **98** [1983] No. 60 702).

[4] Zeveguintzoff, D.; Gourisse, D. (Bull. Soc. Chim. France **1979** I 179/84).

[5] Vasil'eva, A. A.; Gindin, L. M.; Pelina, G. N.; Mal'chikov, G. D. (Izv. Sibirsk. Otd. Akad. Nauk SSSR Ser. Khim. Nauk **1969** No. 6, pp. 40/6; Sib. Chem. J. **1969** 675/9).

Reactions of $[Pd_2Cl_6]^{2-}$

The ion is believed to be implicated in the catalysed isomerism of alkenes by $PdCl_2$–NaCl-glacial CH_3COOH solutions [1]; the intermediates $[Pd_2Cl_5(alkene)]^-$ and $Pd_2Cl_4(alkene)_2$ may be formed. With acetic acid $[Pd_2Cl_5(OOCCH_3)]^-$ is formed [2,3,4].

References:

[1] Davies, N. R. (Australian J. Chem. **17** [1964] 212/8).

[2] Alcock, R. M.; Hartley, F. R.; Rogers, D. E.; Wagner, J. L. (Coord. Chem. Rev. **16** [1975] 59/66).

[3] Alcock, R. M.; Hartley, F. R.; Rogers, D. E.; Wagner, J. L. (J. Chem. Soc. Dalton Trans. **1975** 2189/93).

[4] Hartley, F. R.; Wagner, J. L. (J. Organometal. Chem. **55** [1973] 395/403).

4.5.9.2.1 Salts of $[Pd_2Cl_6]^{2-}$ with Nonmetallic Cations

$[\{((CH_3)_2N)_3C\}_2[Pd_2Cl_6]$ and $[(n-C_3H_7)_3C_3]_2[Pd_2Cl_6]$

Both are cyclopropenium salts. The first salt is made from the tetrafluoroborate of the cation and aqueous $Na_2[PdCl_4]$. It is orange-pink, melting with decomposition at 232°C.

The far-infrared spectrum (20 to 400 cm^{-1}, reproduced in paper) of the solid was measured and also the Raman and infrared spectra in solution. The ^1H NMR spectrum was recorded and gives evidence of cation-anion interaction in CH_2Cl_2 solution.

$[(n-C_3H_7)_3C_3]_2[Pd_2Cl_6]$ forms long, thin orange-pink needles by reaction of $[(n-C_3H_7)_3C_3]BF_4$ and $Na_2[PdCl_4]$ in water. It melts at 105 to 106°C.

The far-infrared spectrum of the solid was measured (20 to 400 cm^{-1}, reproduced in paper), and also the Raman and infrared spectra in solution.

Reference:

Harris, D. C.; Gray, H. B. (Inorg. Chem. **13** [1974] 2250/5).

$(C_{13}H_{21}O)_2[Pd_2Cl_6]$

The X-ray crystal structure of this, the 2,6-ditertbutylpyridinium salt of $[Pd_2Cl_6]^{2-}$, shows the crystals to be monoclinic with a = 12.950, b = 23.597, c = 11.620 Å, β = 103.77°, space group C2/c-C_{2h}^6, Z = 4. The palladium has square-planar coordination with bond parameters (figure reproduced in paper).

Reference:

Kuz'mina, L. G.; Struchkov, Yu. T. (Koord. Khim. **9** [1983] 705/10; Soviet J. Coord. Chem. **9** [1983] 407/11).

$[(N(CH_3)_2COCH_3)_2H_2][Pd_2Cl_6]$. The X-ray crystal structure of this shows the Pd–Cl (terminal) distance to be 2.27 Å and the Pd–Cl (bridge) 2.32 Å, with a Pd–Pd distance of 3.41 Å [4].

$(Bu_4^nN)_2[Pd_2Cl_6]$

This is made from an aqueous solution of $K_2[PdBr_4]$ and $(Bu_4^nN)Cl$ [1] or from "$H_2[PdCl_4]$" in concentrated aqueous solution and the stoichiometric quantity of $(Bu_4^nN)Cl$ [2].

The far-infrared spectrum was measured of the solid (100 to 400 cm^{-1}) and assigned [1] (see table on p. 159), and force constants calculated for the in-plane infrared active $[Pd_2Cl_6]^{2-}$ modes [1]. Electronic spectra for the complex in a propionitrile-2-methyltetrahydrofuran solution (180 to 600 nm) were measured and assigned [2].

With $(Bu_4^nN)Cl$ it gives $(Bu_4^nN)_2[PdCl_4]$ [3].

References:

[1] Adams, D. M.; Chandler, P. J.; Churchill, R. G. (J. Chem. Soc. A **1967** 1272/4).
[2] Mason, W. R.; Gray, H. B. (J. Am. Chem. Soc. **90** [1968] 5721/9).
[3] Goggin, P. L.; Mink, J. (J. Chem. Soc. Dalton Trans. **1974** 1479/83).
[4] Shestakov, A. F.; Makitova, D. D.; Atovmyan, L. O.; Kasochka, O. N. (Koord. Khim. **13** [1987] 949/52; C.A. **107** [1987] No. 183846).

$(Ph_4As)_2[Pd_2Cl_6]$

This is made from an aqueous solution of $K_2[PdCl_4]$ and $(Ph_4As)Cl$, and after recrystallisation gives orange-brown crystals [1]. X-ray data show the unit cell to be triclinic [2].

The far-infrared spectrum of the solid was measured for 100 to 400 cm^{-1} and assignments proposed (see table on p. 159); force constants for the in-plane infrared active $[Pd_2Cl_6]^{2-}$ modes were calculated [1]. The far-infrared spectrum of the solid was also measured (800 to 400 cm^{-1}, reproduced in paper) [2], and a band at 186 cm^{-1} previously assigned to a PdCl$_2$ deformation [1] was re-assigned to a mode arising from $(Ph_4As)^+$ [3].

References:

[1] Adams, D. M.; Chandler, P. J.; Churchill, R. G. (J. Chem. Soc. A **1967** 1272/4).
[2] Chow, C.; Chang, K.; Willett, R. D. (J. Chem. Phys. **59** [1973] 2629/40, 2630).
[3] Harris, D. C.; Gray, H. B. (Inorg. Chem. **13** [1974] 2250/5).

(AsPh₃Me)₂[Pd₂Cl₆]

This brownish-pink salt is made from an aqueous solution of $K_2[PdCl_4]$ with (AsPh₃Me)Cl in water.

Ionic weight measurements indicated that the anion remains in the dimeric form in nitrobenzene. The molar conductance in this solvent was 53.7 $cm^2 \cdot \Omega^{-1}$ [1]. The electronic absorption (reflectance) spectrum of the solid has been recorded (200 to 700 nm, reproduced in paper), and also the spectra (200 to 600 nm, maxima listed in paper) in nitrobenzene, acetic anhydride, nitromethane, acetone, methanol and acetonitrile have been measured [2].

With excess (AsPh₃Me)Cl in boiling acetone (AsPh₃Me)₂[PdCl₄] is formed [1].

References:

[1] Harris, C. M.; Livingstone, S. E.; Stephenson, N. C. (J. Chem. Soc. **1958** 3697/702).
[2] Harris, C. M.; Livingstone, S. E.; Reece, I. H. (J. Chem. Soc. **1959** 1505/11).

(C₁₉H₃₈NH)₂[PdCl₄]

This n-cetyltrimethylammonium salt (R = C₁₉H₃₈NH) was made from RCl and aqueous $K_2[PdBr_4]$[1] [1].

The far-infrared spectrum was measured (100 to 400 cm^{-1}) and force constants for $[Pd_2Cl_6]^{2-}$ calculated for the in-plane infrared active modes.

Reference:

[1] Adams, D. M.; Chandler, P. J.; Churchill, R. G. (J. Chem. Soc. A **1967** 1272/4).

(Pn₄ⁿN)₂[Pd₂Cl₆]

This was prepared by reaction of n-pentylammonium chloride with $[PdBr_4]^{2-}$ and forms red-brown crystals, melting at 79 to 80°C [1].

The Raman spectrum of the salt in CH_2Cl_2 was measured (200 to 400 cm^{-1}) and the infrared spectra in $CHCl_3$ and CH_2Br_2 (200 to 400 cm^{-1}); the infrared spectrum of the solid was measured over the same range (for data see table on p. 159). Assignments of modes were proposed [1] and more complete assignments with terminal and bridging Pd–Cl stretching force constants calculated [2].

References:

[1] Goggin, P. L. (J. Chem. Soc. Dalton Trans. **1974** 1483/6).
[2] Goggin, P. L.; Mink, J. (Inorg. Chim. Acta **26** [1978] 119/24).

4.5.9.2.2 Alkali Metal Salts

Li₂[Pd₂Cl₆]. The existence of this in solution has been inferred from molecular weight data on LiCl–PdCl₂ solutions in acetic acid, though no complex was isolated. For Li₂[Pd₂Cl₆] + 2LiCl → 2Li₂[PdCl₄] the equilibrium constant K = 0.1M⁻¹ at 25°C was calculated.

Na₂[Pd₂Cl₆]. This species is thought to be present in solutions of PdCl₂ and NaCl in acetic acid, though no salt was isolated [1]. Spectrophotometric studies on the equilibria between NaCl and PdCl₂ in glacial acetic acid at 25°C [2] give values of $K_1 = 0.56 \pm 0.02$ and $K_2 = (2.9 \pm 0.6) \times 10^{-5}$ mol/dm³ for $2Na_2[PdCl_4] \xrightarrow{K_1} Na_2[Pd_2Cl_6] + 2NaCl$ and $Na_2[Pd_2Cl_6] \xrightarrow{K_2} Na_2[Pd_2Cl_5(OOCCH_3)] + NaCl$.

Molecular weight determinations for NaCl with $PdCl_2$ in acetic acid suggested that the dimer $Na_2[Pd_2Cl_6]$ was present in such solutions [1].

References:

[1] Henry, P. M.; Marks, O. W. (Inorg. Chem. **10** [1971] 373/6).
[2] Alcock, R. M.; Hartley, F. R.; Rogers, D. E.; Wagner, J. L. (J. Chem. Soc. Dalton Trans. **1975** 2189/93).

4.5.9.3 Other Binuclear Species

$Li_3[Pd_2Cl_7]$ was misprinted in [1] as $Li_2[Pd_2Cl_7]$ but later corrected to $Li_3[Pd_2Cl_7]$ [2]; it may be present in solutions of LiCl and $PdCl_2$ in acetic acid. On the basis of spectrophotometric data the equilibrium constant K for $Li_2[Pd_2Cl_6] + LiCl \rightleftharpoons Li_3[Pd_2Cl_7]$ was calculated as lying between 0.4 and $5.3 M^{-1}$ [1].

The ion has been briefly mentioned as a possible constituent of solutions of $PdCl_2$ in HCl [3].

$[Pd_2Cl_{6+n}]^{(2+n)-}$. Such species may be formed, according to the findings of spectrophotometric data, from $PdCl_2$ and LiCl in CH_3CN [4].

$[Pd_2Cl_n(H_2O)_{6-n}]^{4-n}$. Such species (formulated in the references without coordinated aqua ligands) may be present in some solutions of $PdCl_2$ in HCl [5, 6, 7]. The pentachloro species has also been mentioned in connection with determination of stability constants for $[PdCl_6]^{2-}$ [7].

References:

[1] Henry, P. M.; Marks, O. W. (Inorg. Chem. **10** [1971] 373/6).
[2] Alcock, R. M.; Hartley, F. R.; Rogers, D. E.; Wagner, J. L. (J. Chem. Soc. Dalton Trans. **1975** 2189/93).
[3] Kaszonyi, A.; Vojtko, J.; Hrusovsky, M. (Collection Czech. Chem. Commun. **43** [1978] 3002/6).
[4] Volchenskova, I. I.; Yatsimirskii, K. B. (Zh. Neorgan. Khim. **18** [1973] 1875/82; Russ. J. Inorg. Chem. **18** [1973] 990/5).
[5] Kaszonyi, A.; Vojtko, J.; Hrusovsky, M. (Collection Czech. Chem. Commun. **45** [1980] 179/86).
[6] Kaszonyi, A.; Vojtko, J.; Hrusovsky, M. (Collection Czech. Chem. Commun. **43** [1978] 3002/6).
[7] Dubinskii, V. I. (Zh. Neorgan. Khim. **16** [1971] 1145/7; Russ. J. Inorg. Chem. **16** [1971] 607/8).

Heterobinuclear Species $[PdCuCl_6]^{2-}$

On the basis of ESR spectra of copper(II)-doped $(Ph_4As)_2[Pd_2Cl_6]$ the existence of this species was postulated.

Reference:

Chow, C.; Chang, K.; Willett, R. D. (J. Chem. Phys. **59** [1973] 2629/40).

4.6 Miscellaneous Multicomponent Systems and Compounds

$Pd_4Bi_5Cl_3$ is a mineral phase from the pyrrhotite from the Nor'ilsk region. Its analysis was reported [2].

The $NaCl-PdCl_2-TeCl_4$ System

The phase diagram of this system is shown in **Fig. 34**, p. 166. It was constructed according to DTA, DTG, X-ray diffraction and IR spectroscopy. The liquidus surface consists of the primary crystallisation fields Na$_2$[PdCl$_4$], NaCl, PdCl$_2$ and TeCl$_4$. Points of invariant equilibria (compositions in mol%) [17]:

E$_1$ at 206°C: L ⇌ Na$_2$[PdCl$_4$] + PdCl$_2$ + TeCl$_4$ (18.0 NaCl, 14.5 PdCl$_2$, 67.5 TeCl$_4$)
E$_2$ at 190°C: L ⇌ Na$_2$[PdCl$_4$] + NaCl + TeCl$_4$ (26.0 NaCl, 10.0 PdCl$_2$, 64.0 PdCl$_4$)
e at 208°C: L ⇌ Na$_2$[PdCl$_4$] + TeCl$_4$ (21.0 NaCl, 10.0 PdCl$_2$, 69.0 TeCl$_4$)

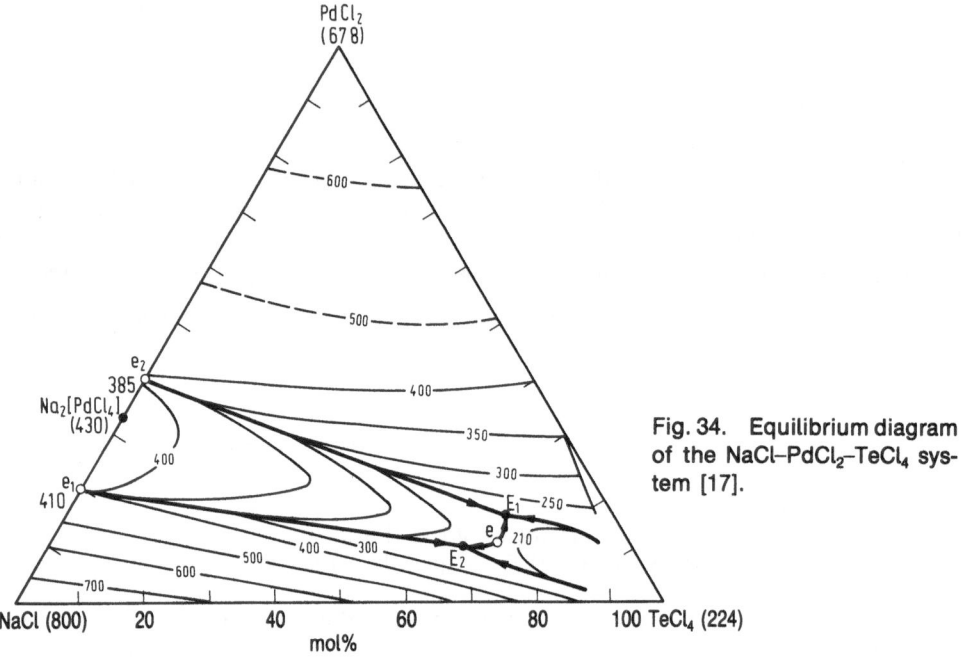

Fig. 34. Equilibrium diagram of the NaCl–PdCl$_2$–TeCl$_4$ system [17].

The KCl–NaCl–PdCl$_2$ System

This system has been studied by DTA, X-ray diffraction and IR spectroscopy. K$_2$[PdCl$_4$]–NaCl(I) and K$_2$[PdCl$_4$]–Na$_2$[PdCl$_4$](II) are stable sections, the others (III, IV, V) are unstable. **Fig. 35** shows the liquidus diagram of the system. The following ternary eutectics were located:

E$_1$ at 335°C: L ⇌ K$_2$[PdCl$_4$] + Na$_2$[PdCl$_4$] + PdCl$_2$
E$_2$ at 388°C: L ⇌ K$_2$[PdCl$_4$] + Na$_2$[PdCl$_4$] + NaCl
E$_3$ at 473°C: L ⇌ K$_2$[PdCl$_4$] + KCl + NaCl

The K$_2$[PdCl$_4$]–NaCl diagram (shown in paper) is of the simple eutectic type, the eutectic point being at 480°C, 54 mol% KCl, 19 mol% NaCl, and 27 mol% PdCl$_2$. **Fig. 36** shows the K$_2$[PdCl$_4$]–Na$_2$[PdCl$_4$] diagram from 0 to 66.67 mol% NaCl. The minimum is at 392°C, and (in mol%) 18 KCl, 49 NaCl, 33 PdCl$_2$. Thermal effects at 330 to 340°C were attributed to the decomposition of K$_2$[PdCl$_4$]–Na$_2$[PdCl$_4$] solid solutions [21]. Phase equilibria in the sub-solidus region of the K$_2$[PdCl$_4$]–Na$_2$[PdCl$_4$] system have been studied by X-ray powder diffraction, DTA and electrical conductivity methods [22].

PdAl$_2$Cl$_8$. This is made by reaction of PdCl$_2$ with Al$_2$Cl$_6$ in an evacuated silica tube at 200°C followed by separation of the products under a temperature gradient [1], or from a PdCl$_2$–Al$_2$Cl$_6$ mixture over a 180 to 110°C temperature gradient [2]. It is remarkably stable in the gas phase under gas-phase transportation conditions [3, 4, 5]. It forms deep red crystals [2] which are very hygroscopic [1].

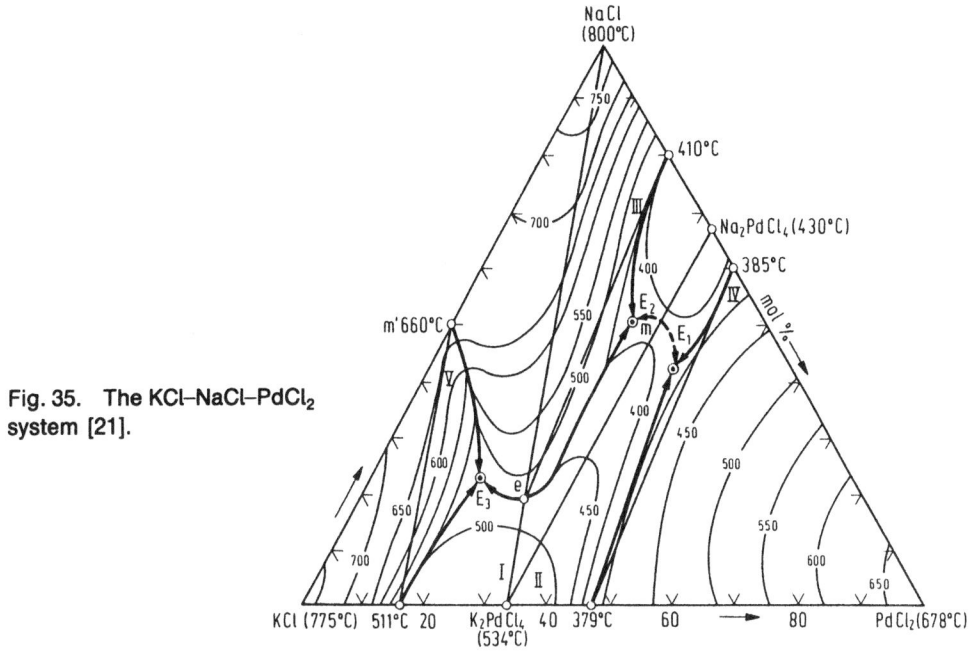

Fig. 35. The KCl–NaCl–PdCl$_2$
system [21].

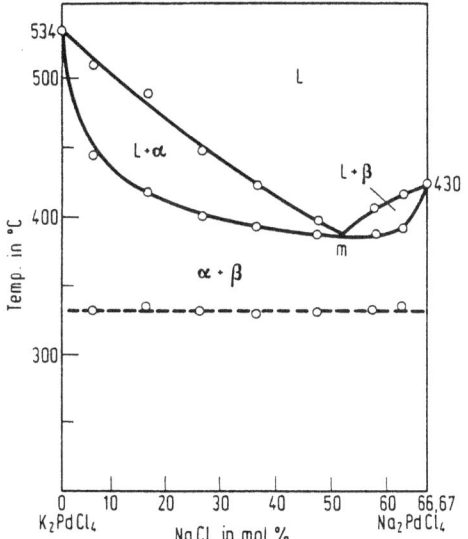

Fig. 36. Partial diagram of the K$_2$[PdCl$_4$]–
Na$_2$[PdCl$_4$] system [21].

The single X-ray crystal structure of the solid, see **Fig. 37**, p. 168, shows the palladium atoms
to have square-planar coordination. The crystals are monoclinic, space group P2$_1$/c-C$_{2h}^5$; Z = 2,
a = 6.583(3), b = 7.342(30), c = 13.033(4) Å, β = 96.40(3)°; measured and calculated densities
2.35(2) and 2.36 g/cm^3, respectively [2]. X-ray powder diffraction data were also listed [1, 6].

The molar magnetic susceptibility is $-(161+6) \times 10^{-6}$ cgsu at 80 K ($\mu_{eff} = 0.22$ BM) and
$-(181 \pm 7) \times 10^{-6}$ cgsu (0.36 BM) at 300 K [7]. Changes in the magnetic susceptibility consequent on addition of PdCl$_2$ to Al$_2$Cl$_6$ have been measured [7].

Thermodynamic data for $PdCl_2(solid) + Al_2Cl_6(gas) \rightarrow PdAl_2Cl_8(gas)$ are $\Delta H_R = 30.2$ kJ/mol and $\Delta S_R = 39.5$ J·deg^{-1}·mol^{-1} [8]. For $PdCl_2(solid) + PdAl_2Cl_8(gas) \rightarrow PdAl_2Cl_{10}(gas)$ $\Delta H^\circ_{298} = 31$ kJ/mol, $\Delta S^\circ_{298} = 22 \pm 8$ J·deg^{-1}·mol^{-1} [9].

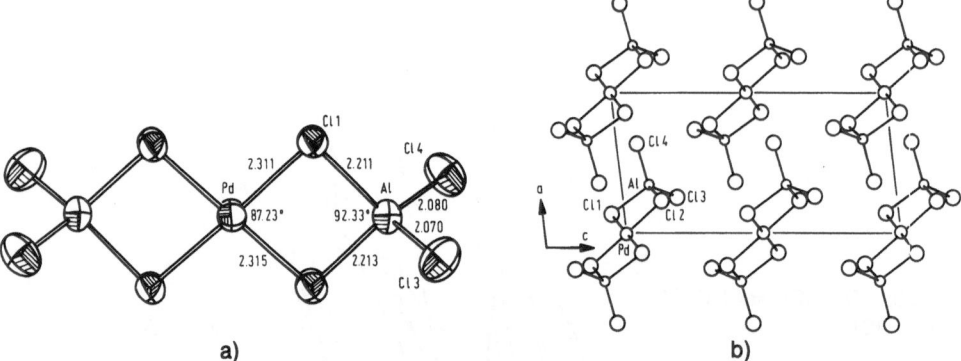

a)   b)

Fig. 37.   X-ray crystal structure of $PdAl_2Cl_8$ [2]; a) the complex anion; b) the unit cell.

Resonance Raman spectra have been measured of gaseous $PdAl_2Cl_8$ [10 to 13] from 0 to 650 cm^{-1} with a variety of laser exciting lines and at temperatures between 600 and 750 K (spectra reproduced in paper). Excitation profiles were obtained and assignments of modes proposed: it seems likely that the gaseous molecule adopts a D_{2h} structure [10].

Electronic absorption spectra of the vapour were measured at temperatures from 600 to 750 K over the range 300 to 750 nm (reproduced in papers) [8, 10]. There is no evidence for formation of the complex in aqueous solution when solutions of $PdCl_2$ in HCl are mixed with Al_2Cl_6 [14].

Mass spectrometry [9, 19] gave evidence for the existence of the species $PdAl_2Cl_6^+$, $PdAlCl_5^+$, $PdAlCl_4^+$, $PdAl_2Cl_8^+$ [9] and for $PdAlCl_4^+$, $PdAl_2Cl_6^+$ and $PdAl_2Cl_7^+$ [19].

$PdAl_2Cl_{10}$. The existence of this species was demonstrated by mass spectrometric observations on a heated reaction mixture of $PdCl_2$ and Al_2Cl_6. For the reaction α-$Pd_2Cl_2(solid) + PdAl_2Cl_8(gas) \rightarrow Pd_2Al_2Cl_{10}(gas)$ the values of $\Delta H^\circ_{298} = 31$ kJ/mol and $\Delta S^\circ_{298} = 22.8 \pm 8$ J·deg^{-1}·mol^{-1} were obtained [9].

$Pd(AlCl_4)_2$ is sometimes referred to in the literature [1, 6] but is identical to $PdAl_2Cl_8$ (see above).

The $NaCl$–$PdCl_2$–$PbCl_2$ System

The liquidus diagram is given in **Fig. 38**.

The $PbCl_2$–$Na_2PdCl_4(S)$ section is quasi-binary. Point e (a van Rijn point) is at $NaCl$ 39, $PbCl_2$ 41, $PdCl_2$ 20 mol%; 378°C. The stable section divides the liquidus diagram into two partial systems: $NaCl$–$PdCl_2$–S and $PbCl_2$–$PdCl_2$–S. The points of four phase equilibria E_1 and E_2 are eutectics (E_1: 355°C, $PbCl_2$ 38.0, $NaCl$ 47.0, $PdCl_2$ 15.0 mol%; E_2: 358°C, $PbCl_2$ 32.0, $NaCl$ 37.0, $PdCl_2$ 31.0 mol%). There are fields of primary crystallisation of $NaCl$, $PbCl_2$, $PdCl_2$ and Na_2PdCl_4 on the liquidus surface [18].

The $AgCl$–$PdCl_2$–$TeCl_4$ System

The liquidus diagram of the system is shown in **Fig. 39**. There is an extensive region of liquid immiscibility (LI $\approx$ 42% of the area of the diagram). The point of four-phase equilibrium E is eutectic (E: 206°C, $AgCl$ 3.0, $TeCl_4$ 94.0, $PdCl_2$ 3.0 mol%). The co-existing phases are: L = $AgCl + PdCl_2 + TeCl_4$ [18].

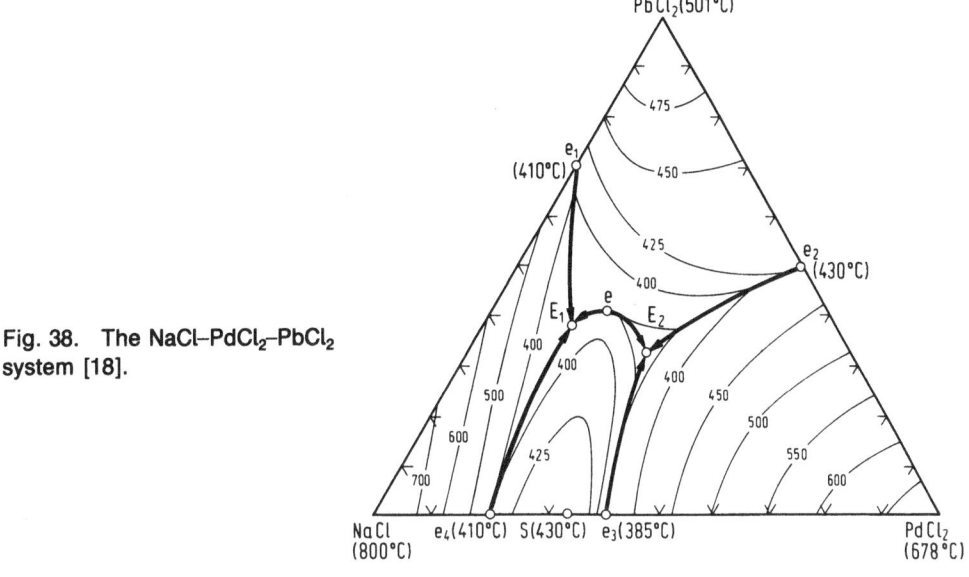

Fig. 38. The NaCl–PdCl$_2$–PbCl$_2$ system [18].

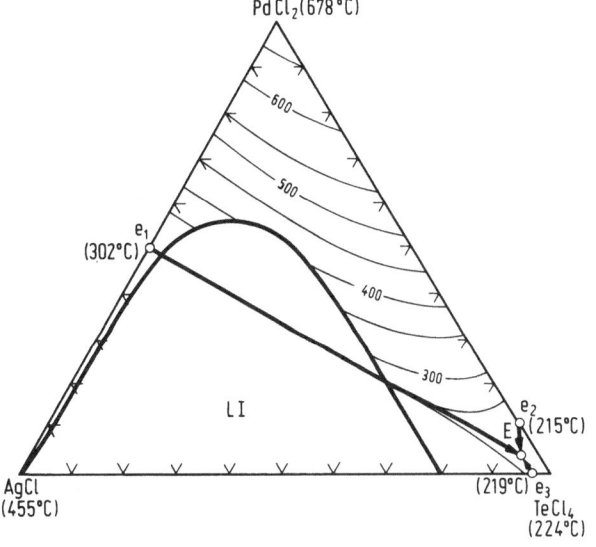

Fig. 39. The AgCl–PdCl$_2$–TeCl$_4$ system [18].

The AgCl–NaCl–PdCl$_2$ System

Fig. 40, p. 170, shows the liquidus diagram of the system. The AgCl–Na$_2$PdCl$_4$(S) section is quasi-binary and the system is of the eutectic type. Point e is a van Rijn point and is at AgCl 55, NaCl 30, PdCl$_2$, 15mol%, 312°C. This stable section divides the liquidus diagram into two partial ones: AgCl–NaCl–S and AgCl–PdCl$_2$–S. At the liquidus surface in the composition range adjacent to the AgCl–NaCl side there are solid solutions which decompose when ~8 to 10% of PdCl$_2$ is added. The points of four-phase equilibria E$_1$ and E$_2$ are eutectics (E$_1$: 310°C, AgCl 53.5, NaCl 36.0, PdCl$_2$ 10.5 mol%; E$_2$: 290°C, AgCl 48.0, NaCl 25.5, PdCl$_2$ 26.5 mol%) [18].

In the AgCl–PdCl$_2$ system there is a eutectic at 50 mol% PdCl$_2$, 302°C [18].

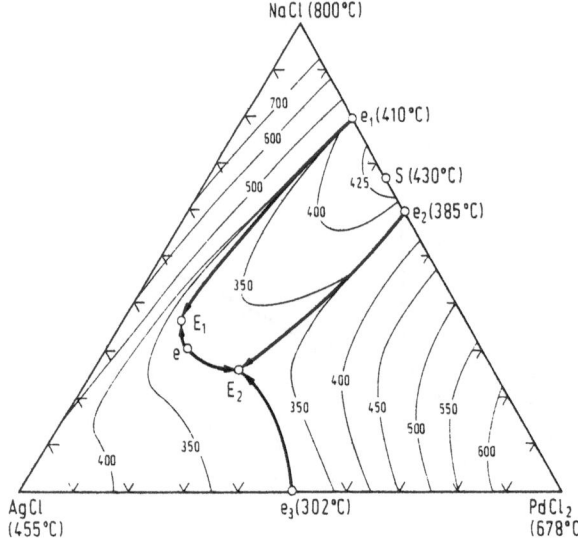

Fig. 40. The AgCl–PdCl$_2$–NaCl system [18].

The Systems PdCl$_2$–PbCl$_2$, PdCl$_2$–FeCl$_3$ and PdCl$_2$–FeCl$_3$–PbCl$_2$

Equilibrium diagrams for these three ternary systems and the AgCl/FeCl$_3$/PbCl$_2$ system which bound the AgCl/FeCl$_3$/PbCl$_2$/PdCl$_2$ system are reproduced in **Fig. 41**. Points of four-phase equilibria, E, are given below (m.p. = melting point, L = liquid) [20].

co-existing phases		m.p. in °C	composition in mol%			
			AgCl	FeCl$_3$	PbCl$_2$	PdCl$_2$
L = AgCl + FeCl$_3$ + PbCl$_2$	E$_1$	170	15.0	61.0	24.0	—
L = AgCl + FeCl$_3$ + PdCl$_2$	E$_2$	140	7.0	88.0	—	5.0
L = AgCl + PbCl$_2$ + PdCl$_2$	E$_3$	290	40.0	—	20.0	40.0
L = FeCl$_3$ + PbCl$_2$ + PdCl$_2$	E$_4$	158	—	68.0	21.0	11.0

Fig. 41. The FeCl$_3$–AgCl–PbCl$_2$–PdCl$_2$ system [20].

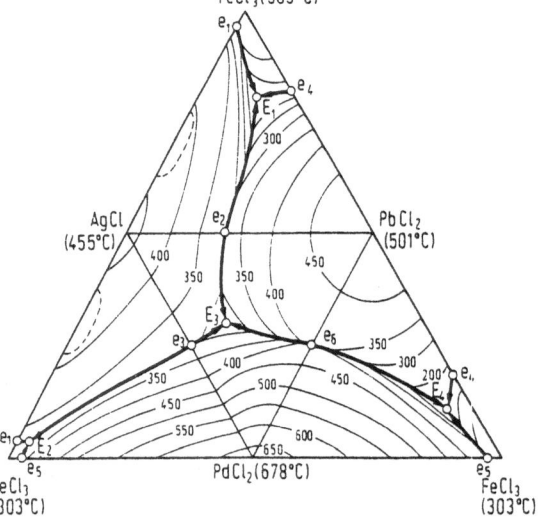

Rb$_4$PdAu$_2$Cl$_{12}$. Cs$_4$PdAu$_2$Cl$_{12}$

Rb$_4$PdAu$_2$Cl$_{12}$ (not Rb$_2$PdAuCl$_7$ as previously reported by Tananaew et al.) is obtained when RbCl, AuCl$_3$ and PdCl$_2$ react in boiling aqueous HCl and the solution is allowed to crystallise. The crystals are cubic, a = 20.53 (±0.02) Å, Z = 16, D$_{exp}$ = 3.88; D$_{calc}$ = 3.894 g/cm^3 [15].

Cs$_4$PdAu$_2$Cl$_{12}$ (not Cs$_2$PdAuCl$_7$ as previously reported by Tananaew et al.) is obtained when CsCl, AuCl$_3$ and PdCl$_2$ react in boiling aqueous HCl and the solution is allowed to crystallise. The crystals are cubic, a = 20.91 (±0.01) Å, Z = 16, D$_{exp}$ = 4.21 g/cm^3 [15]. The specific conductivity of compressed pellets is given as $4.8 \times 10^{-11} \Omega^{-1} \cdot cm^{-1}$ [16].

References:

[1] Papatheodorou, G. N. (Inorg. Nucl. Chem. Letters **10** [1974] 115/8).
[2] Lenhard, W.; Schäfer, H.; Hürter, H.-U.; Krebs, B. (Z. Anorg. Allgem. Chem. **482** [1981] 19/26).
[3] Schäfer, H.; Trenkel, M. (Z. Anorg. Allgem. Chem. **414** [1975] 137/50).
[4] Schäfer, H. (Z. Anorg. Allgem. Chem. **403** [1974] 116/26).
[5] Schäfer, H.; Nowitzki, J. (J. Less-Common Metals **61** [1978] 47/50).
[6] Belt, R. F.; Scott, H. (Inorg. Chem. **3** [1964] 1785/8).
[7] Papatheodorou, G. N. (Inorg. Nucl. Chem. Letters **14** [1978] 249/53).
[8] Papatheodorou, G. N. (J. Phys. Chem. **77** [1973] 472/7).
[9] Florke, U.; Schäfer, H. (Z. Anorg. Allgem. Chem. **459** [1979] 140/4).
[10] Papatheodorou, G. N.; Capote, M. A. (J. Chem. Phys. **69** [1978] 2067/75).

[11] Papatheodorou, G. N.; Capote, M. A. (Proc. Electrochem. Soc. **1978** 78/81; C.A. **89** [1978] No. 119976).
[12] Papatheodorou, G. N. (CONF-780941-8 [1978] 1/63; C.A. **91** [1979] No. 99402).
[13] Papatheodorou, G. N. (NBS-SP-561-1 [1979] 647/78; C.A. **92** [1980] No. 67053).
[14] Kravchuk, L. S.; Stremok, I. P.; Markevich, S. V. (Zh. Neorgan. Khim. **21** [1976] 728/31; Russ. J. Inorg. Chem. **21** [1976] 391/3).
[15] Ferrari, A.; Cavalca, L.; Nardelli, M. (Gazz. Chim. Ital. **85** [1955] 137/44).
[16] Gomm, P. S.; Underhill, A. E. (Inorg. Nucl. Chem. Letters **10** [1974] 309/13).
[17] Safonov, V. V.; Chaban, N. G. (Zh. Neorgan. Khim. **24** [1979] 265/8; Russ. J. Inorg. Chem. **24** [1979] 149/50).
[18] Safonov, V. V.; Mireev, V. A.; Ivnitskaya, R. B. (Zh. Neorgan. Khim. **24** [1979] 3361/3; Russ. J. Inorg. Chem. **24** [1979] 1873/5).
[19] Binnewies, M. (Z. Anorg. Allgem. Chem. **435** [1977] 156/60).
[20] Safonov, V. V.; Mireev, V. A. (Zh. Neorgan. Khim. **24** [1979] 3396/8; Russ. J. Inorg. Chem. **24** [1979] 1894/5).

[21] Safonov, V. V.; Mireev, V. A. (Zh. Neorgan. Khim. **28** [1983] 1779/82; Russ. J. Inorg. Chem. **28** [1983] 1004/6).
[22] Mireev, V. A.; Tsygankov, V. N.; Safonov, V. V. (Zh. Neorgan. Khim. **32** [1987] 454/5; Russ. J. Inorg. Chem. **32** [1987] 252/3).

4.7 Hydrido-chloro Complexes

[HPdCl$_3$]$^-$. It has been proposed that this species is an intermediate in the reaction between [PdCl$_4$]$^{2-}$ and molecular H$_2$ [1].

[HPdCl$_4$]$^-$ (or H[PdCl$_4$]$^-$). This species is said to be one of those extracted into a xylene-trilaurylamine mixture from PdCl$_2$ solutions in HCl [2].

References:

[1] Fasman, A. B.; Golodov, V. A.; Postil'nikov, L. M.; Luk'yanov, A. T.; Baranovskii, B. P.; Darinskii, Y. V. (Izv. Sibirsk. Otd. Akad. Nauk SSSR Ser. Khim. Nauk **1968** No. 3, pp. 144/8; Sib. Chem. J. **1968** 352/5).
[2] Zveguintsoff, D.; Gourisse, D. (Bull. Soc. Chim. France **1979** I 179/84).

4.8 Chloro-aquo Complexes

$[PdCl_5(H_2O)_n]^{3-}$

Some evidence for formation of this species, n being unspecified, was obtained from stopped-flow kinetic experiments on $[PdCl_4]^{2-}$ in M $HClO_4$ in the presence of an excess of Cl^-; a fast reaction attributed to the $[PdCl_3(H_2O)]^--Cl^-$ reaction was observed together with a slower one which may involve this species of higher coordination number [1].

$[PdCl_4(H_2O)_2]^{2-}$. This may be involved in the $[PdCl_4]^{2-}-Cl^--H_2O$ system [1].

$[PdCl_3(H_2O)]^-$. For "$[PdCl_3]^-$" (presumably $[PdCl_3(H_2O)]^-$) see "Palladium" 1942, p. 326.

Electronic Spectra. Electronic spectra have been measured (240 to 700 nm, reproduced in paper) for this species and assigned [2, 5], and the electronic absorption spectrum of a benzene solution of $[PdCl_3(H_2O)]^-$ in association with crystal violet has been measured (420 to 670 nm, reproduced in paper) [3]. Absorption maxima were listed (200 to 500 nm) [4]. Electronic spectra for $[PdCl_3(H_2O)]^-$ were listed (200 to 450 nm) [4].

Kinetic and Thermodynamic Data. The enthalpy ($\Delta H°$) and entropy ($\Delta S°$) of activation for $H_2O + [PdCl_4]^{2-} \rightarrow [PdCl_3(H_2O)]^- + Cl^-$ at 25°C were found to be 50 ± 8 kJ/mol and -54 ± 24 J·deg^{-1}·mol^{-1}, respectively, with $\Delta H° = 41.8 \pm 4$ kJ/mol and $\Delta S° = -63 \pm 12$ J·deg^{-1}·mol^{-1} for the reverse anation reaction [5]. For the aquation reaction values of $\Delta H = 64 \pm 8$ kJ/mol and $\Delta S = -50 \pm 20$ J·deg^{-1}·mol^{-1} for anation 52 ± 8 kJ/mol and -29 ± 12 J·deg^{-1}·mol^{-1} were calculated [6]. The kinetics of formation of $[PdCl_3(H_2O)]^-$ by aquation of $[PdCl_4]^{2-}$ and the reverse anation reactions have been studied at total ionic strength 2.1 M in 2M $HClO_4$ and at 15°C by a stopped-flow method [7]; at 18 to 25°C by a temperature-jump method [6]; from 15 to 35°C by stopped-flow methods [5], and by spectrophotometric and stopped-flow methods [8].

Data from [7, 8] have been critically assessed by [9, 10].

For thermodynamic data on formation and aquation reactions of $[PdCl_3(H_2O)]^-$ see p. 174, for stability constants see p. 96.

The reactions $[PdCl_3(H_2O)]^- + H_2O \rightleftharpoons cis-PdCl_2(H_2O)_2 + Cl^-$ and $[PdCl_3(H_2O)]^- + H_2O \rightleftharpoons trans-PdCl_2(H_2O)_2 + Cl^-$ have been studied by stopped-flow techniques at 25°C, ionic strength 2.1M in 2M $HClO_4$ [7]; at 25°C and at 1M ionic strength using stopped-flow techniques [9]. In general acid hydrolyses are some 10^5 faster for Pd than for Pt chloro complexes, while anations are some 10^4 times faster for Pd than for Pt [9].

The data in [7] have been critically assessed [9, 10].

The pK_a of the reaction $[PdCl_3(H_2O)]^- \rightarrow [PdCl_3(OH)]^{2-}$ is 7 [21]; see also [7].

Chemical Reactions. Kinetics of the reaction of $[PdCl_3(H_2O)]^-$ with ethylenediamine(en) to give $PdCl_2en$ have been measured [11]. The $[PdCl_5(H_2O)_n]^{3-}$ ion may result from reaction of $[PdCl_3(H_2O)]^-$ in 1M $HClO_4$ with excess Cl^- (see above) [1]. It is thought that $[PdCl_3(H_2O)]^-$ may be the active species in the catalytic reduction by CO of Fe^{III} salts in the presence of aqueous $[PdCl_4]^{2-}$ [12], and the mechanism of reaction of $[PdCl_3(H_2O)]^-$ with dienes has been studied [13].

Miscellaneous Data. Sorption of $[PdCl_3(H_2O)]^-$ on γ-Al_2O_3 has been studied by means of sorption isotherms [14]. It is likely that $[PdCl_3(H_2O)]^-$ (given as "$[PdCl_3]^-$" in paper) is generated during electro-reduction of $[PdCl_4]^{2-}$ in 3M H_2SO_4 at a Pd electrode, and that this in turn is reduced to $PdCl_2(H_2O)_2$ [15, 17]. Extraction of $[PdCl_3(H_2O)]^-$ from solutions of $PdCl_2$ in HCl by tri-isobutylphosphine sulphide was investigated [16].

Catalysis by "$[PdCl_3]^-$", again presumably $[PdCl_3(H_2O)]^-$ as the Li salt in the reaction of organic halides with trimethylvinylsilane, has been reported [18]. The $[PdCl_3(H_2O)]^-$ ion may be an important catalytic intermediate for organometallic reactions [1]. Changes in the composition and reactivity of chloropalladates, consequent on addition of organic solvents to their aqueous solutions (including $[PdCl_3(H_2O)]^-$) have been investigated [20].

References:

[1] Cooper, J. C.; Venezky, D. L.; Lorenz, T. (Fundam. Res. Homogeneous Catal. **3** [1979] 847/57, 849, 855).
[2] Elding, L. V.; Olsson, L. F. (J. Phys. Chem. **82** [1978] 69/74).
[3] Khvatkova, Z. M.; Golovina, V. V. (Zh. Analit. Khim. **34** [1979] 2035/9; J. Anal. Chem. [USSR] **34** [1979] 1577/80).
[4] Victori, L.; Tomás, X.; Malgosa, F. (Afinidad **32** [1975] 867/72).
[5] Elding, L. I. (Inorg. Chim. Acta **6** [1972] 683/8).
[6] Vargaftik, M. N.; Igoshin, V. A.; Syrkin, Ya. K. (Izv. Akad. Nauk SSSR Ser Khim. **1972** 1426/8; Bull. Acad. Sci. USSR Div. Chem. Sci. **1972** 1380/2).
[7] Bekker, P. van Z.; Robb, W. (J. Inorg. Nucl. Chem. **37** [1975] 829/34).
[8] Pearson, R. G.; Hynes, M. G. (Kgl. Tek. Høegsk. Handl. No. 285 [1972] 461/9).
[9] Elding, L. I. (Inorg. Chim. Acta **7** [1973] 581/8).
[10] Elding, L. I. (Inorg. Chim. Acta **15** [1975] L9/L11).

[11] de Waal, D. J. A.; Robb, W. (Intern. J. Chem. Kinet. **6** [1974] 309/21).
[12] Golodov, V. A.; Glubokovskikh, N. G.; Taneeva, G. V. (React. Kinet. Catal. Letters **22** [1983] 101/5).
[13] Asfour, H. M.; Green, M. (J. Organometal. Chem. **292** [1985] C25/C27).
[14] Ananin, V. N.; Trokhimetz, A. I. (React. Kinet. Catal. Letters **27** [1985] 129/32).
[15] Zelenskii, M. I.; Kravtsov, V. I. (Elektrokhimiya **6** [1970] 793/8; Soviet Electrochem. **6** [1970] 768/72).
[16] Baba, Y.; Ohshima, M.; Inoue, K. (Bull. Chem. Soc. Japan **59** [1986] 3829/33).
[17] Kravtsov, V. I.; Zelenskii, M. I. (Elektrokhimiya **2** [1966] 1138/43; Soviet Electrochem. **2** [1966] 1042/6).
[18] Chistovalova, N. M.; Akhrem, I. S.; Reshetova, E. V.; Vol'pin, M. E. (Izv. Akad. Nauk SSSR Ser. Khim. **1984** 2342/3; Bull. Acad. Sci. USSR Div. Chem. Sci. **1984** 2139/40).
[19] Rund, J. V. (Inorg. Chem. **9** [1970] 1211/5).
[20] Kaszonyi, A.; Hrusovsky, M.; Ilavsky, J. (Petrochemia **26** [1986] 62/8; C.A. **78** [1987] No. 88386).

$PdCl_2(H_2O)_2$

Both cis and trans isomers have been identified as well as the undifferentiated species; the latter is considered first.

Electronic Spectra. The spectrum of an equilibrium mixture of the cis and trans forms in aqueous solution has been recorded and assigned (220 to 700 nm, reported in paper) [1, 11].

Electronic spectral maxima for $PdCl_2(H_2O)_2$ have been listed (200 to 430 nm) [2].

Miscellaneous Data. The species "$PdCl_2$", presumably $PdCl_2(H_2O)_2$, is thought to be implicated in the electrodeposition of Pd from Cl^-–$[PdCl_4]^{2-}$ solutions in 3 M H_2SO_4 [3, 6]. Values of E_0 for $PdCl_2(H_2O)_2$ have been obtained from automated electrode kinetics measurements [4].

Sorption interactions of $PdCl_2(H_2O)_2$ with γ-Pl_2O_3 have been studied by means of sorption isotherms [5].

For stability constant data see table on p. 94.

cis-PdCl$_2$(H$_2$O)$_2$. Kinetics for the reaction $[PdCl_3(H_2O)]^- + H_2O \rightleftharpoons cis\text{-}PdCl_2(H_2O)_2 + Cl^-$ and the reverse anation reaction at a total ionic strength of 2.1M in 2M $HClO_4$ at 25°C have been studied by a stopped-flow procedure [7]; at 1 M ionic strength at 25°C by stopped-flow methods [8] and, for the aquation and anation reactions, by spectrophotometric and stopped-flow methods [9].

In general acid hydrolysis and halide anation reactions are some 10^4 to 10^5 times faster for the Pd than the Pt analogues [8].

The activation enthalpy and entropy for the reaction $H_2O + [PdCl_3(H_2O)]^- \rightarrow cis\text{-}PdCl_2(H_2O)_2 + Cl^-$ at 25°C are 61 kJ/mol and $-21\,J \cdot deg^{-1} \cdot mol^{-1}$, respectively [7].

For $cis\text{-}PdCl_2(H_2O)_2 + Cl^- \rightarrow [PdCl_3(H_2O)]^- + H_2O$ they are 41.8 ± 4 kJ/mol and $-29 \pm 12\,J \cdot deg^{-1} \cdot mol^{-1}$, respectively [8] or 17.9 kJ/mol and $-138\,J \cdot deg^{-1} \cdot mol^{-1}$ [7]. Possible reasons for the discrepancies in these values have been discussed [10].

The equilibrium $cis\text{-}PdCl_2(H_2O)_2 \rightleftharpoons trans\text{-}PdCl_2(H_2O)_2$ has been studied. At equilibrium the cis species is dominant, the equilibrium constant cis/trans being 2.1 ± 0.3 at 25°C in 1M $HClO_4$ [8].

trans-PdCl$_2$(H$_2$O)$_2$. Kinetics of the reaction $[PdCl_3(H_2O)]^- + H_2O \rightleftharpoons trans\text{-}PdCl_2(H_2O)_2 + Cl^-$ and of the reverse anation have been studied at ionic strength 2.1M in 2M $HClO_4$ at 25°C using a stopped-flow method [7]; at 1M ionic strength at 25°C also by stopped-flow methods [8], and for the aquation and anation reactions by temperature-jump and spectrophotometric techniques [9].

In general the acid hydrolysis and halide anation reactions are some 10^4 to 10^5 times faster for the Pd than the Pt complexes [8].

For the reaction $[PdCl_3(H_2O)]^- + H_2O \rightarrow trans\text{-}PdCl_2(H_2O)_2 + Cl^-$ the activation enthalpy $\Delta H°$ and entropy $\Delta S°$ have been given as 100 ± 4 [9] or 59 kJ/mol [7] and as -50 ± 12 [9] or $-50\,J \cdot deg^{-1} \cdot mol^{-1}$ [7], respectively, while for the reverse reaction $trans\text{-}PdCl_2(H_2O)_2 + Cl^- \rightarrow [PdCl_3(H_2O)]^- + H_2O$ $\Delta H°$ and $\Delta S°$ are given as 54 ± 4 [8] and 57.2 kJ/mol [7] and -37 ± 12 [8] or $-22.6\,J \cdot deg^{-1} \cdot mol^{-1}$ [7].

Possible reasons for the discrepancies in these values have been discussed [10].

For the $cis\text{-}PdCl_2(H_2O)_2 \rightleftharpoons trans\text{-}PdCl_2(H_2O)_2$ equilibrium see above.

References:

[1] Elding, L. I.; Olsson, L. F. (J. Phys. Chem. **82** [1978] 69/74).
[2] Victori, L.; Tomàs, X.; Malgosa, F. (Afinidad **32** [1975] 867/71).
[3] Kravtsov, V. I.; Zelenskii, M. I. (Elektrokhimiya **2** [1966] 1138/43; Soviet Electrochem. **2** [1966] 1042/6).
[4] Harrison, J. A. (Electrochim. Acta **27** [1982] 1123/8).
[5] Ananin, Y. N.; Trokhimetz, A. I. (React. Kinet. Catal. Letters **27** [1965] 129/32).
[6] Zelenskii, M. I.; Kravtsov, V. I. (Elektrokhimiya **6** [1970] 793/8; Soviet Electrochem. **6** [1970] 768/72).
[7] Bekker, P. van Z.; Robb, W. (J. Inorg. Nucl. Chem. **37** [1975] 829/34).
[8] Elding, L. I. (Inorg. Chim. Acta **7** [1973] 581/8).
[9] Pearson, R. G.; Hynes, M. J. (Kgl. Tek. Høgsk. Handl. No. 285 [1972] 461/9).
[10] Elding, L. I. (Inorg. Chim. Acta **15** [1975] L9/L11).

[11] Elding, L. I. (Inorg. Chim. Acta **6** [1972] 647/51).

[PdCl(H₂O)₃]⁺. The electronic spectrum of the aqueous solution has been measured and assigned (220 to 700 nm, reproduced in paper) [1, 6] and electronic spectral maxima for $[PdCl(H_2O)_3]^+$ were listed (200 to 450 nm) [2].

Kinetic data for the reactions 1) $cis\text{-}PdCl_2(H_2O) + H_2O \rightarrow [PdCl(H_2O)_3]^+ + Cl^-$ and 2) $trans\text{-}PdCl_2(H_2O)_2 + H_2O \rightarrow [PdCl(H_2O)_3]^+ + Cl^-$ and for the reverse anation reactions have been measured by stopped-flow methods at 1M ionic strength and at 25°C [3]. Kinetic and rate data for reaction 2) were also obtained by temperature-jump and stopped-flow methods [4]. Similar data, using spectrophotometric and stopped-flow methods at 1M ionic strength, have been obtained for 3) $[PdCl(H_2O)_3]^+ + H_2O \rightarrow [Pd(H_2O)_4]^{2+} + Cl^-$ and the reverse anation reaction 4) $[Pd(H_2O)_4]^{2+} + Cl^- \rightarrow [PdCl(H_2O)_3]^+ + H_2O$. Rate data for reaction 4) were obtained by temperature-jump and stopped-flow methods [4].

Activation enthalpies and entropies $\Delta H°$ and $\Delta S°$ for 2) are 50 ± 4 kJ/mol and -38 ± 12 J $\cdot deg^{-1} \cdot mol^{-1}$ [3]; for 3) 59 ± 8 kJ/mol and 54 ± 24 J$\cdot deg^{-1} \cdot mol^{-1}$ [4] and for 4) 41.8 ± 8 kJ/mol and -25 ± 25 J$\cdot deg^{-1} \cdot mol^{-1}$ [4].

For stability constant data see table on p. 93.

The effect of adding organic solvents to aqueous solutions of chloropalladates, including $[PdCl(H_2O)_3]^+$, on their composition and reactivity towards reductants, has been investigated [7].

References:

[1] Elding, L. I.; Olsson, L. F. (J. Phys. Chem. **82** [1978] 69/74).

[2] Victori, L.; Tomàs, L.; Malgosa, F. (Afinidad **32** [1975] 867/72).

[3] Elding, L. I. (Inorg. Chim. Acta **7** [1973] 583/8).

[4] Elding, L. I. (Inorg. Chim. Acta **6** [1972] 683/8).

[5] Pearson, R. G.; Hynes, M. J. (Kgl. Tek. Høegsk. Handl. No. 285 [1972] 461/9).

[6] Elding, L. I. (Inorg. Chim. Acta **6** [1972] 647/51).

[7] Kaszonyi, A.; Hrusovsky, M.; Ilavsky, J. (Petrochemia **26** [1986] 62/8; C.A. **107** [1987] No. 88386).

4.9 Chloro-hydroxo Complexes

[PdCl₃(OH)]²⁻. The ion has been briefly mentioned as a product of the reaction $[PdCl_3(H_2O)]^- \rightleftharpoons [PdCl_3(OH)]^{2-} + H^+$ and of $[PdCl_4]^{2-} + H_2O \rightleftharpoons [PdCl_3(OH)]^{2-} + H^+ + Cl^-$ [1, 2].

The pK_a of $[PdCl_3(H_2O)]^-$ is 7 [3].

[PdCl₄(OH)]²⁻. The possible formation of this from the $[PdCl_4]^{2-}$–OH radical reaction has been mentioned [4].

[PdCl₂(OH)(H₂O)]⁻. The pK_a of $Pd(H_2O)_2Cl_2$ is 4.3 [3].

[PdCl₂(OH)₂]²⁻. The existence of this ion has been mentioned [3], see also "Palladium" 1942, p. 277.

References:

[1] Grinberg, A. A.; Nikol'skaya, L. E.; Kiseleva, N. V. (Zh. Neorgan. Khim. **1** [1956] 220/4; Russ. J. Inorg. Chem. **1** [1956] 31/5).

[2] Grinberg, A. A.; Kiseleva, N. V. (Zh. Neorgan. Khim. **3** [1958] 1804/9; Russ. J. Inorg. Chem. **3** [1958] 119/26).

[3] Rund, J. V. (Inorg. Chem. **9** [1976] 1211/5).

[4] Broszkiewicz, R. (Intern. J. Radiat. Phys. Chem. **6** [1974] 249/58).

5 Palladium and Bromine

The dibromide is well established, and there is an extensive chemistry of $[PdBr_4]^{2-}$ and, to a lesser extent, of $[PdBr_6]^{2-}$. There seems to be no evidence for the existence of palladium(III) bromide or bromo complexes, nor for $PdBr_4$.

5.1 Palladium Dibromide, $PdBr_2$ (see "Palladium" 1942, p. 287)

Formation and Preparation

The compound is made as a brown powder by the reaction of Br_2-containing HBr on metallic palladium, or in crystalline black-brown form by a chemical transport procedure using the brown powder at 700°C in a sealed tube [1]. It can also be made by chemical transport from Pd and Al_2Br_6 [2], and from palladium and bromine vapour at 500°C [3]. It is insoluble in water but dissolves in HBr to give a solution of "$H_2[PdBr_4]$" [1].

The compound is probably involved during the electrodeposition and anodic dissolution of Pd in bromide electrolytes [4]. The distribution of trace metals during electrodialytic purification of $PdBr_2$ solutions was studied and purification procedures described [5].

References:

[1] Brodersen, K.; Thiele, G.; Gaedcke, H. (Z. Anorg. Allgem. Chem. **348** [1966] 162/7).
[2] Schäfer, H.; Nowitzki, J. (J. Less-Common Metals **61** [1978] 47/50).
[3] Williams, R. C; Gregory, N. W. (J. Phys. Chem. **73** [1969] 623/31).
[4] Nemova, V.A.; Kolpakova, N. A. (Izv. Tomsk. Politekh. Inst. **258** [1976] 48/51; C.A. **89** [1978] No. 13923).
[5] Kozin, L. F.; Grushina, N. V.; Saprykina, T. I.; Lysenko, A. I. (Izv. Akad. Nauk Kaz.SSR Ser. Khim. **28** [1978] 9/14; C.A. **90** [1979] No. 93200).

Physical Properties

Structure, Density. The crystals belong to the monoclinic $P2_1/c$-C_{2h}^5 space group with $Z = 4$; $a = 6.59(2)$, $b = 3.96(2)$, $c = 25.22(3)$ Å, $\beta = 92.6°$. The measured density is 5.38 g/cm³, pyknometric density 5.35 g/cm³. There is planar coordination with the four metal atoms forming a distorted quadrilateral with bridging Br atoms in infinite puckered chains, see **Fig. 42**a and **42**b [1].

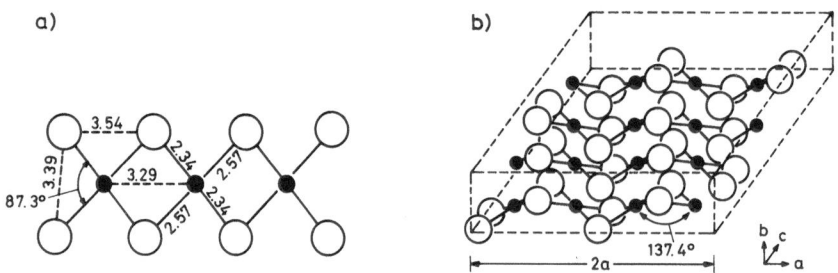

Fig. 42. Structure of $PdBr_2$; a) dimensions in chain, b) the Pd–Br sheet.

Magnetic Properties. The compound is essentially diamagnetic, the small paramagnetism arising from temperature-independent paramagnetism; $\chi_M = +63 \times 10^{-6}$ cgsu [1].

Thermodynamic Functions. For the reactions $PdBr_2$(solid) $\rightarrow$ Pd(solid) + Br_2(gas) from 760 to 970 K, $\ln K = -19840 \, T^{-1} - 3.02 \ln T + 40.124$ and for Pd(solid) + Br_2(gas) $\rightarrow PdBr_2$(gas) from 1090 to 1155 K, $\ln K = 17386 \, T^{-1} - 1.006 \ln T + 13.278$, the data being obtained from transpiration method studies [3].

From these data, ΔH°_{298} for formation of solid $PdBr_2$ from solid Pd and liquid Br_2 is -124 kJ/mol with $S^\circ_{298} = 125 \, J \cdot deg^{-1} \cdot mol^{-1}$. There is evidence for the existence of monomeric, tetrameric and decameric species in the vapour phase [3]. The free energy ΔG° has been calculated as -93.2 kJ/mol for $PdBr_2$ [4]. The lattice energy has been calculated as 2709 kJ/mol [2].

Electronic Spectra. For such spectra of $PdBr_2$ in Br^--containing media, see p. 180.

References:

[1] Brodersen, K.; Thiele, G.; Gaedcke, H. (Z. Anorg. Allgem. Chem. **348** [1966] 162/7).
[2] Karapet'yants, M. Kh. (Zh. Fiz. Khim. **28** [1954] 1136/52, 51; C.A. **1955** 7917).
[3] Williams, R. C.; Greogory, N. W. (J. Phys. Chem. **73** [1969] 623/31).
[4] Karapet'yants, M. Kh. (Zh. Fiz. Khim. **28** [1954] 353/8; C.A. **1955** 5953).

Miscellaneous Properties

From X-ray photoelectronic (ESCA) measurements the $3d_{5/2}$ and $3d_{3/2}$ binding energies were found to be 337.3 ± 0.2 and 342.5 ± 0.2 eV, respectively [1].

The distribution coefficient of $PdBr_2$ between the two immiscible liquid phases in the $AlBr_3$–KBr system has been determined as 1.4 at 110°C [2]. The effect of heat treatment on the H_2-sensitivity of ZnO films loaded with $PdBr_2$ has been explored [7].

The atomic charge on $PdBr_2$ has been calculated by the self-consistency of electronegativity method [3]. Calculation was attempted of a quantitative measure of "softness" of $PdBr_2$, amongst other halides [4]. The effect of ionisation energy on the structures of a number of compounds, including $PdBr_2$, has been discussed [5]. Ab initio calculations have been carried out on PdX_2 compounds, including $PdBr_2$, using an effective core potential [6].

For solid $PdBr_2$, $E_0 = +0.492$ V at 25°C and $+0.061$ V at 800°C; for the aqueous solution, $E_0 = 0.094$ V at 25°C [8].

References:

[1] Kumar, G.; Blackburn, J. R.; Albridge, R. G.; Moddeman, W. E.; Jones, M. M. (Inorg. Chem. **11** [1972] 296/300).
[2] Smith, F. J. (J. Less-Common Metals **97** [1984] 21/6).
[3] Gagarin, S. G. (Zh. Strukt. Khim. **19** [1978] 723/4; J. Struct. Chem. [USSR] **19** [1978] 620/1).
[4] Singh, P. P.; Srivastava, S. K.; Srivastava, A. K. (J. Inorg. Nucl. Chem. **42** [1980] 521/32, 530).
[5] Ahrens, L. H.; Morris, D. F. C. (J. Inorg. Nucl. Chem. **3** [1956] 270/80, 271).
[6] Bäckvall, J. E.; Björkman, E. E.; Pettersson, L.; Siegbahn, P. (J. Am. Chem. Soc. **107** [1985] 7265/7).
[7] Li, W.; Yoneyama, H.; Tamura, H. (Mater. Chem. Phys. **10** [1984] 69/81).
[8] Hamer, W. J.; Malmberg, M. S.; Rubin, B. (J. Electrochem. Soc. **112** [1965] 750/7).

Chemical Reactions

The compound reacts with acetonitrile to give $PdBr_2(NCCH_3)_2$ [1], while diethylenetriaminepentaacetic acid (H_5L) gives $Pd(LH_4)Br \cdot 3H_2O$ [2]. With Al_2Br_6, $PdBr_2$ gives $PdAl_2Br_8$ [3]. Reaction of $PdBr_2$ with (2- and 3-hydroxyalkyl)isocyanides $CNCHRCHR'OH$ ($R = R' = H$, $R = C_2H_5$, $R = H$, $R = CH_3$) and $CN(CH_2)_3OH$ give $[Pd(CNHCHRCHR'O)_4]Br_2$ and $[Pd(CNH(CH_2)_3O)_4]Br_2$ [4].

With ditertiary phosphine sulphides or selenides (L, where L=1,2-ethylene, 1,3-trimethylene, 1,4-tetramethylene, 1,6-hexamethylenebisdiphenylphosphine sulphide or selenide) $PdBr_2 \cdot L$ is formed [5]. With Ph_3PS, $Ph_3PSe(L)$, $Ph_2P(X)CH_2P(X)Ph_2$ (X = S, Se) (L'), $Ph_2P(Se)(CH_2)_3P(Se)Ph_2$ and $(CH_3)_2P(S)P(S)(CH_3)_2$ (L''), $PdBr_2 \cdot L_2$, $2PdBr_2 \cdot L_3$, PdX_2L', PdX_2L'' and $3PdBr_2 \cdot 2L''$ are formed [6]. With 1,4,7-thiadiazonines (L) $PdBr_2$ gives the five coordinate $PdBr_2 \cdot L$ [7]. Reaction of $PdBr_2$ with $P(OPh)_3$ and HBr in dimethylformamide yields $(CH_3)_2NH_2[PdBr_3P(OPh)_3]$ [8] and $PdBr_2(P(OPh)_3)_2$ is also formed [9]. Reactions between $PdBr_2$ and $SnBr_2$ have been studied by [10].

References:

[1] Frye, H.; Kulpan, E.; Vanbrolk, J. (Z. Naturforsch. **20b** [1965] 269).
[2] Grevtsev, A. M.; Zheligovskaya, N. N.; Popov, L. V.; Spitsyn, V. I. (Izv. Akad. Nauk SSSR Ser. Khim **1977** 941/3; Bull. Acad. Sci. USSR Div. Chem. Sci. **26** [1977] 867/9).
[3] Merker, H.-B.; Schäfer, H. (Z. Anorg. Allgem. Chem. **480** [1981] 76/80).
[4] Fehlhammer, W. P.; Bartel, K.; Plaia, U.; Volkl, A.; Liu, A. T. (Chem. Ber. **118** [1985] 2235/54, 2238).
[5] Sandhu, S. S.; Singh, T. (Transition Metal Chem. [Weinheim] **1** [1976] 155/8).
[6] Lobana, T. S.; Sharma, K. (Transition Metal Chem. [Weinheim] **7** [1982] 333/5).
[7] Sandhu, S. S.; Tandon, S. S.; Singh, H. (Transition Metal Chem. [Weinheim] **5** [1980] 333/7).
[8] Nyrkova, A. N.; Kuz'mina, L. G.; Struchkov, Yu. T.; Temkin, O. N. (Koord. Khim. **8** [1982] 82/8; C. A. **96** [1982] No. 134803).
[9] Nyrkova, A. N.; Kukina, G. A.; Temkin, O. N.; Porai-Koshits, M. A. (Koord. Khim. **11** [1985] 954/8; C.A. **103** [1985] No. 170963).
[10] Shukla, S. K. (Chem. Ind. [London] **1965** 511/2).

Applications of $PdBr_2$

The compound has been used as a catalyst for the preparation of phenoxyacetone derivatives from 3-phenoxypropene and alkyl nitrites [1]. It has also been used for selective metal deposition for microelectronic applications [2]. In 8M HBr, $PdBr_2$ (presumably as $[PdBr_4]^{2-}$) can be used as anolyte in $CO-O_2$ fuel cells [3].

References:

[1] Ube Industries, Ltd. (Japan. Kokai Tokkyo Koho 85-123431 [1985]; C.A. **104** [1986] No. 50666).
[2] Eskildsen, S. S.; Soerensen, G. (Nucl. Instrum. Methods Phys. Res. **218** Pt. 1/3 [1983] 485/8; C.A. **100** [1984] No. 112970).
[3] Babcock, Hitachi, K. K. (Japan. Kokai Tokkyo Koho 87-274564 [1987]; C.A. **108** [1988] No. 97939).

5.2 Bromopalladates (see "Palladium" 1942, pp. 287, 308, 315/6, 324, 329/31, 333, 335/8, 340)

General Survey. The coordination chemistry of bromo complexes of palladium resembles that of the chloro species rather more than of the iodopalladates. There is an extensive chemistry of the square-planar $[PdBr_4]^{2-}$ ion, and (as with the chloro and iodo species) the $\mu\mu'$-dibromo bridged $[Pd_2Br_6]^{2-}$ complexes are also known. There are mixed chloro-bromo and bromo-iodo complexes and the aquo species $[PdBr_n(H_2O)_{4-n}]^{2-n}$. There is some rather inconclusive evidence for the existence of palladium(II) bromo complexes with coordination numbers > 4, and electron spin resonance evidence for the transient existence of palladium(I) and palladium(III) bromo complexes of uncertain formulas. There is a relatively large chemistry of $[PdBr_4]^{2-}$.

5.2.1 Bromopalladates(I)

The only evidence for the existence of these comes from ESR data.

Ag$_3$[PdBr$_6$] may be the species generated when AgBr, doped with $PdBr_2$, is irradiated by light at 425 nm at 100 K. An appropriate ESR spectrum was observed.

Reference:

Eachus, R. S.; Graves, R. E.; Olm, M. T. (Phys. Status Solidi A **57** [1980] 429/37).

5.2.2 Tetrabromopalladates, [PdBr$_4$]$^{2-}$

Preparation. The normal method of preparation is reaction of $PdCl_2$ or $[PdCl_4]^{2-}$ with HBr followed by addition of the bromide salt of the appropriate cation.

5.2.2.1 Vibrational Spectra (see table below)

Data have been reviewed on this topic [1, 10]. The Raman spectrum of the aqueous solution has been measured to a high degree of accuracy [2, 4] and bond polarisability derivatives ascertained [2]. Interference between preresonance scattering from an intense electronic transition and resonance scattering from a forbidden electronic transition has been demonstrated in $[PdBr_4]^{2-}$; an excitation profile for the ν_1 mode (given as 192 cm^{-1}) was obtained (599 to 700 nm, reproduced in paper) [3]. The resonance Raman spectrum of $[PdBr_4]^{2-}$ in 2 M HBr was measured (50 to 800 cm^{-1}, reproduced in paper) and the second overtone of ν_1 observed [11]. The Raman spectrum of $[PdBr_4]^{2-}$ in water was measured from 50 to 280 cm^{-1} (reproduced in paper) with 514.5 and 589.9 nm excitation and the excitation profiles for ν_1 and ν_2 given [5].

Raman spectra of $[PdBr_4]^{2-}$ in aqueous solution:

$\nu_1(A_{1g})$	$\nu_2(B_{1g})$	$\nu_4(B_{2g})$	Ref.
189.6 ± 0.7	174.1 ± 0.7	97 ± 4	[2]
188	172	102	[4]
192	177	106	[5]
190			[11]

Frequency numbering as on p. 192; for data on solid $K_2[PdBr_4]$ see table on p. 182.

General valence force field constants and bond polarisability data for $[PdBr_4]^{2-}$ have been calculated by [6] using the data of [2], and force constants calculated with the data of [4] by [4, 7]; the latter data have also been used to calculate modified Urey-Bradley force field

(MUBFF) constants [8] and general valence force field (GVFF) [7]. Values of bond orders were estimated, using the data of [4], for a number of square-planar species including [PdBr$_4$]$^{2-}$ [9]. Mean amplitudes of vibration were also calculated from the data of [4] by [7].

For other force-field data based on measurements of the vibrational spectra of individual salts, see pp. 191/2.

References:

[1] James, D. W.; Nolan, M. J. (Progr. Inorg. Chem. **9** [1968] 195/275, 237).
[2] Bosworth, Y. M.; Clark, R. J. H. (Inorg. Chem. **14** [1975] 170/7).
[3] Zgierski, M. Z. (J. Raman Spectrosc. **6** [1977] 53/6).
[4] Goggin, P. L.; Mink, J. (J. Chem. Soc. Dalton Trans. **1974** 1479/83).
[5] Stein, P.; Miskowski, V.; Woodruff, W. H.; Griffin, J. P.; Werner, K. G.; Gaber, B. P.; Spiro, T. G. (J. Chem. Phys. **64** [1976] 2159/67).
[6] Pandey, A. N.; Sharma, D. K.; Verma, U. P.; Kumar, V. (J. Raman Spectrosc. **6** [1977] 163/4).
[7] Sanyal, N. K.; Verma, D. N.; Dixit, L. (Indian J. Pure Appl. Phys. **14** [1976] 819/22).
[8] Pandey, A. N.; Verma. U. P. (J. Mol. Struct. **42** [1977] 171/9).
[9] Rastogi, V. K.; Kumar, V.; Pandey, A. N. (Indian J. Pure Appl. Phys. **20** [1982] 150/1).
[10] Mink, J.; Goggin, P. L. (Kem. Kozl. **58** [1982] 215/41, 220; C.A. **99** [1983] No. 12876).

[11] Hamaguchi, H.; Harada, I.; Shimanouchi, T. (Chem. Letters [1973] 1049/53).

5.2.2.2 Electronic Spectra

The electronic spectra of [PdBr$_4$]$^{2-}$ have been recorded as follows: 250 to 700 nm in aqueous solution with 0.42 to 2M KBr [1]; 210 to 500 nm in aqueous Br$^-$ [18]; in water-dioxan mixtures 200 to 500 nm [2]; in water in the presence of LiBr 280 to 530 nm [3, 19]; in 2M HBr 350 to 650 nm [4] and 200 to 500 nm [20]; of PdBr$_2$ and LiBr in water-n-butanol 220 to 460 nm [5]; in water with varying amounts of Br$^-$ 230 to 400 nm [6]; in acetic anhydride, nitrobenzene, nitromethane, acetone, methanol, acetonitrile 200 to 600 nm (spectrum in methanol only reproduced in paper) [7]; in acetonitrile with (Bu$_4$N)Br 250 to 450 nm [8]; in nitrobenzene alone and in the presence of (Et$_4$N)Br 400 to 600 nm [9]; in dioxan from 180 to 600 nm [10]; in water from 450 to 650 nm [11]; in 1 M HClO$_4$ from 210 to 600 nm [12]; in water with or without added Br$^-$ 200 to 500 nm, also in water over the same range as a function of time [13]; in water from 230 to 400 nm [14]. All the above have spectra recorded in the papers.

There are also listed data, but with no reproduced spectra, for [PdBr$_4$]$^{2-}$ generated from [Pd(H$_2$O)$_4$]$^{2+}$ and Br$^-$ 215 to 400 nm [15]; for [PdBr$_4$]$^{2-}$ in 0.1 to 4M HBr 200 to 350 nm [16]; with excess Br$^-$ in acetone 300 to 700 nm [17]; for [PdBr$_4$]$^{2-}$ in excess Br$^-$ in aqueous solution 200 to 500 nm [21].

Magnetic circular dichroism (MCD) spectra have been measured (350 to 650 nm, reproduced in paper) in 2M HBr [4].

For other electronic spectral data see under individual salts in the following sections.

References:

[1] Rush, R. M.; Martin, D. S.; LeGrand, R. G. (Inorg. Chem. **14** [1975] 2543/50).
[2] Kushnikov, Y. A.; Vozdvizhenskii, V. F.; Zinevich, L. N.; Fasman, A. H. (Zh. Neorgan. Khim. **11** [1966] 2193/6; Russ. J. Inorg. Chem. **11** [1966] 1175/7).

[3] Srivastava, S. C.; Newman, L. (Inorg. Chem. **6** [1967] 762/5).

[4] McCaffery, A. J.; Schatz, P. N.; Stephens, P. J. (J. Am. Chem. Soc. **90** [1968] 5730/5).

[5] Nyrkova, A. N.; Kaliya, O. L.; Temkin, O. N.; Flid, R. M.; Vorob'eva, N. K. (Zh. Neorgan. Khim. **20** [1975] 3017/21; Russ. J. Inorg. Chem. **20** [1975] 1669/71).

[6] Shlenskaya, N. I.; Biryukov, A. A. (Vestn. Mosk. Univ. Ser. II Khim. **19** No. 3 [1964] 65/8; C.A. **61** [1964] 9043).

[7] Harris, C. M.; Livingstone, S. E.; Reece, I. H. (J. Chem. Soc. **1959** 1505/11).

[8] Klotz, P.; Feldberg, S.; Newman, L. (Inorg. Chem. **12** [1973] 164/8).

[9] Harris, C. M.; Livingstone, S. E.; Reece, I. H. (Australian J. Chem. **10** [1957] 282/6).

[10] Ivanova, S. N.; Chernyaeva, A. P.; Gindin, L. M.; Chernobrov, A. S. (Izv. Sibirsk. Otd. Akad. Nauk SSSR Ser. Khim. Nauk **1978** No. 6, pp. 74/8; C.A. **90** [1979] No. 77223).

[11] Stein, P.; Miskowski, V.; Woodruff, W. H.; Griffin, J. P.; Werner, K. G.; Gaber, B. P.; Spiro, T. G. (J. Chem. Phys. **64** [1976] 2159/67).

[12] Elding, L. I. (Inorg. Chim. Acta **6** [1972] 647/54).

[13] Grinberg, A. A.; Kiseleva, N. V. (Zh. Neorgan. Khim. **3** [1958] 1804/9; Russ. J. Inorg. Chem. **3** No. 8 [1958] 119/26).

[14] Srivastava, S. C.; Newman, L. (Inorg. Chem. **5** [1966] 1506/10).

[15] Biryukov, A. A.; Shlenskaya, V. I. (Vestn. Mosk. Univ. Ser. II Khim. **19** No. 5 [1964] 81/6; C.A. **62** [1965] 4890).

[16] Vorlicek, J.; Dolezal, J. (Z. Anal. Chem. **262** [1972] 365/7).

[17] Lobaneva, O. A.; Kononova, M. A.; Davydova, M. K. (Dokl. Akad. Nauk SSSR **194** [1970] 133/5; Dokl. Phys. Chem. Proc. Acad. Sci. USSR **190/195** [1970] 669/70).

[18] Kutyukov, G. G.; Fasman, A. B.; Lyuts, A. E.; Kushnikov, Yu. A.; Vozkvizhenskii, V. F.; Golodov, V. A. (Zh. Fiz. Khim. **40** [1966] 1468/71; Russ. J. Phys. Chem. **40** [1966] 798/802).

[19] Elding, L. I.; Olsson, L. F. (J. Phys. Chem. **82** [1978] 69/74).

[20] Hamaguchi, H.; Harada, I.; Shimanouchi, T. (Chem. Letters [1973] 1049/54).

[21] Gray, H. B.; Ballhausen, C. J. (J. Am. Chem. Soc. **85** [1963] 260/5).

5.2.2.3 Stability Constants

The data are collected in the table on p. 182. A useful survey of earlier data is given in [2] and [5]; there are also older data from potentiometric measurements [11]. The position of Br^- in a number of $[PdX_4]^{2-}$ and other complexes in stability orders has been briefly discussed [13].

From standard electrode potential measurements the expression $\log K\ [PdBr_4^{2-}] = 3.56 + 2863.39/T$ has been deduced, e.g., 13.16 at 25°C, 9.04 at 100°C. From these data ΔH_{298}° was given as -56.0 ± 0.8 kJ/mol and ΔS_{298} as 67 ± 4 J·deg^{-1}·mol^{-1} [14].

References:

[1] Elding, L. I. (Inorg. Chim. Acta **6** [1972] 647/51).

[2] Morer, X. T.; Companys, L. V.; Carbonell, A. T. (Afinidad **31** [1974] 745/50).

[3] Shchukarev, S. A.; Lobaneva, O. A.; Ivanova, M. A.; Kononova, M. A. (Zh. Neorgan. Khim. **9** [1964] 1503/4).

[4] Grinberg, A. A.; Gel'fman, M. I.; Kiseleva, N. V. (Zh. Neorgan. Khim. **12** [1967] 1171/5; Russ. J. Inorg. Chem. **12** [1967] 620/2).

[5] Companys, L. V.; Morer, X. T.; Carbonell, A. T. (Afinidad **31** [1974] 559/65).

Stability constants for $[PdBr_n(H_2O)_{4-n}]^{2-n}$:

	method	t in °C	total ionic strength (M)	K (M^{-1})	log K (M^{-1})	log β	Ref.
K$_1$	a)	25	1	$(1.47 \pm 0.06) \times 10^5$	5.17*)	5.17 ± 0.02	[1]
	a)	25	1	$(3.009 \pm 0.055) \times 10^5$	5.478 ± 0.008	5.478 ± 0.008	[2]
	a)	20	0.8	$2.34 \times 10^{4*}$	4.37	4.37	[3]
	b)			6.25×10^6	6.79*)	6.79*)	[4]
K$_2$	a)	25	1	$(1.80 \pm 0.01) \times 10^4$	4.25*)	9.42 ± 0.04	[1]
	a)	25	1	$(1.823 \pm 0.062) \times 10^4$	4.261 ± 0.015	9.739 ± 0.015	[2]
	a)	20	0.8	$1.2 \times 10^{4*}$	4.08		[3]
K$_3$	a)	25	1	$(2.00 \pm 0.1) \times 10^3$	3.3*)	12.72 ± 0.06	[1]
	a)	25	1	$(1.800 \pm 0.301) \times 10^3$	3.255 ± 0.067		[5]
	a)	20	0.8	$6.16 \times 10^{3*}$	3.79		[3]
	b)	20	0.43			11.6*)	[12]
	b)	20	0.60			11.28*)	[12]
K$_4$	a)	25	1	163 ± 6	2.21*)	14.94 ± 0.08	[1]
	a)	25	1	209.8 ± 18	2.322 ± 0.037	15.7 ± 0.5	[5]
	a)	20	0.8	3610*)	3.50		[3]
	a)	20	0.5		2.14		[6]
	a)	25	0.5	162 ± 2	2.21*)		[7]
	a)	10	1.0	316*)	2.50		[8]
	a)	25	1.0	200*)	2.30		[8]
	a)	45	1.0	144*)	2.16		[8]
	a)	20	1.0	5.8**)	0.76*)		[9]
	a)	20	1.0	170 ± 8	2.23*)		[10]
	b)	20	0.43			13.39*)	[12]
	b)	20	0.60			13.41*)	[12]

a) Spectrophotometric method; b) solubility method; *) calculated values by author from data in reference; **) in aqueous butanol.

[6] Biryukov, A. A.; Shlenskaya, V. I. (Vestn. Mosk. Univ. Ser. II Khim. **19** No. 5 [1964] 81/8).

[7] Shlenskaya, V. I.; Biryukov, A. A. (Vestn. Mosk. Univ. Ser. II Khim. **19** No. 3 [1964] 65/8).

[8] Shlenskaya, V. I.; Biryukov, A. A. (Zh. Neorgan. Khim. **11** [1966] 54/60; Russ. J. Inorg. Chem. **11** [1966] 28/31).

[9] Nyrkova, A. N.; Kaliya, O. L.; Temkin, O. N.; Flid, R. M.; Vorob'eva, N. K. (Zh. Neorgan. Khim. **20** [1975] 3017/21; Russ. J. Inorg. Chem. **20** [1975] 1669/71).

[10] Gulko, A.; Schmuckler, G. (J. Inorg. Nucl. Chem. **25** [1973] 603/7).

[11] Grinberg, A. A.; Kiseleva, N. V.; Gel'fman, M. I. (Dokl. Akad. Nauk SSSR **153** [1963] 1327/9; Dokl. Chem. Proc. Acad. Sci. USSR **148/153** [1963] 1025/7).

[12] Popovicheva, N. K.; Biryukov, A. A.; Shlenskaya, V. I (Zh. Neorgan. Khim. **9** [1964] 1482/3; Russ. J. Inorg. Chem. **9** [1964] 803/4).

[13] Ahrland, S. (Acta Chem. Scand. **10** [1956] 723/6).

[14] Nikolaeva, N. M.; Pogodina, L. P. (Ivz. Sibirsk. Otd. Akad. Nauk SSSR Ser. Khim. Nauk **1980** No. 1, pp. 130/4; C. A. **93** [1980] No. 15634).

5.2.2.4 Thermodynamic Data

For early measurements see [1] and a brief review in [2]. On the basis of the stability constant data of [3] and calorimetric data the following were obtained for the stepwise enthalpy, free energy and entropy changes [4]:

log β	ΔH°_{298} (kJ/mol)	ΔG°_{298} (kJ/mol)	(J·mol^{-1}·deg^{-1})
5.17 ± 0.02	-21.3 ± 0.2	-29.5 ± 0.1	27.2 ± 1
9.42 ± 0.04	—	-24.2 ± 0.3	—
12.72 ± 0.06	—	-18.8 ± 0.4	—
14.94 ± 0.08	—	-12.7 ± 0.7	—

ΔH° for the stepwise formation of $[PdBr_4]^{2-}$ for the fourth stage of complex formation has been calculated as -18.0 kJ/mol and ΔS as -14.6 J·deg^{-1}·mol^{-1} [5].

References:

[1] Goldberg, R. N.; Hepler, L. G. (Chem. Rev. **68** [1968] 229/52; 241).

[2] Ashcroft, S. J.; Mortimer, C. T. (Thermochemistry of Transition Metal Complexes, Academic, New York 1970, p. 362).

[3] Elding, L. I. (Inorg. Chim. Acta **6** [1962] 647/51).

[4] Ryhl, T. (Acta Chem. Scand. **26** [1972] 2961/2).

[5] Shlenskaya, V. I.; Biryukov, A. A. (Zh. Neorgan. Khim. **11** [1966] 28/31; Russ. J. Inorg. Chem. **11** [1966] 28/31).

5.2.2.5 Electrochemical Behaviour

Polarographic reduction of $[PdBr_4]^{2-}$ at the dropping mercury electrode has been studied in HCl, HBr, H_2SO_4 and $HClO_4$ [1,2]. Reduction to the metal occurs. The diffusion coefficient of $[PdBr_4]^{2-}$ (1×10^{-3} M in 1 M HBr) is 1.03×10^{-5} cm^2/sec [1,2]. The first step in the electroreduction of $[PdBr_4]^{2-}$ is the dissociation of two bromo ligands [3,10]. Because of a reaction with mercury the polarographic reduction of $[PdBr_4]^{2-}$ in 1 M KBr begins without any applied potential – i. e., the dissolution wave of mercury is followed by the reduction wave of $[PdBr_4]^{2-}$ [4].

The electrochemistry of $[PdBr_4]^{2-}$ has been studied using computer-controlled methods, and results are briefly described [5]. The standard potential of $[PdBr_4]^{2-}$ as a function of temperature was determined by [6, 7]; at 50°C it is 0.527 V and at 150°C it is 0.464 V [7].

At pH = 2 in 1 M Br$^-$ E_0 is 0.55 V at 20°C and 0.548 V at 60°C [8]. For $[PdBr_6]^{2-}/[PdBr_4]^{2-}$ in IM KBr solution, E_0 is given as +0.994 V by [11] and as +0.7070 V by [12].

The anodic dissolution of Pd in aqueous 0.5 M KBr solution has been studied [9].

For thermodynamic data derived from electrochemical data see p. 183.

References:

[1] Kravtsov, V. I.; Shereshevskaya, I. I. (Elektrokhimiya **7** [1971] 99/102; Soviet Electrochem. **7** [1971] 91/4).

[2] Kravtsov, V. I.; Shereshevskaya, I. I. (Elektrokhimiya **5** [1969] 985/8; Soviet Electrochem. **5** [1969] 926/8).

[3] Kravtsov, V. I. (Elektrokhimiya **6** [1970] 275/7; Soviet Electrochem. **6** [1970] 265/7).

[4] Hirota, M.; Umezawa, Y.; Nakamura, M.; Fujiwara, S. (J. Inorg. Nucl. Chem. **33** [1971] 2617/21).

[5] Harrison, J. A. (Electrochim. Acta **27** [1982] 1123/8).

[6] Nikolaeva, N. M.; Pogodina, L. P. (Izv. Sibirsk. Otd. Akad. Nauk SSSR Ser. Khim. Nauk **1980** No. 1, pp. 130/4; C.A. **93** [1980] No. 15634).

[7] Nikolaeva, N. M.; Tsvelodub, D. D.; Erenburg, A. M. (Izv. Sibirsk. Otd. Akad. Nauk SSSR Ser. Khim. Nauk **1978** No. 3, pp. 44/7; C.A. **89** [1978] No. 206412).

[8] Fasman, A. B.; Kutyukov, G. G.; Sokol'skii, D. V. (Zh. Neorgan. Khim. **10** [1965] 1338/43; Russ. J. Inorg. Chem. **10** [1965] 727/30).

[9] Sorokin, I. N.; Alekhin, A. P.; Lavrishchev, V. P. (Zh. Fiz. Khim. **48** [1974] 3067/70; Russ. J. Phys. Chem. **48** [1974] 1793/5).

[10] Nemova, V. A.; Kolpakova, N. A. (Izv. Tomsk. Politekh. Inst. **258** [1976] 48/51; C.A. **89** [1978] No. 13923).

[11] Grinberg, A. A.; Shamsiev, A. M. (Zh. Obshch. Khim. **12** [1942] 55/72; C.A. **1943** 1915).

[12] Dubinskii, V. L. (Zh. Neorgan. Khim. **16** [1971] 1145/7; Russ. J. Inorg. Chem. **16** [1971] 607/8).

5.2.2.6 Extraction and Separation of $[PdBr_4]^{2-}$

The mechanism of $PdBr_2$ extraction by $[(n-C_8H_{17})_4N]Br$ or $[(n-C_8H_{17})_3NH]Br$ in dioxan has been studied spectrophotometrically [1]. Extraction of $[PdX_4]^{2-}$ (X = Cl, Br, I) by trioctylammonium salts, for instance $(n-C_8H_{17})_3NH^+$, was measured, and extraction constants and ligand hydration energies were correlated [2]. The extraction of $[PdBr_4]^{2-}$ from $CHCl_3-CCl_4$ mixtures with methylene-blue in H_2SO_4 has been studied [3]. High-voltage electrophoresis has been used to separate $[PdBr_4]^{2-}$ from other halo-species in aqueous solution [4]. Transport of $[PdBr_4]^{2-}$ through $Br^--S_2O_3^{2-}$-toluene emulsion membranes using K^+-dicyclohexano-18-crown-6 as carrier is reported [5].

References:

[1] Ivanova, S. N.; Chernyaeva, A. P.; Gindin, L. M.; Chernobrov, A. S. (Izv. Sibirsk. Otd. Akad. Nauk SSSR Ser. Khim. Nauk **1978** No. 6, pp. 74/8; C.A. **90** [1979] No. 77223).

[2] Ivanova, S. N.; Gindin, L. M.; Chernyaeva, A. P.; Chernobrov, A. S.; Shaidurova, N. N. (Izv. Sibirsk. Otd. Akad. Nauk SSSR Ser. Khim. Nauk **1978** No. 6, pp. 68/74; C.A. **90** [1979] No. 110710).

[3] Tarayan, V. M.; Mikaelyan, D. A. (Arm. Khim. Zh. **26** [1973] 720/6; C.A. **80** [1974] No. 90761).

[4] Meier, H.; Zimmerhackl, E.; Albrecht, W.; Bösche, D.; Hecker, W.; Menge, P.; Unger, E.; Zeitler, G. (Mikrochim. Acta **1971** 811/20).

[5] Izatt, R. M.; Clark, G. A.; Christensen, J. J. (Separ. Sci. Technol. **22** [1987] 691/9).

5.2.2.7 Chemical Reactions

Reactions with Hydrogen

The kinetics of reduction of $[PdBr_4]^{2-}$ by H_2 in aqueous solution were studied.

Reference:

Fasman, A. B.; Golodov, V. A.; Pustyl'nikov, L. M.; Luk'yanov, A. T.; Baranovskii, B. P.; Darinskii, Y. V. (Izv. Sibirsk. Otd. Akad. Nauk SSSR Ser. Khim. Nauk **1968** No. 3, pp. 144/8; Sib. Chem. J. **1968** 302/5).

Reaction with Nitrogen Donors

The kinetics of reaction of $[PdBr_4]^{2-}$ with 1,10-phenanthroline (phen) have been studied: the product is $Pd(phen)Br_2$, and $[PdBr_3(H_2O)]^-$ is also involved in the reaction scheme [1]. Reaction of $[PdBr_4]^{2-}$ with $[Pd(NH_3)_4]^{2+}$ yields $[PdBr_4][Pd(NH_3)_4]$ [2]. Reaction with 4,4'-bis(ethoxycarbonyl)-3,3', 5,5'-tetramethyldipyrromethane or 3,4-bis(ethoxycarbonyl)-5-chloro-3', 4,5'-trimethyldipyrromethane (HL) gave PdBrL (LH) [3]. With benzimidazole (L), "$H_2[PdBr_4]$" in alcohol gives $(LH)_2[PdBr_4]$ [4], and carbohydrazide $H_2NNHCONHNH_2$ (L) gives PdL_2Br_2 [5]. Reaction of $[PdBr_4]^{2-}$ with N-methyl-2,2'-bipyridylium ion (L) gives $PdLBr_3$ [6]. With alanine (LH), $[PdBr_4]^{2-}$ gives $K[Pd(L)Br_2]$ [7].

References:

[1] Rund, J. V. (Inorg. Chem. **13** [1974] 738/40).

[2] Grinberg, A. A.; Nikol'skaya, L. E.; Kiseleva, N. V. (Zh. Neorgan. Khim. **1** [1956] 220/4; Russ. J. Inorg. Chem. **1** No. 2 [1956] 31/5).

[3] March, F. C.; Fergusson, J. E.; Robinson, W. T. (J. Chem. Soc. Dalton Trans. **1972** 2069/76).

[4] Kukushkin, Y. N.; Sedova, G. N.; Pyzhova, L. Y.; Vlasova, R. A.; Garnovskii, A. D. (Zh. Neorgan. Khim. **24** [1979] 2265/7; Russ. J. Inorg. Chem. **24** [1979] 1257/8).

[5] Ivanov, M. G.; Kalinichenko, I. I. (Zh. Neorgan. Khim. **29** [1984] 1237/40; Russ. J. Inorg. Chem. **29** [1984] 708/10).

[6] Dholakis, S.; Gillard, R. D.; Wimmer, F. L. (Inorg. Chim. Acta **69** [1983] 179/81).

[7] Chernova, N. N.; Shakhova, I. P.; Kukushkin, Y. N. (Zh. Neorgan. Khim. **21** [1976] 3027/30; Russ. J. Inorg. Chem. **21** [1976] 1671/2).

Reactions with Oxygen Donors; Aquation and Hydrolysis

Kinetics of the aquation of $[PdBr_4]^{2-}$ to $[PdBr_3(H_2O)]^-$ have been studied [1, 2]. The hydrolysis of $[PdBr_4]^{2-}$ to "$Pd(OH)_2$" at room temperature occurs at pH = 7.5 [3]. The rate of decomposition of H_2O_2 by $[PdBr_4]^{2-}$ has been measured [4].

References:

[1] Vargaftik, M. N.; Igoshin, V. A.; Syrkin, Ya. K. (Izv. Akad. Nauk SSSR Ser. Khim. **1972** 1426/8; Bull. Acad. Sci. USSR Div. Chem. Sci. **1972** 1380/2).

[2] Elding, L. I. (Inorg. Chim. Acta **6** [1972] 683/8).

[3] Pshenitsyn, N. K.; Ginzburg, S. I. (Izv. Sekt. Platiny **28** [1954] 213/28; C. A. **1956** 722).

[4] Grinberg, A. A.; Kukushkin, Y. N.; Vlasova, R. A. (Zh. Neorgan. Khim. **13** [1968] 2177/83; Russ. J. Inorg. Chem. **13** [1968] 1126/9).

Reactions with Halogen Donors

The $[PdBr_4]^{2-}/Br^-$ exchange rate has been measuerd and is much faster than that for $[PtBr_4]^{2-}/Br^-$ [1, 2]. With Cl^-, $[PdCl_nBr_{4-n}]^{2-}$ complexes are formed [3, 4, 5], and with I^-, $[PdCl_nI_{4-n}]^{2-}$ [6, 7]; for equilibria constants for these systems see pp. 203 and 215.

References:

[1] Grinberg, A. A.; Nikol'skaya, L. E.; Kiseleva, N. V. (Zh. Neorgan. Khim. **1** [1956] 220/4; Russ. J. Inorg. Chem. **1** No. 2 [1956] 31/5).

[2] Grinberg, A. A. (Chem. Zvesti **13** [1959] 201/3).

[3] Srivastava, S. C.; Newman, L. (Inorg. Chem. **5** [1966] 1506/10).

[4] Feldberg, S.; Klotz, P.; Newman, L. (Inorg. Chem. **11** [1972] 2860/5).

[5] Klotz, P.; Feldberg, S.; Newman, L. (Inorg. Chem. **12** [1973] 164/8).

[6] Srivastava, S. R.; Newman, L. (Inorg. Chem. **11** [1972] 2855/9).

[7] Srivastava, S. R.; Newman, L. (Inorg. Chem. **6** [1967] 762/5).

Reactions with Sulphur and Selenium Donors

Reaction of $S(OPr^{iso})_2$ (L) with $(NH_4)_2[PdBr_4]$ in THF gives PdL_2Br_2 [1]. The rates and activation energy parameters of halide ligand substitution in $[PdBr_4]^{2-}$ in HBr by N-alkyl substituted thioureas have been evaluated [2]. The formation of thiourea complexes from $[PdBr_4]^{2-}$ in aqueous solutions has been studied spectrophotometrically [3], and also the kinetics and mechanism of the substitution reaction of thiourea and selenourea with $[PdBr_4]^{2-}$ [4]. With hydrogenated thiamines (L), PdL_2Br_2 complexes were formed [5]. The open-chain 1,3-bis(methylselenoethylseleno)propane (L) gives Pd_2LBr_4 with $[PdBr_4]^{2-}$ [6]. With N-methyl O-ethylthiocarbonate (L), PdL_2Br_2, $[PdL_3Br]Br$ and $[PdL_4]Br_2$ are formed [7].

References:

[1] Wenschuh, E.; Schleif, P.; Kolbe, A. (Z. Chem. [Leipzig] **19** [1979] 414).

[2] Bekker, P. van Z. (Inorg. Chim. Acta **31** [1978] 109/15).

[3] Shlenskaya, V. I.; Biryukov, A. A.; Moskovkina, E. M. (Zh. Neorgan. Khim. **11** [1966] 600/5; Russ. J. Inorg. Chem. **11** [1966] 325/8).

[4] Bekker, P. van Z.; Robb, W. (Intern. J. Chem. Kinet. **7** [1975] 87/98).

[5] Hadjiliadis, N.; Markopoulos, J. (J. Chem. Soc. Dalton Trans. **1981** 1635/44).

[6] Levason, W.; McAuliffe, C. A.; Murray, S. G. (J. Chem. Soc. Dalton Trans. **1976** 269/71).

[7] Faraglia, G.; Sindellari, L.; Zarli, B. (Inorg. Chim. Acta **48** [1981] 247/53).

Reactions with Carbon Donors

Reaction of 2-methylenebicyclo(2,2,1)heptane and $[PdBr_4]^{2-}$ gives the π-allylnorcamphane (R) species RPdBr [1]. The kinetics of reaction of alkyl iodides RI (R = Me, Et, Me_2CHCH_2) with $[PdBr_4]^{2-}$ in aqueous $HClO_4$ have been measured [2]. With 1,3-dienes, $[PdBr_4]^{2-}$ in aqueous solution reacts to give a Pd-containing complex which, on treatment with dimethylglyoxime, yields butenylethers [3]. The kinetics of the oxidation of olefins by $[PdBr_4]^{2-}$ in aqueous solutions have been studied [4], as have those of the reduction of $[PdBr_4]^{2-}$ in 10% Br^- aqueous solution by CO [5, 6]. The catalysis of reduction of organic substrates such as 1,4-naphthoquinone by CO in the presence of $[PdBr_4]^{2-}$ has been studied kinetically [6]. Oxidation of CO by $[PdBr_4]^{2-}$ as a catalyst has been studied [7]. The $[PdBr_4]^{2-}$ ion will effect a chirality transfer from optically active alkylsilanes to π-allyl complexes [8]. Reduction of $[PdBr_4]^{2-}$ by CO is promoted by O_2, and this was studied in aqueous media at 25°C [9].

References:

[1] Castanet, Y.; Petit, F. (J. Chem. Res. S **1982** 238/9).
[2] Litvinenko, S. L.; Zamashchikov, V. V.; Rudakov, E. S.; Strel'tsov, V. I. (Ukr. Khim. Zh. **47** [1981] 186/8; Soviet Progr. Chem. **47** No. 2 [1981] 72/4).
[3] Lawrence, R. V.; Ruff, J. K.; Taylor, R. C. (J. Chem. Soc. Chem. Commun. **1976** 9/10).
[4] Osipov, A. M.; Likholobov, V. A.; Ermeeva, N. M. (Izv. Vysshikh Uchebn. Zavedenii Khim. Khim. Tekhnol. **15** [1972] 1485/8; C.A. **78** [1973] No. 83590).
[5] Fasman, A. B.; Kutyukov, G. G.; Sokol'skii, D. V. (Dokl. Akad. Nauk SSSR **158** [1964] 1176/9; Dokl. Phys. Chem. Proc. Acad. Sci. USSR **154/159** [1964] 958/61).
[6] Kutyukov, G. G.; Fasman, A. B.; Lyuts, A. E.; Kushnikov, Yu. A.; Vozdvizhenskii, V. F.; Golodov, V. A. (Zh. Fiz. Khim. **40** [1966] 1468/71; Russ. J. Phys. Chem. **40** [1966] 798/802).
[7] Zakumbaev, G. D.; Noskova, N. F.; Konaev, E. N.; Sokol'skii, D. N. (Dokl. Akad. Nauk SSSR **156** [1964] 1386/8; Dokl. Chem. Proc. Acad. Sci. USSR **154/159** [1964] 644/6).
[8] Hayash, T.; Konishi, M.; Kumada, M. (J. Chem. Soc. Chem. Commun. **1983** 736/7).
[9] Golodov, V. A.; Kuksenko, E. L. (C₁ Mol. Chem. **1** [1984] 109/13; C.A. **103** [1985] No. 43430).

Reactions with Phosphorus Donors and with Metals

Reaction of "$H_2[PdBr_4]$" with L (L = $Ph_2PCH_2CHMeCN$, $Ph_2P(CH_2)_2CN$ and $Ph_2P(o\text{-}C_6H_4CN)$) gives trans-$PdBr_2L_2$ [1]. Reaction of $PdBr_2$ with $LiBr_2$ (presumably giving $[PdBr_4]^{2-}$) and $P(OPh)_3$ with $(Me_2NH_2)^+$ gives $(Me_2NH_2)[PdBr_3(P(OPh)_3)]$ [2]. With $(\pi\text{-}C_3H_5)HgBr$ and $[PdBr_4]^{2-}$, $(\pi\text{-}C_3H_5)PdBr \cdot HgBr_2$ is formed [3].

References:

[1] Habib, M.; Trujillo, H.; Alexander, C. A.; Storhoff, B. N. (Inorg. Chem. **24** [1985] 2344/9).
[2] Nyrkova, A. N.; Kuz'mina, L. G.; Struchkov, Yu. T.; Temkin, O. N. (Koord. Khim. **8** [1982] 82/8; C.A. **96** [1982] No. 134803).
[3] Nesmeyanov, A. N.; Rubezhov, A. Z. (J. Organometal. Chem. **164** [1978] 259/73).

5.2.2.8 Bromopalladates of Nonmetals

A number of these is known, though it is sometimes difficult to obtain them free of $[Pd_2Br_6]^{2-}$ salts.

"$H_2[PdBr_4]$" (see "Palladium" 1942, p. 287). The free acid has been made by dissolving $PdBr_2$ in 40% HBr [1, 17] or from "$H_2[PdCl_4]$" and conc. HBr [2]. A brief reference has been made to isolation at -5°C of a solid hydrate of $H_2[PdBr_4]$ from a solution of $PdBr_2$ in HBr [17].

$(RH)_2[PdBr_4]$. A number of these has been made, all by the same general method, and the temperature of decomposition and the Pd–Br stretch ν_{Pd-Br} determined (see table on p. 188). The salts are obtained by dissolving $PdBr_2$ in 40% HBr and the calculated quantity of (RH)Br added. On cooling the red-brown products crystallise out. They are soluble in water, alcohol and acetone but insoluble in $CHCl_3$ or ether [2]. On heating they give trans-PdR_2Br_2. Differential thermal analysis curves were reproduced for the anilinium complex (20 to 320°C) [2].

The pyridinium salt $(pyH)_2[PdBr_4]$ has also been reported, prepared by dissolving $PdBr_2$ in conc. HBr and adding pyridine. The infrared spectrum (4000 to 250 cm^{-1}, reproduced in paper) showed a band at 315 cm^{-1} (unconvincingly assigned to v_{Pd-Br}) [3]. The dark brown material was said to decompose in water. The same properties were described for $(RH)_2[PdBr_4]$ (R = 2,4,6-collidine), made in similar fashion to $(pyH)_2[PdBr_4]$; for this, v_{Pd-Br} was assigned to an infrared band at 310 cm^{-1} (spectra from 4000 to 250 cm^{-1}, reproduced in paper) [3].

Properties of $(RH)_2[PdBr_4]$ [2]:

R	v_{Pd-Br} (cm^{-1})	$T^{a)}$ (°C)	$T^{b)}$(°C)
methylamine	255, 240	225	250
½ ethylenediamine	253	240	320
piperidine	250	235	315
pyridine	252	180	290
γ-picoline	246, 220	190	260
aniline	252	160	270
p-toluidine	251	180	275
m-nitroaniline	255, 240	175	295
quinoline	250	170	250
½ 1,10-phenanthroline	238	105	175

a) onset of decomposition; b) peak decomposition temperature

$(C_7H_6N_2H)_2[PdBr_4]$. This is made from $H_2[PdBr_4]$ and benzimidazole (L) in ethanol. It has a v_{Pd-Br} stretch at 250 cm^{-1} in the infrared. The compound decomposes at 240°C to give cis-PdL_2Br_2, and at 250 to 300°C it gives trans-PdL_2Br_2 [4].

$(Me_2NH_2)_2[PdBr_4]$ (misprinted as $[(CH_3)_2NH_2][PdBr_4]$ in abstract). This is made by oxidative dissolution of Pd metal in benzyl bromide and dimethylformamide at 150°C. It melts at 148 to 150°C [19].

The X-ray crystal structure of the complex showed the crystals to belong to the orthorhombic CmCa-D_{2h}^{18} space group, Z = 4; a = 18.457(4), b = 7.320(2), c = 9.844(2) Å with a density of 2.604 g/cm^3. The crystals are isomorphous with those of $(NH_4)_2[CuCl_4]$. The Pd–Br distances in the planar $[PdBr_4]^{2-}$ anions are 2.434(?) and 2.446(4) Å. The infrared spectrum was measured [19].

$[(CH_3)_2NH_2]_2[PdBr_4]\cdot[(CH_3)_2NH_2]Br$. This was obtained as a light-red material from the mother liquors from the preparation of $[(CH_3)_2(NH_2)_2]_2[PdBr_4]$ (see above). It melts at 125°C [19].

$(Me_3NH)_2[PdBr_4]$. This is made by treatment of Pd metal with HBr at 80°C in the presence of oxygen, $CuCl_2$ and trimethylamine. The product can be recrystallised from ethanol. The infrared spectrum from 4000 to 1600 cm^{-1} was listed but not reproduced [5]. It is also made by reaction of Pd metal with Me_3NO and $CHBr_3$ at 80°C [20].

(Me₃NOH)₂[PdBr₄]. This is made by reaction of Pd metal with Me_3NO and $CHBr_3$ at 80°C [20].

(Et₄N)₂[PdBr₄]. This brownish-pink salt is made by treating $(Et_4N)_2[Pd_2Br_6]$ with excess $(Et_4N)Br$ in boiling acetone. It is insoluble in water and acetone but fairly soluble in nitrobenzene [6]. The conductance in nitrobenzene at 25°C is 60.5 $cm^2 \cdot \Omega^{-1} \cdot mol^{-1}$ [7]. Electronic spectra in a variety of solvents (nitrobenzene, acetic anhydride, nitromethane, acetone, methanol, acetonitrile) were listed (200 to 550 nm; spectrum in methanol reproduced 200 to 600 nm); the spectrum of the solid was also measured (200 to 800 nm, reproduced in paper) [8].

(Bu₄ⁿN)₂[PdBr₄]. This is a red-brown material, melting at 128 to 129°C; no preparation was given but it was probably made by reaction of $[PdCl_4]^{2-}$ in excess HBr with $(Bu_4^nN)Br$ [14]. Attempts to make the salt, however, by the above method gave, according to an earlier report, $(Bu_4^nN)_2[PdBr_6]$ and no $(Bu_4^nN)_2[PdBr_4]$ [16].

The infrared spectrum of the solid was measured (40 to 300 cm^{-1}) and bands assigned as follows (numbering scheme as on p. 192): $\nu_6(E_u) = 243$; $\nu_5(A_{2u}) = 114$; $\nu_7(E_u) = 104$ cm^{-1} [14].

(C(NH₂)₃)₂[PdBr₄]. This is made from $PdBr_2$, guanidinium bromide and HBr, and forms brown crystals. The X-ray powder pattern was measured (reproduced in paper). From ¹H NMR data, studies on the motion of the guanidinium ions and dipolar-quadrupolar cross-relaxation in the crystals were made [18].

[N(CH₃)₂(C₆H₅CH₂)₂]₂[PdBr₄]. This is made by oxidative dissolution of metallic Pd in a benzyl bromide-dimethylformamide-benzene solvent at 130 to 150°C. It forms dark brown crystals, melting at 198 to 199°C.

The X-ray crystal structure shows the crystals to belong to the $B2_1/b$-C_{2h}^5 space group with $Z = 2$; $a = 18.876(5)$, $b = 11.094(2)$, $c = 9.430(2)$ Å, $\gamma = 58.06(2)°$; density 1.751 g/cm^3. The structure is shown in the paper. In the planar $[PdBr_4]^{2-}$ anion the Pd–Br distances are 2.441 and 2.447 Å [19].

(C₅H₅NOCH₂CH₂Br)₂[PdBr₄]. This is made by reaction of Pd metal with C_5H_5NO and $BrCH_2CH_2Br$ at 90°C [20].

The X-ray crystal structure (see figure in the paper) shows the crystals to be triclinic, space group P1-C_1^1; $Z = 1$; $a = 9.586(2)$, $b = 7.523(4)$, $c = 9.340(2)$ Å; $\alpha = 88.68(2)°$, $\gamma = 112.54(2)°$. The mean Pd–Br distance in the square-planar anion is 2.436(1) Å. Infrared spectra were measured and assignments proposed for the 200 to 4000 cm^{-1} range [20].

[(C₈H₁₇)₃NH]₂[PdBr₄] is apparently the species extracted from trioctylaminium bromide and $[PdBr_4]^{2-}$ in CH_2Cl_2 or C_6H_5Me at 25°C. Extraction constants and hydration energies for $[(C_8H_{17})_3NH]_2[PdX_4]$ (X = Cl, Br, I) were determined [9].

[(n-C₈H₁₇)₄N]₂[PdBr₄]. This is thought to be present in solutions of $PdBr_2$ in $(n-C_8H_{17})_4NBr$. The electronic spectrum in dioxan was measured from 180 to 600 nm (reproduced in paper); in a deficiency of $(n-C_8H_{17})_4NBr$, the dimer $[(n-C_8H_{17})_4N]_2[Pd_2Br_6]$ is formed in dioxan solution [15].

(RH)₂[PdBr₄]. The imipramine [5(3-dimethylaminopropyl) 10,11-dihydro-5-dibenzyl azepine] salt $(C_{19}H_{24}N_2H)_2[PdBr_4]$ has been made and its electronic spectra measured (300 to 800 nm, reproduced in paper) [10].

Salts with R = diantipyrylmethane, diantipyrylmethylmethane, diantipyrylpropylmethane and diantipyrylphenylmethane have been reported and the diantipyrylpropylmethane complex

190

$(C_{26}H_{30}N_4O_2H)_2[PdBr_4]$ was used as a gravimetric reagent for the metal. The salts are made by dissolving $Pd(OH)_2$ in 1M HBr and adding the pyrazolone [11].

Thiazole salts $(LH)_2[PdBr_4]$ have anti-tumour properties [12].

$(SO_2)_2[PdBr_4]$. This salt has been briefly reported [13].

References:

[1] Kukushkin, Yu. N.; Sedova, G. N.; Vlasova, R. A.; Pyzhova, L. Ya. (Zh. Neorgan. Khim. **24** [1979] 733/8; Russ. J. Inorg. Chem. **24** [1979] 409/12).

[2] Rush, R. M.; Martin, D. S.; LeGrand, R. G. (Inorg. Chem. **14** [1975] 1479/83).

[3] Doadrio, A.; Craciunescu, D.; Ghirvu, G. (Rev. Chim. Minerale **12** [1975] 129/3).

[4] Kukushkin, Yu. N.; Sedova, G. N.; Pyzhova, L. Ya.; Vlasova, R. A. (Zh. Neorgan. Khim. **24** [1979] 2265/7; Russ. J. Inorg. Chem. **24** [1979] 1257/8).

[5] Letuchii, Ya. A.; Lavrent'ev, I. P.; Khidekel, M. L. (Izv. Akad. Nauk SSSR Ser. Khim. **28** [1979] 888/90; Bull. Acad. Sci. USSR Div. Chem. Sci. **28** [1979] 832/4).

[6] Harris, C. M.; Livingstone, S. E.; Reece, I. H. (Australian J. Chem. **10** [1957] 282/6).

[7] Harris, C. M.; Livingstone, S. E.; Stephenson, N. C. (J. Chem. Soc. **1958** 3697/3702).

[8] Harris, C. M.; Linvingstone, S. E.; Reece, I. H. (J. Chem. Soc. **1959** 1505/11).

[9] Ivanova, S. N.; Gindin, L. M.; Chernyaeva, A. P.; Chernobrov, A. S.; Shaidurova, N. N. (Izv. Sibirsk. Otd. Akad. Nauk SSSR Ser. Khim. Nauk **1978** No. 6, pp. 68/74; C.A. **90** [1979] No. 110710).

[10] Dembinski, B. (Chem. Anal. [Warsaw] **28** [1983] 161/5).

[11] Akimov, V. K.; Dzhishkariani, G. I.; Busev, A. I.; Emel'yanova, I. A. (Zh. Analit. Khim. **29** [1974] 2112/6; J. Anal. Chem. [USSR] **29** [1974] 1814/7).

[12] Doadrio, A.; Craciunescu, D.; Ghirvu, C. (Inorg. Perspect. Biol. Med. **1** [1978] 223/31).

[13] Letuchii, Ya. A. (Fiz. Khim. Protsessy Gazov. Kondens. Fazakh **1979** 41; C.A. **93** [1980] No. 60170).

[14] Goggin, P. L.; Mink, J. (J. Chem. Soc. Dalton Trans. **1974** 1479/83).

[15] Ivanova, S. N.; Chernyaeva, A. P.; Gindin, L. M.; Chernobrov, A. S. (Izv. Sibirsk. Otd. Akad. Nauk SSSR Ser. Khim. Nauk **1978** No.6, pp. 74/8; C.A. **90** [1979] No. 77223).

[16] Mason, W. R.; Gray, H. B. (J. Am. Chem. Soc. **90** [1968] 5721/9).

[17] Brodersen, K.; Thiele, G.; Gaedcke, H. (Z. Anorg. Allgem. Chem. **348** [1966] 162/7).

[18] Furukawa, V.; Gima, S.; Nakamura, D. (Ber. Bunsenges. Physik. Chem. **89** [1985] 863/75, 864).

[19] Makitova, D. D.; Korableva, L. G.; Krasochka, O. N.; Lavrent'ev, I. P.; Atovmyan, L. O.; Khidekel', M. L. (Koord. Khim. **12** [1986] 1262/9; Soviet J. Coord. Chem. **12** [1986] 729/36).

[20] Makitova, D. D.; Letuchii, Ya. A.; Krasochka, O. N.; Roshchupkina, O. S.; Lavrent'ev, I. P.; Atovmyan, L. O.; Khidekel', M. L. (Koord. Khim. **14** [1988] 394/404; C.A. **108** [1988] No. 230909).

5.2.2.9 Tetrabromopalladates of Alkali Metals and Ammonium

For the sodium salt see "Palladium" 1942, p. 308.

$K_2[PdBr_4]$ and $K_2[PdBr_4] \cdot 2H_2O$

Preparation. The anhydrous salt can be made from Pd metal and a mixture of conc. (48%) HBr and HNO_3 followed by addition of KBr [1, 2]. The product is dark red [1]. It is also made from $PdCl_2$ in conc. HBr with KBr [3], from $K_2[PdCl_4]$ with excess HBr [4]; from $H_2[PdCl_4]$ and conc. HBr followed by addition of KOH to pH = 2 when red crystals form [5]; from $PdBr_2$, HBr and KBr [6, 11] to give red-brown crystals [6]; or from $K_2[PdCl_4]$ and 48% HBr to give brown needles [10].

The dihydrate has been claimed [7] as a brown material made by a literature method which in fact refers to a preparation of $K_2[PdBr_6]$ [8].

Structure. The anhydrous salt is tetragonal, Z = 1, a = 7.403, b = 4.293 Å, while the brown dihydrate is A-centred rhombic, Z = 4; a = 9.567, b = 13.855, c = 8.357 Å [9].

References:

[1] Livingstone, S. E. (Syn. Inorg. Metal-Org. Chem. **1** [1971] 1/7).
[2] Harris, C. M.; Livingstone, S. E.; Reece, I. H. (J. Chem. Soc. **1959** 1505/11).
[3] Hamaguchi, H.; Harada, I.; Shimanouchi, T. (Chem. Letters [Tokio] [1973] 1049/54).
[4] Sharpe, A. G. (J. Chem. Soc. **1950** 3444/50).
[5] Rush, R. M.; Martin, D. S.; LeGrand, R. G. (Inorg. Chem. **14** [1975] 2543/50).
[6] Ito, K.; Nakamura, D.; Kurita, Y.; Ito, K.; Kubo, M. (J. Am. Chem. Soc. **83** [1961] 4526/8).
[7] Yamada, S. (J. Am. Chem. Soc. **73** [1951] 1182/4).
[8] Gutbier, A.; Krell, A. (Ber. Deut. Chem. Ges. **38** [1905] 2385/9).
[9] Marchee, W. G. J.; van Rosmalen, G. M. (J. Cryst. Growth **39** [1977] 358/61).
[10] Rund, J. V. (Inorg. Chem. **13** [1974] 738/40).

[11] Perry, C. H.; Athans, D. P.; Young, E. F.; Durig, J. R.; Mitchell, B. R. (Spectrochim. Acta A **23** [1967] 1137/47; 39).

Vibrational Spectra. Data on $K_2[PdBr_4]$ and the $[PdBr_4]^{2-}$ ion have been reviewed [1]. The most comprehensive data are from [2, 18] for Raman spectra and [3, 16] for infrared data. The infrared spectrum of the salt was reproduced in the paper (20 to 400 cm^{-1}); spectra for both room and liquid nitrogen temperatures were given [16]. The resonance Raman spectrum of $K_2[PdBr_4]$ has been measured (50 to 1000 cm^{-1}, reproduced in paper) and the fourth overtone observed of ν_1 [18].

There is a brief mention of the ν_6 stretching mode (264 cm^{-1}) [6] and a comparison of M–X stretches in the infrared spectra of $[MX_4]^{2-}$ (M = Pd, Pt; X = Cl, Br) [7].

A number of force constant studies has been carried out, mostly using the data of [3]. Thus, Urey-Bradley (UBFF) [3], general valence (GVFF) [9] and modified valence (MVFF) [10] force field calculations were carried out. The study in [9], based on vibrational data from [3, 4, 16], refers to "$PdBr_4$", but clearly $[PdBr_4]^{2-}$ is the species under consideration. There are also valence force field calculations [8] using the data of [16] and similar calculations [17] using the data of [3] and [16].

Mean amplitudes of vibration for "$PdBr_4$" (presumably $[PdBr_4]^{2-}$) were calculated [11], using the data of [3, 4, 16], and also [17] using the data of [3] and [16], and [10] using the data of [16]. The Pd–Br force constant was related to ligand field parameters for a number of square-planar complexes, including $[PdBr_4]^{2-}$ [12]. No source data were quoted for other force constant studies on $[PdBr_4]^{2-}$ [13, 14]. Compliance constants for $[PdBr_4]^{2-}$ in $K_2[PdBr_4]$ using the data of [4] have been calculated [15], and Coriolis coupling constants [17] using the data of [3] and [16]. Potential energy distributions were also calculated [10] using the data of [16].

Vibrational spectra of $K_2[PdBr_4]$:

$\nu_1(A_{1g})$	$\nu_2(B_{1g})$	$\nu_3(A_{2u})$	$\nu_4(B_{2g})$	$\nu_5(A_{2u})$	$\nu_6(E_u)$	$\nu_7(E_u)$	Ref.
187.4	166.6		125.4				[2]
192	165	130	125		260	140	[3]
187	167		125				[4]
				106	254	136	[5]
					260	140	[16]*)
190			124				[18]

Frequency numbering as in [1] but with assignments of [2] and [3]; in [1]. – For Raman spectra of $[PdBr_4]^{2-}$ in aqueous solution see on p. 179. – *) The B_{1u} out-of-plane bending mode was found at 130 cm^{-1} [16].

References:

[1] James, D. W.; Nolan, M. J. (Progr. Inorg. Chem. **9** [1968] 195/275, 237).

[2] Bosworth, Y. M.; Clark, R. J. H. (Inorg. Chem. **14** [1975] 170/7).

[3] Hendra, P. J. (J. Chem. Soc. A **1967** 1298/1301).

[4] Hendra, P. J. (Nature **212** [1966] 179).

[5] Adams, D. M.; Morris, D. M. (Nature **208** [1965] 283).

[6] Hendra, P. J.; Sadasivan, N. (Spectrochim. Acta **21** [1965] 1271/5).

[7] Adams, D. M.; Chandler, P. J. (J. Chem. Soc. Commun. **1966** 69).

[8] Tranquille, M.; Forel, M. T. (J. Chim. Phys. **68** [1971] 471/4).

[9] Thirugnanasambandam, P.; Mohan, S. (Indian J. Pure Appl. Phys. **12** [1974] 206/9).

[10] Durig, J. R.; Nagarajan, G. (Monatsh. Chem. **100** [1969] 1960/72).

[11] Thirugnanasambandam, P.; Mohan, S. (Indian J. Pure Appl. Phys. **13** [1975] 398/401).

[12] Hancock, R. D.; Evers, A. (J. Inorg. Nucl. Chem. **35** [1973] 2558/61).

[13] Thyagarajan, G.; Subhedar, M. K. (Indian J. Pure Appl. Phys. **12** [1974] 309/11).

[14] Sanyal, N. K.; Dixit, L.; Pandey, A. N.; Singh, H. S.; Singh, B. P. (Indian J. Pure Appl. Phys. **10** [1972] 493/4).

[15] Raja, S. V. K.; Savariraj, G. V. (Acta Ciencia Indica **2** [1975] 180/4).

[16] Perry, C. H.; Athans, D. P.; Young, E. F.; Durig, J. R.; Mitchell, B. R. (Spectrochim. Acta **A 23** [1967] 1137/47, 46).

[17] Cyvin, S. J.; Cyvin, B. N.; Müller, A.; Krebs, B. (Z. Naturforsch. **23a** [1968] 479/81).

[18] Hamaguchi, H.; Harada, I.; Shimanouchi, T. (Chem. Letters [1973] 1049/54).

Electronic Spectra. The polarised single-crystal spectra of $K_2[PdBr_4]$ have been measured from 200 to 700 nm (reproduced in paper) for various crystal orientations and dimensions, and assignments proposed for the bands [1]. Single-crystal absorption spectra of $K_2[PdBr_4] \cdot 2H_2O$ have also been measured at 50 K from 300 to 800 nm (not reproduced in paper), assigned and a value of 21600 cm^{-1} for 10 Dq calculated [2]. The reflectance spectrum of anhydrous $K_2[PdBr_4]$ was measured and assigned (200 to 600 nm) [3], and the dichroism reflectance spectrum of

$K_2[PdBr_4] \cdot 2H_2O$ (200 to 500 nm) with the spectrum reproduced in the paper [4]. The electronic spectrum of solid $K_2[PdBr_4]$ (reproduced from 200 to 800 nm in paper) has been measured [5].

References:

[1] Rush, R. M.; Martin, D. S.; LeGrand, R. G. (Inorg. Chem. 14 [1975] 1543/50).
[2] Tuszynski, W.; Gliemann, G. (Z. Naturforsch. 34a [1979] 211/9).
[3] Gray, H. B.; Ballhausen, C. J. (J. Am. Chem. Soc. 85 [1963] 260/5).
[4] Yamada, S. (J. Am. Chem. Soc. 73 [1951] 1182/4).
[5] Harris, C. M.; Livingstone, S. E.; Reece, I. H. (J. Chem. Soc. 1959 1505/11).

Nuclear Quadrupole Resonance (NQR). This has been reviewed for $[PdBr_4]^{2-}$ and other transition metal complexes [1]. The NQR spectrum of $K_2[PdBr_4]$ was measured at liquid nitrogen temperatures (no resonance observed), and at -73 and $27°C$, at which temperatures resonances at 129.77 and 129.34 mHz were observed. The covalent character of the Pd–Br bond was estimated as 0.4 and the net charge 0.4 [2]. The data have been compared with those for $K_2[PtBr_4]$ [2, 3].

X-Ray Photoelectron Spectroscopy (ESCA). The ESCA spectrum of $K_2[PdBr_4]$ shows binding energies of 337.9 eV ($3d_{5/2}$), 69.6 eV (Br 3d) and 293.2 eV (K $2p_{3/2}$) [4] or of 342.8 ± 0.1 ($3d_{5/2}$) and 337.5 ± 0.1 ($3d_{3/2}$) eV [5].

References:

[1] van Bronswyk, W. (Struct. Bonding [Berlin] 7 [1970] 87/113).
[2] Ito, K.; Nakamura, D.; Kurita, Y.; Ito, K.; Kubo, M. (J. Am. Chem. Soc. 83 [1961] 4526/8).
[3] Machmer, P. (Z. Naturforsch. 24b [1969] 193/9).
[4] Nefedov, V. I.; Zakharova, I. A.; Moiseev, I. I.; Porai-Koshits, M. A.; Vargaftik, M. N.; Belov, A. P. (Zh. Neorgan. Khim. 18 [1973] 3264/8; Russ. J. Inorg. Chem. 18 [1973] 1738/40).
[5] Kumar, G.; Blackburn, J. R.; Albridge, R. G.; Moddeman, W. E.; Jones, M. M. (Inorg. Chem. 11 [1972] 296/300).

Miscellaneous Properties of $K_2[PdBr_4]$. The refractive index of single crystals of $K_2[PdBr_4]$ shows that $n_c(n_E) = 1.5804$ and $n_a(n_D) > 1.74$ at the Na-D lines in 1-bromonaphthalene and CH_2I_2, so the crystals are dichroic [1]. The molar refraction of $K_2[PdBr_4]$ is 47.85 [2]. The effect on electrical conductivity, viscosity and density on water-dioxan mixtures consequent on addition of $K_2[PdBr_4]$ to the solvent has been measured [3]. The pseudo-lattice energy of $K_2[PdBr_4]$ has been calculated [4], and the effective charge on Pd in $K_2[PdBr_4]$ calculated on the basis of an electrostatic model [5]. The ionic character of the Pd–Br bond has also been calculated as 0.42 [6].

The anaphylactoid properties of $K_2[PdBr_4]$ towards cats and guinea pigs have been assessed [7], as has its antiphage activity against bacteriophages [8, 9].

Thermal decomposition of $K_2[PdBr_4]$ under H_2 has been investigated, and the decomposition characteristics were compared with those of $K_2[PdCl_4]$ and the corresponding Pt^{II} complexes [10].

References:

[1] Rush, R. M.; Martin, D. S.; LeGrand, R. G. (Inorg. Chem. 14 [1975] 2543/50).
[2] Shamsiev, A. Sh.; Obel'chenko, P. F. (Tr. Sredneaziat. Gos. Univ. [2] No. 33 [1952] 57/63; C.A. 1956 3830).

[3] Golodov, V. A.; Fasman, A.; Roganov, V. V.; Enker, K. P. (Zh. Neorgan. Khim. **15** [1970] 236/9; Russ. J. Inorg. Chem. **15** [1970] 121/3).

[4] de Jonge, R. M. (J. Inorg. Nucl. Chem. **38** [1976] 1821/6).

[5] de Jonge, R. M. (J. Inorg. Nucl. Chem. **38** [1976] 1283/6).

[6] Batsanov, S. S. (Zh. Neorgan. Khim. **9** [1964] 323/7; Russ. J. Inorg. Chem. **9** [1964] 722/5).

[7] Tomilets, V. A.; Zakharova, I. A. (Farmakol. Toksikol. [Moscow] **42** [1979] 170/3; C. A. **91** [1979] No. 13476).

[8] Zakharova, I. A.; Moshkovskii, Yu. Sh.; Suraikina, T. I.; Fonshtein, L. M. (Koord. Khim. **2** [1976] 1642/5; Soviet J. Coord. Chem. **2** [1976] 1263/6).

[9] Zakharova, I. A.; Moshkovskii, Yu. S.; Suraikina, T. I.; Fonshtein, L. M. (Dokl. Akad. Nauk SSSR **222** [1975] 1229/31; Dokl. Biochem. **222** [1975] 252/3).

[10] Marchée, W. G. J.; van Rosmalen, G. M.; Hakvoort, G. (Thermochim. Acta **25** [1978] 91/9).

Applications for $K_2[PdBr_4]$. The use of $K_2[PdBr_4]$ with a polyoxyethylene nonionic surfactant to improve the antistatic properties of silver halide photographic materials has been described.

Reference:

Anonymous (Res. Discl. **238** [1984] 44/6; C. A. **100** [1984] 200799).

$(NH_4)_2[PdBr_4]$

This red-brown material is made by dissolution of $PdBr_2$ in 40% HBr followed by addition of the stoichiometric quantity of NH_4Br [1, 2]. The salt is also formed from Pd metal, HBr and NH_4Br with oxygen [3].

The salt crystallises in the tetragonal space group $P4/mmm–D_{4h}^1$, Z=1; a=7.5258(9), c=4.446(4) Å. The Pd–Br distance in the square-planar anion is 2.438(c) Å [5]. It is isostructural with $(NH_4)_2[PdCl_4]$ [4].

The infrared spectrum of the salt has been measured, ν_6 at 254 cm^{-1}, ν_7 at 169 cm^{-1}, and the B_{1u} out-of-plane bend at 140 cm^{-1} [2]. The ν_{Pd-Br} stretch lies at 257 cm^{-1} in the infrared; thermal decomposition starts at 225°C and reaches a peak at 310°C [1]. The infrared stretches in the N- and N–H region in $(NH_4)_2[PdBr_4]$ with 2% deuteration have been measured at 77 and 298 K [4].

References:

[1] Kukushkin, Yu. N.; Sedova, G. N.; Vlasova, R. A.; Pyzhova, L. Ya. (Zh. Neorgan. Khim. **24** [1979] 733/8; Russ. J. Inorg. Chem. **24** [1979] 409/12).

[2] Perry, C. H.; Athans, D. P.; Young, E. F.; Durig, J. R.; Mitchell, B. R. (Spectrochim. Acta A **23** [1967] 1137/47, 41).

[3] Letuchii, Ya. A. (Fiz. Khim. Protsessy Gazov. Kondens. Fazakh **1979** 41; C. A. **93** [1980] No. 60170).

[4] Knop, O.; Oxton, I. A.; Falk, M. (Can. J. Chem. **57** [1979] 404/23, 414, 420).

[5] Larina, T. B.; Bol'shakova, L. D.; Surazhskaya, M. A.; Skubochkina, E. F.; Koz'min, P. A. (Zh. Neorgan. Khim. **33** [1988] 807; C. A. **109** [1988] No. 214359).

$Rb_2[PdBr_4]$

This is made as needle-shaped crystals by adding RbBr to a solution of $Na_2[PdBr_4]$ [1] or from $PdBr_2$ and HBr with RbBr [2].

The infrared spectrum was measured at room and liquid nitrogen temperatures (20 to 400 cm^{-1}, reproduced in paper). Assignments were: $\nu_6 = 258$ cm^{-1}, $\nu_7 = 135$ cm^{-1} (numbering as on p. 192) and the out-of-plane B_{1u} bend was found at 125 cm^{-1} [2].

References:

[1] Bekker, P. van Z.; Robb, W. (Intern. J. Chem. Kinet. **7** [1975] 87/98).
[2] Perry, C. H.; Athans, D. P.; Young, E. F.; Durig, J. R.; Mitchell, B. R. (Spectrochim. Acta A **23** [1967] 1137/47, 40).

$Cs_2[PdBr_4]$

This red salt is made from $K_2[PdBr_4]$ and CsBr in 6 M HBr [1] or from $PdBr_2$, HBr and CsBr [2].

The infrared spectrum of the salt was measured at room and liquid nitrogen temperatures (20 to 450 cm^{-1}, reproduced in paper). The ν_6 stretch was found at 249 cm^{-1} and ν_7 at 130 cm^{-1} (numbering as on p. 192) with the B_{1u} out-of-plane bend at 114 cm^{-1} [2].

The electronic spectrum of the salt, doped in $Cs_2[ZrBr_6]$ at 2 K, was measured and assigned (reproduced from 400 to 550 nm in paper); evidence for the possible presence of a Jahn-Teller effect in one transition was presented. Molecular orbital calculations were made on the data [1].

References:

[1] Harrison, T. G.; Patterson, H. H.; Hau, M. T. (Inorg. Chem. **15** [1976] 3018/24).
[2] Perry, C. H.; Athans, D. P.; Young, E. F.; Durig, J. R.; Mitchell, B. R. (Spectrochim. Acta A **23** [1967] 1137/47, 40).

For salts $[PdL_4][PdBr_4]$ (e.g., $L = NH_3$, CH_3NH_2), see under $[PdL_4]^{2+}$.

5.2.3 Bromopalladates(II) with Coordination Numbers >4

Spectrophotometric data suggest that $[PdBr_4]^{2-}$ in excess Br^- gives $[PdBr_n]^{2-n}$ where $n>4$; such species are more stable than the corresponding ones for $[PdCl_n]^{2-n}$ (see p. 192) [1].

Absorption spectra of $(Et_4N)_2[PdBr_4]$ in solution with excess $(Et_4N)Br$ give evidence for the existence of an unstable $[PdBr_6]^{4-}$ species in solution [2, 3]; aqueous solutions of $K_2[PdBr_4]$ show, however, no significant shifts or intensity changes when extra Br is added [3].

References:

[1] Grinberg, A. A.; Kiseleva, N. V. (Zh. Neorgan. Khim. **3** [1958] 1804/9; Russ. Inorg. Chem. **3** No. 8 [1958] 119/26).
[2] Harris, C. M.; Livingstone, S. E.; Reece, I. H. (Australian J. Chem. **10** [1957] 282/6).
[3] Harris, C. M.; Livingstone, S. E.; Reece, I. H. (J. Chem. Soc. **1959** 1505/11).

5.2.4 Bromopalladates(III)

The only evidence for the existence of these is from electron spin resonance (ESR) data.

$Ag_3[PdBr_6]$. This substance may be formed, together with $Ag_5[PdBr_6]$, when $PdBr_2$ doped into AgBr is irradiated with light at 425 nm at 100 K. The ESR spectrum was measured (reproduced in paper).

Reference:

Eachus, R. S.; Graves, R. E.; Olm, M. T. (Phys. Status Solidi A **57** [1980] 429/37).

5.2.5 Hexabromopalladates [PdBr$_6$]$^{2-}$

See "Palladium" 1942, pp. 287, 316, 324, 331/3, 335.

The normal method of preparation is to oxidise [PdBr$_4$]$^{2-}$ with Br$_2$.

Vibrational Spectra are considered under K$_2$[PdBr$_6$], see below.

Stability Constants. The values of $k_5 = 3.04 \times 10^3$ and $k_6 = 4.3 \times 10^{-2}$ were determined by a redox-potentiometric method for [PdBr$_6$]$^{2-}$; the values are some 10^4 times lower than for the corresponding [PtBr$_6$]$^{2-}$ values and were originally quoted as the "instability constants" (for which $k_5 = 3.29 \times 10^{-4}$ and $k_6 = 2.3 \times 10^{-3}$) [1].

Redox Potential. The value $E_0 = 0.7070$ V has been given for [PdBr$_6$]$^{2-}$ [1] and, in 1M aqueous KBr solution, as +0.994 V for [PdBr$_6$]$^{2-}$/[PdBr$_4$]$^{2-}$ [2].

References:

[1] Dubinskii, V. I. (Zh. Neorgan. Khim. **16** [1971] 1145/7; Russ. J. Inorg. Chem. **16** [1971] 607/8).
[2] Grinberg, A. A.; Shamsiev, A. M. (Zh. Obshch. Khim. **12** [1943] 55/72; C.A. **1943** 1915).

5.2.5.1 (MeNH$_3$)$_2$[PdBr$_6$]

The salt was made from Pd metal with Br$_2$, HBr and methylammonium bromide. Structural phase transitions were studied by nuclear quadrupole resonance (NQR) of ^{81}Br; such a transition was observed at 108 K. The NQR frequency of ^{81}Br in the compound was 174.60 ± 0.05 at 77 K, 174.02 ± 0.05 at 190 K and 173.33 ± 0.05 mHz at 300 K.

Reference:

Kume, Y.; Ikeda, R.; Nakamura, D. (J. Magn. Resonance **33** [1979] 331/44, 338).

5.2.5.2 Alkali Metal and Ammonium Salts of [PdBr$_6$]$^{2-}$

K$_2$[PdBr$_6$]

Preparation. This is made by dissolving PdO in HBr and treating the solution with Br$_2$ and KBr; crystals separate on cooling to 0°C [1], or by oxidising a concentrated solution of K$_2$[PdBr$_4$] with bromine [2]. It forms black crystals. The salt is only slightly soluble in water [2].

Structure. The lattice constant is 10.25 Å; the salt has the K$_2$[PtCl$_6$] structure [1].

Vibrational Spectra. The Raman and infrared spectra of solid K$_2$[PdBr$_6$] have been measured [3].

$\nu_1(A_{1g})$	$\nu_2(E_g)$	$\nu_3(F_{1u})$	$\nu_4(F_{1u})$	$\nu_5(F_{2g})$	$\nu_6(F_{2g})$	Ref.
198	176	253	130	100		[3]

Force constants were calculated on the basis of the data from [3] on a Urey-Bradley force field (UBFF) [3, 4, 6], general valence force-field (GVFF) [3 to 6, 18], orbital valence force field (OVFF) [4, 6], modified Urey-Bradley force field (MUBFF) [6], modified orbital valence force field (MOVFF) [6].

Coriolis coupling constants (using the GVFF results) and potential energy distributions for the fundamental modes were established (from the data of [3]) [4] and vibrational mean amplitudes at 298 K (also from the data of [3]) [4, 5]. "Progressive rigidity constants" for

$Cs_2[PdBr_6]$ have been given (using the data of [3, 7], see p. 196) and reduced partition functions for $[PdBr_6]^{2-}$ have been calculated for 200 to 2000 K from spectroscopic data from v_3 and v_4 [8]. Compliance constants for $[PdBr_6]^{2-}$ from the data of [3] have been calculated [9]. Coriolis coupling constants, using the data of [3], were determined [4]. Constrained force constants have been calculated by [19] using the data of [3].

CNDO molecular orbital calculations have been carried out on $[PdBr_6]^{2-}$ (using the data of [3]) [14].

Nuclear Quadrupole Resonance (NQR) and Miscellaneous Data. Resonances were seen at 205.34, 202.43 and 201.81 mHz/s at liquid nitrogen temperatures, −73 and 27°C, respectively. The covalent character of the Pd–Br bond was estimated as 0.63 and the net charge on Pd as 0.22 [10], see also [16]. These data have been discussed with reference to other hexabrome species $K_2[MBr_6]$ (M = W, Re, Os, Ir, Pt) [12]. Data were also correlated with the ionic character of the bond in this and other salts [11].

The X-ray electron spectrum of $K_2[PdBr_6]$ has been measured and the electronic structure of $[PdBr_6]^{2-}$ discussed on the basis of the results [13, 15].

The ionic character of the Pd–Br bond in $K_2[PdBr_6]$ has been calculated [16].

Chemical Reactions. With dithiomalonamide (HL), $[PdBr_6]^{2-}$ in HBr gives the red mixed-valence species $Pd(HL)_2Br_3$ [17].

References:

[1] Sharpe, A. G. (J. Chem. Soc. **1953** 4177/9).

[2] Dubinskii, V. I. (Zh. Neorgan. Khim. **16** [1971] 1145/7; Russ. J. Inorg. Chem. **16** [1971] 607/8).

[3] Hendra, P. J.; Park, P. J. D. (Spectrochim. Acta A **23** [1967] 1635/40).

[4] Rai, S. N.; Thakur, S. N.; Rai, D. K. (Proc. Indian Acad. Sci. A **74** [1971] 243/54).

[5] Mohan, S.; Mukunthan, A. (Indian J. Phys. B **56** [1982] 91/5).

[6] Labonville, P.; Ferraro, J. R.; Wall, M. C.; Basile, L. J. (Coord. Chem. Rev. **7** [1972] 257/87).

[7] Kovrikov, A. B. (Zh. Prikl. Spektrosk. **15** [1971] 81/5; J. Appl. Spectrosc. [USSR] **15** [1971] 899/902).

[8] Kotaka, M.; Shono, T.; Ikuta, E.; Kakihana, H. (Bull. Res. Lab. Nucl. React. [Tokyo Inst. Technol.] **3** [1978] 31/7; C.A. **90** [1979] No. 210438).

[9] Raja, S. V. K.; Savaryraj, G. A. (Acta Ciencia Indica **2** [1976] 180/4).

[10] Ito, K.; Nakamura, D.; Kurita, Y.; Ito, K.; Kubo, M. (J. Am. Chem. Soc. **83** [1961] 4526/8).

[11] Kubo, M.; Nakamura, D. (Colloq. Intern. Centre Natl. Rech. Sci. [Paris] No. 191 [1970] 117/23; C.A. **77** [1972] No. 81770).

[12] Machmer, P. (Z. Naturforsch. **24b** [1969] 193/9).

[13] Peshchevitskii, B. I.; Zemskov, S. V.; Sadovskii, A. P.; Kravtsova, E. A.; Mit'kin, V. N. (Koord. Khim. **5** [1979] 1838/45; Soviet J. Coord. Chem. **5** [1979] 1433/9).

[14] Sakaki, S.; Kato, H. (Bull. Chem. Soc. Japan **46** [1973] 2227/9).

[15] Peshchevitskii, B. I.; Sadovskii, A. P.; Zemskov, S. V.; Kravtsova, E. A.; Mit'kin, V. N.; Shipachev, V. A.; Nefedov, V. I. (Tezisy Dokl. 13th Vses. Chugaevskoe Soveshch. Khim. Kompleksn. Soedin., Moscow 1978, pp. 311; C.A. **90** [1979] No. 112 666).

[16] Batsanov, S. S. (Zh. Neorgan. Khim. **9** [1964] 1323/7; Russ. J. Inorg. Chem. **9** [1964] 722/5).

[17] Bret, J.-M.; Castan, P.; Laurent, J.-P. (J. Chem. Soc. Dalton Trans. **1984** 1975/80).

[18] Mohan, S.; Gunasekaran, G.; Ravikumar, K. G.; Govindarajan, S. (Oriental J. Chem. **1** [1985] 11/15).

[19] Geldard, J. F.; McDowell, H. K. (Spectrochim. Acta A **43** [1987] 439/45).

NH₄, Rb and Cs Salts

(NH₄)₂[PdBr₆]. This is made by warming PdO in HBr with Br_2 followed by addition of ammonium ion; crystals appear on cooling [1].

The infrared spectrum of the $[NH_3D]^+$ probe-ion in $(NH_4)_2[PdBr_6]$ (2% deuteriated) has been measured in the 2400 to 1200 cm^{-1} region (spectra reproduced in paper) from 10 to 90 K and a phase transition located near 50 K [2].

Rb₂[PdBr₆]. This is made from PdO, conc. HBr and Br_2 with RbBr; the salt crystallises at 0°C. The lattice constant is 10.38 Å [1].

Cs₂[PdBr₆]. This is made from PdO, HBr and Br_2 with heating and addition of CsBr, and crystallises at 0°C [1].

The lattice constant is 10.62 Å [1]. The compliance (force) constants for the anion in this salt have been calculated [3] using the data, apparently, for the vibrational spectra of $K_2[PdBr_6]$ given in [4].

[Pd(NH₃)₄][PdBr₄]. This is made from $[Pd(NH_3)_4]^{2+}$ and $[PdBr_4]^{2-}$. The polarised single-crystal electronic spectrum was measured at 6 and 300 K from 250 to 800 nm (reproduced in paper) [5].

[Pd en₂][PdBr₄]. This is made by reaction of $[Pd\ en_2]^{2+}$ and $[PdBr_4]^{2-}$. The polarised single-crystal electronic absorption spectrum was measured at 6 and 300 K from 250 to 800 nm (reproduced in paper) [5].

References:

[1] Sharpe, A. G. (J. Chem. Soc. **1953** 4177/9).
[2] Knop, O.; Oxton, I. A.; Westerhans, W. J.; Falk, M. (J. Chem. Soc. Faraday Trans. II **77** [1981] 811/32, 819).
[3] Keng, F. T.; Kovrikov, A. B. (Zh. Prikl. Spektrosk. **15** [1971] 81/5; J. Appl. Spectrosc. [USSR] **15** [1971] 899/902).
[4] Hendra, P. J.; Park, P. J. D. (Spectrochim. Acta A **23** [1967] 1635/40).
[5] Rodgers, M. L.; Martin, D. S. (Polyhedron **6** [1987] 225/54, 226, 250).

5.2.6 Binuclear Bromopalladates

5.2.6.1 The [Pd₂Br₆]²⁻ Ion

As with the corresponding chloro (p. 159) and iodo (p. 212) systems this species exists in equilibrium with $[PdBr_4]^{2-}$, and salts of the anion can be isolated with organic cations.

Factors affecting the conformation of the $[Pd_2Br_6]^{2-}$ anion in its complexes have been discussed [12].

Vibrational Spectra. Raman and infrared data on $(Pn_4^nN)_2[Pd_2Br_6]$ in solution [1] and infrared spectra of $R_2[Pd_2Br_6]$ in the solid state ($R = Pn_4^nN$ [1], $R = Et_4N$ [2]) have been given (see following table). Force constant calculations have been performed on the in-plane vibrations [2, 3] and potential energy distributions of these modes calculated [3].

Electronic Spectra. The first studies were on $(Et_4N)_2[Pd_2Br_6]$ in nitromethane, nitrobenzene, acetic anhydride, acetone, methanol and acetonitrile (200 to 500 nm, data listed but not reproduced in paper) [4]; there are listed data (200 to 500 nm) for $(Et_4N)_2[Pd_2Br_6]$ dissolved in methanol with some band assignment [5], and for solutions in acetone and in methanol (200 to 500 nm, data listed) [9]; of $(Bu_4^nN)_2[Pd_2Br_6]$ in acetonitrile from 160 to 250 nm (reproduced in paper) with band assignments [6]; 180 to 600 nm for the trioctylammonium salt in dioxan (reproduced in paper) [7].

Experimental and theoretically calculated spectra of $[Pd_2Br_6]^{2-}$ have been compared (200 to 650 nm, reproduced in paper) [8]. The effect of solvents on the electronic spectra of the $(Et_4N)^+$ and $(Bu_4^nN)^+$ salts of $[Pd_2Br_6]^{2-}$ has also been discussed [10]. Spectra of $[Pd_2Br_6]^{2-}$ have been used for the analysis of Pd [11].

Raman and infrared data for $R_2[Pd_2Br_6]^{2-}$ salts:

	$R = Et_4N$ [2]	$R = Pn_4^nN$ [1]		$R = Et_4N$ [2]	$R = Pn_4^nN$ [1]
$\nu_1(A_g)\nu_{MBr}$		262	$\nu_{10}(B_{1u})\delta_{MBr}$	88[1]	
$\nu_2(A_g)\nu_{MBr}$		194	$\nu_{12}(B_{2u})\nu_{MBr}$	266[1]	257 257
$\nu_3(A_g)\delta_{MBr}$		115	$\nu_{13}(B_{2u})\nu_{MBr}$	177[1]	178[2] 179[1]
$\nu_4(A_g)\delta_{MBr}$		83	$\nu_{14}(B_{2u})\delta_{MBr}$	65[1]	
$\nu_6(B_{1g})\nu_{MBr}$		253	$\nu_{16}(B_{3u})\nu_{MBr}$	262[1]	264 266[1]
$\nu_7(B_{1g})\nu_{MBr}$		173	$\nu_{17}(B_{3u})\nu_{MBr}$	192[1]	192[2] 193[1]
$\nu_9(B_{1u})\delta_{MBr}$		113[2] 113[1]	$\nu_{18}(B_{3u})\delta_{MBr}$	117[1]	

[1] Solids; otherwise solution data in CH_2Cl_2 or [2] in CH_2Cl_2–$CHCl_3$; numbering as in [1].

Molecular Orbital Calculations. Calculations using an extended Hückel method on $[Pd_2Cl_6]^{2-}$ and $[Pd_2Br_6]^{2-}$ permit prediction of the folding angle in various salts of these anions [12].

References:

[1] Goggin, P. L. (J. Chem. Soc. Dalton Trans. **1974** 1483/6).

[2] Adams, D. M.; Chandler, P. J.; Churchill, R. G. (J. Chem. Soc. **1967** 1272/4).

[3] Goggin, P. L.; Mink, J. (Inorg. Chim. Acta **26** [1978] 119/24).

[4] Harris, C. M.; Livingstone, S. E.; Reece, I. H. (J. Chem. Soc. **1959** 1505/11).

[5] Lobaneva, O. A.; Kononova, M. A.; Kunaeva, N. T.; Davydova, M. K. (Zh. Neorgan. Khim. **17** [1972] 3011/4; Russ. J. Inorg. Chem. **17** [1972] 1583/5).

[6] Batiste, J. L. H.; Rumfeldt, R. (Can. J. Chem. **52** [1974] 174/81).

[7] Ivanova, S. N.; Chernyaeva, A. P.; Gindin, L. M.; Chernobrov, A. S. (Izv. Sibirsk. Otd. Akad. Nauk SSSR Ser. Khim. Nauk **1978** No. 6, pp. 74/8; C.A. **90** [1979] No. 77223).

[8] Baranovskii, V. I.; Davydova, M. K.; Panina, N. S.; Panin, A. I. (Koord. Khim. **2** [1976] 409/15; Soviet J. Coord. Chem. **2** [1976] 308/13).

[9] Lobaneva, O. A.; Kononova, M. A.; Davydova, M. K. (Dokl. Akad. Nauk SSSR **194** [1970] 133/5; Dokl. Phys. Chem. Proc. Acad. Sci. USSR **190/195** [1970] 669/70).

[10] Lobaneva, O. A.; Kononova, M. A.; Davydova, M. K. (Vestn. Leningr. Univ. Fiz. Khim. **1971** No. 4, pp. 149/50; C.A. **76** [1972] No. 92298).

[11] Sednev, Yu. M.; Rakhman'ko, E. M.; Starobinets, G. L.; Gulevitch, A. L. (Zh. Analit. Khim. **40** [1985] 2216/9; C.A. **104** [1986] No. 141174).

[12] Shestakov, A. F.; Makitova, D. D.; Atovmyan, L. O.; Krasochka, O. N. (Koord. Khim. **13** [1987] 949/52; Soviet J. Coord. Chem. **13** [1987] 538/42).

5.2.6.2 Salts of [Pd$_2$Br$_6$]$^{2-}$

(Et$_4$N)$_2$[Pd$_2$Br$_6$]. This is made by reaction of K$_2$[PdBr$_4$] in aqueous solution with (Et$_4$N)Br [1, 2] and forms brown plates after recrystallisation from 1:1 acetone/ethanol. Both ebullioscopic and cryoscopic measurements show it to be dimeric in solution [1,2]. The conductance in nitrobenzene is 60.5 cm$^2 \cdot \Omega^{-1} \cdot$ mol^{-1} at 25°C [2].

Raman data for solutions in CH$_2$Cl$_2$ (50 to 500 cm^{-1}) and infrared data for solutions in CH$_2$Cl$_2$ or CH$_2$Cl$_2$-CHCl$_3$ (100 to 300 cm^{-1}) have been measured [4]; for band assignment and force constant data see above and table on p. 199. Far-infrared spectra were measured for the solid (50 to 300 cm^{-1}) [7]. Electronic absorption spectra were measured from 200 to 600 nm in nitromethane, acetonitrile, acetone, acetic anhydride and methanol (data listed but not reproduced) and also of the solid (200 to 800 nm, reproduced in paper) [3]. The electronic spectrum of a single crystal was also measured (200 to 700 nm, data listed) [8].

The density of the crystalline (Et$_4$N)$_2$[Pd$_2$Br$_6$] is 2.13 g/cm^3. The solid is red-brown and is slightly soluble in water and in alcohol, but more soluble in acetone. The aqueous solutions are greenish and the organic solutions are red-yellow; alcohol causes gradual reduction to the metal [8].

(Pn$_4^n$N)$_2$[Pd$_2$Br$_6$] is made from [PdBr$_4$]$^{2-}$ and n-pentylammonium bromide followed by recrystallisation from diethyl ether/CH$_2$Cl$_2$. It is deep red, melting from 101 to 102°C. The Raman spectrum of the solution in CH$_2$Cl$_2$ or in CH$_2$Cl$_2$/CHCl$_3$ and the infrared spectrum of the solid were measured (see table on p. 199) [4].

(Bu$_4^n$N)$_2$[Pd$_2$Br$_6$]. This is made from "H$_2$[PdBr$_6$]" and excess conc. HBr and (Bu$_4^n$N)Br added. The electronic spectrum was measured from 300 to 700 nm (data listed in paper) and the bands assigned [9].

((n-C$_8$H$_{17}$)$_4$N)$_2$[Pd$_2$Br$_6$]. This tetraoctylammonium salt is involved in the extraction of PdBr$_2$ from solutions containing (n-C$_8$H$_{17}$)$_4$NBr; in the presence of excess of the latter in dioxan, [(n-C$_8$H$_{17}$)$_4$N]$_2$[Pd$_2$Br$_6$] is converted to [(n-C$_8$H$_{17}$)$_4$N]$_2$[PdBr$_4$]. The electronic spectrum of the solution in dioxan was measured from 180 to 600 nm, reproduced in paper [10].

((C$_5$H$_5$NO)C$_2$H$_4$Br)$_2$[Pd$_2$Br$_6$]. Details of the X-ray crystal structure of this have been noted; the Pd–Br (terminal) distance is 2.26 Å and the Pd–Br (bridge) distance 2.32 Å. The Pd–Pd distance is 3.35 Å [12].

(C$_5$H$_5$NOCH$_2$CH$_2$Br)$_2$[Pd$_2$Br$_6$]. This is made by reaction of Pd metal with C$_5$H$_5$NO and BrCH$_2$CH$_2$Br at 90°C [13].

The X-ray crystal structure (see figure in the paper) shows the crystals to be monoclinic, space group P2$_1$/a-C$_{2h}^5$, Z = 4; a = 19.904(5), b = 18.143(5), c = 8.024(6) Å; α = β = 90°, γ = 66.32(2)°. The mean Pd–Br terminal distance is 2.409(3) Å and the mean Pd–Br bridge distance is 2.452(4) Å. Infrared spectra were measured (200 to 4000 cm^{-1}, reproduced in paper) and assignments proposed [13].

[(C$_5$H$_5$NO)$_2$H]$_2$[Pd$_2$Br$_6$]. This is formed as a deep red compound by treatment of Pd metal with HBr and pyridine-N-oxide at 82°C in the presence of oxygen [5] or from Pd and pyridine-N-oxide in CHBr$_3$, heated for 20 h at 95°C [6, 11]. The salt forms dark red crystals, m.p. 127 to 128°C, which are soluble in acetone, CHCl$_3$ and methanol but insoluble in water, ether, CCl$_4$ and pentane [6].

The X-ray crystal structure shows the salt to be monoclinic, space group B2/b-C$_{2h}^6$, Z = 4, a = 21.218(6), b = 20.097(8), c = 7.044(4) Å, γ = 97.84(2)°, calculated density 2.412 g/cm^3.

The structure shows a centrosymmetric binuclear anion, see **Fig. 43**, mean Pd–Br^1 = 2.398, Pd–Br^2 = 2.445, Pd–Br^3 = 2.405, Pd–Br$^{2'}$ = 2.452 Å; the PdBr^{2}Pd angle is 92.9°, the Br^{1}PdBr^2 angle 89.9° [6].

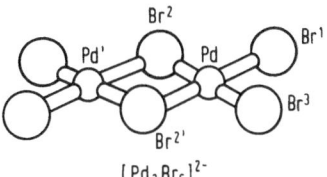

$[Pd_2Br_6]^{2-}$

Fig. 43.   X-ray crystal structure of $[(C_5H_5NO)_2H]^+[Pd_2Br_6]$, showing the structure of the anion and cation [6].

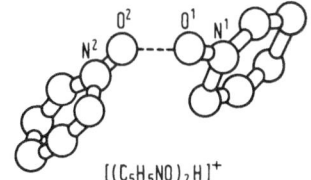

$[(C_5H_5NO)_2H]^+$

Infrared data were listed for the salt [5, 6, 11].

(pyOH)$_2$[Pd$_2$Br$_6$], [ROCH$_2$CH$_2$Br)$_2$[Pd$_2$Br$_6$]. For preparation and properties of these compounds see [13].

[N(CH$_3$)$_2$(C$_6$H$_5$CH$_2$)$_2$]$_2$[Pd$_2$Br$_6$]·(N(CH$_3$)$_2$H$_2$)·Br. This (with Pd misprinted as Pb in the paper) was isolated as dark red crystals as a by-product in the preparation of [N(CH$_3$)$_2$(C$_6$H$_5$)$_2$]$_2$[PdBr$_4$] by oxidative dissolution of Pd metal in a benzyl bromide-benzene-dimethylformamide mixture at 150°C. It melts at 178 to 180°C [11].

References:

 [1] Harris, C. M.; Livingstone, S. E.; Reece, I. H. (Australian J. Chem. **10** [1957] 282/6).
 [2] Harris, C. M.; Livingstone, S. E.; Stephenson, N. C. (J. Chem. Soc. **1958** 3697/3702).
 [3] Harris, C. M.; Livingstone, S. E.; Reece, I. H. (J. Chem. Soc. **1959** 1505/11).
 [4] Goggin, P. L. (J. Chem. Soc. Dalton Trans. **1974** 1483/6).
 [5] Letuchii, Ya. A.; Lavrent'ev, I. P.; Khidekel', M. L. (Izv. Akad. Nauk SSSR Ser. Khim. **1979** 888/91; Bull. Acad. Sci. USSR Div. Chem. Sci. **28** [1979] 832/4).
 [6] Letuchii, Ya. A.; Lavrent'ev, I. P.; Khidekel', M. L.; Krasochka, O. N.; Makitova, D. D.; Atovmyan, L. O. (Izv. Akad. Nauk SSSR Ser. Khim. **1978** 1902/3; Bull. Acad. Sci. USSR Div. Chem. Sci. **27** [1978] 1673/4).
 [7] Adams, D. M.; Chandler, P. J. (J. Chem. Soc. A **1967** 1272/4).
 [8] Lobaneva, O. A.; Kononova, M. A.; Davydova, M. K. (Dokl. Akad. Nauk SSSR **194** [1970] 133/5; Dokl. Phys. Chem. Proc. Acad. Sci. USSR **100/195** [1970] 669/70).
 [9] Mason, W. R.; Gray, H. B. (J. Am. Chem. Soc. **90** [1968] 5721/9).
[10] Ivanova, S. N.; Chernyaeva, A. P.; Gindin, L. M.; Chernobrov, A. S. (Izv. Sibirsk. Otd. Akad. Nauk SSSR Ser. Khim. Nauk **1978** No. 8, pp. 74/8; C.A. **90** [1979] No. 77223).

[11] Makitova, D. D.; Korableva, L. G.; Krasochka, O. N.; Lavrent'ev, I. P.; Atovmyan, L. O.; Khidekel', M. L. (Koord. Khim. **12** [1986] 1262/9; Soviet J. Coord. Chem. **12** [1986] 729/36).
[12] Shestakov, A. F.; Makitova, D. D.; Atovmyan, L. O.; Krasochka, O. N. (Koord. Khim. **13** [1987] 949/52; Soviet J. Coord. Chem. **13** [1987] 538/42).
[13] Makitova, D. D.; Letuchii, Ya. A.; Krasochka, O. N.; Roshchupkina, O. S.; Lavrent'ev, I. P.; Atovmyan, L. O.; Khidekel', M. L. (Koord. Khim. **14** [1988] 394/404; C.A. **108** [1988] No. 230909).

5.2.7 Ternary Bromopalladates

PdAl$_2$Br$_8$. The equilibrium constant for PdBr$_2$ + Al$_2$Br$_6$ ⇌ PdAl$_2$Br$_8$ has been determined from 573 to 723 K; for the process ΔH°_{298} = 34.7 kJ/mol and ΔS_{298} = 35.1 J·deg^{-1}·mol^{-1} [1]. The species is presumably involved in the chemical transport of PdBr$_2$ by Al$_2$Br$_6$ [2]. Resonance Raman spectra of the compound have been measured [3].

References:

[1] Merker, H.-B.; Schäfer, H. (Z. Anorg. Allgem. Chem. **480** [1981] 76/80).
[2] Schäfer, H.; Nowitzki, J. (J. Less-Common Metals **61** [1978] 47/50).
[3] Papatheodorou, G. N. (NBS-SP-561-1 [1979] 647/86).

5.2.8 Aquo-bromo Complexes, [PdBr$_n$(H$_2$O)$_{4-n}$]$^{2-n}$

[PdBr$_3$(H$_2$O)]$^-$. For stability constant data see p. 181 and table on p. 182.

The electronic spectrum in 1 M HClO$_4$ has been recorded (210 to 600 nm, reproduced in paper) [1, 13] and that of CV[PdBr$_3$(H$_2$O)] (CV = crystal violet; see below) in benzene (420 to 620 nm, reproduced in paper) [2] and in dioxan (200 to 500 nm, reproduced in paper) [4]. The electronic spectrum has also been calculated (200 to 500 nm, reproduced in paper) [3].

Rate constants for the reaction [PdBr$_4$]$^{2-}$ → [PdBr$_3$(H$_2$O)]$^-$ + Br$^-$ have been measured at 15, 25 and 35°C, and rate constants and thermodynamic data for the anation of [PdBr$_3$(H$_2$O)]$^-$ to give [PdBr$_4$]$^{2-}$ were measured [5, 10]. There are also data on anation of trans-PdBr$_2$(H$_2$O)$_2$ to the species [11,12]. Rate constants for reaction of R$_2$ (R = Me, Et, Me$_2$CHCH$_2$) with [PdBr$_3$(H$_2$O)]$^-$ have been measured [6]. Kinetics of reaction of [PdBr$_4$]$^{2-}$ with 1,10-phenanthroline to give Pd phen Br$_2$ have been measured; [PdBr$_3$(H$_2$O)]$^-$ is involved in the reaction [7]. The species has been shown to be the most active one for the catalysis of reduction by CO of 1,4-naphthoquinone, and the kinetics of the reaction established [9]. It is also active in the hydration of acetylene [8].

For solid CV[PdBr$_3$(H$_2$O)] see below.

PdBr$_2$(H$_2$O)$_2$. For stability constant data see p. 181 and table on p. 182.

Electronic spectra (210 to 600 nm, reproduced in paper) have been measured in 1 M HClO$_4$ [1, 13]; assignments of electronic bands proposed [4] and the electronic spectrum calculated (200 to 600 nm, reproduced in paper) [3].

The species is active in the hydration of acetylene; the kinetics and mechanism of the reaction were studied [8].

cis- and trans-PdBr$_2$(H$_2$O)$_2$. Rate constants for the reaction

$$[PdBr(H_2O)_3]^+ + Br^- \rightarrow cis\text{-}PdBr_2(H_2O)_2,$$
$$\text{for } cis\text{-}PdBr_2(H_2O)_2 + Br^- \rightarrow [PdBr_3(H_2O)]^-$$
$$\text{and for } trans\text{-}PdBr_2(H_2O)_2 + Br^- \rightarrow [PdBr_3(H_2O)]^-$$

have been studied by stopped-flow methods, as well as the equilibria between cis- and trans-isomers of PdBr$_2$(H$_2$O)$_2$. At equilibrium the cis species is dominant, the equilibrium constant at 25°C being 6.0 ± 0.8 at ionic strength 1M [11, 12].

[PdBr(H$_2$O)$_3$]$^+$. For stability constant data see p. 181 and table on p. 182.

The electronic spectrum has been measured in 1M $HClO_4$ (210 to 600 nm, reproduced in paper) [1,13]; assignments of bands made [4] and the spectrum (200 to 600 nm, reproduced in paper) calculated [3].

The kinetics of the aquation of $[PdBr(H_2O)_3]^+$ to $[Pd(H_2O)_4]^{2+}$ have been studied [5] and of its anation to cis-$PdBr_2(H_2O)_2$ [11,12]. The species is active in the hydration of acetylene; the kinetics and mechanism of the reaction have been studied [8].

Isolated Salts

CV[PdBr$_3$(H$_2$O)]. This is made from crystal violet (CV) and $PdCl_2$ dissolved in H_2SO_4 with excess Br$^-$. The salt is sparingly soluble in water but readily extracted into benzene, in which solvent the electronic absorption spectrum (420 to 620 nm, reproduced in paper) was measured [2].

HPdBr$_3$·dioxan·2H$_2$O. This may contain the $[PdBr_3(H_2O)]^-$ ion; it is made from dioxan and palladium carbonyl bromide [14].

References:

[1] Elding, L. I. (Inorg. Chim. Acta **6** [1974] 647/51).
[2] Khvatkova, Z. M.; Golovina, V. V. (Zh. Analit. Khim. **34** [1979] 2035/9; J. Anal. Chem. [USSR] **34** [1977] 1577/80).
[3] Companys, L. V.; Morer, X. T.; Carbonell, A.T. (Afinidad **31** [1974] 843/51).
[4] Kushnikov, Yu. A.; Vozdvizhenskii, V. F.; Zinevich, L. N.; Fasman, A. B. (Zh. Neorgan. Khim. **11** [1966] 2193/6; Russ. J. Inorg. Chem. **11** [1966] 1175/7).
[5] Elding, L. I. (Inorg. Chim. Acta **6** [1972] 683/8).
[6] Litvinenko, S. L.; Zamashchikov, V.V.; Rudakov, E. S.; Strel'tsov, V. I. (Ukr. Khim. Zh. **47** [1981] 186/8; Soviet Progr. Chem. **47** No. 2 [1981] 72/4).
[7] Rund, J. V. (Inorg. Chem. **13** [1974] 738/40).
[8] Sokol'skii, D. V.; Segizbaeva, S. S.; Dorfman, Y. A. (Zh. Org. Khim. **6** [1970] 893/7; J. Org. Chem. [USSR] **6** [1970] 899/902).
[9] Kutyukov, G. G.; Fasman, A. B.; Lyuts, A. E.; Kushnikov, Yu. A.; Vozdvizhenskii, V. F.; Golodov, V.A. (Zh. Fiz. Khim. **40** [1966] 1468/71; Russ. J. Phys. Chem. **40** [1966] 798/802).
[10] Vargaftik, M. N.; Igoshin, V. A.; Syrkin, Ya. K. (Izv. Akad. Nauk SSSR Ser. Khim. **1972** 1426/8; Bull. Acad. Sci. USSR Div. Chem. Sci. **1972** 1380/2).

[11] Elding, L. I. (Inorg. Chim. Acta **7** [1973] 581/8).
[12] Elding, L. I. (Inorg. Chim. Acta **15** [1975] L9/L11).
[13] Elding, L. I.; Olsson, L. F. (J. Phys. Chem. **82** [1978] 69/74).
[14] Golodov, V.A.; Fasman, A. B.; Vozdvizhenskii, V. F.; Kushnikov, Yu. A.; Roganov, V. V. (Zh. Neorgan. Khim. **11** [1968] 3314/6; Russ. J. Inorg. Chem. **11** [1968] 1705/7).

5.2.9 Chloro-bromo Complexes

Monomeric Species

[PdCl$_n$Br$_{4-n}$]$^{2-}$. Reaction of Cl$^-$ with $[PdBr_4]^{2-}$ gives the above species, for each of which resolved electronic spectra were reproduced in the paper (230 to 410 nm). Values of log K were as follows [1]:

$$\log K_1 \text{ for } \frac{[PdCl_3Br^{2-}][Cl^-]}{[PdCl_4^{2-}][Br^-]} = 1.55 \pm 0.05$$

$$\log K_2 \text{ for } \frac{[PdCl_2Br_2^{2-}][Cl^-]}{[PdCl_3Br^-][Br^-]} = 1.09 \pm 0.07$$

$$\log K_3 \text{ for } \frac{[PdClBr_3^{2-}][Cl^-]}{[PdCl_2Br_2^{2-}][Br^-]} = 0.95 \pm 0.07$$

$$\log K_4 \text{ for } \frac{[PdBr_4^{2-}][Cl^-]}{[PdClBr_3^{2-}][Br^-]} = 0.55 \pm 0.05$$

Later values using new data for M ionic strength gave $\log K_1 = 1.40 \pm 0.02$, $\log K_2 = 1.06 \pm 0.02$, $\log K_3 = 0.72 \pm 0.02$, $\log K_4 = 0.27 \pm 0.03$ [2, 3]. The relative distributions of species as a function of the ligand ratio R was also given, see **Fig. 44**.

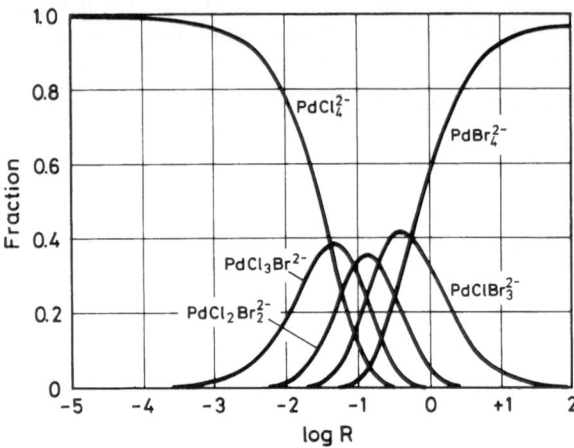

Fig. 44. Relative distribution of species as a function of ligand ratio R [2, 3].

Reaction of $p\text{-}CH_3C_6H_4SO_3Na$ and $Na_2[PdCl_2Br_2]$ with $HgCl_2$ at 100°C gives p, p'-bitolyl [4]. Both $[PdCl_3Br]^{2-}$ and $[PdCl_2Br_2]^{2-}$ have been shown to be active as catalysts for the oxidation of CO [5].

$(Bu_4^nN)_2[PdCl_2Br_2]$. The charge-transfer electronic spectrum of this compound, made from $PdCl_2$ and molten $(Bu_4^nN)Br$, has been recorded [6].

References:

[1] Srivastava, S. C.; Newman, L. (Inorg. Chem. 5 [1966] 1506/10).
[2] Feldberg, S.; Klotz, P.; Newman, L. (Inorg. Chem. **11** [1972] 2860/5).
[3] Klotz, P.; Feldberg, S.; Newman, L. (Inorg. Chem. **12** [1973] 164/8).
[4] Garves. K. (J. Org. Chem. **35** [1970] 3273/5).
[5] Zakumbaeva, G. D.; Noskova, N. F.; Konaev, E. N.; Sokol'skii, D. V. (Dokl. Akad. Nauk SSSR **156** [1964] 1386/8; Dokl. Chem. Proc. Acad. Sci. USSR **154/159** [1964] 644/6).
[6] Islam, N.; Aroof, A. M.; Islam, M. R. (Indian J. Chem. A **20** [1981] 1089/92).

Binuclear Species

$(Et_4N)_2[Pd_2Cl_4Br_2]$. The electronic spectrum of this has been briefly mentioned, and the suggestion made that the bromo ligands occupy the bridging positions.

Reference:

Lobaneva, O. A.; Kononova, M. A.; Davydova, M. K. (Zh. Neorgan. Khim. **20** [1975] 2648/50; Russ. J. Inorg. Chem. **20** [1975] 1577/8).

For **bromo-iodo complexes** $[PdBr_nI_{4-n}]^{2-}$, see p. 216.

6 Palladium and Iodine

6.1 Palladium Iodides

General Remarks. The di-iodide is well established and, on account of its relative insolubility, has long been used in the gravimetric determination of the metal. There seems to be no evidence for iodides of higher oxidation state, although the $[PdI_6]^{2-}$ anion is known as the caesium salt.

6.1.1 Palladium Di-iodide PdI$_2$ (see "Palladium" 1942, p. 288)

This exists in three modifications, the α, β and γ. The latter is probably the form normally obtained by addition of I$^-$ to palladium(II) solutions.

Preparation. The most explicit procedures for the three modifications involve the prior formation of the black γ-form made as crystals by controlled precipitation at 140°C with 5% aqueous HI in sealed tubes from an unspecified palladium(II) salt (presumably PdCl$_2$). Heating this form with dilute HI in sealed tubes above 600°C gives black needles of the α-modification, a metastable form of which can also be obtained from the β-form by reversible transformation at 20°C. The β-modification is made from γ-PdI$_2$ in sealed tubes with dilute HI at temperatures below 600°C, and forms deep red needles or plates [1].

Reaction of aqueous PdCl$_2$ solution in the presence of a little HCl with aqueous KI gives the black PdI$_2$; an excess of KI must be avoided [2], and a similar method involves the use of "palladium nitrate" in HNO$_3$, with NaI [3].

Anhydrous PdI$_2$ can also be made from aqueous palladium(II) perchlorate and a stoichiometric amount of aqueous NaI; the precipitate is then refluxed with 0.1M aqueous NaI solution [12]. Another method is to heat Pd and Al$_2$I$_6$ together at 400°C [13].

It appears likely that the compound is anhydrous [3]. Colloidal PdI$_2$ can be made from Pd-containing electrolytes or from HCl solutions containing PdCl$_2$ with iodide and a gelatin foam, and this is said to constitute a method for separating Pd from other platinum-group metals. The kinetics of extraction of PdI$_2$ by this method were measured [4].

Physical Properties

The normal γ-form is black, insoluble in water, alcohol and ether but soluble in KI solution. It does not melt below 700°C nor develop a significant vapour pressure [2]. The solubility product is $(7 \pm 3) \times 10^{-32}$ M^3 at 25°C [17].

Crystal Structures. The α-form is orthorhombic, space group Pnmn-D_{2h}^{12}, Z = 2; a = 6.69, b = 8.00, c = 3.80 Å, experimental density 6.06, calculated 5.88 g/cm^3. In the α-form tetragonal PdI$_4$ units form side-by-side chains; Pd–I = 2.60 Å, Pd–Pd = 3.80 Å, I–I at 3.80 and 3.55 Å; the structure undergoes extensive modification at higher temperatures [1]. The β-form is monoclinic, space group P2$_1$/c-C_{2h}^5, Z = 4, a = 6.69(1), b = 8.60(1), c = 6.87(1) Å, β = 103.5(3)°, calculated density 6.229 g/cm^3 at 300°C [1, 5]. There are layers of flat Pd$_2$I$_6$ units (see **Fig. 45**, p. 206) which are cross-linked via two I atoms to give octahedrally coordinated Pd with two long Pd–I bonds (3.29 and 3.49 Å) in addition to those shown in **Fig. 46**, p. 206.

Though no structural X-ray data have been given for the γ-form its d-spacing patterns were recorded [5].

Thermodynamic Data and Magnetic Properties. For a review of early thermodynamic values see [6]. The best data derive from dissociation pressure measurements made by a flow method between 500 and 544°C and by static methods between 350 and 640°C. These give the following values: $\Delta H^\circ_{298} = -58 \pm 8$ kJ/mol from solid Pd and solid I_2; $\Delta H^\circ_{298} = -121 \pm 8$ kJ/mol from solid Pd and gaseous I_2; $\Delta S^\circ_{298} = -138 \pm 9$ J·deg^{-1}·mol^{-1}; $S^\circ = +159 \pm 8$ J·deg^{-1}·mol^{-1} [2].

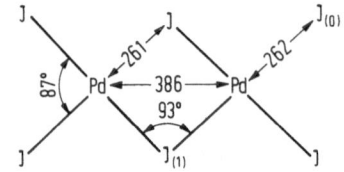

Fig. 45. The Pd_2I_6 layer in β-PdI_2 [5].

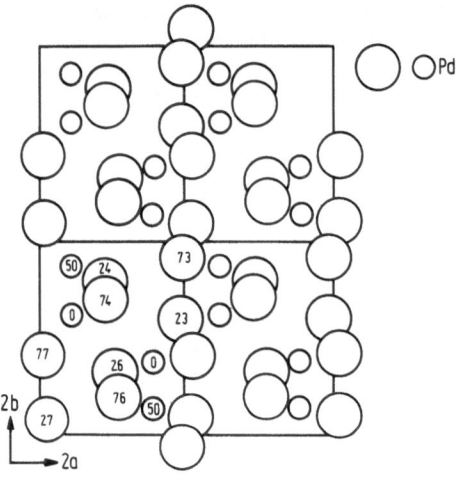

Fig. 46. The β-PdI_2 structure (001 layer, figures indicate z-parameters) [5].

An early estimate of ΔH°_{298} was -50 kJ/mol [7]. The E° value for PdI_2 has been calculated as -0.351 V for the aqueous solution at 25°C, as 0.325 ± 0.17 V for the solid at 25°C up to -0.082 V at 827°C [8]. The lattice energy has been calculated as 2637 kJ/mol [9].

The compound (β-form) is diamagnetic; corrected molar susceptibility from 90 to 300 K: $\chi_m = 69.1 \times 10^{-6}$ cgsu [5]. Its diamagnetism has been briefly discussed [11].

Spectroscopic Data. The electronic absorption spectrum by diffuse reflectance of the β-form gave a band at 645 nm, assigned to a d-d transition [5].

The ESCA spectrum gave binding energies of 341.9 ± 0.1 eV for $3d_{3/2}$ and 336.6 ± 0.1 eV for $3d_{5/2}$ [10].

Effective Charge. The effective charge on Pd atoms in PdI_2 was calculated using the self-consistent electronegativity method [14]. Attempts have been made to evaluate "softness" parameters for PdI_2 as a Lewis base [15, 16].

References:

[1] Thiele, G.; Brodersen, K.; Kreuse, E.; Holle, B. (Naturwissenschaften **54** [1967] 615).

[2] Shchukarev, S. A.; Tolmacheva, T. A.; Pazukhina, T. L. (Zh. Neorgan. Khim. **9** [1964] 2507/10; Russ. J. Inorg. Chem. **9** [1964] 1354/6).

[3] Horner, D. E.; Mailen, J. C.; Bigelow, H. R. (J. Inorg. Nucl. Chem. **39** [1977] 1645/9).

[4] Borisikhina, V. I.; Anan'ina, N. P.; Mokrushin, S. G. (Zh. Prikl. Khim. **48** [1975] 2354/7; J. Appl. Chem. [USSR] **48** [1975] 2439/42).

[5] Thiele, G.; Brodersen, K.; Kruse, E.; Holle, B. (Chem. Ber. **101** [1968] 2771/5).

[6] Goldberg, R. N.; Hepler, L. G. (Chem. Rev. **68** [1968] 229/52, 241).

[7] Karapet'yants, M. K. (Zh. Fiz. Khim. **30** [1956] 593/60).

[8] Hamer, W. J.; Malmberg, M. S.; Rubin, B. (J. Electrochem. Soc. **112** [1965] 750/5).

[9] Karapet'yants, M. K. (Zh. Fiz. Khim. **28** [1954] 1136/52).

[10] Kumar, G.; Blackburn, J. R.; Albridge, R. G.; Moddeman, W. E.; Jones, M. M. (Inorg. Chem. **11** [1972] 296/300).

[11] Morris, D. F. C.; Ahrens, L. H. (J. Inorg. Nucl. Chem. **3** [1956] 263/9).

[12] Elding, L. I.; Olsson, L. F. (Inorg. Chem. **16** [1977] 2789/94).

[13] Schäfer, H.; Nowitzki, J. (J. Less-Common Metals **61** [1978] 47/50).

[14] Gagarin, S. G. (Zh. Strukt. Khim. **19** [1978] 723/4; J. Struct. Chem. [USSR] **19** [1978] 620/1).

[15] Singh, P. P. (J. Sci. Ind. Res. [India] **42** [1983] 140/8).

[16] Singh, P. P.; Srivastava, S. M.; Srivastava, A. K. (J. Inorg. Nucl. Chem. **42** [1980] 521/32).

[17] Elding, L. I.; Olsson, L. F. (Inorg. Chim. Acta **117** [1986] 9/16).

Chemical Reactions

Thermal gravimetric analysis (TGA) has been carried out on PdI_2: from 20 to 84°C water is lost, then decomposition to Pd and I_2 occurs from 365 to 651°C [8].

The solubility of PdI_2 has been studied because of the importance of the compound as a gravimetric reagent for palladium. The solubility products (K_{sp}) of a number of "insoluble" halides have been correlated with electronegativities, including PdI_2 [4] and the values gives as 3.73×10^{-16} at 25°C from potentiometric data [5], and in 1M $NaClO_4$, as 1.85×10^{-17} at 24°C [6]. The solubilities of PdI_2 in aqueous nitrate [6, 7] and perchlorate solutions at 25°C were determined as a function of ionic strength; it is 3.6×10^{-8} M soluble in 2M HNO_3 at 29°C and 5.9×10^{-8} M soluble at 46°C [6]. The solubility in aqueous iodide has also been studied [14]; in water it is estimated to have a solubility lower than 2×10^{-3} g/l [15].

The equilibrium constant PdI_2(liquid) + Al_2I_6(gas) $\rightleftharpoons$ $PdAl_2I_8$(gas) has been studied using a flow method [9]. Heterogeneous isotope-exchange reactions between $CH_3^{131}I$ and PdI_2 have been studied [10]. With I^-, $[PdI_4]^{2-}$ [1, 14], $[PdI_3(H_2O)]^-$ [14] and $[Pd_2I_6]^{2-}$ [2, 14] are formed.

Reaction of a suspension of PdI_2 with 2-hydroxyethylisocyanide gives oxazolidin-2-ylidene palladium complexes [11]; with pyridine or substituted pyridine (L), PdL_2I_2 complexes are formed [3].

Applications of PdI_2

The use of PdI_2 as a histochemical stain has been demonstrated [12] though it appears that in vivo PdI_2 is remarkably unstable [13]. The use of self-indicating PdI_2 in palladium analysis has been studied [15]. The iodide has been used catalytically in the presence of O_2 with CO for the carbonylation of acetylenes [16].

References:

[1] Yamada, S.; Tsuchida, R. (Bull. Chem. Soc. Japan **26** [1953] 489/93).
[2] Harris, C. M.; Livingstone, S. E.; Stephenson, N. C. (J. Chem. Soc. **1958** 3697/702).
[3] Lau, P.-S.; House, J. E. (J. Inorg. Nucl. Chem. **36** [1974] 207/8).
[4] Clifford, A. F. (J. Am. Chem. Soc. **79** [1957] 5404/7).
[5] Ackermann, G. (Z. Chem. [Leipzig] **22** [1982] 341).
[6] Horner, D. E.; Mailen, J. C.; Bigelow, H. R. (J. Inorg. Nucl. Chem. **39** [1977] 1645/9).
[7] Mailen, J. C.; Horner, D. E. (Nucl. Technol. **33** [1977] 260/3).
[8] Dupuis, T.; Duval, C. (Anal. Chim. Acta **4** [1959] 615/22).
[9] Merker, H.-B.; Schäfer, H. (Z. Anorg. Allgem. Chem. **480** [1981] 76/80).
[10] Galiba, I.; Latzkovits, L.; Gal, D. (Magy. Kem. Folyoirat **67** [1961] 323/4).

[11] Bartel, K.; Fehlhammer, W. P. (Angew. Chem. Intern. Ed. Engl. **13** [1974] 599/600).
[12] Knospe, C. (Acta Histochem. **72** [1983] 77/84).
[13] Anghileri, L. J. (Acta Isotopica **9** [1969] 347/56; C.A. **74** [1971] No. 1858).
[14] Elding, L. I.; Olsson, L. F. (Inorg. Chem. **16** [1977] 2789/94).
[15] Rhoda, R. N.; Atkinson, R. H. (Anal. Chem. **28** [1956] 535/6).
[16] Chiusoli, G. P.; Costa, M.; Pergreffi, P.; Reverberi, S.; Salerno, G. (Gazz. Chim. Ital. **115** [1985] 691/6).

6.1.2 Palladium Tetra-iodide PdI_4(?)

Although there is one reference to this in "Palladium" 1942, p. 289, no recent work seems to have been done on it, and its existence seems doubtful.

6.2 Iodopalladates (see "Palladium" 1942, pp. 289, 308, 317, 324, 333, 335)

Although several salts of $[PdI_4]^{2-}$ are reported in the early literature salts of the ion are curiously elusive, and in many cases the formation in the solid state at least of $[Pd_2I_6]^{2-}$ seems to be favoured. There is only one brief report of a salt of the $[PdI_6]^{2-}$ ion (see p. 212).

Early thermochemical data on iodopalladates have been summarised [1].

6.2.1 Tetra-iodopalladates

The normal method of preparation is reaction of $PdCl_2$ or other palladium(II) species with excess I^- followed by appropriate precipitation.

6.2.1.1 The Tetra-iodopalladate(II) ion $[PdI_4]^{2-}$

Spectra. Although it is stated that attempts to measure the Raman spectrum of $[PdI_4]^{2-}$ in aqueous solution failed owing to instability of the solution [2] there is a report of an antiresonance Raman spectrum of $[PdI_4]^{2-}$ in aqueous solution, whereby the ν_1 (A_{1g}) vibration was found at 141 cm^{-1}, the ν_4 (B_{2g}) mode at 126 cm^{-1} and the ν_2 (B_{1g}) deformation mode at 80 cm^{-1}. These could be re-assigned as ν_2 (B_{1g}) at 126 cm^{-1} and ν_4 (B_{2g}) at 80 cm^{-1} in line with the frequency numbering for other $[PdX_4]^{2-}$ species. Excitation profiles of these bands were given over the range 400 to 750 nm [3].

Electronic absorption spectra of $[PdI_4]^{2-}$ in aqueous solution have been given from 400 to 750 nm [3], 280 to 520 nm in 1M I^- [4]; 350 to 620 nm in 0.06 to 0.6 M I^- [5], 240 to 550 nm in 1M I^- [6]; in $CHCl_3$ (300 to 750 nm) [13], all with the spectra reproduced in the papers. Electronic spectra without reproduced data have also been given [7, 8].

Stability Constants. Stability of $[PdI_4]^{2-}$ in aqueous solutions has been studied spectro-photometrically and at ionic strength I: $\log K_1 = 5.50 \pm 0.08$, $\log K_2 = 4.32$, $\log K_3 = 3.15$, $\log K_4 = 1.97 \pm 0.05$ at 20°C [8]; $\log K_1$ and $\log K_4$ values have also been given as 4.95 and 2.92, respectively, with $\log K_2$ calculated as 4.27 and $\log K_3$ as 3.60 [9]. The value of $360 \pm 60\,M^{-1}$ for K_4 has also recently been determined spectrophotometrically for aqueous solutions [10], and potentiometric data also given [11, 12]. At 25°C and 1M ionic strength, $\log \beta_4 = (2 \pm 1) \times 10^{28}\,M^{-4}$ [14].

Redox Data. Redox potentials of $[PdI_4]^{2-}$ for various concentrations and pH have been determined for 10 to 60°C; for 0.01M $[PdI_4]^{2-}$, E_0 is 0.18 V at 10°C and 0.212 V at 60°C. From redox data the trans influence of ligands in square-planar palladium(II) complexes has been shown to lie in the order $CN^- > $ thiourea $ > CO > NH_3 \approx SCN^- > I^- > NO_2^- > Br^- > Cl^-$ [12].

References:

[1] Goldberg, R. N.; Hepler, L. G. (Chem. Rev. **68** [1968] 229/52, 241).
[2] Goggin, P. L.; Mink, J. (J. Chem. Soc. Dalton Trans. **1974** 1479/83).
[3] Stein, P.; Miskowski, V.; Woodruff, W. H.; Griffin, J. P.; Werner, K. G.; Gaber, B. P.; Spiro, T. G. (J. Chem. Phys. **64** [1976] 2159/67, 2163).
[4] Srivastava, S. C.; Newman, L. (Inorg. Chem. **6** [1967] 762/5).
[5] Morrow, J. J.; Markham, J. J. (Anal. Chem. **36** [1964] 1159).
[6] Srivastava, S. C.; Newman, L. (Inorg. Chem. **11** [1972] 2855/9).
[7] Misovic, J. D.; Gligorijevic, V. Z. (Glasnik Hem. Drustva Beograd **39** [1975] 599/606; C.A. **85** [1976] No. 13336).
[8] Victori, L.; Tomas, X.; Duarte, O. (Afinidad **34** [1977] 107/10).
[9] Shchukarev, S. A.; Lobaneva, O. A.; Kononova, M. A. (Vestn. Leningr. Univ. Fiz. Khim. **20** [1965] 149/50; C.A. **62** [1965] 13928).
[10] Elding, L. I.; Olsson, L. F. (Inorg. Chem. **16** [1977] 2789/94).

[11] Grinberg, A. A.; Kiseleva, N. V.; Gel'fman, M. I. (Dokl. Akad. Nauk SSSR **153** [1963] 1327/9; Dokl. Chem. Proc. Acad. Sci. USSR **148/153** [1963] 1025/7).
[12] Fasman, A. B.; Kutyukov, G. G.; Sokol'skii, D. V. (Zh. Neorgan. Khim. **10** [1965] 1338/44; Russ. J. Inorg. Chem. **10** [1965] 727/30).
[13] Dembinski, B. (Chem. Anal. [Warsaw] **28** [1983] 161/5).
[14] Elding, L. I.; Olsson, L. F. (Inorg. Chim. Acta **117** [1986] 9/16).

Chemical Reactions

Since it is not clear whether $[PdI_4]^{2-}$ is the only species present in solution (see p. 208) reactions are assumed to proceed from $[PdI_4]^{2-}$, normally generated from Pd^{II} salts and excess iodide. The kinetics and mechanism of the $2[PdI_4]^{2-} \rightleftharpoons [Pd_2I_6]^{2-} + 2I^-$ reaction have been studied; a dimeric transition state is probably involved with double iodo bridges, and three reaction pathways linking $[PdI_4]^{2-}$ and $[PdI_3(H_2O)]^-$ [1], see p. 215. The kinetics and mechanism of reaction of the assumed $[PdI_4]^{2-}$ with 1,10-phenanthroline have been measured; $[PtI_3(H_2O)]^-$ is probably involved also. The product is $Pd(phen)I_2$ [2]. Kinetics have also been investigated of the reaction of $[PdI_4]^{2-}$ (misprinted as $[PbI_4]^{2-}$ at one point in abstract) with ethylene and propylene [3], and of $[PdI_4]^{2-}$ with CO [13]. Kinetics of the $[Pd(H_2O)_4]^{2+}/I^-$ reaction have been measured [14].

Reaction of $[PdI_4]^{2-}$ (as constituted from $[PdCl_4]^{2-}$ and I^-) with 1,3-bis(methylselenoethyl-seleno) propane (L) gives Pd_2LI_4 [4]; and bis(3-dimethylarsinopropyl)but-3-enylarsine (L') gave PdI_2L' [5]; cis-1,2-bis(o-methylthiophenylthio)ethene (L) gives M_2LI_4 [6]; o-$(Ph_2Sb)_2C_6H_4$ (L') gives $PdL'I_2$ [7]. Reaction of "$K_2[PdI_4]$" with N,N-ethylenedimorpholine (EDM) gives the paramagnetic $Pd(EDM)I_2$ [8].

Extraction of $[PdI_4]^{2-}$. A number of studies of extraction of Pd as $[PdI_4]^{2-}$ have been reported, though in no case has it been established that $[PdI_4]^{2-}$ is the extracted species. Correlations between heats of hydration and extraction constants of $[PdI_4]^{2-}$ and other Pd complexes have been made [9]. Extraction in the presence of dimedrol as $[PdI_4]^{2-}$ has been claimed [10]. Mixtures of alcohols and cyclohexanone have been used to extract $[PdI_4]^{2-}$ and extraction coefficients determined [11].

Applications. A Pd-iodo complex, presumably $[PdI_4]^{2-}$, has been used as a means of the photometric determination of iodide [12].

References:

[1] Elding, L. I.; Olsson, L. F. (Inorg. Chem. **16** [1977] 2789/94).
[2] Rund, J. V. (Inorg. Chem. **13** [1974] 738/40).
[3] Osipov, A. M.; Likholobov, V. A.; Bremeeva, N. M. (Izv. Vysshikh Uchebn. Zavedenii Khim. Khim. Tekhnol. **15** [1972] 1485/8; C.A. **78** [1973] No. 83590).
[4] Levason, W.; McAuliffe, C. A.; Murray, S. G. (J. Chem. Soc. Dalton Trans. **1976** 269/71).
[5] Ashmawy, F. M.; Benn, F. R.; McAuliffe, C. A.; Watson, D. G.; Hill, W. E.; Perry, W. D. (Inorg. Chim. Acta **97** [1984] 25/9).
[6] McAuliffe, C. A.; Murray, S. G. (Inorg. Nucl. Chem. Letters **12** [1976] 897/8).
[7] Levason, W.; McAuliffe, C. A.; Murray, S. G. (Inorg. Nucl. Chem. Letters **12** [1976] 849/53).
[8] Lott, A. L.; Rasmussen, P. G. (J. Am. Chem. Soc. **91** [1969] 6502/3).
[9] Ivanova, S. N.; Gindin, L. M.; Chernyaeva, A. P.; Bezzubenko, A. A.; Mogilevkina, M. F.; Marohkina, L. Ya. (Izv. Sibirsk. Otd. Akad. Nauk SSSR Ser. Khim. Nauk **1974** No. 2, pp. 3/7; C.A. **81** [1974] No. 17347).
[10] Shesterova, I. P.; Talipov, S. T.; Ismailova, V. K. (Izv. Vysshikh Uchebn. Zavedenii Khim. Khim. Tekhnol. **18** [1975] 748/50; C.A. **83** [1975] No. 172124).

[11] Golub, A. M.; Pomerants, G. V. (Ukr. Khim. Zh. **31** [1965] 104/12; C.A. **62** [1965] 13822).
[12] Novak, J.; Slama, I. (Collection Czech. Chem. Commun. **37** [1972] 2907/10).
[13] Fasman, A. B.; Kutyukov, G. G.; Sokol'skii, D. V. (Zh. Neorgan. Khim. **10** [1965] 1338/44; Russ. J. Inorg. Chem. **10** [1965] 727/30).
[14] Elding, L. I.; Olsson, L. F. (Inorg. Chim. Acta **117** [1986] 9/16).

6.2.1.2 Tetra-iodopalladates of Nonmetals

No free acid has been isolated (see "Palladium" 1942, p. 289). There are, however, a few salts with organic cations though most seem to be incompletely characterised.

$R_2[PdI_4]$. The imipraminium salt $(C_{19}H_{24}N_2H)_2[PdI_4]$ (imipraminium = 5(3-dimethyl-o-amino-propyl)-10,11-dihydro-5H-dibenzylazepine) salt has been reported by reaction of the base, $PdCl_2$ and iodide ion, but it is not clear whether it was isolated. The electronic spectrum was measured in $CHCl_3$ (300 to 750 nm, reproduced in paper) [1]. A number of species $(RH)_2[PdI_4]$ (R = pyrazolone derivatives such as antipyrine, diantipyrylmethane, diantipyrylmethylmethane, diantipyrylpropylmethane) were made in 0.5 M KI and in 0.5 to 2 M H_2SO_4; such species can be used for extraction of Pd [2], as is also the case for the trioctylammonium salt

$[(C_8H_{17})_3NH]_2[PdI_4]$, which can be extracted in dichloroethane or toluene. The extraction constant and ligand hydration energies were correlated for $[PdX_4]^{2-}$ (X = Cl, Br, I) [3]. Salts of polyethyleneglycoloxonium cations have also been used in extraction studies of $[PdI_4]^{2-}$ [4].

References:

[1] Dembinski, B. (Chem. Anal. [Warsaw] **28** [1983] 161/5).
[2] Dzhishkariani, G. I.; Akinov, V. K.; Busev, A. I. (Sobshch. Akad. Nauk Gruz.SSR **73** [1974] 337/40; C.A. **81** [1974] No. 32925).
[3] Ivanova, S. N.; Gindin, L. M.; Chernyaeva, A. P.; Chernobrov, A. S.; Shaidurova, N. N. (Izv. Sibirsk. Otd. Akad. Nauk SSSR Ser. Khim. Nauk **1978** No. 6, pp. 68/74; C.A. **90** [1979] No. 110710).
[4] Ziegler, M.; Winkler, H. (Z. Anal. Chem. **195** [1963] 241/3).

6.2.1.3 Alkali-Metal Tetra-iodopalladates

Although early references ("Palladium" 1942, p. 317) refer to a potassium salt of $[PdI_4]^{2-}$ the evidence for their existence in the solid state has not been satisfactorily demonstrated; a similar situation exists for $[PtI_4]^{2-}$ [1]. The existence in solution of a $[PdI_4]^{2-}-[Pd_2I_6]^{2-}$ equilibrium has been demonstrated [2] (see p. 212) and it seems to be easier to isolate salts of the dimer than of the monomer.

$K_2[PdI_4]$. Although solid $K_2[PdI_4]$ is said to be formed as black dichroic rhombic tablets from $K_2[PdCl_4]$ in solution and hot KI solution, no analytical data were presented [3]. A later paper claims that $K_2[PdI_4]$ has not been isolated [4], though it is also said to be formed in solution when $PdCl_2$ reacts with excess KI [5]. It is said that a sixfold excess of KI is needed to convert PdI_2 to $K_2[PdI_4]$ in solution [6].

The crystals claimed as $K_2[PdI_4]$ are dichroic, being brown-black or green-black by polarised light; the dichroic electronic absorption spectra were measured for the solid from 200 to 500 nm (reproduced in paper) [3].

The molar refraction of $K_2[PdI_4]$ has been reported as 78.51 [7].

$[K(DB18C6)]_2[PdI_4]$. This is made from dibenzo-18-crown-6 (DB18C6) and KI with $PdCl_2$. The electronic spectrum was measured (300 to 700 nm, reproduced in paper) [9].

$Cs_2[PdI_4]$. This is said to be a thermal decomposition product of $Cs_2[PdI_6]$; it will decompose on further heating to CsI, Pd and I_2 [8].

References:

[1] Corain, B.; Poë, A. J. (J. Chem. Soc. A **1967** 1318/22).
[2] Elding, L. L.; Olsson, L. E. (Inorg. Chem. **16** [1977] 2789/94).
[3] Yamada, S.; Tsuchida, R. (Bull. Chem. Soc. Japan **26** [1953] 489/93).
[4] Rund, J. V. (Inorg. Chem. **13** [1974] 738/40).
[5] Lott, A. L.; Rasmussen, P. G. (J. Am. Chem. Soc. **91** [1969] 6502/3).
[6] Rhoda, B. N.; Atkinson, R. H. (Anal. Chem. **28** [1956] 535/6).
[7] Shamsiev, A. S.; Obel'chenko, P. F. (Tr. Sredneaziat Gos. Univ. [4] No. 33 [1952] 57/63; C.A. **1956** 3830).
[8] Sinram, D.; Brendel, C.; Krebs, B. (Inorg. Chim. Acta **64** [1982] L131/2).
[9] Poladyan, V. E.; Burtnenko, L. M.; Avlasovich, L. M.; Andrianov, A. M. (Zh. Neorgan. Khim. **32** [1987] 737/40; Russ. J. Inorg. Chem. **32** [1987] 413/5).

212

6.2.2 Higher Iodopalladates

A search for $[PdI_5]^{3-}$ or $[PdI_6]^{4-}$ proved fruitless [1], but $[PdI_6]^{2-}$ has been briefly reported as the caesium salt [2].

$Cs_2[PdI_6]$. Despite an earlier report [1] that $M_2^I[PdI_6]$ (M = Na, K, NH_4, Rb, Cs) did not exist, there is a brief report of the preparation of $Cs_2[PdI_6]$, made from $[PdCl_6]^{2-}$ and a large excess of CsI. No analytical data were given. The compound is dark red and X-ray powder studies show it to be of the $K_2[PtCl_6]$ type with a unit cell constant of 11.311(3) Å. Thermal decomposition data show two decomposition steps apparently via $Cs_2[PdI_4]$, the latter decomposing to Cs^I, Pd and I_2 [2]. The redox potential of the reaction $[PdI_6]^{2-} + 2e^- \rightarrow [PdI_4]^{2-} + 2I$ is given as $+0.482$ V [3].

Reaction of $[PdCl_6]^{2-}$ with KI and dithiomalonamide (HL) gives $[Pd^{II}(HL)_2][Pd^{IV}(HL)_2I_2]I_4$ and $[Pd^{II}(HL)_2][Pd^{IV}I_6]$, presumably via $[PdI_6]^{2-}$ as an intermediate [4]. There is some spectrophotometric evidence too for the existence of $[PdI_6]^{2-}$ [5].

References:

[1] Ketelaar, J. A. A.; van Walsem, J. F. (Rec. Trav. Chim. **57** [1938] 963/6).
[2] Sinram, D.; Brendel, C.; Krebs, B. (Inorg. Chim. Acta **64** [1982] L131/2).
[3] Grinberg, A. A.; Shamsiev, A. S. (Zh. Obshch. Khim. **12** [1942] 55/72, 62).
[4] Bret, J.-M.; Castan, P.; Laurent, J.-P. (J. Chem. Soc. Dalton Trans. **1984** 1975/80).
[5] Misovic, J. D.; Grigorijevic, V. Z. (Glasnik Hem. Drustva Beograd. **39** [1975] 599/606).

6.2.3 Binuclear Iodopalladates

6.2.3.1 The $[Pd_2I_4]^{?-}$ Ion

The lithium and caesium salts have been isolated of this anion; their diamagnetism suggests the dimeric structure [2].

$Li[PdI_2] \cdot 2H_2O$. $Cs[PdI_2] \cdot 2H_2O$. These were prepared by reaction of $[Pd(CO)Cl]_n$ in HCl with LiI and Cs^I under a CO atmosphere. They are soluble in alcohols and acetone in the presence of I^-, and the Li salt decomposes at 120°C. They are active catalysts for acetylene carbonylation and are probably best formulated as containing $[Pd_2I_4]^{2-}$ [1].

It has been estimated that E for $[PdI_4]^{2-}/[Pd_2I_4]^{2-} \ll 0.45$ V and E_0 for $[Pd_2I_4]^{2-}/[PdI_4]^{2-} \gg 0.15$ V [2].

References:

[1] Bruk, L. G.; Temkin, O. N.; Gonchanova, Z. V.; Utsenko, I. S.; Flid, V. R. (React. Kinet. Catal. Letters **9** [1978] 303/8).
[2] Temkin, O. N.; Bruk, L. G. (Usp. Khim. **52** [1983] 206/43; Russ. Chem. Rev. **52** [1983] 117/37, 129).

6.2.3.2 The $[Pd_2I_6]^{2-}$ Ion

The existence of this ion in aqueous solutions of PdI_2 in excess aqueous iodide has been demonstrated; in iodide concentrations between 0.1 and 0.02M the predominant species are $[PdI_4]^{2-}$ and $[Pd_2I_6]^{2-}$. There are likely to be three parallel reaction paths for formation of

$[Pd_2I_6]^{2-}$ from $[PdI_4]^{2-}$ and $[PdI_3(H_2O)]^-$: $PdI_2(\text{solid}) + I^- + H_2O \rightleftharpoons [PdI_3(H_2O)]^-$ and $[PdI_3(H_2O)]^- + I^- \rightleftharpoons [PdI_4]^{2-} + H_2O$ (fast); $[PdI_4]^{2-} + [PdI_3(H_2O)]^- \rightleftharpoons [Pd_2I_6]^{2-} + H_2O + I^-$ (slow); $2[PdI_3(H_2O)]^- \rightleftharpoons [Pd_2I_6]^{2-} + 2H_2O$ (slow); $PdI_2(\text{solid}) + 2I^- \rightleftharpoons [PdI_4]^{2-}$, $2PdI_2(\text{solid}) + 2I^- \rightleftharpoons [Pd_2I_6]^{2-}$. Thermodynamic and rate data were established. In saturated solutions the main species is $[PdI_4]^{2-}$ with only some 6% of the dimer [1].

The structure of the anion is likely to be similar to that found by X-ray structure determination for $(C_5H_5NOH)_2[Pd_2Br_6]$, see Fig. 41, p. 201.

The electronic absorption spectrum of $[Pd_2I_6]^{2-}$ in aqueous solution was reported (250 to 450 nm, reproduced in paper) [1], and in acetone (of $[(C_2H_5)_4N]_2[Pd_2I_6]$, 300 to 800 nm, reproduced in paper) [2]. The electronic structure and charge-transfer spectra (200 to 600 nm, calculated electronic spectrum reproduced in paper) were given [3], and the spectrum in acetone of $[(C_2H_5)_4N]_2[Pd_2I_6]$ from 350 to 840 nm was given (no spectrum reproduced in paper) and assigned [4].

For Raman and infrared data on $[Pd_2I_6]^{2-}$ in solution see table below.

6.2.3.3 Salts of $[Pd_2I_6]^{2-}$

$[(C_2H_5)_4N]_2[Pd_2I_6]$. This black crystalline material is made from PdI_2 in boiling ethanol with $[(C_2H_5)_4N]I$ and excess NaI in the same solvent [2, 5]. It has a conductance in $C_6H_5NO_2$ of $60.6\ cm^2 \cdot \Omega^{-1} \cdot mol^{-1}$ at 25°C [5]. The reflectance electronic absorption spectrum has been measured (200 to 800 nm, reproduced in paper) [2]; for Raman and infrared data see table below. Force constant calculations for the anion were made on the basis of far-infrared data for the solid [6]. The salt is slowly reduced to Pd metal by alcohol [4].

$(Bu_4^nN)_2[Pd_2I_6]$. This is made by reaction of $[PdCl_4]^{2-}$ in water with excess KI followed by slow addition of $(Bu_4^nN)I$, and the product recrystallised from $CH_2Cl_2-(C_2H_5)_2O$ [7].

Raman and infrared spectra were measured for the solid and solution (table below).

Force-constant calculations were made for the in-plane vibrations of $[Pd_2I_6]^{2-}$ [8] using the data of [7].

Raman and infrared data for $R_2[Pd_2I_6]^{2-}$ salts:

	R = (C₂H₅)₄N [6]	R = Bu₄ⁿN [7]		R = (C₂H₅)₄N [6]	R = Bu₄ⁿN [7]
ν_1 (A_g) ν_{MI}^s	—	219a)	ν_9 (B_{1u}) δ_{MI}	85*)	— 91*)
ν_2 (A_g) ν_{MI}	—	143a)	ν_{12} (B_{2u}) ν_{MI}	222*)	218 214*)
ν_3 (A_g) δ_{MI}	—	—	ν_{13} (B_{2u}) ν_{MI}	144*)	— 133*)
ν_4 (A_g) δ_{MI}	—	—	ν_{16} (B_{3u}) ν_{MI}	220*)	218 220*)
ν_6 (B_{1g}) ν_{MI}	—	—	ν_{17} (B_{3u}) ν_{MI}	132*)	140 143*)
ν_7 (B_{1g}) ν_{MI}	—	130a)	ν_{18} (B_{3u}) δ_{MI}	96*)	— —

*) Solids, otherwise solution data in CH_2Cl_2. – a) Raman data, otherwise infrared; frequency numbering as in reference [7].

$(Me_3PhN)_2[Pd_2I_6]$. This is made from PdI_2 and excess NaI with $(Me_3PhN)I$ in hot 90% ethanol. The salt is black and crystalline, with a conductance in $C_6H_5NO_2$ of $60.6\ cm^2 \cdot \Omega^{-1} \cdot mol^{-1}$ at 25°C [5].

214

References:

[1] Elding, L. I.; Olsson, L. F. (Inorg. Chem. **16** [1977] 2789/94).

[2] Harris, C. M.; Livingstone, S. E.; Reece, I. H. (J. Chem. Soc. **1959** 1505/11).

[3] Baranovskii, V. I.; Davydova, M. K.; Panina, N. S.; Panin, A. I. (Koord. Khim. **2** [1976] 409/15; Soviet J. Coord. Chem. **2** [1976] 308/13).

[4] Lobaneva, O. A.; Kononova, M. A.; Davydova, M. K. (Dokl. Akad. Nauk SSSR **194** [1970] 133/5; Dokl. Phys. Chem. Proc. Acad. Sci. USSR **190/195** [1970] 669/70).

[5] Harris, C. M.; Livingstone, S. E.; Stephenson, N. C. (J. Chem. Soc. **1958** 3697/702).

[6] Adams, D. M.; Chandler, P. J.; Churchill, R. G. (J. Chem. Soc. A **1967** 1272/4).

[7] Goggin, P. L. (J. Chem. Soc. Dalton Trans. **1974** 1483/6).

[8] Goggin, P. L.; Mink, J. (Inorg. Chim. Acta **26** [1978] 119/24).

6.2.4 Ternary Iodopalladates

Pd_5AlI_2. This is made from Pd_2Al and I_2 at 600°C. It forms dark plates. The X-ray crystal structure shows it to belong to the I4/mmm-D_{4h}^{17} space group, Z = 2, a = 4.052, c = 19.559 Å, density (X-ray) 8.405 g/cm³. In the tetragonal structure ordered Pd/Al layers alternate with iodine layers, Pd–Pd = 2.670(1), Pd–Al = 2.865(1), Pd–I = 3.575(2) Å; see **Fig. 47** [1].

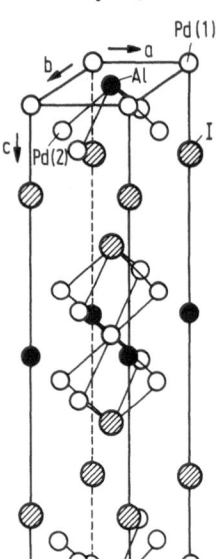

Fig. 47.   X-ray crystal structure of Pd_5AlI_2 [1].

The electrical conductivity is $2 \times 10^{-6} \, \Omega^{-1} \cdot cm^{-1}$ in one direction and $1.5 \times 10^{-1} \, \Omega^{-1} \cdot cm^{-1}$ in the perpendicular direction, the higher conductivity being parallel to the layers [1].

$Pd_5Al_2I_8$. Study of the reaction PdI_2(liquid) + Al_2I_6(gas) $\rightleftharpoons$ $PdAl_2I_8$(gas) from 583 to 773 K gave an equilibrium constant with a ΔH_{298}° value of 54 kJ/mol and $\Delta S_{298}^{\circ} = 42 \, J \cdot deg^{-1} \cdot mol^{-1}$ [2].

References:

[1] Merker, H.-B.; Schäfer, H.; Krebs, B. (Z. Anorg. Allgem. Chem. **462** [1980] 49/56).

[2] Merker, H.-B.; Schäfer, H. (Z. Anorg. Allgem. Chem. **480** [1981] 76/80).

6.2.5 Iodo-aquopalladates

Three species $[PdI_n(H_2O)_{3-n}]^{2-n}$ are known though only one seems to have been isolated.

$[PdI_3(H_2O)]^-$. This is formed as a minor species when PdI_2 is dissolved in excess aqueous KI [1]. Kinetic data show that the species is involved in the reaction of aqueous "$[PdI_4]^{2-}$" with 1,10-phenanthroline [2]. The log K_4 value for $[PdI_3(H_2O)]^-/[PdI_4]^{2-}$ is given as 1.97 ± 0.05 at 20°C [3]; see also p. 209. The electronic absorption spectrum of $(CV)[PdI_3(H_2O)]$ (CV = crystal violet cation) in benzene has been measured from 420 to 520 nm (reproduced in paper) [4]. The log β_3 value has been given as $(6 \pm 4) \times 10^{25} M^{-3}$ [5].

$(CV)[PdI_3(H_2O)]$ (CV = crystal violet cation). This salt, sparingly soluble in water but readily extracted by benzene, is formed from $PdCl_2$ dissolved in H_2SO_4 with excess I^- and crystal violet [4].

$[PdI_2(H_2O)_2]$. The log K_3 value for $PdI_2(H_2O)_2/[PdI_3(H_2O)]^-$ is 3.15 at 20°C (see also p. 209); log K_1 for $[Pd(H_2O)_4]^{2+}/[PdI(H_2O)_3]^-$ is 5.50 [3].

$[PdI(H_2O)_3]^-$. The log K_2 value for $[PdI(H_2O)_3]^+/[PdI_2(H_2O)]$ is 4.32 at 20°C (see also p. 209) [3]. Rate constants have been measured for the reaction of $[Pd(H_2O)_4]^{2+}$ with I^- to give $[PdI(H_2O)_3]^+$, and also for aquation of $[Pd^I(H_2O)_3]^+$ [5].

References:

[1] Elding, L. I.; Olsson, L. F. (Inorg. Chem. **16** [1977] 2789/94).
[2] Rund, J. V. (Inorg. Chem. **13** [1974] 738/40).
[3] Victori, L.; Tomas, X.; Duarte, O. (Afinidad **34** [1977] 107/10).
[4] Zhvatkova, Z. M.; Golovina, V. V. (Zh. Analit. Khim. **34** [1979] 2035/9; J. Anal. Chem. [USSR] **34** [1979] 1577/80).
[5] Elding, L. I.; Olsson, L. F. (Inorg. Chim. Acta **117** [1986] 9/16).

6.3 Mixed Halo-iodo Complexes

Chloro-iodido Complexes $[PdCl_nI_{4-n}]^{2-}$. Spectrophotometric studies on $[PdCl_4]^{2-}/I^-$ mixtures gave the following formation constants (log K): 3.95 ± 0.05 for $[PdCl_3I]^{2-}$ from $[PdCl_4]^{2-}$; 4.1 ± 0.2 for $[PdCl_2I_2]^{2-}$ from $[PdCl_3I]^{2-}$; 2.8 ± 0.03 for $[PdClI_3]^{2-}$ from $[PdCl_2I_2]^{2-}$; and 1.30 ± 0.05 for $[PdI_4]^{2-}$ from $[PdClI_3]^{2-}$. The relative distribution of the complexes is shown in **Fig. 48**.

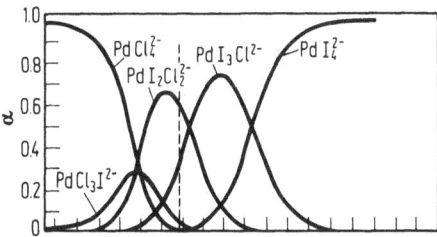

Fig. 48. Relative distribution of palladium chloro-iodo complexes (R = ratio $[I^-]/[Cl^-]$, α = mole fraction of named complex) [1].

In the chloro-iodo system the formation constants are somewhat higher than those calculated from a purely statistical model [1].

Bromo-iodo Complexes [PdBr$_n$I$_{4-n}$]$^{2-}$. Spectrophotometric studies of the [PdBr$_4$]$^{2-}$–I$^-$ reaction gave the following formation constants (log K): 2.75 ± 0.05 for [PdBr$_3$I]$^{2-}$ from [PdBr$_4$]$^{2-}$; 3.00 ± 0.15 for [PdBr$_2$I$_2$]$^{2-}$ from [PdBr$_3$I]$^{2-}$; 1.70 ± 0.15 for [PdBrI$_3$]$^{2-}$ from [PdBr$_2$I$_2$]$^{2-}$, and 0.80 ± 0.05 for [PdI$_4$]$^{2-}$ from [PdBrI$_3$]$^{2-}$. The absorption spectrum of each complex was resolved (280 to 530 nm, reproduced in paper) [2], and the relative distribution of the complexes are shown in **Fig. 49**.

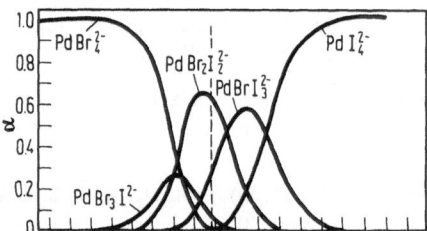

Fig. 49. Relative distribution of palladium bromo-iodo complexes (R = ratio [I$^-$]/[Br$^-$], α = mole fraction of named complex).

The constants for the bromo-iodo system are relatively close to those predicted on the basis of a purely statistical model [1].

References:

[1] Srivastava, S. C.; Newman, L. (Inorg. Chem. **11** [1972] 2855/9).
[2] Srivastava, S. C.; Newman, L. (Inorg. Chem. **6** [1967] 762/5).

7 Palladium and Sulphur

7.1 Phase Diagram

A section of the phase diagram for the Pd–S system is shown in **Fig. 50**.

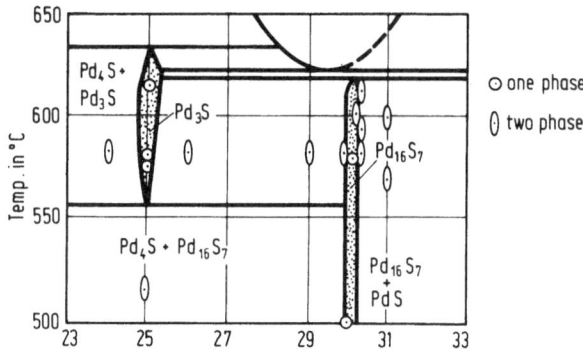

Fig. 50.   Pd–S diagram from 23
to 33 mol% S and 500 to 650°C.

Reference:

Matković, P.; El-Boragy, M.; Schubert, K. (J. Less-Common Metals **50** [1976] 165/76).

7.2 Binary Palladium Sulphides

General References:

Schoellhorn, R., Solvated Intercalation Compounds of Layered Chalcogenide and Oxide
Bronzes, Intercalation Chem. **1982** 315/60.

Tanaka, S., Physical Properties and Charge Density Waves in Transition Metal Dichalcogenides, Kotai Butsuri **14** [1979] 546/58.

Rao, S. G. V.; Shafer, M. W., Intercalation in Layered Transition Metal Dichalcogenides, Phys.
Chem. Mater. Layered Struct. **6** [1979] 99/199.

Rouxel, J., Alkali Metal Intercalation Compounds of Transition Metal Chalcogenides: TX$_2$, TX$_3$
and TX$_4$ Chalcogenides, Phys. Chem. Mater. Layered Struct. **6** [1979] 201/50.

Shatynski, S. R., The Thermochemistry of Transition Metal Sulphides, Oxid. Metals **11** [1977]
307/20.

Schubert, K., On the Binding in Phases of T^{10}–B^6 Mixtures, Acta Cryst. B **33** [1977] 2631/9.

Vandenberg-Voorhoeve, J. M., Structural and Magnetic Properties of Layered Chalcogenides
of the Transition Elements, Phys. Chem. Mater. Layered Struct. **4** [1976] 423/57.

Mills, K. C., Thermodynamic Data for Inorganic Sulphides, Selenides and Tellurides, Butterworths, London 1974.

Pd$_4$S (see "Palladium" 1942, p. 292). The compound is prepared by heating a stoichiometric
mixture of PdS (obtained from Pd and S at 500 to 800°C) and palladium at 600°C for several
days [2]. Crystal quality was improved by annealing at 450°C [3]. The crystals are tetragonal,
space group P$\bar{4}$2$_1$c-D$_{2d}^4$, a = 5.1147, c = 5.5903 Å, D$_{exp}$ = 10.27 g/cm^3 at 25°C, Z = 2 [2, 3]. Powder
photograph data [2] and a table of observed and calculated intensities [3] are reproduced in
papers. In Pd$_4$S each palladium is surrounded by 2 sulphur atoms (Pd–S = 2.34, 2.48 Å) and 10
palladium atoms (Pd–Pd = 2.78 to 3.10 Å), each sulphur is surrounded by 8 palladium atoms

[3]. The magnetic susceptibility has been measured over the temperature range -183 to $450°C$ [$\psi_g \times 10^6 = 0.20$ (150°C), 0.01 (300°C), 0.06 (450°C)]; at room temperature and below the susceptibility is field strength dependent [$\psi_g \times 10^6$ at $-183°C = 19.9$ (4015 Φ), 17.5 (4700 Φ) and 16.4 (5110 Φ)] [2]. The electrical resistivity ϱ is $3.20 \times 10^{-4}\,\Omega \cdot cm$ at 20°C with $d\varrho/dt = 1.05 \times 10^{-6}\,\Omega \cdot cm \cdot deg^{-1}$ [4]. The heat capacity of Pd_4S, determined by adiabatic calorimetry at 5 to 350 K is of the usual sigmate shape without transitions or anomalies. Values of the heat capacity Cp, entropy S° and Gibbs energy function $-(G°-H_0°)/T$ at 298.15 K are 27.48, 43.18 and 23.16 cal·mol^{-1}·K^{-1}, respectively [5]. Equilibrium constants K have been measured over the temperature range ~ 620 to 795 K for the reactions $\frac{4}{3}PdS + H_2 \rightleftharpoons \frac{1}{3}Pd_4S + H_2S$ and $Pd_4S + H_2 \rightleftharpoons 4Pd + H_2S$. Values for $\Delta G°$ are $12200 \pm 300 - (13.4 \pm 0.5)T$; ± 130 cal/mol at 613 to 798 K and $9600 \pm 400 - (7.7 \pm 0.6)T$; ± 90 cal/mol at 638 to 795 K, respectively. Equilibrium constants are tabulated and plots of log K vs reciprocal temperature are reproduced in the paper [6].

Pd₃S (see "Palladium" 1942, p. 292). The compound is prepared by melting, quenching and grinding a stoichiometric mixture of Pd and S, then annealing the powder at 600°C for "some days" in an evacuated silica tube. The Pd_3S phase is stable between 554 and 623°C. The X-ray crystal structure was determined using crystals obtained by the transport reaction using small amounts of bromine as transport agent. The crystals are orthorhombic, probable space group Ama2-C_{2v}^{16}, a = 6.088, b = 5.374, c = 7.453 Å, $D_{obs} = 9.58$, $D_{calc} = 9.57$ g/cm³, Z = 4. Tables of powder pattern data, positional parameters, temperature factors, structure factors and interatomic distances are reproduced in the paper. The crystal structure is illustrated in **Fig. 51**. The shortest Pd-S and Pd-Pd distances are 2.28 and 2.75 Å, respectively [7]. The compound is a metallic phase but is not superconducting [8].

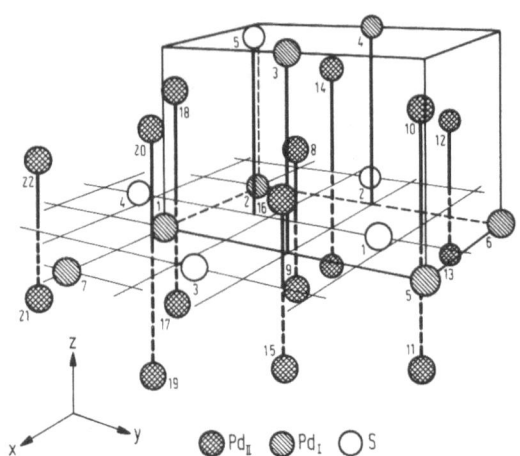

Fig. 51. A perspective view of the crystal structure of Pd₃S. The parallelepiped is a quarter of the unit cell: a/2, b, c/2 [7].

Pd₃.₁S to Pd₂.₈S. Samples with compositions in this range are thought to be impure Pd_3S contaminated with Pd_4S or $Pd_{2.2}S$ [7]. X-ray data for the supposed $Pd_{2.8}S$ phase, prepared by quenching samples of composition Pd_3S from 600°C, are tabulated in a paper [2].

"Pd₂.₂S" is found in samples of fused Pd/S mixtures either slowly cooled or quenched from 600°C [2]. It is now thought to be identical to $Pd_{32}S_{14}$ [9]. The powder photograph has been indexed as cubic with a = 8.9300 Å. X-ray data are tabulated in the paper. Magnetic susceptibility, χ_g, ranges from 0.27×10^{-6} (at $-183°C$) to 0.12×10^{-6} (at 450°C) [2]. Electrical resistivity: $\varrho = 10.00 \times 10^{-4}\,\Omega \cdot cm$ at 20°C, $d\varrho/dt = 1.80 \times 10^{-6}\,\Omega \cdot cm \cdot deg^{-1}$ [4]. Superconductivity occurs below 1.63 K [8].

Pd$_{32}$S$_{14}$ or Pd$_{16}$S$_7$(Pd$_{2.3}$S) is prepared by melting together the appropriate proportions of palladium and sulphur, grinding the product and annealing at 600°C for one week. The crystals are cubic, space group T$_d^3$–I$\bar{4}$3m, a = 8.954(2) Å. There are 32 Pd and 14 S atoms to the unit cell, Pd–Pd contacts occur at 2.79 and 2.94 Å, Pd–S bonds range from 2.27 to 2.49 Å [9]. A second X-ray diffraction study confirms the space group and reports a = 8.93 Å. The structure is reproduced in the paper [1].

Pd$_{2.5}$S to Pd$_2$S. Samples with compositions in this range are thought to be impure forms of Pd$_{2.2}$S contaminated with Pd$_4$S or PdS. Powder photograph data for "Pd$_{2.25}$S" annealed at 300°C are reproduced in the paper [2].

Pd$_2$S. A brown-black product of this stoichiometry is obtained on reduction of K$_2$PdCl$_4$, with excess NaHSO$_2$·CH$_2$O·2H$_2$O ('Rongalite'). It is amorphous to X-rays but is not a 1:1 mixture of PdS and metallic palladium [10].

References:

[1] Matković, P.; El-Boragy, M.; Schubert, K. (J. Less-Common Metals **50** [1976] 165/76).
[2] Grønvold, F.; Røst, E. (Acta Chem. Scand. **10** [1956] 1620/34).
[3] Grønvold, F.; Røst, E. (Acta Cryst. **15** [1962] 11/3).
[4] Fischmeister H. (Acta Chem. Scand. **13** [1959] 852/3).
[5] Grønvold, F.; Westrum, E. F.; Radebaugh, R. (J. Chem. Eng. Data **14** [1969] 205/7).
[6] Niwa, K.; Yokokawa, T.; Isoya, T. (Bull. Chem. Soc. Japan **35** [1962] 1543/5).
[7] Røst, E.; Vestersjø, E. (Acta Chem. Scand. **22** [1968] 819/26).
[8] Raub, Ch. J.; Zachariasen, W. H.; Geballe, T. H.; Matthias, B. T. (J. Phys. Chem. Solids **24** [1963] 1093/100).
[9] Rømming, C.; Røst, E. (Acta Chem. Scand. A **30** [1976] 425/8).
[10] Kukushkin, Yu. N.; Marshak, E. M. (Zh. Neorgan. Khim. **16** [1971] 266/7; Russ. J. Inorg. Chem. **16** [1971] 138/9).

PdS (see "Palladium" 1942, p. 292). Palladium(II) sulphide deposits from aqueous solutions as a dihydrate PdS·2H$_2$O which loses water to form PdS at 150°C. The dihydrate has been obtained by treatment of aqueous H$_2$PdCl$_4$/Na$_2$CO$_3$ (or NaOH) solutions with H$_2$S followed by acidification (pH = 0) [1]. The anhydrous form has been prepared by heating a mixture of a palladium salt, thiourea and concentrated H$_2$SO$_4$ at ~190°C [2] or by treatment of PdCl$_2$ with Li$_2$S in nonaqueous solution [3]. Other routes to PdS are thermal decomposition of cis-PdCl$_2$(SEt$_2$)$_2$ [4] or Pd(NCS)$_2$(PPh$_3$)$_2$ [5] and treatment of K$_2$[PdCl$_4$] with 1 mol K$_2$S$_2$O$_3$ [6]. The dihydrate [7, 8] and the anhydrous compound [9] have been used for the gravimetric determination of palladium. The thermogravimetric analysis of the dihydrate has been reported, thermal decomposition curves are reproduced in papers [9, 10]. Redetermined lattice parameters for tetragonal PdS are a = 6.4287, c = 6.6082 Å, powder photograph data are tabulated in the paper [11]. In PdS the Pd atoms adopt a structure of the β-W (eg. V$_3$Si) type with the sulphur atoms occupying one half of the type II tetrahedral sites [12]. The X-ray crystal structure of PdS has been re-investigated using single crystals isolated from a reaction of EuPd$_3$S$_4$ with ~1at% Br$_2$ in a sealed, evacuated silica tube at 900 K for 12 d. The crystals are tetragonal, P4$_2$/m-C$_{4h}^2$ with a = 6.429(2), c = 6.611(2) Å. Z = 8, D$_{exp}$ = 6.728 g/cm^3. In the structure, see **Fig. 52**, p. 220, the square-planar PdS$_4$ units are more regular than previously reported [31]. Distortion in the PdS structure is greater than for PtS but less than for CuS [13]. The compound is diamagnetic, $\psi_{mol} \times 10^6 = -21$ (80 K), -23 (295 K) cgs units [14]. The X-ray emission and 2pS X-ray photoelectron spectra of PdS have been recorded and analysed [15]. The electron structure of PdS has been analysed within the framework of the cluster model using the Xα-scattered wave method [16, 17]. Energy levels and charge distributions for

"$[PdS_6]^{10-}$" cluster has been calculated using SCF-SW-Xα method. Calculated binding energies for "$[PdS_6]^{10-}$" and "$[PdS_4]^{6-}$" have been compared with corresponding features in XPS spectrum of PdS [18]. The relationship between the electronic structure of transition metal sulphides, including PdS, and their ability to catalyse the hydrodesulphurisation of dibenzothiophene has been investigated [3, 19].

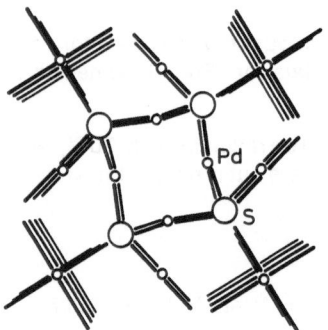

Fig. 52.   Structure of PdS [31].

Thermodynamic data for PdS at 25°C are $\Delta H_f^\circ = -35.8$, $\Delta F_f^\circ = -34.4$ kcal/mol, $S^\circ = 12$ gbs/mol [20]. Thermal decomposition of PdS (in air) involves formation of $PdSO_4$ which subsequently breaks down to PdO and palladium metal, a DTA curve is reproduced in the paper [21]. The thermal stability and thermodynamics of dissociation of PdS over the temperature range 700 to 1200 K have been investigated [22]. For thermodynamic data on the reaction $\frac{4}{3}PdS + H_2 = \frac{1}{3}Pd_4S + H_2S$, see Pd_4S, p. 214.

Low electrical resistivity and a positive temperature coefficient of resistivity, not quantified in the paper, are taken to indicate metallic character for PdS [23]. PdS particles do not aggregate on ageing [24]. The photoelectrochemical properties have been investigated [32].

Oxidation with $NaClO$ in alkaline solution or $KMnO_4$ in acid solution has been used to separate Ru (as RuO_4) from RuS/PdS mixtures [25]. Sorption of H_2S by PdS has been investigated, data are plotted in paper [26]. The exchange of Pd^{2+} ions with thin layers of PdS on cellulose has been investigated [27]. The kinetics [28] and thermodynamics [29] of the reaction between PdS and H_2SO_4 have been studied. Corrosion of Ti/Pd alloys by concentrated H_2SO_4 leads to buildup of surface palladium as PdS rather than metallic palladium [30].

References:

[1] Pittwell, L. R. (Nature **207** [1965] 1181/2).
[2] Pshenitsyn, N. K.; Prokof'eva, I. V. (Zh. Neorgan. Khim. **3** [1958] 996/1001; Russ. J. Inorg. Chem. **3** No. 4 [1958] 249/57; C.A. **1958** 16969).
[3] Pecoraro, T. A.; Chianelli, R. R. (J. Catal. **67** [1981] 430/45).
[4] Kochetkova, A. P.; Sveshnikova, L. B. (Zh. Neorgan. Khim. **20** [1975] 210/3; Russ. J. Inorg. Chem. **20** [1975] 114/6).
[5] Sodnomov, B. G.; Polovnyak, V. K.; Troitskaya, A. D. (Koord. Khim. **6** [1980] 1241/3; Soviet J. Coord. Chem. **6** [1980] 627/9).
[6] Ryabchikov, D. I. (Omagiu Raluca Ripan, Bucuresti 1966, pp. 485/93; C.A. **67** [1967] No. 39696).
[7] Sant, S. B.; Chow, A.; Beamish, F. E. (Anal. Chem. **33** [1961] 1257/60).
[8] Taimni, I. K.; Salaria, G. B. S. (Anal. Chim. Acta **11** [1954] 329/38).
[9] Champ, P.; Fauconnier, P.; Duval, C. (Anal. Chim. Acta **6** [1952] 250/8).
[10] Taimni, I. K.; Tandon, S. N. (Anal. Chim. Acta **22** [1960] 553/7).

[11] Grønvold, F.; Røst, E. (Acta Chem. Scand. **10** [1956] 1620/34).

[12] Smirnova, N. L. (Zh. Strukt. Khim. **1** [1960] 342/5; J. Struct. Chem. [USSR] **1** [1960] 317/20).

[13] Ahrens, L. H.; Morris, D. F. C. (J. Inorg. Nucl. Chem. **3** [1956] 270/80).

[14] Hulliger, F. (J. Phys. Chem. Solids **26** [1965] 639/45).

[15] Mikhailova, S. S.; Nemnonov, S. A.; Minin, V. I.; Maksyutov, F. B.; Galakhov, V. R. (Izv. Akad. Nauk SSSR Ser. Fiz. **40** [1976] 439/42; Bull. Acad. Sci. USSR Phys. Ser. **40** No. 2 [1976] 186/9; C. A. **85** [1976] No. 12126).

[16] Gagarin, S. G.; Kovtun, A. P.; Krichko, A. A.; Sachenko, V. P. (Kinetika Kataliz **20** [1979] 1138/45; Kinet. Catal. [USSR] **20** [1979] 935/41).

[17] Gagarin, S. G.; Gubskii, A. L.; Kovtun, A. P.; Krichko, A. A.; Sachenko, V. P. (Kinetika Kataliz **23** [1982] 585/90; Kinet. Catal. [USSR] **23** [1982] 491/6).

[18] Harris, S. (Chem. Phys. **67** [1982] 229/37).

[19] Harris, S.; Chianelli, R. R. (J. Catal. **86** [1984] 400/12).

[20] McDonald, J. E.; Cobble, J. W. (J. Phys. Chem. **66** [1962] 791/4).

[21] Timonova, R. I.; Ivashentsev, Ya. I. (Zh. Neorgan. Khim. **22** [1977] 2303/4; Russ. J. Inorg. Chem. **22** [1977] 1246/7).

[22] Tagirov V. K.; Kazenas, E.; Zviadadze, G. N.; Pavlyuchenko, N. M. (Izv. Akad. Nauk SSSR Metally **1981** No. 3, pp. 49/53; Russ. Met. **1981** No. 3, pp. 33/6; C. A. **95** [1981] No. 50 514).

[23] Fischmeister, H. (Acta Chem. Scand. **13** [1959] 852/3).

[24] Rudnev, N. A.; Malofeeva G. I. (Zh. Analit. Khim. **19** [1964] 151/5; J. Anal. Chem. [USSR] **19** [1964] 139/43).

[25] Konečny, C. (Collection Czech. Chem. Commun. **23** [1958] 1999/2004).

[26] Rudnev, N. A.; Malofeeva, G. I. (Zh. Neorgan. Khim. **6** [1961] 1885/90; Russ. J. Inorg. Chem. **6** [1961] 963/6).

[27] Schüssler, H. D.; Herrmann, G. (Radiochim. Acta **13** [1970] 65/9).

[28] Barkan, V. Sh.; Greiver, T. N. (Zh. Prikl. Khim. **50** [1977] 1380/4; J. Appl. Chem. [USSR] **50** [1977] 1324/8).

[29] Greiver, T. N.; Barkan, V. Sh. (Zh. Prikl. Khim. **50** [1977] 1384/7; J. Appl. Chem. [USSR] **50** [1977] 1328/31).

[30] Tomashov, N. D.; Chernova, G. P.; Chukalovskaya, T. V. (Zashch. Metal. **6** [1970] 3/8; Prot. Metals [USSR] **6** [1970] 1/5; C. A. **72** [1970] No. 106497).

[31] Brese, N. E.; Squattrito, P. J.; Ibers, J. A. (Acta Cryst. C **41** [1985] 1829/30).

[32] Folmer, J. C. W.; Turner, J. A.; Parkinson, B. A. (J. Solid State Chem. **68** [1987] 28/37).

PdS$_2$ (see "Palladium" 1942, p. 294). This compound is a derivative of palladium(II) (Pd^{2+}/S$_2^{2-}$). Attempted preparation by heating PdS with excess sulphur at 300 to 500°C for six months gave a product contaminated with PdS. However, pure material was obtained as a black crystalline powder by heating PdCl$_2$ with excess sulphur at 450°C for 4 d in an evacuated silica tube [1]. Amorphous PdS$_2$ prepared by passing H$_2$S over dry (NH$_4$)$_2$PdCl$_6$ at 110°C for 2 h can be converted to a poorly crystallised form by annealing at 700°C for 5 d [2]. Thermal decomposition of Pd(NCS)$_2$(PEt$_3$)$_2$ also affords PdS$_2$ [3].

A high pressure orthorhombic form of PdS$_2$ has been obtained by heating the first form or an intimate mixture of the elements at 1450°C and 63 kbar pressure for 5 min, then quenching [4]. Thermogravimetric analysis data are tabulated [2].

Crystals of the first orthorhombic form [1, 5], probable space group Pbca–D$_{2d}^{15}$ [5], have a = 5.460, b = 5.541, c = 7.531 Å, Z = 4 [1, 5] or a = 5.457(2), b = 5.542(2), c = 7.534(2) Å [2]. Powder photograph data for a sample annealed at 450°C [1] and observed and calculated intensities [5] are tabulated in papers. Each Pd atom is coordinated tetragonally by 4 S atoms

(Pd–S = 2.30 Å), sulphur atoms are bonded in pairs (S–S = 2.13 Å) [5]. The structure of the high pressure form is derived from that of pyrite by elongation of the Z axis. Lattice parameters are a = 5.51, b = 5.56, c = 7.16 Å, Z = 4, D_{calc} = 5.17, D_{pyk} = 4.92 ± 0.1 g/cm^3 [4]. Polymorphism of PdS_2 has been investigated over the temperature range 400 to 1350°C at 75 kbar pressure [6]. Thermodynamic data for PdS_2 are ΔH_f° = – 39.0, ΔF_f° = – 37.1 kcal/mol, S° = 18 gbs/mol [7], ΔG° = – 49 700 + 41.7 T cal/mol [8]. Resistivity ϱ measurements indicate that PdS_2 is nonmetallic, ΔE = 0.7 to 0.8 eV, ϱ ≈ 100 Ω · cm, Seebeck coefficient α = 240 μV/K [9]. Auger spectra and electron diffraction studies on the nature of the interaction between sulphur and palladium surfaces indicate formation of PdS_2 as one of the products [10].

References:

[1] Grønvold, F.; Røst, E. (Acta Chem. Scand. **10** [1956] 1620/34).
[2] Passaretti, J. D.; Kaner, R. B.; Kershaw, R.; Wold, A. (Inorg. Chem. **20** [1981] 501/3).
[3] Sodnomov, B. G.; Polovnyak, V. K.; Troitskaya, A. D. (Koord. Khim. **6** [1980] 1241/3; Soviet J. Coord. Chem. **6** [1980] 627/9).
[4] Munson, R. A.; Kaspar, J. S. (Inorg. Chem. **8** [1969] 1198/9).
[5] Grønvold, F.; Røst, E. (Acta Cryst. **10** [1957] 329/31).
[6] Larchev, V. N.; Popova, S. V. (Izv. Akad. Nauk SSSR Neorgan. Materialy **14** [1978] 775/6; Inorg. Materials [USSR] **14** [1978] 611/2).
[7] McDonald, J. E.; Cobble, J. W. (J. Phys. Chem. **66** [1962] 791/4).
[8] Niwa, K.; Yokokawa, T.; Isoya, T. (Bull. Chem. Soc. Japan **35** [1962] 1543/5).
[9] Hulliger, F. (J. Phys. Chem. Solids **26** [1965] 639/45).
[10] Sutyagina, A. A.; Perepelitsa, V. A.; Semenenko, M. N. (Zh. Fiz. Khim. **57** [1983] 681/4; Russ. J. Phys. Chem. **57** [1983] 408/10).

7.3 Ternary Palladium Sulphides

General References:

Bronger, W., Ternary Transition Metal Chalcogenides with Framework Structures and the Characterisation of their Bonding by Magnetic Properties, Pure Appl. Chem. **57** [1985] 1363/72.

Bronger, W., Ternary Sulphides: Model Compounds for the Correlation of Crystal Structure and Magnetic Properties, Angew. Chem. Intern. Ed. Engl. **20** [1981] 52/62; Angew. Chem. **93** [1981] 12/23.

7.3.1 Alkali-Metal Palladium Sulphides

Na_2PdS_2 is prepared by heating Na_2S_2 with $PdCl_2(NH_3)_2$ at 900°C for 18 h. The crystals which are red to transmitted light and blue to reflected light are orthorhombic, space group $Cmc2_1$–C_{2v}^{12}, a = 3.539(2), b = 10.411(4), c = 10.886(4) Å. The structure has·square-planar PdS_4 units connected laterally in one dimension by sulphide bridges, interatomic distances are Pd–S = 2.35, 2.39; Pd–Pd = 3.54; S–S = 3.54, 3.15 Å [1].

$K_2Pd_3S_4$ is prepared by heating K_2CO_3, palladium or $PdCl_2(NH_3)_2$ and sulphur at 800 to 900°C [2], by heating stoichiometric amounts of K_2S and PdS at 600 to 800°C [2] or by heating K_2CO_3 with powdered palladium in a stream of H_2S at 750 to 850°C [3]. Syntheses are performed under anhydrous, anaerobic conditions. Samples prepared at 800°C or at 730°C over a period of days afford orthorhombic crystals, space group Fddd–D_{2h}^{24}, a = 10.652, b = 25.614, c = 6.095 Å, Z = 8

[3]. The structure is shown in **Fig. 53**. A second form prepared at 850°C was indexed as hexagonal with $a = 6.117$ and $c = 25.575$ Å, $Z = 4$ but did not afford single crystals suitable for further work [3]. The compound is nonconductive with permeability $\varepsilon' \approx 8$ [4].

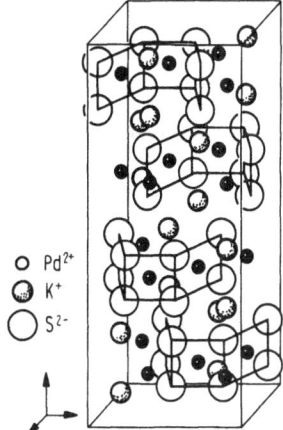

Fig. 53. Unit cell of $K_2Pd_3S_4$ [3].

O Pd^{2+}
◉ K^+
◯ S^{2-}

$Rb_2Pd_3S_4$ is prepared by heating Rb_2CO_3, palladium or $PdCl_2(NH_3)_2$ and sulphur at 800 to 900°C [2], by heating stoichiometric amounts of Rb_2S and PdS at 600 to 800°C [2] or by heating Rb_2CO_3 with powdered palladium in a stream of H_2S at 750 to 850°C [3]. The compound exists in two crystalline forms. One form is isotypic with the orthorhombic $K_2Pd_3S_4$, space group Fddd–D_{2h}^{24}, $a = 10.715$, $b = 26.623$, $c = 6.114$ Å, the other is isotypic with the monoclinic $Cs_2Pd_3S_4$, space group C2/m–C_{2h}^3, $a = 6.224$, $b = 13.320$, $c = 6.117$ Å, $\beta = 120.58°$, $Z = 2$ [3]. The compound is essentially diamagnetic, $\chi_{mol}^{corr} \times 10^6 = 37$ (90 and 195 K) and 32 (295 K) thus confirming square-planar palladium(II) [2] and is nonconductive with permeability $\varepsilon' \approx 8$ [4].

$Cs_2Pd_3S_4$ is prepared by heating Cs_2CO_3, $PdCl_2(NH_3)_2$ and sulphur under nitrogen at 800 to 900°C, by heating Cs_2S and PdS in stoichiometric amounts at 600 to 800°C [2] or by heating Cs_2CO_3 and palladium powder in a stream of H_2S at 750 to 850°C [3].

The crystals are monoclinic, space group C2/m–C_{2h}^3, $a = 6.275$, $b = 14.02$, $c = 6.166$ Å, $\beta = 120.6°$, $Z = 2$, $D_{X\text{-ray}} = 5.07$, $D_{pyk} = 4.93$ g/cm^3 [2, 5]. The structure is shown in **Fig. 54**, p. 224 [5]. The compound is diamagnetic $\chi_{mol}^{corr} \times 10^6 = 36$ (90 K), 34 (195 K), 32 (295 K) [2] and nonconductive with permeability $\varepsilon' \approx 8$ [4].

References:

[1] Bronger, W.; Günther, O.; Huster, J.; Spangenberg, M. (J. Less-Common Metals **50** [1976] 49/55).

[2] Bronger, W.; Eyck, J.; Rüdorff, W.; Stössel, A. (Z. Anorg. Allgem. Chem. **375** [1970] 1/7).

[3] Huster, J.; Bronger, W. (J. Solid State Chem. **11** [1974] 254/60).

[4] Gerstenhauer, E.; Vits, P. (Medd. Svenska Tekn. Vetenskapsakad. Finland No. 31 [1977] 160/81).

[5] Bronger, W.; Huster, J. (J. Less-Common Metals **23** [1971] 67/72).

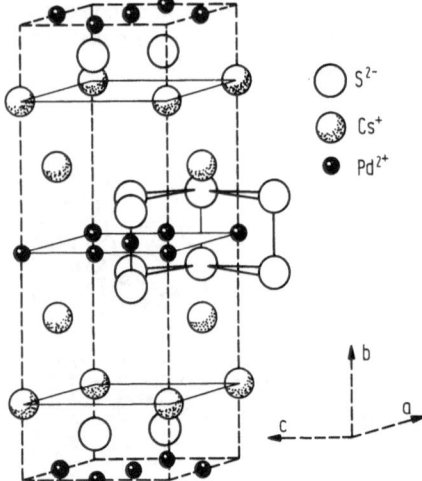

Fig. 54. The $Cs_2Pd_3S_4$ structure [5].

7.3.2 Barium–Palladium Sulphide $BaPdS_2$

$BaPdS_2$ is prepared by treatment of $BaPd(CN)_4 \cdot 4H_2O$ with dry H_2S or S. The crystals are orthorhombic, space group $Cmcm–D_{2h}^{17}$ with a = 6.783(1), b = 10.634(2), c = 5.627(1) Å, Z = 4. $BaPdS_2$ is isotypic with $BaNiO_2$, containing infinite zigzag chains $_\infty^1[PdS_{4/2}]^{2-}$ in which the palladium atoms are planar coordinated by the sulphur ligands. A projection and a perspective view of the structure along [100] are reproduced.

Reference:

Huster, J.; Bronger, W. (J. Less-Common Metals **119** [1986] 159/65).

7.3.3 Rare-Earth Metal/Palladium Sulphides

MPd_3S_4 (M = rare earth). These sulphide bronzes are prepared by heating the elements together at 1125 K in a sealed evacuated silica tube for 21 d with grinding every fourth day [1, 2]. X-ray and neutron diffraction powder data establish the $NaPt_3O_4$ type structure shown in **Fig. 55** for $LaPd_3S_4$ [1] and the other examples listed in the table below [2]. Plots of the lattice parameter a vs. crystal radius of M^{III} [2] and of reciprocal molar susceptibility vs. temperature for M = Tb, Sm [1, 2], Pr and Ho [2] are reproduced. Magnetic data given in the table below are consistent with the formulation $[M^{III}Pd_3^{II}S_4] + e^-$ [2].

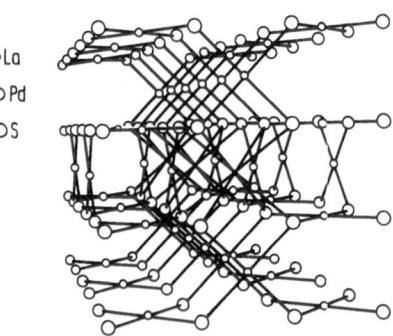

o La
O Pd
O S

Fig. 55. Sketch of the $LaPd_3S_4$ structure.

LaPd$_3$S$_4$ in pressed form displays metallic conductivity and provides the first example of a metallic Pd bronze [2]. A plot of electrical conductivity vs. temperature is reproduced for LaPd$_3$S$_4$ [3].

Structural data for the series MPd$_3$S$_4$:

compound	a in Å	M–S in Å	Pd–S in Å	S–S in Å	crystal radius*) of M^{III}/Å
LaPd$_3$S$_4$	6.7398(1)	2.9184(1)	2.3829(1)	3.3699(1)	1.300
CePd$_3$S$_4$	6.7130(1)	2.9068(1)	2.3734(1)	3.3565(1)	1.283
PrPd$_3$S$_4$	6.6989(1)	2.9007(1)	2.3684(1)	3.3495(1)	1.266
SmPd$_3$S$_4$	6.6692(1)	2.8878(1)	2.3579(1)	3.3346(1)	1.219
EuPd$_3$S$_4$	6.675(3)	2.890(1)	2.360(1)	3.338(2)	1.206
YPd$_3$S$_4$	6.6349(1)	2.8730(1)	2.3458(1)	3.3175(1)	1.159
HoPd$_3$S$_4$	6.6346(1)	2.8729(1)	2.3457(1)	3.3173(1)	1.155
ErPd$_3$S$_4$	6.6240(1)	2.8683(1)	2.3419(1)	3.3120(1)	1.144

*) Crystal radii for coordination number 8.

Other relevant radii are: Pd^{II} = 0.625; S^{-II} = 1.70 Å.

Magnetic data for ternary rare-earth metal sulphides:

compound	Θ_p in K	μ_{eff} exptl. data	μ_{eff} calc. data (M^{III})
PrPd$_3$S$_4$	—	3.66	3.58
SmPd$_3$S$_4$	—	1.66*)	1.55*)
TbPd$_3$S$_4$	−9	9.64	9.74
HoPd$_3$S$_4$	−10	10.44	10.60

*) At 260 K.

References:

[1] Keszler, D. A.; Ibers, J. A. (Inorg. Chem. **22** [1983] 3366/7).
[2] Keszler, D. A.; Ibers, J. A.; Mueller, M. H. (J. Chem. Soc. Dalton Trans. **1985** 2369/73).

7.3.5 Other Heavy-Metal Palladium Sulphides

Tl$_2$Pd$_3$S$_2$ is prepared by heating a mixture of TlS and palladium metal at 900 to 1000°C for 24 h then annealing at 600°C for 4 d. The crystals have Shandite-type rhombohedral unit cell a = 5.748 Å, α = 60.52(3)° [1].

TlPd$_2$S$_3$ is formed as a crystalline compound by coprecipitation of thallium with PdS [2, 3, 4]. Coprecipitation and calculated X-ray diffraction diagrams are reproduced in papers and data on coprecipitation are tabulated [2, 3]. Palladium is quantitatively precipitated from Tl$_2$SO$_4$/H$_2$SO$_4$ solution as TlPd$_2$S$_3$ [5].

Nb$_2$PdS$_6$ is obtained in crystalline form by heating the elements at 1075 K for 3 weeks in a sealed evacuated silica tube. The crystal cell dimensions are a = 9.99(1), b = 3.30(1), c = 11.69(1) Å, β = 114.7(1)° [12].

Ta_2PdS_6 is obtained in a similar manner. The crystals are monoclinic, space group C_{2h}^3–$I2/m$ with $a = 9.960(13)$, $b = 3.271(5)$, $c = 11.696(16)$ Å, $\beta = 114.51(4)°$, $Z = 2$ [12]. The structure is similar to that reported for Ta_2PdSe_6 (see p. 246).

Ni–Pd–S System. X-ray diffraction, electron microprobe analysis and DTA have been used to plot sections of the Ni–Pd–S phase diagram at <50 mol% S. A ternary eutectic of Pd 66, Ni 18 and S 16% (m.p. 450°C) was composed of $(Pd_xNi_{1-x})_4S$, Ni_3S_2 and Ni–Pd solid solution [6].

Co–Pd–S System. It has been investigated at 1000, 800, 600 and 400°C but no ternary phase was discovered. The system is characterised by a broad sulphide melt field which recedes to Pd-rich portions at 600°C and disappears at >500°C [11].

Cu–Pd–S System. X-ray diffraction, electron microprobe analysis and DTA have been used to plot sections of the Cu–Pd–S phase diagram at <50 mol% S. A ternary eutectic of Pd 70, Cu 20 and S 10% (m.p. 560°C) was composed of $(Pd_xCu_{1-x})S$, Cu_2S and Cu–Pd solid solution [6]. Part of the Cu–Pd–S phase diagram has been determined and is reproduced in the paper. X-ray data have been reported for $Pd_{12}Cu_4S_7$. The crystals have cubic symmetry, space group T_d^3–$I\bar{4}3$ m, $a = 8.874(2)$ Å [7].

Ag_2Pd_3S is prepared by heating a briquetted mixture of the elements in an evacuated silica tube at 630°C for several days [8, 9]. The resistivity at 20°C is $\varrho = 1.93 \times 10^4 \, \Omega \cdot cm$ and $d\varrho/dt = +0.22 \times 10^{-6} \, \Omega \cdot cm \cdot deg^{-1}$ [8]. Thermoanalytical, microscopical and X-ray investigation of the Ag–Pd–S system are reported. Ag_2Pd_3S forms a eutectic with Ag_2S at ~700°C [8].

$Ru_{4.5}Pd_{4.5}S_8$. A product of this composition has been obtained by dry synthesis from the pure elements, and is thought to consist of Ru + C2 phase and an unidentified component. The C2 phase had $a = 5.6100$ Å and was thought to be RuS_2, possibly with some Pd substitution [10].

Rh_8PdS_8. Prepared from pure elements by dry synthesis. X-ray pattern corresponds to homogeneous ϱ-phase with $a = 9.9301$ Å [10].

References:

[1] Zabel, M.; Wandinger, S.; Range, K. J. (Z. Naturforsch. **34 b** [1979] 238/41).

[2] Rudnev, N. A.; Malofeeva, G. I. (Talanta **11** [1964] 531/42).

[3] Rudnev, N. A.; Malofeeva, G. I. (Zh. Neorgan. Khim. **6** [1961] 1885/90; Russ. J. Inorg. Chem. **6** [1961] 963/6).

[4] Rudnev, N. A. (Sovrem. Metody Anal. **1965** 262/73; C. A. **64** [1966] 5821).

[5] Rudnev, N. A.; Malofeeva, G. I. (Analiz Blagorodn. Metal. **1965** 27/36; C.A. **66** [1967] No. 121788).

[6] Zviadadze, G. N.; Gulyanitskaya, Z. F.; Pavlyuchenko, N. M.; Blagoveshchenskaya, N. V. (Izv. Akad. Nauk SSSR Metally **1982** No. 5, pp. 53/6; Russ. Met. **1982** No. 5, pp. 40/3; C.A. **98** [1983] No. 7232).

[7] Matković, P.; El-Boragy, M.; Schubert, K. (J. Less-Common Metals **50** [1976] 165/76).

[8] Fischmeister, H. (Acta Chem. Scand. **13** [1959] 852/3).

[9] Raub, E.; Wullhorst, B.; Plate, W. (Z. Metallk. **45** [1954] 533/7).

[10] Knop, O. (Can. J. Chem. **41** [1963] 1832/3).

[11] Karup-Moeller, S.; Makovicky, E. (Econ. Geol. **81** [1986] 1049/55; C. A. **105** [1986] 176107).

[12] Keszler, D. A.; Squattrito, P. J.; Brese, N. E.; Ibers, J. A.; Maoyu, S.; Jiaxi, L. (Inorg. Chem. **24** [1985] 3063/7).

7.3.6 Miscellaneous Pd–S Compounds

Na₂[Pd(SH)₄] is obtained as a rather unstable brown powder by adding a solution of sodium hydrosulphide to Na_2PdCl_4.

Reference:

Pittwell, L. R. (Nature **207** [1965] 1181/2).

[NH₄]₂PdS₁₁·2H₂O (see "Palladium" 1942, p. 324). An improved synthesis involving addition of aqueous K_2PdCl_4 solution to a solution of ammonium polysulphide, prepared from $(NH_4)_2S$ and free sulphur, has been described [1, 2]. The complex forms dark red water-insoluble crystals which decompose on dissolution in warm organic solvents. The crystals are tetragonal, space group $P4/mnc–D_{4h}^6$, a = 11.065(12), c = 6.862(15) Å, Z = 2, D_{obs} = 2.09, D_{calc} = 2.10 g/cm³. The Pd atoms are linked by S_6 chains in a three-dimensional array, there are no discrete PdS_{12}^{2-} groups. Sulphur absences, usually from the middle of chains, reduce the ideal composition PdS_{12}^{2-} to PdS_{11}^{2-}. Bond lengths are Pd–S = 2.32; S–S ≈ 2.01 to 2.06 Å. A stereo view of the structure is reproduced in **Fig. 56** [1].

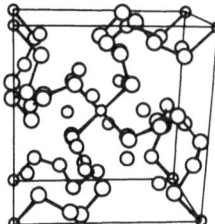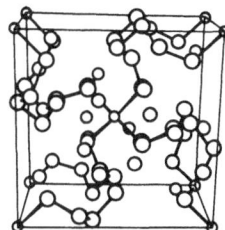

Fig. 56. Stereo view of the unit cell contents (down c) showing possible Pd–Pd linkages via S_6 chains. Pd atoms are shown as small circles, S atoms as large circles, NH_4^+ and H_2O as intermediate (nonbonded) circles [1].

Reaction with cyanide anions in DMF leads to oxidation of Pd^{II} to Pd^{IV} by sulphur radical ions and formation of $[Pd(CN)_6]^{2-}$ [1]. The complex does not melt but changes from red to black at 165 to 167°C. The infrared spectrum below 500 cm⁻¹ shows ν(S–S) = 480, 440 cm⁻¹ (weak) and additional bands at 360 (broad) and 300 cm⁻¹ [2].

K₂PdS₁₁·xH₂O is similarly prepared as a red solid using K_2S in place of $(NH_4)_2S$ [1].

References:

[1] Haradem, P. S.; Cronin, J. L.; Krause, R. A.; Katz, L. (Inorg. Chim. Acta **25** [1977] 173/9).
[2] Krause, R. A.; Kozlowski, A. W.; Cronin, J. L. (Inorg. Syn. **21** [1982] 12/6).

[NR₄]₂[Pd(MoS₄)₂] (R = Et, Pr). These complex salts are prepared by treatment of K_2PdCl_4 in $H_2O/MeCN$ solution with $(NH_4)_2MoS_4$ followed by (NR₄)Br and crystallise spontaneously from solution [1, 2].

[PPh₄]₂[Pd(MoS₄)₂] is similarly prepared using (PPh₄)Br [1].

The $[Pd(MoS_4)_2]^{2-}$ anion is red in colour, the electronic spectrum shows λ_{max} = 21400, 26300, 28700, 32000, 36500 and 37300 cm⁻¹; infrared spectrum: ν(Mo=S) ≈ 510, 495; ν(Mo–S) ≈ 450, 435; ν(Pd–S) ≈ 312, 295 cm⁻¹, bending modes ≈ 203, 185, 173 and 169 cm⁻¹ [1]. The anion undergoes 2 one-electron reductions – one reversible, one irreversible. $E_{1/2}$ = –1.09

and −1.24 V, respectively. A cyclic voltammogram for [NEt₄]₂[Pd(MoS₄)₂] is reproduced and reduction potentials are tabulated [1, 2].

[NEt₄]₂[Pd(WS₄)₂]. [PPh₄]₂[Pd(WS₄)₂]. These salts are prepared by methods similar to those used for the corresponding molybdenum containing species [1, 2]. The anion is rust-coloured, electronic spectrum: λ_{max} 25300, 27400, 32200, 35000, 39700 and 44400 cm^{-1}; infrared spectrum: ν(W–S) $\approx$500, 490; ν(W–S) $\approx$440; ν(Pd–S) $\approx$310, 295 cm^{-1}, bending modes $\approx$190, 170, 165 and 155 cm^{-1} [1]. The anion undergoes 2 one-electron reductions – one reversible, the other irreversible. $E_{1/2}=-1.05$ and -1.28 V, respectively [1, 2]. A cyclic voltammogram for [NEt₄]₂[Pd(WS₄)₂] is reproduced [1] and reduction potentials are tabulated [1, 2].

References:

[1] Callahan, K. P.; Piliero, P. A. (Inorg. Chem. **19** [1980] 2619/26).
[2] Callahan, K. P.; Piliero, P. A. (J. Chem. Soc. Chem. Commun. **1979** 13/4).

7.4 Quaternary Palladium Sulphides

K₂Pd₃TiS₆ is obtained as a blue-black powder by heating K₂Pd₃S₄ with TiS₂ at 700°C.

K₂Pd₃SnS₆ is obtained as a red powder by heating K₂Pd₃S₄ with SnS₂ at 500 to 600°C. The crystals are hexagonal, a = 7.09, c = 13.02 Å, Z = 2; D_{X-ray} 4.16 g/cm³.

K₂Pd₃PtS₆ is obtained as a grey-blue powder by heating K₂Pd₃S₄ with PtS₂ at 600°C. The crystals are hexagonal-rhombic, space group D_{3d}^5–R$\overline{3}$m, a = 6.96, c = 19.31 Å, Z = 3, D_{X-ray} 4.83 g/cm^{-3}.

Reference:

Rüdorff, W.; Stössel, A.; Schmidt, V. (Z. Anorg. Allgem. Chem. **357** [1968] 264/72).

7.5 Palladium Sulphoxylate Pd₂SO₂(?)

This may be the product of reaction of Rongalite (sodium formaldehydesulphoxylate, NaHSO₂·CH₂O·2H₂O), with K₂[PdCl₄] in solution in acetate buffer. It is dark brown.

Reference:

Kukushkin, Yu. N.; Marshak, E. M. (Zh. Neorgan. Khim. **16** [1971] 266/7; Russ. J. Inorg. Chem. **16** [1971] 138/9).

7.6 Palladium Sulphito Complexes (see "Palladium" 1942, pp. 308/9, 324)

A number of these have been characterised, all of palladium(II). In most of them it seems likely that there is Pd–S rather than Pd–O bonding.

7.6.1 Unsubstituted Complexes

Examples of the [Pd(SO₃)₄]$^{6-}$ and [Pd(SO₃)₂]$^{2-}$ ions are known.

[Pd(SO₃)₄]$^{6-}$. The standard reduction potential E_0 of 0.058(5) V has been determined for this ion potentiometrically in 0.5 M KNO₃ and 0.1 M Na₂SO₃ solution, and log $\beta_4 = 29.1 \pm 0.2$ l^4·mol^{-4} calculated for it [1].

7.6.1.1 Tetrasulphito Complexes

Na$_6$[Pd(SO$_3$)$_4$]·2H$_2$O. This is made from PdCl$_2$, NaHSO$_3$ and free SO$_2$ in dilute acetic acid. Crystals were obtained by a gel-crystallisation method [2]. It can also be made from PdCl$_2$ in hydroxide media with SO$_2$ gas [3].

The tetragonal crystals belong to the space group I4$_1$/a-C$_{4h}^6$, a = b = 16.488(1), c = 10.663(2) Å; Z = 8; measured density 2.69 g/cm^3, calculated density 2.75 g/cm^3. The structure is made up of isolated, slightly distorted square-planar [Pd(SO$_3$)$_4$]$^{6-}$ units stacked into columns (see **Fig. 57**). There are two crystallographically independent anions (Pd–S = 2.316(1) Å in one and Pd–S = 2.341(1) Å in the other) with all sulphite ligands S-bonded to palladium. The Pd–Pd distances are large (5.33 Å), the columnar stacking arising from peripheral coordination of sodium ions to different O atoms [2].

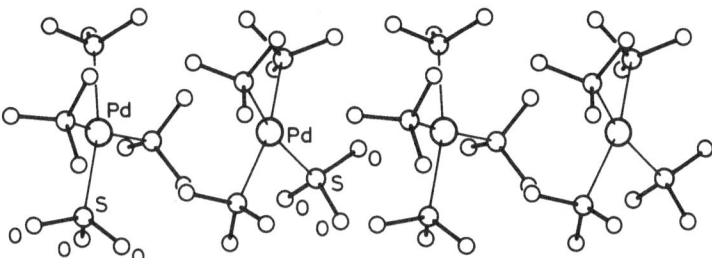

Fig. 57. Stacking of [Pd(SO$_3$)$_4$]$^{6-}$ anions in Na$_6$[Pd(SO$_3$)$_4$]·2H$_2$O [2].

The infrared spectrum of the salt was measured from 2000 to 400 cm^{-1}; ν_{SO} appears at 1114, 1083, 1054 and 974 cm^{-1} [3].

7.6.1.2 Disulphito Complexes

Na$_2$[Pd(SO$_3$)$_2$]·H$_2$O. This is made by reaction of Pd(SO$_3$)(H$_2$O)$_3$ with Na$_2$SO$_3$·7H$_2$O.

It is yellow; the molar conductance at 25°C is 191 Ω^{-1}·cm^2 [4].

K$_2$[Pd(SO$_3$)$_2$]·H$_2$O. This is made by reaction of PdCl$_2$ and K$_2$S$_2$O$_5$ in hot water or by reaction of Pd(SO$_3$)(H$_2$O)$_3$ with K$_2$SO$_3$ in water (this gives the purer product). It is yellow [4]. Crystals have been made of the salt by a diffusion method using a solution of K$_2$SO$_3$ and K$_2$S$_2$O$_5$ with PdCl$_2$ in HCl [5].

The X-ray crystal structure analysis shows the crystals to be orthorhombic with a = 7.037(3), b = 14.749(6), c = 7.441(3) Å, space group Pnma-D$_{2v}^{16}$ or Pn2$_1$a-C$_{2v}^9$, Z = 4, measured density 3.05 g/cm^3, calculated density 3.12 g/cm^3. The palladium atoms have a slightly distorted square-planar environment with two cis-coordinated S atoms (Pd–S = 2.241(1) Å) and two cis-coordinated O atoms (2.121(3) Å), such that the sulphite ligands function as bidentate ligands bridging two Pd atoms through one S and one O atom, see figure in the paper [5].

The infrared spectrum was measured (400 to 4000 cm^{-1}, reproduced in paper); the S–O stretches ν_{SO} are at 1157, 1099, 1056, 933 and 904 cm^{-1} [3].

The dehydrated compound reacts with aqueous ammonia forming cis-K$_2$[Pd(SO$_3$)$_2$(NH$_3$)$_2$] ·2H$_2$O [2].

230

References:

[1] Hancock, R. D.; Finkelstein, N. P.; Evers, A. (J. Inorg. Nucl. Chem. **39** [1977] 1031/4).

[2] Messer, D.; Breitinger, D. K.; Haegler, W. (Acta Cryst. B **37** [1981] 19/23).

[3] Newman, G.; Powell, D. (Spectrochim. Acta **19** [1963] 213/24).

[4] Earwicker, G. A. (J. Chem. Soc. **1960** 2620/6).

[5] Messer, D.; Breitinger, D.; Haegler, W. (Acta Cryst. B **35** [1979] 815/8).

7.6.2 Substituted Sulphito Complexes

7.6.2.1 Sulphito-aquo Complex

$Pd(SO_3)(H_2O)_3$. This is made by reaction of $PdCl_2$ with Ag_2SO_3 in water followed by evaporation, or from $Pd(OH)_2$ suspended in acetone with SO_2 gas. It forms orange needles. Freezing point depression and molecular weight measurements suggest that it is monomeric in water [1].

It slowly decomposes at room temperature to Pd and H_2SO_4. With ammonia it gives $Pd(SO_3)(NH_3)(H_2O)_2$, $Pd(SO_3)(NH_3)_2(H_2O)$ and $Pd(SO_3)(NH_3)_3$, while SO_3^{2-} gives $[Pd(SO_3)_2]^{2-}$ [1].

Conductimetric titrations with ammonia demonstrated its reaction with one, two and three molecules of NH_3, respectively [1].

7.6.2.2 Sulphito-ammine Complexes

$Pd(SO_3)(NH_3)_3$. This is made by reaction of $Pd(SO_3)(H_2O)_3$ and ammonia or from $PdCl_2(NH_3)_2$, NH_4OH and Na_2SO_3. It is white [1].

The X-ray crystal structure shows the complex to be monoclinic, space group $P2_1/c\text{-}C_{2h}^5$; $Z = 4$; $a = 7.84(3)$, $b = 6.96(3)$, $c = 13.77(10)$ Å; $\beta = 126.7(5)°$; density by flotation 2.68 g/cm³, calculated density 2.62 g/cm³. The metal has square-planar coordination and the sulphito ligand is S-bonded; $Pd–S = 2.254(6)$ Å, mean $Pd–N = 2.11(2)$ Å, mean $S–O = 1.50(2)$ Å, mean OSO angle $= 108.6(1.1)°$, see **Fig. 58** [2].

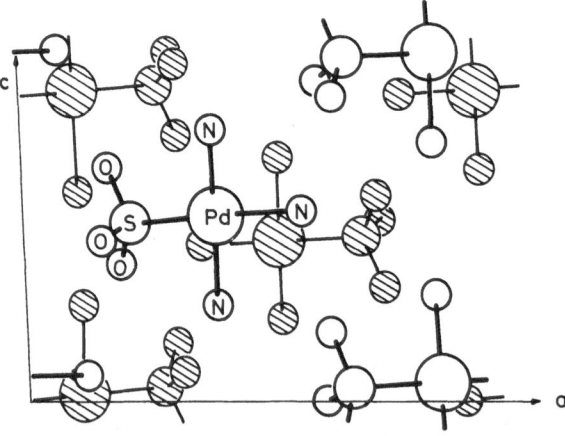

Fig. 58.   Part of the unit cell of $Pd(SO_3)(NH_3)_3$ [2].

The infrared spectrum (400 to 4000 cm^{-1}) shows v_{SO} at 1157, 1074, 983, 913, 838 and 813 cm^{-1} [3] or at 1068 and 976 cm^{-1} with v_{PdS} at 253 cm^{-1} [6]; in the Raman spectrum, v_{SO} lies at 1090, 1075 and 978 cm^{-1} [6].

Pd(SO$_3$)(NH$_3$)$_2$(H$_2$O). This yellow material can be made by reaction of PdCl$_2$(NH$_3$)$_2$ with Ag$_2$SO$_3$, from Pd(SO$_3$)(H$_2$O)$_3$ and NH$_4$OH or from Pd(SO$_3$)(NH$_3$)$_3$ and water [1].

The compound is stable at 100°C but decomposes at 150°C. In HCl it gives PdCl$_2$(NH$_3$)$_2$ [1].

The infrared spectrum (400 to 4000 cm^{-1}) showed bands at 1079, 1008 and 847 cm^{-1}; it was suggested that monodentate S-bonded sulphite is present in the complex [3].

trans-Na$_2$[Pd(SO$_3$)$_2$(NH$_3$)$_2$]·6H$_2$O. This is made by reaction of [Pd(NH$_3$)$_4$]Cl$_2$ and Na$_2$SO$_3$ [1]. It forms pale yellow tablets, with molar conductance in water of 220 Ω^{-1}·cm^2 at 25°C [1].

The X-ray crystal structure shows the salt to be triclinic, space group P$\bar{1}$-C$_i^1$, Z=1, a = 8.92(2), b = 6.52(2), c = 7.07(2) Å, α = 73.5(2)°, β = 102.5(2)°, γ = 81.6(2)°; density by flotation 2.01(4) g/cm^3, calculated density 2.010 g/cm^3. There is a trans arrangement of S-bonded sulphito ligands as shown in **Fig. 59**. The structure consists of square-planar Pd(SO$_3$)$_2$(NH$_3$)$_2$ units at the corners of each unit cell with distorted NaO$_6$ octahedra in the middle of the cell [4].

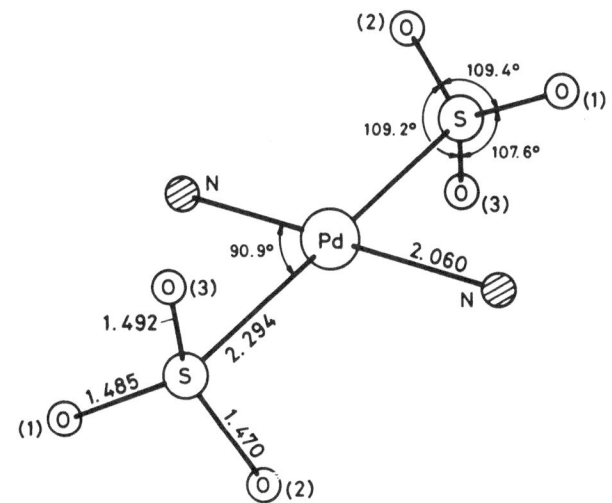

Fig. 59. X-ray crystal structure of the trans-[Pd(SO$_3$)$_2$(NH$_3$)$_2$]$^{2-}$ anion in trans-Na$_2$[Pd(SO$_3$)$_2$(NH$_3$)$_2$]·6H$_2$O [4].

The infrared spectrum of the salt (misprinted in the text at one point as cis-K$_2$[Pd(SO$_3$)$_2$(NH$_3$)$_2$]) from 400 to 4000 cm^{-1} was measured; v_{SO} lies at 1093 to 1056, 995, 977 and 958 cm^{-1} [3].

With CO, Pd metal is formed. With ethylenediamine (en) it gives Na$_2$[Pd(SO$_3$)$_2$en] [1].

cis-K$_2$[Pd(SO$_3$)$_2$(NH$_3$)$_2$]·2H$_2$O. Reaction of K$_2$[Pd(SO$_3$)$_2$] with aqueous ammonia gives salt as almost colourless needles which readily lose their water of crystallisation. The molar conductance is 260 Ω^{-1}·cm^2 at 25°C [1].

The infrared spectrum (400 to 4000 cm^{-1}) shows v_{SO} at 1074, 986 and 958 cm^{-1}; it was suggested that monodentate S-bonded sulphito groups are present [3]. With CO, Pd metal is formed [1].

trans-[Pd(SO₃)₂(NH₃)₂][Pd(NH₃)₄]. This is made as pale yellow needles by reaction of [Pd(NH₃)₄]Cl₂ with Na₂SO₃·7H₂O, or from Na₂[Pd(SO₃)₂(NH₃)₂] and PdCl₂(NH₃)₂. The molar conductance in water is 285 Ω⁻¹·cm² [1].

cis-[Pd(SO₃)₂(NH₃)₂][Pd(NH₃)₄]. This is made by reaction of [Pd(NH₃)₄]Cl₂ with cis-K₂[Pd(SO₃)₂(NH₃)₂], and forms short, pale yellow prisms. The molar conductance in water is 280 Ω⁻¹·cm² [1].

Pd(SO₃)(NH₃)(H₂O)₂. This is made by reaction of Pd(SO₃)(H₂O)₃ in water with aqueous ammonia, and has a low solubility in water [1].

The infrared spectrum was measured (400 to 4000 cm⁻¹); ν_{SO} absorbs at 1118, 1067 and 1003 cm⁻¹, and it was suggested that a monodentate S-bonded sulphito ligand is present in the complex [3].

7.6.2.3 Sulphito-ethylenediamine Complex Na₂[Pd(SO₃)₂en]

This is made by reaction of Na₂[Pd(SO₃)₂(NH₃)₂] with ethylenediamine (en). It forms yellow crystals [1].

7.6.2.4 Sulphito-chloro Complex

K₃[Pd((SO₃)₂H)Cl₂]. This is made from K₂[PdCl₆] in aqueous solution with K₂CO₃ after treatment with SO₂ gas. The salt forms yellow needles [5].

The crystals are orthorhombic, space group Cmc2₁-C²⁰², a=15.083(3), b=10.391(4), c=7.002(4) Å; Z=4; measured density 2.69 g/cm³, calculated density 2.757 g/cm³. The structure is of the type found in the isostructural Pt complex. The Pd atom is square planar with cis chloro ligands, Pd–Cl = 2.384(1) Å, the other two positions being occupied by the two S atoms Pd–S = 2.254(1) Å of two SO₃²⁻ units joined by a short, symmetric and almost linear hydrogen bond (O–O 2.396(4) Å) thus forming a six-membered chelate ring, see **Fig. 60** and figure in the paper [5].

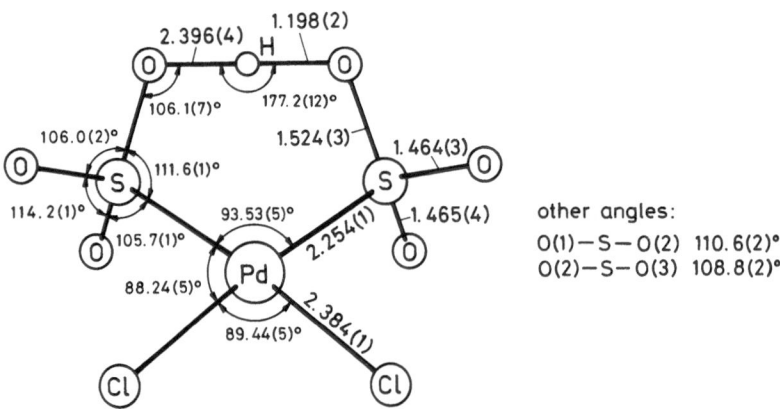

other angles:
O(1)—S—O(2) 110.6(2)°
O(2)—S—O(3) 108.8(2)°

Fig. 60.   The [Pd(SO₃)₂HCl₂]³⁻ anion [5]; distances in Å.

K₃[Pd((SO₃)₂H)Br₂]. This is made by reaction of PdBr₂ and K₂CO₃ in aqueous solution with SO₂. It is yellow [5].

X-ray crystal structure analysis showed the salt to be isotypic with $K_3[Pd((SO_3)_2H)Cl_2]$; orthorhombic, space group $Cmc2_1\text{-}C^{12}_{2v}$, $Z = 4$, $a = 15.29(3)$, $b = 10.51(2)$, $c = 7.13(2)$ Å [5].

References:

[1] Earwicker, G. A. (J. Chem. Soc. **1960** 2620/6).
[2] Spinnler, M. A.; Becka, L. N. (J. Chem. Soc. A **1967** 1194/9).
[3] Newman, G.; Powell, D. B. (Spectrochim. Acta **19** [1963] 213/24).
[4] Capparelli, M. V.; Becka, L. N. (J. Chem. Soc. A **1969** 260/5).
[5] Liehr, G.; Breitinger, D. K.; Raidel, M. (Acta Cryst. C **41** [1985] 304/6).
[6] Hall, J. P.; Griffith, W. P. (Inorg. Chim. Acta **48** [1981] 65/71).

7.7 Palladium Sulphates and Sulphato Complexes

As is frequently the case with sulphates and sulphato complexes of the platinum-group metals the species reported are often ill-defined.

There is a brief review of early work on palladium sulphates and sulphato complexes.

Reference:

Ginzburg, S. I.; Chalisova, N. N. (Zh. Neorgan. Khim. **13** [1968] 1239/45; Russ. J. Inorg. Chem. **13** [1968] 648/51).

7.7.1 Palladium Sulphate (see "Palladium" 1942, p. 295)

$PdSO_4$. This is made by dehydration of $PdSO_4 \cdot 2H_2O$ at 250°C and drying in vacuum over KOH [1, 2]. It is brown [1]. It can also be obtained from PdO and $K_2S_2O_7$ with heat [3], and from $PdCl_2$ in boiling concentrated H_2SO_4 [4], and can be made by heating PdS in air [2].

It forms brown crystals which are apparently deliquescent, though it is said to be only sparingly soluble [2].

The aqueous solution is hydrolysed between pH = 1.7 and 1.17 [5].

The X-ray powder diffraction diagram was recorded [2]. The infrared spectrum (700 to 1200 cm^{-1}, reproduced in paper) shows ν_{SO} at 1160, 1105, 1035 and 996 cm^{-1} [1]; values of 1155, 1105 and 966 cm^{-1} were also given [3]. A polymeric structure with bridging sulphato groups was proposed for the material on the basis of its infrared spectrum [1, 3].

Differential thermal analysis for $PdSO_4$ has been carried out (thermogram for 300 to 900°C reproduced in paper) [2].

For electronic spectra of $PdSO_4$ see p. 236; for electrochemical data see p. 236; for its chemical reactions see p. 234.

$PdSO_4 \cdot 2H_2O$. This is made by reaction of Pd sponge with an H_2SO_4–HNO_3–water mixture (1:2:2 by volume) [1, 2]. It is a greenish brown solid giving a deep red solution [1]. It may be present in the solute after anodic dissolution of Cu–Pd alloys in H_2SO_4–Na_2SO_4 electrolytes [6].

The infrared spectrum was measured (700 to 1400 cm^{-1}, reproduced in paper); ν_{SO} lies at 1235, 1089, 995 and 960 cm^{-1} [1].

$PdSO_4$ in Solution. This is made from finely divided Pd in 9 to 60% H_2SO_4 at 80 to 150°C with oxygen under pressure [7].

234

References:

[1] Eskenazi, R.; Raskovan, J.; Levitus, R. (J. Inorg. Nucl. Chem. **28** [1966] 521/6).
[2] Timonova, R. I.; Ivashentsev, Ya. I. (Zh. Neorgan. Khim. **22** [1977] 2303/4; Russ. J. Inorg. Chem. **22** [1977] 1246/7).
[3] Jasim, F.; Jameel, I. (Thermochim. Acta **86** [1985] 155/62).
[4] Chernyaeva, A. P.; Ivanova, S. N.; Gindin, L. M.; Ginzburg, S. I.; Chalisova, N. N.; Chernobrov, A. S. (Izv. Sibirsk. Otd. Akad. Nauk SSSR Ser. Khim. Nauk **1974** 49/52; C.A. **81** |1974] No. 17348).
[5] Pshenitsyn, N. K.; Ginzburg, S. I. (Izv. Sekt. Platiny No. 28 [1954] 213/28; C.A. **1956** 722).
[6] Kukushkin, Yu. N.; Maslov, E. I.; Simanova, S. A.; Alashkevich, V. P.; Vorotnikov, M. V. (Zh. Prikl. Khim. **44** [1971] 2441/6; J. Appl. Chem. [USSR] **44** [1971] 2512/7).
[7] Stowe, S. C.; Dow Chemical Co. (U.S. 3425801 [1969]; C.A. **70** [1969] No. 69682).

Extraction of PdSO₄

The extraction of "PdSO$_4$" (the exact nature of the extracted species seems uncertain) from aqueous solution by tetra-octylammonium bisulphate in toluene and by octanoic acid in decane has been explored. It appears that $[Pd(H_2O)_{4-n}(OH)_n]^{2-n}$ complexes are involved rather than inner-sphere sulphato complexes [1].

Biological Activity of PdSO₄

Intravenously injected PdSO$_4$ in unanesthetised rats produced immediate cardiovascular reaction [2]. The toxicity of PdSO$_4$ is higher than that of PdCl$_2 \cdot 2H_2O$ [3].

References:

[1] Chernyaeva, A. P.; Ivanova, S. N.; Gindin, L. M.; Ginzburg, S. I.; Chalisova, N. N.; Chernobrov, A. S. (Izv. Sibirsk. Otd. Akad. Nauk SSSR Ser. Khim. Nauk **1974** 49/52; C.A. **81** [1974] No. 17348).
[2] Wiester, M. J. (Environ. Health Persp. **12** [1975] 41/4).
[3] Holbrook, D. J.; Washington, M. E.; Leake, H. B.; Brubaker, P. E. (Environ. Health Persp. **10** [1975] 95/101).

Reactions of Palladium Sulphates

With $(CH_3)_2SO$, PdSO$_4 \cdot 2H_2O$ gives $PdSO_4[(CH_3)_2SO]_2$, and 1,10-phenanthroline (phen) gives Pd(SO$_4$)phen [1]. With 1,2,3-benzotriazole, PdSO$_4$ in aqueous solution gives $[Pd(C_6H_4N_3)_2]SO_4$ and $[Pd(C_6H_4N_3)(C_6H_4NHN_2)]_2SO_4$ [2].

Kinetics of the reaction of PdSO$_4$ in H$_2$SO$_4$ with solutions of alkanes were measured. Pd and SO$_2$ are formed and the alkane undergoes sulphonation, oxidation and dehydrogenation reactions [3]. Acid solutions of PdSO$_4$ absorb C$_2$H$_4$ to give an unstable Pd(C$_2$H$_4)_2$ complex [4]. In acetic acid solution, PdSO$_4$ with CO gives Pd, CO$_2$ and acetic anhydride [5]. Kinetics of the oxidation of benzene-1,4-diol by PdSO$_4$ in solution [6], and the rates of oxidation of cyclohexane and $(CH_3)_3CH$ by PdSO$_4$ in H$_2$SO$_4$ with Al$_2$(SO$_4)_3$ have been measured [7]. The kinetics of oxidation of alkanes by PdSO$_4$ and nitronium ions in 80 to 100% H$_2$SO$_4$ has been reviewed [8]. The effectiveness of PdSO$_4$ in acid for oxidation of alkanes has been compared with that of Pd^{II} in H$_3$PO$_4$–BF$_3$ media, and it was concluded that the high acidity rather than any intrinsic properties of SO$_4^{2-}$ was one of the main factors responsible for the reaction [9].

Oxidation of cyclohexane to benzenesulphonic acid and of cycloheptane to the tropylium ion by PdSO$_4$ in H$_2$SO$_4$ has been observed [10, 11]; there is a review of PdSO$_4$ oxidations [12]. A

[Pd(OSO$_3$H$_2$)] complex species may be involved in the reaction of PdSO$_4$ in H$_2$SO$_4$ with alkanes [13]. Toluene is oxidised to benzoic acid in 20% sulphuric acid and benzene to biphenyl [14].

References:

[1] Eskenazi, R.; Raskovan, J.; Levitus, R. (J. Inorg. Nucl. Chem. **28** [1966] 521/6).
[2] Wilson, R. F.; Wilson, L. E.; Baye, L. J. (J. Am. Chem. Soc. **78** [1956] 2370/1).
[3] Rudakov, E. S.; Zamashchikov, V. V.; Belyaeva, N. P.; Rudakova, R. I. (Zh. Fiz. Khim. **47** [1973] 2732; Russ. J. Phys. Chem. **47** [1973] 1542/3).
[4] Shitova, N. B.; Matveev, K. I.; Kuznetsova, L. I. (Izv. Sibirsk. Otd. Akad. Nauk SSSR Ser. Khim. Nauk **1973** No. 1, pp. 25/30; C.A. **78** [1973] No. 152113).
[5] Moiseev, I. I.; Vargaftik, M. N.; Gentosh, O. I.; Zhavoronkov, N. M.; Kalechits, I. V.; Pazderskii, Yu. A.; Stromnova, T. A.; Shcherbakova, L. S. (Dokl. Akad. Nauk SSSR **237** [1977] 645/7; Dokl. Phys. Chem. Proc. Acad. Sci. USSR **232/237** [1977] 1119/21).
[6] Coe, J. S.; Rispoli, P. L. (J. Chem. Soc. Dalton Trans. **1976** 2215/8).
[7] Rudakov, E. S.; Lutsyk, A. I.; Rudakova, R. I.; Volkova, L. K. (Dopov. Akad. Nauk Ukr. RSR Ser. B Geol. Khim. Biol. Nauk **1977** No. 3, pp. 234/7; C.A. **87** [1977] No. 5104).
[8] Rudakov, E. S.; Lutsyk, A. I. (Strukt. Reakts. Sposobn. Org. Soedin. Mekh. Reacts. **1980** 69/101; C.A. **96** [1982] No. 67919).
[9] Rudakov, E. S.; Lutsyk, A. I.; Rudakova, R. I. (Kinetika Kataliz **18** [1977] 525; Kinet. Catal. [USSR] **18** [1977] 441).
[10] Rudakov, E. S.; Rudakova, R. I. (Dokl. Akad. Nauk SSSR **218** [1974] 1377/80; Dokl. Chem. Proc. Acad. Sci. USSR **218** [1974] 747/50).

[11] Rudakov, E. S.; Zamashchikov, V. V.; Lutsyk, A. I. (Teor. Eksperim. Khim. **12** [1976] 474/81; Theor. Exptl. Chem. [USSR] **12** [1976] 361/8, 363).
[12] Rudakov, E. S. (Izv. Sibirsk. Otd. Akad. Nauk SSSR Ser. Khim. Nauk **1980** No. 3, pp. 161/71; C.A. **93** [1980] No. 237980).
[13] Rudakov, E. S.; Rudakova, R. I. (Dopov. Akad. Nauk Ukr. SSR Ser. B Geol. Khim. Biol. Nauk. **1979** No. 1, pp. 46/9; C.A. **90** [1979] No. 137055).
[14] Davidson, J. M.; Triggs, C. (Chem. Ind. [London] **1966** 457).

Uses of PdSO$_4$

The application of an aqueous solution of PdSO$_4$ with Na$_2$MoO$_4$ as a colorimetric reagent for determination of the hydrogen content of gases was evaluated but found to be less effective than a PdCl$_2$-methylene-blue mixture [1]. A solution of PdSO$_4$ adsorbed onto alumina will absorb oxides of nitrogen [2]. A mixture of PdSO$_4$ and (NH$_4$)$_2$MoO$_4$ solution adsorbed on to silica gel, after drying, forms an effective filter to remove CO from cigarette smoke [3]. The efficiency of PdSO$_4$–H$_2$SO$_4$ mixtures adsorbed onto columns for gas-chromatographic determinations of aldehydes, ketones and saturated hydrocarbons has been assessed [4]. Aqueous solutions of PdSO$_4$ can be used to convert alkenes to aldehydes or ketones [5].

References:

[1] Silverman, L.; Bradshaw, W. (Anal. Chim. Acta **15** [1956] 31/42).
[2] Kitagawa, T.; Kobe Steel Works Ltd. (Fr. 1388242 [1965]; C.A. **63** [1965] 11054).
[3] Osawa, K. (Japan. 75-101598 [1975]; C.A. **83** [1975] No. 203978).
[4] Black, M. S.; Rehg, W. R.; Sievers, R. E.; Brooks, J. J. (J. Chromatog. **142** [1973] 809/22, 815).
[5] Stowe, S. C.; Dow Chemical Co. (U.S. 3425801 [1969]; C.A. **70** [1969] No. 69682).

7.7.2 Sulphato Complexes

7.7.2.1 Unsubstituted Species

The $[Pd(SO_4)_4]^{6-}$ and $[Pd(SO_4)_2]^{2-}$ Ions

Although no salts of this seem to have been isolated there are some physical data on aqueous solutions of the supposed ions.

$[Pd(SO_4)_4]^{6-}$. The promoting effect of oxygen on the oxidation of CO by $[PdX_4]^{n-}$ ($X = Cl^-$, Br^-, SO_4^{2-}) in aqueous media at 298 K has been measured [1].

$[Pd(SO_4)_2]^{2-}$. **Electronic Absorption Spectra.** Electronic spectra of Pd^{II} at various concentrations of sulphate ion have been measured (180 to 350 nm, reproduced in paper) [2]. The electronic spectra of $[Pd_2Cl_6]^{2-}$ and of $PdCl_2$ in $NaHSO_4$–$KHSO_4$ and K_2SO_4–$ZnSO_4$ melts have been measured at 77 and 300 K (300 to 500 nm) and assignments for an unformulated square-planar complex made [3]. The electronic spectra of aqueous $Pd(NO_3)_2$ in various media including SO_4^{2-} have been recorded (300 to 500 nm) [4].

Stability Constants. From polarographic measurements at a Norbide (B_4C) electrode log β_2 was calculated as 3.16 ± 0.15 [2].

Electrochemical Reduction. Polarographic reduction of $[Pd(SO_4)_2]^{2-}$ at a Norbide (B_4C) electrode showed that $E_{1/2}$ varied from $0.179 \pm 0.021\,V$ for $10^{-4}\,M$ Pd^{II} in $0.2\,M$ $HClO_4$ at zero sulphate concentration to $0.043 \pm 0.013\,V$ at $1.8\,M$ SO_4^{2-} concentration; $\Delta E_{1/2}$ is given by $\Delta E_{1/2} = (0.103 \pm 0.014) + (0.065 \pm 0.014)\log[SO_4^{2-}]$ according to [2].

Reduction of $PdSO_4$ (misprinted frequently in paper as $PbSO_4$) occurs in molten Li_2SO_4 at 580°C at 0.518 V in molten Na_2SO_4 at 580°C at 0.546 V and in molten K_2SO_4 at 0.441 V [5]. A potentiometric study has been made of $PdSO_4$ and also of $[PdCl_4]^{2-}$ in H_2SO_4 using Pd electrodes made reversible with respect to Pd complexes by depositing Pd on a Pt wire [6].

References:

[1] Golodov, V. A.; Kuksenko, E. L. (C_1 Mol. Chem. **1** [1984] 109/13).
[2] Jackson, E.; Pantony, D. A. (J. Appl. Electrochem. **1** [1971] 783/91, 287).
[3] Duffy, J. A.; MacDonald, W. J. D. (Phys. Chem. Glasses **12** [1971] 87/90).
[4] Jörgensen, C. K.; Parthasarathy, V. (Acta Chem. Scand. A **32** [1978] 957/62).
[5] Johnson, K. E.; Laitinen, H. A. (J. Electrochem. Soc. **110** [1963] 314/8).
[6] Kravtsov, V. I.; Simakova, I. V. (Vestn. Leningr. Univ. Fiz. Khim. **1969** No. 4, pp. 124/30; C.A. **72** [1970] No. 125686).

Miscellaneous Sulphates

$PdSO_4 \cdot 7PdO \cdot 6H_2O$ may be the formula of the material obtained by hydrolysis at pH 1 to 2 of palladium sulphates [1, 2].

References:

[1] Pshenitsyn, N. K.; Ginzburg, S. I. (Izv. Sekt. No. 28 [1954] 213/28; C.A. **1956** 722).
[2] Ginzburg, S. I.; Chalisova, N. N. (Zh. Neorgan. Khim. **13** [1968] 1239/45; Russ. J. Inorg. Chem. **13** [1968] 648/51).

7.7.2.2 Substituted Sulphato Complexes

7.7.2.2.1 Sulphato-aquo and -oxo Complexes

Unspecified palladium(II) sulphato-aquo complexes, possibly $Pd(SO_4)(H_2O)_2$, are involved in the anodic dissolution of Pd alloys in H_2SO_4 [1].

$Na_4[PdO(SO_4)]_2$. Some evidence for formation of this by heating a mixture of PdO and $Na_2S_2O_8$ above 155°C was presented. The X-ray powder diffraction diagram was recorded [2].

7.7.2.2.2 Sulphato-ammine Complexes

$Pd(SO_4)(NH_3)_2$. This is made by heating $Pd(SO_4)(NH_3)_2 \cdot H_2O$ to 120°C for several days. It very slowly reacts with water to give hydrolysis products [3].

The infrared spectrum was measured (700 to 1300 cm^{-1}, reproduced in paper); ν_{SO} is at 1195, 1110, 1035 and 960 cm^{-1}, and it was suggested that the sulphate has a bridging role in the complex [3].

$Pd(SO_4)(NH_3)_2 \cdot H_2O$. This is made by reaction of trans-$PdCl_2(NH_3)_2$ with Ag_2SO_4 in water. It is a yellow solid, rapidly decomposing in hot water.

At 120°C it slowly loses water to give $Pd(SO_4)(NH_3)_2$ [3].

The infrared spectrum (700 to 1300 cm^{-1}, reproduced in paper) shows ν_{SO} at 1140 to 1110, 1050 to 1030 and at 970 cm^{-1}; it was suggested that monodentate coordinated SO_4^{2-} was present [3].

$Pd(SO_4)(NH_3)_2(H_2O)$. This is made by reaction of $PdCl_2(NH_3)_2$ and Ag_2SO_3. It gave a clear yellow solution which slowly deposited $Pd(OH)_2$. The molar conductance of a freshly prepared solution was $196\,\Omega^{-1} \cdot cm^2$ [4].

$(NH_4)(H_3O)[Pd_3(SO_4)_4(NH_3)_6]$. This yellow material is made by reaction of trans-$Pd(NH_3)_2(NO_2)_2$ with conc. H_2SO_4 and subsequent precipitation with iso butyric acid [5].

The infrared spectrum was measured (400 to 4000 cm^{-1}, reproduced in paper) $\nu_{SO} = 1202$ to 1195, 1123 to 1110, 980 to 945 cm^{-1}, and on the basis of this a polymeric structure involving chelating bidentate and bridging bidentate sulphato ligands was proposed.

The X-ray photoelectronic (ESCA) spectrum was measured and binding energies of 338.5 eV (Pd $3d_{5/2}$), 169.5 eV (S $2p_{3/2}$), 401.2 and 400.3 eV (N1s) were measured [5].

References:

[1] Kukushkin, Yu. N.; Maslov, E. I.; Simanova, S. A.; Alashkevich, V. P.; Vorotnikov, M. V. (Zh. Prikl. Khim. **44** [1971] 2441/6; J. Appl. Chem. [USSR] **44** [1971] 2512/7).
[2] Jasim, F.; Jameel, I. (Thermochim. Acta **86** [1985] 155/62).
[3] Eskenazi, R.; Raskovan, J.; Levitus, R. (J. Inorg. Nucl. Chem. **28** [1966] 521/6).
[4] Earwicker, G. (J. Chem. Soc. **1960** 2620/6).
[5] Abashkin, V. E.; Muraveiskaya, G. S.; Estaf'eva, O. N.; Shchelokov, R. N. (Koord. Khim. **6** [1980] 590/3; Soviet J. Coord. Chem. **6** [1980] 278/80).

7.7.2.2.3 Sulphato-amino Complexes

$Pd(SO_4)py_2 \cdot H_2O$. This is made by reaction of trans-$PdCl_2py_2$ in water with Ag_2SO_4. It is orange [1].

The infrared spectrum was measured from 700 to 1300 cm^{-1}; ν_{SO} lies at 1235, 1125, 1020 and 930 cm^{-1}, and the complex is believed to contain bridging SO_4^{2-} ligands [1].

Pd(SO$_4$)phen. This is prepared by heating Pd(SO$_4$)phen·H$_2$O at 150°C for several days, or from 1,10-phenanthroline and PdSO$_4$·2H$_2$O in H$_2$SO$_4$ [1].

The infrared spectrum was measured (700 to 1350 cm^{-1}, reproduced in paper); ν_{SO} lies at 1240, 1125, 1040 to 1015 and at 955 cm^{-1}. It is thought to contain chelating SO_4^{2-} ligands [1].

Pd(SO$_4$)L·2H$_2$O (L = dihydrazinophthalazine). The complex is said to be formed by reaction of PdCl$_2$ in HCl with the sulphate of the ligand [2].

The complex is yellow-brown. Thermal analysis (thermogram reproduced from 20 to 800°C in paper) found that water is lost between 50 and 200°C; decomposition is more accentuated from 210 to 510°C and PdO is formed above 510°C [2].

References:

[1] Eskenazi, R.; Raskovan, J.; Levitus, R. (J. Inorg. Nucl. Chem. **28** [1966] 521/6).
[2] Grecu, I.; Curea, E. (Rev. Roumaine Chim. **13** [1968] 781/6).

7.7.2.2.4 Sulphato Dimethylsulphoxide Complex Pd(SO$_4$)[(CH$_3$)$_2$SO]$_2$

Reaction of dimethylsulphoxide in 2,2-dimethyloxypropane with PdSO$_4$·2H$_2$O gives this hygroscopic yellow complex.

The infrared spectrum (700 to 1350 cm^{-1}) shows ν_{SO} at 1220, 1135, 1020 and 970 to 940 cm^{-1}.

Reference:

Eskenazi, R.; Raskovan, J.; Levitus, R. (J. Inorg. Nucl. Chem. **28** [1966] 521/6).

7.8 Fluorosulphonato (FSO$_3^-$) Complexes

Pd(SO$_3$F)$_2$. This is made from Pd metal and excess BrOSO$_2$F at 110°C over a two-week period. The product is a light purple, hygroscopic solid.

Raman and infrared data were measured from 300 to 1500 cm^{-1} and assigned, and the diffuse reflectance electronic spectrum measured and assigned (10 Dq is 11770 cm^{-1}). It is paramagnetic at room temperature ($\mu_{eff} = 3.39$ BM). It is likely to have a polymeric structure with the FSO$_3^-$ group functioning as a tridentate ligand.

With S$_2$O$_6$F$_2$ it gives Pd[Pd(SO$_3$F)$_6$].

Pd[Pd(SO$_3$F)$_6$] (or Pd(SO$_3$F)$_3$). This is made by reaction of S$_2$O$_6$F$_2$ (peroxydisulphuryl fluoride) with Pd. It is a dark brown, hygroscopic solid, thermally stable to 180°C.

The Raman and infrared spectra were measured and assigned (200 to 1600 cm^{-1}). It is paramagnetic ($\mu_{eff} = 3.45$ BM from 107 to 334 K, following the Curie-Weiss law with $\Theta = 10 \pm 2$ K).

On the basis of the spectroscopic data it is regarded as Pd^{II}[Pd^{IV}(SO$_3$F)$_6$].

Reference:

Lee, K. C.; Aubke, F. (Can. J. Chem. **55** [1977] 2473/7).

239

7.9 Chlorosulphonate ($ClSO_3^-$) Complex $Na_2[Pd(SO_3Cl)_4]$

This is made from $Na_2[PdCl_4]$ and excess HSO_3Cl below 40°C.

It is brown; the infrared spectrum was measured from 200 to 1200 cm^{-1} ($\nu_{SO} = 1180$, 1140 and 1080 cm^{-1}; $\nu_{SCl} = 445$ cm^{-1}). The electronic spectrum was measured (400 to 500 nm) and assigned.

Reference:

Zaidi, S. A. A.; Saidi, S. R. A.; Siddiqui, Z. A.; Khan, T. A.; Shakia, M. (Syn. React. Inorg. Metal-Org. Chem. **16** [1986] 491/7).

7.10 Thiosulphato ($S_2O_3^{2-}$) Complexes

Few of these are known for palladium. It is likely, by analogy with the sulphito complexes, that bonding occurs via the sulphur atom on the monodentate form and via sulphur and oxygen in the bidentate mode.

$K_2[Pd(S_2O_3)_2]$. The preparation was not given [1] but is perhaps similar to that for $Na_2[Pt(S_2O_3)_2]$ in which $Na_2S_2O_3$ and $Na_2[PtCl_4]$ are reacted together [2].

The infrared spectrum shows ν_{SO} at 1210 and 1165 cm^{-1}, assigned to asymmetric S–O stretches, and a band at 1025 cm^{-1} assigned to a symmetric S–O stretch. On the basis of these data it was suggested that Pd–S bonding occurred in the complex [1].

$[Pd(S_2O_3)_4]^{6-}$. Although it is not clear what the evidence for the existence of this complex is $E_0 = -0.116(5)$ V has been measured for it in 0.1 M $Na_2S_2O_3$ solution, and log $\beta_4 = 35.0 \pm 0.2$ $l^4 \cdot mol^{-4}$ calculated [3].

$Pd(S_2O_3)(NH_3)_3$. No preparation was given for this complex, but the infrared spectrum showed a symmetric S–O stretch at 1010 cm^{-1} and an asymmetric stretch at 1160 cm^{-1}. It was suggested that the complex contains Pd–S bonds [2].

$K_2[Pd(S_2O_3)_2phen]$. No preparation for this complex was reported. The infrared spectrum shows bands assigned to asymmetric S–O stretches at 1175 and 1145 cm^{-1} and a band at 1010 cm^{-1} assigned to a symmetric S–O stretch. Bonding involving donation of the sulphur atoms to Pd was suggested [2].

References:

[1] Costamagna, J. A.; Levitus, R. (J. Inorg. Nucl. Chem. **28** [1966] 1116/8).
[2] Ryabchikov, D. I. (Compt. Rend. Acad. Sci. URSS [2] **27** [1940] 349/52; C. A **1940** 7773).
[3] Hancock, R. D.; Finkelstein, N. P.; Evers, A. (J. Inorg. Nucl. Chem. **39** [1977] 1031/4).

Gmelin Handbook
Pd Main Vol. B 2

8 Palladium and Selenium

8.1 Phase Diagram

A phase diagram based on differential thermal analysis and calorimetric data has been published for the Pd/Se system, see **Fig. 61**. Transformation enthalpies for palladium selenides have been tabulated.

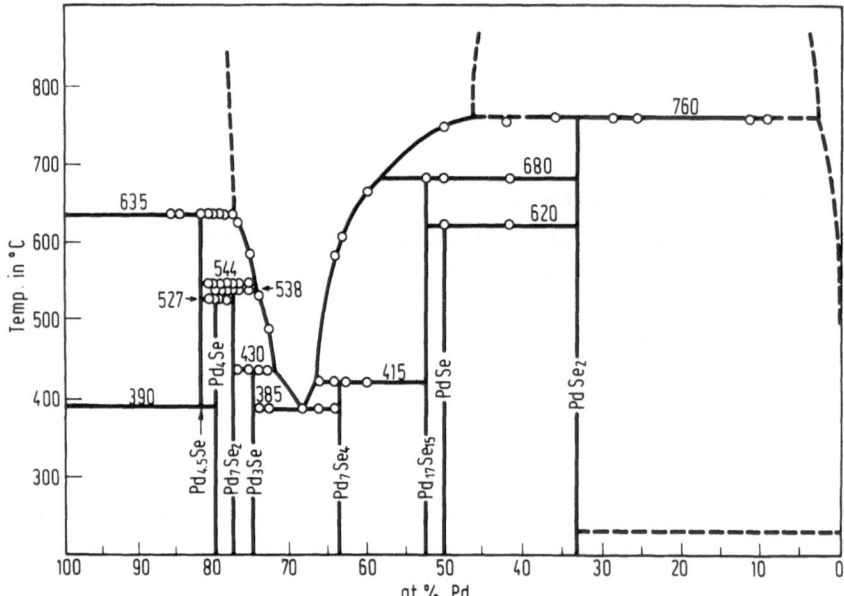

Fig. 61. The palladium–selenium phase diagram.

The phase diagram of the Pd/Se system based on differential thermal analysis and calorimetric data, see Fig. 59 [1], has been confirmed by differential thermal analysis, metallography and X-ray diffraction [2]. Transformation enthalpies for palladium selenides have been tabulated [1]. The solubility limit of selenium in palladium at 590°C has been determined as ~1.5 at% [2].

References:

[1] Olsen, T.; Røst, E.; Grønwold, F. (Acta Chem. Scand. A **33** [1979] 251/6).

[2] Takabatake, T.; Ishikawa, M.; Jorda, J. L. (J. Less-Common Metals **134** [1987] 79/89).

8.2 Binary Palladium Selenides

General References:

Schoellhorn, R., Solvated Intercalation Compounds of Layered Chalcogenide and Oxide Bronzes, Intercalation Chem. **1982** 315/60.

Tanaka, S., Physical Properties and Charge Density Waves in Transition Metal Dichalcogenides, Kotai Butsuri **14** [1979] 546/58.

Rao, S. G. V.; Shafer, M. W., Intercalation in Layered Transition Metal Dichalcogenides, Phys. Chem. Mater. Layered Struct. **6** [1979] 99/199.

Rouxel, J., Alkali Metal Intercalation Compounds of Transition Metal Chalcogenides: TX_2, TX_3 and TX_4 Chalcogenides, Phys. Chem. Mater. Layered Struct. **6** [1979] 201/50.

Schubert, K., On the Binding in Phases of $T^{10}-B^6$ Mixtures, Acta Cryst. B **33** [1977] 2631/9.

Vandenberg-Voorhoeve, J. M., Structural and Magnetic Properties of Layered Chalcogenides of the Transition Elements, Phys. Chem. Mater. Layered Struct. **4** [1976] 423/57.

Mills, K. C., Thermodynamic Data for Inorganic Sulphides, Selenides and Tellurides, Butterworths, London 1974.

Pd_7Se, Pd_6Se. Alloys of this approximate composition (85 to 87 at% Pd) have a phase which is superconducting at 0.66 K but only very poor X-ray powder patterns could be obtained [2].

$Pd_9Se_2(Pd_{4.5}Se)$. A phase of this composition has been identified in the Pd/Se system, the front reflections of a high-temperature X-ray exposure are tabulated in the paper [1].

Both DTA and metallography confirm composition close to Pd_9Se_2 [20].

Pd_4Se (see "Palladium" 1942, p. 296). Described as a brittle, grey metallic-coloured compound prepared by heating PdSe and metallic palladium in the correct proportions at 500 to 800°C for a "few days" [3]. Crystals, formed by annealing at 450°C for 45 d, are tetragonal, probable space group $P\bar{4}2_1c$-D_{2d}^4, a = 5.2324, c = 5.6470 Å, Z = 2, D_{pyk} = 10.74 g/cm³ [3, 4]. A later paper gives a = 5.230, c = 5.643 Å [20]. Powder photograph data [3] observed and calculated structure factors [4] and a view of the structure along the c axis are reproduced in papers; see **Fig. 62**. Each palladium is coordinated by 2 selenium atoms (Pd–Se = 2.46, 2.49 Å) and 10 palladium atoms (Pd–Pd = 2.76 to 3.12 Å). The selenium atoms have a body-centred arrangement and each is surrounded by 8 palladium atoms [4]. The compound is essentially diamagnetic, $\varphi_g \times 10^6$ = 0.18 (150°C), 0.04 (300°C), 0.00 (450°C), below room temperature the magnetic susceptibility is field-strength dependent (data tabulated in paper) [3]. Pd_4Se was reported not to superconduct above 1.5 K [5] but was later shown to be superconducting at 0.42 K [2].

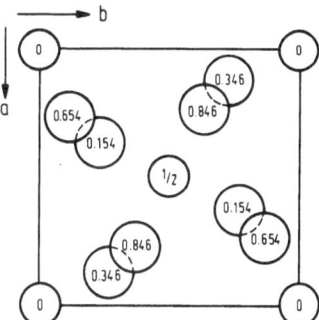

Fig. 62. The structure of Pd_4Se viewed along the c axis.
Large circles indicate Pd and small circles Se atoms.
Figures in the circles give the z parameters of the atoms [4].

The heat capacity of Pd_4Se has been determined by adiabatic calorimetry (5 to 350 K) and found to be of the usual sigmate shape, without transitions or anomalies. Heat capacity

$C_p = 32.75$, entropy $S° = 47.83$, Gibbs energy function $-(G°-H°)/T$ 24.84 cal·mol^{-1}·K^{-1} at 298.15 K [6]. The presence of a high-temperature modification above 515°C was indicated by DTA [20].

Pd$_7$Se$_2$ (Pd$_{3.5}$Se) is one of three new phases found in the Pd/Se system [1, 7] and forms twinned crystals of monoclinic symmetry with lattice constants a = 9.462, b = 5.354, c = 5.501 Å and β = 86.50° [1]. X-ray powder data (d, hkl and I$_0$ values) are tabulated. The phase is stable to 538°C [1]. A more recent paper gives revised lattice data, space group P2$_1$/a-C$_{2h}^5$, a = 9.441, b = 5.370, c = 5.495 Å, β = 93.61° [20].

Pd$_3$Se. A phase of this approximate composition has been identified in the Pd/Se system [1, 7]. It was subsequently identified as Pd$_{34}$Se$_{11}$ [20]. X-ray powder data (d and I$_0$ values) are tabulated. Peritectic decomposition occurs at 430°C [1].

Pd$_{34}$Se$_{11}$ has been obtained from a 24.5 at% selenium sample heat at 1150°C, then annealed at 400°C for 18 d. The crystals are monoclinic, space group P2$_1$/n-C$_{2h}^5$, a = 21.413, b = 5.504, c = 12.030 Å, β = 99.44°, Z = 2. Superconductivity occurs below 2.66 K, the direct current magnetic conductivity has been plotted as a function of temperature 0 to 300 K [20].

Pd$_{2.8}$Se. X-ray powder photographs of samples in the composition range Pd$_4$Se–Pd$_{1.1}$Se contain reflections attributed to a phase of composition Pd$_{2.8}$Se [3].

Pd$_{2.5-1.1}$Se. Phases in this composition range are thought to contain varying proportions of Pd$_{2.8}$Se and Pd$_{1.1}$Se. All display superconductivity below ~1 to 2.5 K [5]. Low-angle reflections from a powder photograph of Pd$_{2.5}$Se (annealed at 450°C) are tabulated [3].

Pd$_7$Se$_4$(Pd$_{1.75}$Se) is obtained by fusing Pd and Se in the correct proportions, crushing the product and annealing at 350°C [1]. Crystals annealed at 405°C for two weeks were found to be orthorhombic, primitive type with a = 5.381, b = 6.873, c = 10.172 Å and D = 9.39 ± 0.05 g/cm^3; X-ray powder data (d, hkl and I$_0$ values) are tabulated [1]. A separate investigation gave the space group D$_2^3$-P2$_1$22$_1$ with a = 6.863, b = 5.375, c = 10.162 Å [7].

"Pd$_{1.1}$Se". Several papers describe a product of this composition made by heating PdSe with palladium metal. However, the stoichiometry was later amended to **Pd$_9$Se$_8$** and finally **Pd$_{17}$Se$_{15}$** (see below). X-ray power photograph data have been recorded for "Pd$_{1.1}$Se", the unit cell is primitive cubic with a = 10.604 Å. The compound is diamagnetic, $\chi_g \times 10^6 = 0.07$ (−183°C), 0.05 (−78°C), 0.02 (20°C), 0.00 (150°C), −0.01 (300°C), −0.01 (450°C) cgs units (data plotted in paper) [3].

"Pd$_9$Se$_8$". An early incorrect formulation for Pd$_{17}$Se$_{15}$, the structure was described as simple cubic, a = 10.62 Å [8]. A high-temperature (20 to 940°C) X-ray study of "Pd$_9$Se$_8$" gave an average linear thermal expansion coefficient of 9.8×10^{-6} deg^{-1} and a volume expansion coefficient of 29.5×10^{-6} deg^{-1} at 20 to 625°C [9].

Pd$_{17}$Se$_{15}$. This phase has been obtained by heating weighed quantities of the elements in evacuated sealed quartz tubes at 600°C for 10 d, then quenching in ice water [10]. A preparation involving annealing at 350°C has also been mentioned. At 680°C Pd$_{17}$Se$_{15}$ transforms peritectically into PdSe$_2$ and a palladium-rich melt [1].

Pd$_{17}$Se$_{15}$ has been shown by X-ray diffraction methods to be identical to "Pd$_{1.1}$Se" and "Pd$_9$Se$_8$" (see above), and isostructural with Rh$_{17}$S$_{15}$. The crystals are cubic, space group Pm3m-O$_h^1$ or possibly P$\bar{4}$3m-T$_d^1$ or P432-O^1 with a = 10.606 Å and Z = 2. The unit cell is illustrated in **Fig. 63** [11]. The structure has been refined by electron diffraction using thin polycrystalline films deposited on NaCl surfaces [12]. Pd$_{17}$Se$_{15}$ is not a superconductor [2]. A plot of electrical resistivity vs. temperature (0 to 300 K) is reproduced in a paper [10].

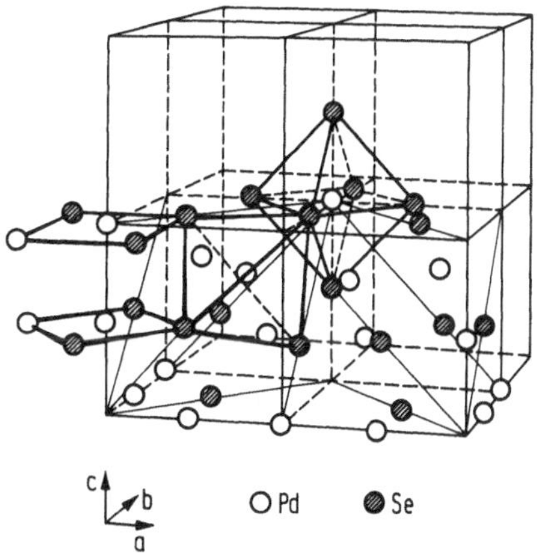

Fig. 63. The unit cell of $Pd_{17}Se_{15}$. The atoms in two octants and some needed to complete the four nonequivalent Se coordination figures are shown [11].

PdSe (see "Palladium" 1942, p. 296). This palladium(II) compound adopts a tetragonal B 34 (PdS type) structure with a = 6.727, c = 6.912 Å [8] or a = 6.733 ± 0.001, c = 6.918 ± 0.001 Å [1]. Polarisation curves for PdSe in NaOH solution have been plotted, electrochemical decomposition of these solutions proceeds rapidly to completion under conditions where oxidation occurs. The degree of decomposition of PdSe during cathodic leaching is very low because of the low overvoltage of H evolution on Pd and PdSe [13]. Gravimetric determination of palladium as PdSe has been described [14, 15]. The solubility product of PdSe has been calculated to be 4.0×10^{-74} [16]. The melting point is >300°C [17]. The kinetics [18] and thermodynamics [19] of the reaction between PdSe and H_2SO_4 below 300°C have been investigated. Values for S^{298} and ΔH^{298} for PdSe are calculated to be 12.7 cal·deg^{-1}·mol^{-1} and −14.0 kcal/mol, respectively [16].

References:

[1] Olsen, T.; Røst, E.; Grønvold, F. (Acta Chem. Scand. A **33** [1979] 251/6).
[2] Raub, C. J.; Compton, V. B.; Geballe, T. H.; Matthias, B. T.; Maita, J. P.; Hull, G. W. (J. Phys. Chem. Solids **26** [1965] 2051/7).
[3] Grønvold, F.; Røst, E. (Acta Chem. Scand. **10** [1956] 1620/34).
[4] Grønvold, F.; Røst, E. (Acta Cryst. **15** [1962] 11/3).
[5] Matthias, B. T.; Geller, S. (Phys. Chem. Solids **4** [1958] 318/9).
[6] Grønvold, F.; Westrum, E. F.; Radebaugh, R. (J. Chem. Eng. Data **14** [1969] 205).
[7] Matković, T.; Schubert, K. (J. Less-Common Metals **59** [1978] P57/P63).
[8] Schubert, K.; Breimer, H.; Burkhardt, W.; Günzel, E.; Haufler, R.; Lukas, H. L.; Vetter, H.; Wegst, J.; Wilkens, M. (Naturwissenschaften **44** [1957] 229/30).
[9] Kjekshus, A. (Acta Chem. Scand. **14** [1960] 1623/6).
[10] Kjekshus, A. (Acta Chem. Scand. **27** [1973] 1452/4).

[11] Geller, S. (Acta Cryst. **15** [1962] 713/21).
[12] Avilov, A. S.; Imamov, R. M. (Izv. Akad. Nauk SSSR Neorgan. Materialy **12** [1976] 1295/6; Inorg. Materials [USSR] **12** [1976] 1076/7).
[13] Baeshov, A.; Ugorets, M. Z.; Semina, O. I.; Nagumanov, P. N. (Zh. Prikl. Khim. **57** [1984] 1520/4; J. Appl. Chem. [USSR] **57** [1984] 1411/5).
[14] Taimni, K.; Rakshpal, R. (Anal. Chim. Acta **25** [1961] 438/47).
[15] Raoot, S.; Vaidya, V. G. (Ind. J. Chem. A **16** [1978] 996/8).
[16] Buketov, E. A.; Ugorets, M. Z.; Pashinkin, A. S. (Zh. Neorgan. Khim. **9** [1964] 526/9; Russ. J. Inorg. Chem. **9** [1964] 292/4).

[17] Dolivo-Dobrovol'skii, V. V. (Khim. Anal. Tsvetn. Redk. Metal. **1964** 144/55; C. A. **63** [1965] 2668).
[18] Barkan, V. Sh.; Greiver, T. N. (Zh. Prikl. Khim. **50** [1977] 1380/4; J. Appl. Chem. [USSR] **50** [1977] 1324/8).
[19] Greiver, T. N.; Barkan, V. Sh. (Zh. Prikl. Khim. **50** [1977] 1384/7; J. Appl. Chem. [USSR] **50** [1977] 1328/31).
[20] Takabatake, T.; Ishikawa, M.; Jorda, J. L. (J. Less-Common Metals **134** [1987] 79/89).

PdSe$_2$ (see "Palladium" 1942, p. 296). This compound is a derivative of palladium(II) (Pd^{2+}/Se$_2^{2-}$). Crystals formed by annealing at 650°C for 50 d [1] adopt a distorted pyrite-type orthorhombic structure, space group Pbca–D$_{2h}^{15}$, a = 5.71, b = 5.79, c = 7.65 Å [2], a = 5.741, b = 5.866, c = 7.691 Å, D$_{obs}$ = 6.77 g/cm^3, Z = 4 [1, 3] or a = 5.73 ± 0.02, b = 5.84 ± 0.02, c = 7.65 ± 0.03 Å [4]. Powder photograph data for a sample annealed at 450°C are tabulated [1, 3] and a projection of the structure on (001) is reproduced [3]. Each Pd atom is coordinated in square-planar manner to 4 selenium atoms (Pd–Se = 2.44 Å), these in turn are bonded in pairs (Se–Se = 2.36 Å) [3]. Polymorphism has been investigated over the temperature range 400 to 1350°C at 75 kbar pressure [5].

The compound is diamagnetic, $\chi_{mol} \times 10^6 = -68$ cgs units [6], $\chi_g \times 10^6 = -0.26$ to -0.27 cgs units over range -183 to -450°C, a plot of effective moment vs. temperature (0 to 800 K) is reproduced [1]. PdSe$_2$ is a nonmetallic conductor [6] and does not superconduct at temperatures down to 1.5 K [7]; energy gap $\Delta E = 0.4$ eV, resistivity $\varrho \approx 1\,\Omega \cdot$cm at room temperature, Seebeck coefficient $\alpha \approx 500$ µV/K [6].

References:

[1] Grønvold, F.; Røst, E. (Acta Chem. Scand. **10** [1956] 1620/34).
[2] Schubert, K.; Breimer, H.; Burkhardt, W.; Günzel, E.; Haufler, R.; Lukas, H. L.; Vetter, H.; Wegst, J.; Wilkens, M. (Naturwissenschaften **44** [1957] 229/30).
[3] Grønvold, F.; Røst, E. (Acta Cryst. **10** [1957] 329/31).
[4] Avilov, A. S.; Imamov, R. M. (Izv. Akad. Nauk SSSR Neorgan. Materialy **12** [1976] 1295/6; Inorg. Materials [USSR] **12** [1976] 1076/7).
[5] Larchev, V. N.; Popova, S. V. (Izv. Akad. Nauk SSSR Neorgan. Materialy **14** [1978] 775/6; Inorg. Materials [USSR] **14** [1978] 611/2).
[6] Hulliger, F. (J. Phys. Chem. Solids **26** [1965] 639/45).
[7] Matthias, B. T.; Geller, S. (Phys. Chem. Solids **4** [1958] 318/9).

8.3 Ternary Palladium Selenides

General Reference:

Bronger, W., Ternary Transition Metal Chalcogenides with Framework Structures and the Characterisation of Their Bonding by Magnetic Properties, Pure Appl. Chem. **57** [1985] 1363/72.

PdSSe is prepared by sintering pressed powder of the component elements at 500°C. It is isomorphous with PdS$_2$ and PdSe$_2$, diamagnetic and nonmetallic; lattice constants a = 5.595, b = 5.713, c = 7.672 (±0.005) Å; energy gap $\Delta E = 0.65$ eV, resistivity $\varrho \approx 20\,\Omega \cdot$cm at room temperature, Seebeck coefficient ~500 µV/K, magnetic susceptibility $\chi \times 10^6 = -56$ cgs units [1].

PdXSe (X = As, Sb, Bi). These cobaltite-type compounds for which few preparative details are reported [2] have a symmetry lower than that of the pyrite structure. They are superconductors below ~1.0 K. Lattice constants a = 6.092 (PdAsSe), 6.323 (PdSbSe) and 6.432 Å (PdBiSe) [3, 4].

The magnetic susceptibility χ_M of PdAsSe has been given as 14×10^{-6} cgs units at 80 K and -5×10^6 cgs units at 295 K [5].

References:

[1] Hulliger, F. (J. Phys. Chem. Solids **26** [1965] 639/45).
[2] Hulliger, F. (U.S. 3295931 [1967]).
[3] Hulliger, F.; Müller, J. (Phys. Letters **5** [1963] 226/7).
[4] Hulliger, F. (Nature **198** [1963] 382/3).
[5] Hulliger, F. (Helv. Phys. Acta **35** [1962] 535/7).

8.3.1 Alkali-Metal Selenides

$K_2Pd_3Se_4$ is prepared by fusing K_2CO_3 with palladium and selenium at 850°C [1, 2] or by heating K_2Se and PdSe (ratio 1:3) at 600 to 800°C in an evacuated quartz ampoule [2]. The crystals are orthorhombic, space group Fddd–D_{2h}^{24} (isotypic with $K_2Pd_3S_4$), a = 10.954, b = 26.196, c = 6.317 Å [1]. The compound is a nonconductor with a permeability $\varepsilon' \approx 8$ [3].

$Rb_2Pd_3Se_4$ is prepared by fusing Rb_2CO_3 with palladium and selenium at 850°C [1, 2] or by heating Rb_2Se and PdSe (ratio 1:3) at 600 to 800°C in an evacuated quartz ampoule [2]. The crystals are orthorhombic, space group Fddd-D_{2h}^{24} (isotypic with $K_2Pd_3Se_4$), a = 11.076, b = 27.092, c = 6.351 Å [1]. The compound is a nonconductor with a permeability $\varepsilon' \approx 8$ [3].

$Cs_2Pd_3Se_4$ is prepared by fusing Cs_2CO_3 with palladium and selenium at 850°C [1, 2] or by heating Cs_2Se and PdSe (ratio 1:3) at 600 to 800°C in an evacuated quartz ampoule [2]. The crystals are monoclinic, space group C2/m-C_{2h}^3 (isotypic with $Cs_2Pd_3S_4$), a = 6.493, b = 14.204, c = 6.414 Å, β = 120.41° [1]. The compound is a nonconductor with a permeability $\varepsilon' \approx 8$ [3].

References:

[1] Huster, J.; Bronger, W. (Z. Naturforsch. **29b** [1974] 594/5).
[2] Bronger, W.; Eyck, J.; Rüdorff, W.; Stössel, A. (Z. Anorg. Allgem. Chem. **375** [1970] 1/7).
[3] Gerstenhauer, E.; Vits, P. (Medd. Svenska Tekn. Vetenskapsakad. Finland No. 31 [1977] 160/81; C.A. **93** [1980] No. 35616).

8.3.2 Other Ternary Selenides

$Pd_3Tl_2Se_2$ is prepared by heating together TlSe and palladium metal. It has a Shandite-type structure with rhombohedral unit cell, a = 5.949(1) Å, α = 58.06(3)° [1].

$PdTl_2Se_2$ is prepared by the reaction of TlSe with palladium at 1100 K over a period of 4 weeks and forms moisture-sensitive needle-shaped crystals with a metallic lustre. The crystals are orthorhombic, space group Pbam-D_{2h}^9 with a = 11.355(9), b = 6.056(4), c = 3.691(2) Å, Z = 2. The crystal structure is of a new type in which palladium and selenium form infinite linear chains of edgesharing PdSe$_4$ rectangles (Pd–Se = 2.483 Å) which are separated from each other by thallium atoms. Thallium is coordinated by eight selenium atoms (Tl–Se = 3.55 Å) in a distinctly polar configuration and has one palladium neighbour (Pd–Tl = 2.92 Å) [5].

Pd₃Pb₂Se₂ is prepared by heating the requisite elements together. It adopts a Shandite-type structure with rhombohedral unit cell, a = 5.944(1) Å, α = 59.44(3)° [1].

Nb₂Pd₃Se₈ is obtained as small needle crystals by heating the elements together in a sealed silica tube for 21 d at 650 to 600°C. The crystals are orthorhombic, space group Pbam-D_{2h}^9 with a = 15.074(6), b = 10.573(4), c = 3.547(2) Å, Z = 2. The structure is illustrated in the paper, bond distances and angles are tabulated. The compound is a metallic semiconductor, a plot of the temperature dependence of the single-crystal conductivity along the needle axis c is reproduced in the paper [6].

Nb₃Pd₀.₇₂Se₇ is prepared by heating the combined elements in the ratio Nb : Pd : Se = 4 : 1 : 10 from 300 to 725 K at 15 deg/h then at 1125 K for 288 h. The crystals are monoclinic, space group C 2/m-C_{2h}^3 with a = 12.803(2), b = 3.413(1), c = 21.058(2) Å, β = 95.47(1)°, Z = 4. The structure is illustrated in the paper [9].

Nb₂Pd₀.₇₁Se₅ is obtained as small rod-shaped crystals by heating a combination of the elements mixed in the ratio Nb : Pd : Se = 2 : 1 : 5 at 1100 K for 288 h in a sealed evacuated silica tube. The crystals are monoclinic, space group C 2/m-C_{2h}^3 with a = 12.788(6), b = 3.391(1), c = 15.416(6) Å, β = 101.48(2)°, Z = 4. The structure is illustrated in the paper. The compound is a metallic conductor [7].

Nb₂PdSe₆ is obtained in crystal or powder form by heating the elements in an appropriate ratio at 1075 K in a sealed evacuated silica tube for 21 d. The crystals are monoclinic, space group C 2/m-C_{2h}^3, a = 10.368(1), b = 3.362(1), c = 12.132(2) Å, β = 113.42(2)° (single crystal) or a = 10.357(5), b = 3.362(2), c = 12.141(6) Å, β = 113.48(5)° [8].

Ta₂Pd₃Se₈ is obtained as needle-shaped crystals by heating a stoichiometric mixture of the elements at 1000 K for 10 d in a sealed evacuated silica tube. The crystals are orthorhombic, space group Pbam-D_{2h}^9 with a = 15.152(2), b = 10.631(2), c = 3.540(1) Å, Z = 2. The structure is identical to Nb₂Pd₃Se₈ [7].

Ta₂PdSe₆. Crystalline material was obtained from the elements by heating at 1050 K for 10 d in a sealed evacuated silica tube. A technique for production of a homogeneous powder form is also described. The crystals are monoclinic, space group C 2/m-C_{2h}^3 with a = 10.423(6), b = 3.375(2), c = 12.196(6) Å, β = 113.68(2)°, Z = 2. The structure is illustrated in the paper. A simple valence description $Ta^V Pd^{II} S^{-II}$ is proposed [8].

Co₂Ta₄PdSe₁₂ is obtained by heating the elements in a sealed evacuated silica tube for 10 d, then 3 weeks at 1000 to 900 K. The crystals are monoclinic, space group C 2/m-C_{2h}^3 with a = 12.951(3), b = 3.413(1), c = 19.277(5) Å, β = 110.37(2)°, Z = 2. The structure is illustrated in the paper [7].

(Cu, Pd)₂Se is one component of the brittle phase obtained when a mixture of Cu, Se and Pd is smelted at 1200 to 1300°C for 2 h [2].

Pd₁₋ₓMₓSe₂ (M = Rh, Ru, Co, Ni). Members of this group of systems prepared at ambient pressure consist of mixtures of PdSe₂ and pyrite phases. Compositions with x > 0.4 (Ru) or 0.3 (Rh, Co, Ni) prepared at 1000°C and 50 kbar pressure give single-phase regions crystallising with the pyrite structure [3, 4].

References:

[1] Zabel, M.; Wandinger, S.; Range, K. J. (Z. Naturforsch. **34b** [1979] 238/41).
[2] Greiver, T. N. (Tsvetn. Metal. **38** [1965] 28/33; Soviet J. Non-Ferrous Metals **6** No. 1 [1965] 27/32; C.A. **62** [1965] 12805).

[3] Carre, D.; Avignant, D.; Collins, R. C.; Wold, A. (Inorg. Chem. **18** [1979] 1370/2).

[4] Avignant, D.; Carre, D.; Collins, R.; Wold, A. (Mater. Res. Bull. **14** [1979] 553/7).

[5] Klepp, K. O. (J. Less-Common Metals **107** [1985] 147/53).

[6] Keszler, D. A.; Ibers, J. A. (J. Solid State Chem. **52** [1984] 73/9).

[7] Keszler, D. A.; Ibers, J. A.; Maoyu, S.; Jiaxi, L. (J. Solid State Chem. **57** [1985] 68/81).

[8] Keszler, D. A.; Squattrito, P. J.; Brese, N. E.; Ibers, J. A.; Maoyu, S.; Jiaxi, L. (Inorg. Chem. **24** [1985] 3063/7).

[9] Keszler, D. A.; Ibers, J. A. (J. Am. Chem. Soc. **107** [1985] 8119/27).

8.4 Quaternary Palladium Selenides

PdPS$_x$Se$_{(1-x)}$ (x = 0 to 1). These products are obtained by heating the requisite elements or binary compounds at ~900 to 1000°C and 20 to 65 kbar pressure. They are semiconductors and have X-ray diffraction patterns that can be indexed on the basis of orthorhombic cells.

Reference:

Bither, T. A. (U.S. 3761572 [1973]).

9 Palladium and Tellurium

9.1 Phase Diagram

Pd–Te alloys as shown in **Fig. 64**, constructed from heating and cooling curves and X-ray diffraction data [1], are discussed in [1, 2]. At room temperature the solubility of PdTe in $PdTe_2$ does not exceed 2% and that of $PdTe_2$ in PdTe does not exceed 4%, alloys with 50 to 66.7 at% Pd are mixtures of PdTe and Pd_2Te (eutectic point 490°C). Above 66.7 at% Pd various compound mixtures are stable over narrow composition ranges: $Pd_2Te/Pd_{2.5}Te$ (66.7 to 71.5 at% Pd), $Pd_{2.5}Te/Pd_3Te$ (71.5 to 75 at% Pd) and Pd_3Te/Pd_4Te (75 to 80 at% Pd) [1].

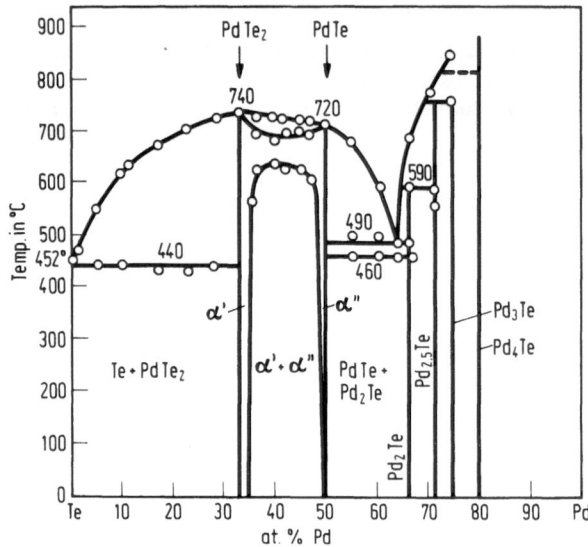

Fig. 64. Pd–Te phase diagram [1].

An early conclusion that the composition region $PdTe-PdTe_2$ was a two-phase field and that the phases did not have extended ranges of homogeneity [3] has been shown to be correct only for temperatures below 640°C. Above this temperature there is a continuous solid solution between 50.0 and 66.7 at% Te [1].

The isothermal lattice parameter vs. composition curves for the $PdTe-PdTe_2$ region are reproduced in [2]. The Te vapour pressures of liquid and solid Pd–Te alloys have been determined by an isopiestic method at 875 to 1350 K and between 35 and 75% Te. Activities and partial molar enthalpies of Te were derived [4].

References:

[1] Medvedeva, Z. S.; Klochko, M. A.; Kuznetsov, V. G.; Andreeva, S. N. (Zh. Neorgan. Khim. **6** [1961] 1737/40; Russ. J. Inorg. Chem. **6** [1961] 886/7).
[2] Kjekshus, A.; Pearson, W. B. (Can. J. Phys. **43** [1965] 438/49).
[3] Grønvold, F.; Røst, E. (Acta Chem. Scand. **10** [1956] 1620/34).
[4] Ipser, H. (Z. Metallk. **73** [1982] 151/8).

9.2 Binary Palladium Tellurides

General References:

Schoellhorn, R., Solvated Intercalation Compounds of Layered Chalcogenide and Oxide Bronzes, Intercalation Chem. **1982** 315/60.

Tanaka, S., Physical Properties and Charge Density Waves in Transition Metal Dichalcogenides, Kotai Butsuri **14** [1979] 546/58.

Rao, S. G. V.; Shafer, M. W., Intercalation in Layered Transition Metal Dichalcogenides, Phys. Chem. Mater. Layered Struct. **6** [1979] 99/199.

Rouxel, J., Alkali Metal Intercalation Compounds of Transition Metal Chalcogenides: TX_2, TX_3 and TX_4 Chalcogenides, Phys. Chem. Mater. Layered Struct. **6** [1979] 201/50.

Schubert, K., On the Binding in Phases of $T^{10}-B^6$ Mixtures, Acta Cryst. B **33** [1977] 2631/9.

Vandenberg-Voorhoeve, J. M., Structural and Magnetic Properties of Layered Chalcogenides of the Transition Elements, Phys. Chem. Mater. Layered Struct. **4** [1976] 423/57.

Mills, K. C., Thermodynamic Data for Inorganic Sulphides, Selenides and Tellurides, Butterworths, London 1974.

Pd_4Te is obtained by heating PdTe with palladium metal (ratio 1:3) at 500 to 800°C for a "few days", single crystals were obtained in a sample of gross composition Pd_6Te which had been heated at 550°C for 50 d. The crystals have primitive cubic symmetry with $a = 12.674$ Å and $D_{pyc} = 11.40$ g/cm³. There is evidence of a sub-cell with $a' = 3.168$ Å. The structure can be regarded as a body-centred arrangement of palladium atoms with 1:5 substituted by tellurium in an ordered manner. Powder photograph data for a sample annealed at 500°C with excess Pd are tabulated [1].

According to a more recent report Pd_4Te and Pd_3Te are both mixtures containing Pd_4Te and another palladium/selenium phase [2]. The compound is diamagnetic, $\chi_g \times 10^6 = 0.51$ (300°C), 0.14 (450°C) cgs units; below 150°C susceptibility is field-strength dependent, data obtained at -183, -78, 20 and 150°C are tabulated in the paper [1]. Electron diffraction data have been reported for $Pd_{3.5}Te$, a palladium-deficient form of the Pd_4Te structure [3].

Pd_xTe (x = 3, 2.5, 2). Alloys containing 71 at% Pd have X-ray powder patterns indicative of 3 phases Pd_3Te, $Pd_{2.5}Te$ and Pd_2Te of which only Pd_3Te can be isolated [4]. Intensity data and d-spacings for low-angle reflections on powder photographs of Pd_3Te, $Pd_{2.5}Te$ and Pd_2Te are tabulated [1]. The compounds Pd_3Te and Pd_2Te are amongst products precipitated when $PdCl_2/Na_2TeO_3$ solutions in aqueous HCl are reduced with hydrazine sulphate [5, 6]. The compound Pd_3Te is superconducting at 0.76 K, alloys containing 71 at% Pd have a transition temperature beginning at 0.40 K [4].

$Pd_{20}Te_7$ ($Pd_{2.85}Te$). The alloy of this composition has a rhombohedral structure isotypic with that of $Pd_{20}Sb_7$. Lattice constants are $a = 11.797(1)$, $c = 11.172(2)$ Å [7].

Pd_9Te_4 ($Pd_{2.25}Te$). This phase crystallises in a 1M 36.16 structure, space group C_{2h}^5-$P2_1/c$, $a = 7.458$, $b = 13.938$, $c = 8.839$ Å, $\beta = 91.97°$. The structure is related to that of Pd_5Sb_2 and is therefore a variant of the W family with replacement, partial occupancy and inner deformation [8].

Pd_5Te_3 ($Pd_{1.67}Te$) formed as a polycrystalline layer by co-depositing elemental Pd and Te vapours in the appropriate ration on an NaCl plate. The crystals are rhombohedral, $a = 11.677$, $c = 11.05$ Å, $Z = 8$ [9].

Pd$_3$Te$_2$ (Pd$_{1.5}$Te) is similarly prepared, the crystals were reported to be orthorhombic, a = 9.05, b = 13.54, c = 7.55 Å, Z = 9 [9]. However, a later report describes Pd$_3$Te$_2$ as being isotypic with Rh$_3$Te$_2$ and having an orthorhombic space group D$_{2h}^{17}$–Amam with a = 7.900(2), b = 12.687(3), c = 3.858(1) Å [10].

Pd$_{1.1}$Te is prepared by heating the elements together in the correct proportions, crystals are hexagonal, structure type B8$_1$ with a = 4.152, c = 5.671 Å. The compound is a superconductor below 4.07 K [4].

PdTe (see "Palladium" 1942, p. 297). Palladium(II) telluride has been prepared by heating 1:1 mixtures of the elements for a "few days" at 500 to 800°C [1], in evacuated silica tubes at 850°C [14] or 800°C for 2 h [15], or at 1000°C for 20 h followed by annealing at 800 to 500°C for 28 h [16]. It has also been obtained by adding solutions of palladium and tellurium in aqua regia to a solution of hydrazine hydrochloride in aqueous ammonia [17], by reduction of PdCl$_2$/Na$_2$TeO$_3$ solutions in aqueous HCl with hydrazine sulphate [5, 6] or by treatment of Na$_2$Te/PdCl$_2$ in 0.5 M HCl with SO$_2$ [6, 18]. Thermal decomposition of PdTe$_2$ also yields PdTe [16].

The crystals are hexagonal (NiAs type), a = 4.1521, c = 5.6719 Å [1] or a = 4.127 ± 0.004, c = 5.663 ± 0.005 Å [14]. X-ray data indicate that PdTe is nonstoichiometric, at least below 500°C, a plot of lattice parameters vs. composition (Te/Pd atomic ratio 0.8 to 1.20) is reproduced [12]. A plot of magnetic susceptibility vs. temperature (0 to 800 K) is reproduced, the hyperbolic form of the curve indicates that the Curie-Weiss law is approximately satisfied but the estimated magnetic moments are low, u_p < 1 BM [12]. De Haas-van-Alphen oscillations were not observed [19]. Graphical representations of changes in electrical resistivity and absolute thermoelectric power with temperature (0 to 280 K) are reproduced [12]. PdTe becomes superconducting at 2.3 K [20 to 22], 3.85 K [4] or ~4.0 to 4.2 K [12] depending upon the method of preparation and hence the exact stoichiometry of the phase.

The thermal conductivity has been measured (25 to ~400°C) by the Kohlrausch method [23]. The heat capacity (5 to 350°C) shows the normal sigmoidal temperature dependence with no anomalies, data are tabulated and plotted in the paper [15]. Low-temperature heat capacities for samples in 0 and 1550 G magnetic fields have been measured and plotted as a function of T^2 [4]. The melting point of PdTe has been given as 720°C [11]. The third-law entropy of PdTe at 298.15 K is 21.42 cal · gfw^{-1} · K^{-1} (gfw = g per formula weight) [15].

Gravimetric determination of palladium as PdTe [24] and the solubility product of PdTe (10^{-79}) [25] have been reported. The kinetics [26] and thermodynamics [13] of the reaction of PdTe with H$_2$SO$_4$ have been studied.

PdTe$_x$ (x = 1.02, 1.04, 1.06, 1.08). These phases prepared by heating the elements together in the correct proportions sealed in evacuated silica tubes are all superconductors. Transition temperatures are 2.56 K (x = 1.02), 2.11 K (x = 1.04, 1.06) and 1.88 K (x = 1.08). All adopt hexagonal NiAs (B8$_1$) type structures, lattice constants are given in the following table.

	PdTe$_{1.00}$	PdTe$_{1.02}$	PdTe$_{1.04}$	PdTe$_{1.06}$	PdTe$_{1.08}$
a in Å	4.152	4.144	4.143	4.138	4.135
c in Å	5.670	5.661	5.659	5.652	5.647

The low-temperature heat capacity of PdTe$_{1.04}$ in 0 and 1550 G magnetic fields has been measured and plotted as a function of T^2 when T = 0 to 8 K [4].

References:

[1] Grønvold, F.; Røst, E. (Acta Chem. Scand. **10** [1956] 1620/34).
[2] Chernyaev, I. I.; Zheligovskaya, N. N.; Borisenkova, M. K.; Subbotina, N. A. (Zh. Neorgan. Khim. **13** [1968] 1620/5; Russ. J. Inorg. Chem. **13** [1968] 848/50).

[3] Khar'kin, V. S.; Imamov, R. M.; Semiletov, S. A. (Kristallografiya **14** [1969] 907/10; Soviet Phys.-Cryst. **14** [1969] 779/81).

[4] Raub, Ch. J.; Compton, V. B.; Geballe, T. H.; Matthias, B. T.; Maita, J. P.; Hull, G. W. (J. Phys. Chem. Solids **26** [1965] 2051/7).

[5] Segui-Cros, M.; Triché, H. (Compt. Rend. **251** [1960] 1127/9).

[6] Segui-Cros, M. (Ann. Fac. Sci. Univ. Toulouse Sci. Math. Sci. Phys. [4] **25** [1961] 165/208).

[7] Wopersnow, W.; Schubert, K. (J. Less-Common Metals **51** [1977] 35/44).

[8] Matković, P.; Schubert, K. (J. Less-Common Metals **58** [1978] P39/P46).

[9] Khar'kin, V. S.; Imamov, R. M.; Semiletov, S. A. (Izv. Akad. Nauk SSSR Neorgan. Materialy **4** [1968] 1801/2; Inorg. Materials [USSR] **4** [1968] 1571/2).

[10] Matkovic, P.; Schubert, K. (J. Less-Common Metals **52** [1977] 217/20).

[11] Medvedeva, Z. S.; Klochko, M. A.; Kuznetsov, V. G.; Andreeva, S. N. (Zh. Neorgan. Khim. **6** [1961] 1737/40; Russ. J. Inorg. Chem. **6** [1961] 886/7).

[12] Kjekshus, A.; Pearson, W. B. (Can. J. Phys. **43** [1965] 438/49).

[13] Greiver, T. N.; Barkan, V. Sh. (Zh. Prikl. Khim. **50** [1977] 1384/7; J. Appl. Chem. [USSR] **50** [1977] 1328/31).

[14] Groeneveld-Meijer, W. O. J. (Am. Mineralogist **40** [1955] 646/57).

[15] Grønvold, F.; Thurmann-Moe, T.; Westrum, E. F.; Chang, E. (J. Chem. Phys. **35** [1961] 1665/9).

[16] Subbotina, N. A.; Zheligovskaya, N. N.; Pashinkin, A. S. (Zh. Neorgan. Khim. **11** [1966] 2393/4; Russ. J. Inorg. Chem. **11** [1966] 1283).

[17] Kulifay, S. M. (J. Am. Chem. Soc. **83** [1961] 4916/9).

[18] Segui-Cros, M. (Bull. Soc. Chim. France **1960** 451/2).

[19] Jan, J. P.; Pearson, W. B.; Springford, M. (Can. J. Phys. **42** [1964] 2357/8).

[20] Matthias, B. T. (Phys. Rev. [2] **92** [1953] 874/6).

[21] Matthias, B. T. (Phys. Rev. [2] **90** [1953] 487).

[22] Buckel, W. (Naturwissenschaften **42** [1955] 451/8).

[23] Okhotin, A. S. (Teploenerg. Akad. Nauk SSSR Energ. Inst. No. 2 [1960] 99/110; C.A. **57** [1962] 6645).

[24] Raoot, S.; Vaidya, V. G. (Ind. J. Chem. A **16** [1978] 996/8).

[25] Buketov, E. A.; Ugorets, M. Z.; Pashinkin, A. S. (Zh. Neorgan. Khim. **9** [1964] 526/9; Russ. J. Inorg. Chem. **9** [1964] 292/4).

[26] Barkan, V. Sh.; Greiver, T. N. (Zh. Prikl. Khim. **50** [1977] 1380/4; J. Appl. Chem. [USSR] **50** [1977] 1324/8).

Pd_xTe_2 (x = 1.05, 1.25, 1.50, 1.75). Phases with these compositions, obtained by fusing the elements together in the correct proportions, are superconducting, transition temperatures are 1.77 K (x = 1.05), 2.20 K (x = 1.25), 2.21 K (x = 1.5) and 2.25 K (x = 1.75) [1].

$PdTe_2$ (see "Palladium" 1942, p. 297). The compound is prepared by heating PdTe with tellurium at ~500 to 800°C in vacuum [2], by fusing the elements together in vacuum [1], by reduction of $PdCl_2/Na_2TeO_3$ in aqueous HCl solution with hydrazine sulphate [3], by reaction of Pd/Te solutions in aqua regia with $N_2H_4 \cdot 2HCl$ in aqueous ammonia [4] or by treatment of $PdCl_2/Na_2Te$ mixtures in 0.5 M HCl with SO_2 [5].

The crystals are trigonal, $Cd(OH)_2$, i.e., C6-type, space group $P\bar{3}ml$-C_{3v}^1, a = 4.0365, c = 5.1262 Å [2, 6], a = 4.036, c = 5.132 Å [7] or a = 4.028 ± 0.003, c = 5.118 ± 0.004 Å [8]. X-ray data for a single crystal, grown at 800°C using the Bridgeman technique, confirmed the space group $P\bar{3}ml$-C_{3v}^1 and gave lattice constants a = 4.035(1), c = 5.129(1) Å with D_{pyk} = 8.33 ± 0.03 and D_{calc} = 8.30 g/cm³. The structure is illustrated in the paper [9]. Thermal expansion of $PdTe_2$

is anisotropic (high temperature X-ray study) expansion along the a-axis over the temperature range 20 to 574°C is linear with the thermal expansion coefficient $\beta_a = 12.0 \times 10^{-6}$ deg^{-1}, along the c axis expansion is linear only over the temperature range 20 to 284°C with $\beta_c = 30 \times 10^{-6}$ deg^{-1}. Lattice constants, unit cell volumes and interatomic distances at 20 to 574°C are tabulated and plotted in the paper [10]. The polymorphism of PdTe$_2$ has been investigated over the temperature range 400 to 1350°C at 75 kbar pressure [11]. A plot of lattice parameters vs. composition (Te/Pd atomic ratio 1.80 to 2.10) is reproduced [12]. The magnetic susceptibility of PdTe$_2$ has been plotted as a function of temperature (0 to 800 K), the hyperbolic form of the curve indicates that the Curie-Weiss law is approximately obeyed but the estimated magnetic moment is low, $\mu_p < 1$ BM. De Haas-van-Alphen oscillations are observed in the magnetic susceptibility of PdTe$_2$ [12].

The electronic structure has been determined (Korringa-Kohn-Rostoker method), unlike other metal dichalcogenides of similar structure PdTe$_2$ has eigenfunctions which show strong metal-ligand mixing [13]. The de Haas-van-Alphen frequencies have been calculated [13] and agree well with experimental results [13, 14, 15]. Frequency branches from 0.1 to 4.5 kT were observed in the three principal planes suggesting a complicated Fermi surface [15]. Plots of the electronic structure and electronic density states of PdTe$_2$ are reproduced in the paper [13]. Calculations using relativistic linear muffin-tin orbitals establish that the band structure of PdTe$_2$ is separated into 3 regions: low-lying Te 5s bands, conduction bands formed by Pd 4 d and Te 5p states, and high-lying bands formed by Pd 5p, Te 6s and Te 5 d states. Details of the Fermi surface are reported and results are used to explain most of the observed de Haas-van-Alphen frequencies [16]. Angle resolved UV photoelectronic spectra from the valence bands of PdTe$_2$ have been measured and empirical band structures along high symmetry directions of the Brillouin zone have been determined. Representative HeI valence band photoelectron spectra and empirical band structure for PdTe$_2$ are reproduced in the paper [17]. The reflectance spectrum (1 to 12.5 eV [18] and 1 to 6 eV [19]) of PdTe$_2$ has been reported.

The self-consistent band structures and density-of-states functions of PdTe$_2$ have been calculated using the LMTO-ASA method. Results indicate that PdTe$_2$ is a metal with a complex Fermi surface and a relatively large electron density of states at the Fermi level. They also show that d orbitals of the metal atoms are strongly hybridised with the 5p orbitals of the tellurium atoms forming mainly covalent bonds. The self-consistent band structure, total and partial l-decomposed density of states functions, and composite energy band picture for PdTe$_2$ are all reproduced in the paper. Data are compared with related results for NiTe$_2$ and PtTe$_2$. A detailed comparison of the results with previously measured photo-emission spectra and de Haas-van-Alphen measurements shows that there is substantial agreement between calculated band structures and experimental data [27].

A complete data set for the phonon dispersion curves of PdTe$_2$ along the [ζ00], [ζζ0] and [00ζ] symmetry directions have been measured and are plotted in the paper. The lattice dynamics for PdTe$_2$ can be modelled reasonably well using axial symmetric Born-von Karman interactions out to seventh nearest neighbours together with shell-model contributions for the intersandwich and intrasandwich Te–Te nearest neighbours. The latter primarily influence the shapes of the low energy optic branches for which a dip in the vicinity of the first Γ point was observed [28].

PdTe$_2$ is a superconductor below 1.53 K [1] or ~1.8 K [12], Hall effect and resistance data have been reported [1]. Binding and conduction character of compounds with CdI$_2$ structure including PdTe$_2$ have been discussed [20, 21]. The electrical resistivity and absolute thermoelectric power of PdTe$_2$ single crystals have been measured and are plotted as a function of temperature (0 to 280 K) [12]. Intercalation with pyridine raises the critical temperature from 1.45 to 1.65 K [22].

The heat capacity of $PdTe_2$ has been measured over the temperature range 5 to 350 K, data are tabulated in the paper, the entropy of $PdTe_2$ at 298.15 K is 30.25 cal$\cdot$gfw$^{-1}\cdot$K^{-1} (gfw = g per formula weight) [23]. Dissociation pressures of $PdTe_2$ as a function of reciprocal temperature are tabulated and plotted, pressure of Te_2 vapour ranges from 3.95 (582°C) to 13.2 (660°C) Torr [24]. $PdTe_2$ melts at 740°C [25, 26].

$PdTe_{2.1}$ adopts trigonal (C6 type) structure, a = 4.037, c = 5.128 Å; it is superconducting below 1.89 K [7].

$PdTe_{2.3}$ adopts trigonal (C6 type) structure, a = 4.037, c = 5.127 Å; it is superconducting below 1.85 K [7].

References:

[1] Guggenheim, J.; Hulliger, F.; Müller, J. (Helv. Phys. Acta **34** [1961] 408/10).
[2] Grønvold, F.; Røst, E. (Acta Chem. Scand. **10** [1956] 1620/34).
[3] Segui-Cros, M.; Triché, H. (Compt. Rend. **251** [1960] 1127/9).
[4] Kulifay, S. M. (J. Am. Chem. Soc. **83** [1961] 4916/9).
[5] Segui-Cros, M. (Bull. Soc. Chim. France **1960** 451/2).
[6] Furuseth, S.; Selte, K.; Kjekshus, A. (Acta Chem. Scand. **19** [1965] 257/8).
[7] Raub, C. J.; Compton, V. B.; Geballe, T. H.; Matthias, B. T.; Maita, J. P.; Hull, G. W. (J. Phys. Chem. Solids **26** [1965] 2051/7).
[8] Groeneveld-Meijer, W. O. J. (Am. Mineralogist **40** [1955] 646/57).
[9] Lyons, A.; Schleich, D.; Wold, A. (Mater. Res. Bull **11** [1976] 1155/9).
[10] Kjekshus, A.; Grønvold, F. (Acta Chem. Scand. **13** [1959] 1767/74).

[11] Larchev, V. I.; Popova, S. V. (Izv. Akad. Nauk SSSR Neorgan. Materialy **14** [1978] 775/6; Inorg. Materials [USSR] **14** [1978] 611/2).
[12] Kjekshus, A.; Pearson, W. B. (Can. J. Phys. **43** [1965] 438/49).
[13] Myron, H. W. (Solid State Commun. **15** [1974] 395/8).
[14] Jan, J.-P.; Pearson, W. B.; Springford, M. (Can. J. Phys. **42** [1964] 2357/8).
[15] Dunsworth, A. E. (J. Low-Temp. Phys. **19** [1975] 51/7).
[16] Jan, J. P.; Skriver, H. L. (J. Phys. F **7** [1977] 1719/20).
[17] Orders, P. J.; Liesegang, J.; Leckey, R. C. G.; Jenkin, J. G.; Riley, J. D. (J. Phys. F **12** [1982] 2737/53).
[18] Sobolev, V. V.; Donetskikh, V. I. (Izv. Akad. Nauk SSSR Neorgan. Materialy **8** [1972] 688/92; Inorg. Materials [USSR] **8** [1972] 599/603; C.A. **77** [1972] No. 26775).
[19] Sobolev, V. V.; Syrbu, N. N.; Popov, Yu. V. (Khim. Svyaz. Poluprov. Tverd. Telakh **1966** 229/39; C.A. **67** [1967] No. 38055).
[20] Hulliger, F. (Helv. Phys. Acta **33** [1960] 959/61).

[21] Hulliger, F. (Helv. Phys. Acta **34** [1961] 303/4).
[22] Gamble, F. R.; Di Salvo, F. J.; Klemm, R. A.; Geballe, T. H. (Science **168** [1970] 568/70).
[23] Westrum, E. F.; Carlson, H. G.; Grønvold, F.; Kjekshus, A. (J. Chem. Phys. **35** [1961] 1670/6).
[24] Subbotina, N. A.; Zheligovskaya, N. N.; Pashinkin, A. S. (Zh. Neorgan. Khim. **11** [1966] 2393/4; Russ. J. Inorg. Chem. **11** [1966] 1283).
[25] Medvedeva, Z. S.; Klochko, M. A.; Kuznetsov, V. G.; Andreeva, S. N. (Zh. Neorgan. Khim. **6** [1961] 1737/40; Russ. J. Inorg. Chem. **6** [1961] 886/7).
[26] Hoffman, E.; MacLean, W. H. (Econ. Geol. **71** [1976] 1461/8; C.A. **86** [1977] No. 93158).
[27] Guo, G. Y.; Liang, W. Y. (J. Phys. C **19** [1986] 5365/80).
[28] Finlayson, T. R.; Reichardt, W.; Smith, H. G. (Phys. Rev. [3] B **33** [1986] 2473/80).

9.3 Ternary Palladium Tellurides

PdSeTe is reported to adopt the CdI_2 (C6) type structure, $a = 3.90$, $c = 4.98$ Å. It is metallic and shows weak, essentially temperature-independent paramagnetism [1].

PdSbTe. Cobaltite phase, $a = 6.533$ Å, superconducting below $\sim$1.2 K [2, 3].

PdBiTe. Cobaltite phase, $a = 6.656$ Å, symmetry lower than pyrite, superconducting below $\sim$1.2 K [2, 3].

Phase relations of michenerite and merenskyite in the Pd–Bi–Te system have been explored. Michenerite forms an extensive but narrow solid solution field by substitution between Bi and Te from $Pd_{0.99}Bi_{0.79}Te_{1.22}$ to $Pd_{0.95}Bi_{1.11}Te_{0.94}$. Merenskyite forms a more extensive solid solution field and varies from $PdTe_2$ to $Pd_{1.05}Te_{1.34}Bi_{0.61}$ [4].

References:

[1] Hulliger, F. (J. Phys. Chem. Solids **26** [1965] 639/45).
[2] Hulliger, F.; Müller, J. (Phys. Letters **5** [1963] 226/7).
[3] Hulliger, F. (Nature **198** [1963] 382/3).
[4] Hoffman, E.; MacLean, W. H. (Econ. Geol. **71** [1976] 1461/8; C.A. **86** [1977] No. 93158).

$(Cu, Pd)_2Te$ is one component of the brittle phase obtained when a mixture of Cu, Te and Pd is smelted at 1200 to 1300°C for 2 h, then cooled to room temperature [1].

Ag_2Pd_2Te. Evidence for a phase ot this composition in the Ag/Pd/Te system has been provided by X-ray data. The phase, which is stable below 425°C is tetragonal (PbFCl type structure) with $a = 6.24$, $c = 11.56$ Å. X-ray phase analysis data are tabulated for a range of Ag/Pd/Te compositions [2].

$Pd_{1-x}Rh_xTe_2$ ($x = 0$ to 1). Polycrystalline samples, obtained by heating stoichiometric quantities of the elements in evacuated silica tubes at 600°C, have been subjected to 1000°C and 60 kbar pressure for 1 h. The compounds with $x < 0.4$ were found to crystallise at ambient pressure with the CdI_2 structure, those with $x > 0.6$ transformed to the pyrite structure at 1000°C and 60 kbar pressure. Lattice parameters and densities are given in the tables below according to [3].

Lattice parameters and densities for phases prepared at 600°C and ambient pressure:

	a_o in Å	c_o in Å	D_{exp} in g/cm^3	D_{calc} in g/cm^3
$PdTe_2$	4.036(2)	5.130(2)	8.29(3)	8.29
$Pd_{0.9}Rh_{0.1}Te_2$	4.029(2)	5.147(2)	8.25(3)	8.29
$Pd_{0.8}Rh_{0.2}Te_2$	4.018(2)	5.160(2)	8.27(3)	8.30
$Pd_{0.7}Rh_{0.3}Te_2$	4.008(2)	5.174(2)	8.31(3)	8.32
$Pd_{0.6}Rh_{0.4}Te_2$	3.999(2)	5.192(2)	8.27(3)	8.32

Lattice parameters and densities for phases prepared at 1000°C and 60 kbar pressure:

	a_o in Å	D_{exp} in g/cm^3	D_{calc} in g/cm^3
$Pd_{0.4}Rh_{0.6}Te_2$	6.502(2)	8.65(3)	8.68
$Pd_{0.3}Rh_{0.7}Te_2$	6.492(2)	8.71(3)	8.72
$Pd_{0.2}Rh_{0.8}Te_2$	6.478(2)	8.77(3)	8.77

	a in Å	D_{exp} in g/cm^3	D_{calc} in g/cm^3
$Pd_{0.1}Rh_{0.9}Te_2$	6.463(2)	8.83(3)	8.82
$RhTe_2$	6.443(2)	8.84(3)	8.89

References:

[1] Greiver, T. N. (Tsvetn. Metal. **38** [1965] 28/33; Soviet J. Non-Ferrous Metals **6** No. 1 [1965] 27/32; C.A. **62** [1965] 12805).

[2] Chernyaev, I. I.; Zheligovskaya, N. N.; Borisénkova, M. K.; Subbotina, N. A. (Zh. Neorgan. Khim. **13** [1968] 1620/25; Russ. J. Inorg. Chem. **13** [1968] 848/50).

[3] McCarron, E.; Korenstein, R.; Wold, A. (Mater. Res. Bull. **11** [1976] 1457/62).

10 Palladium and Boron

See "Palladium" 1942, p. 298

General References:

Matkovich, V. I., Boron and Refractory Borides, Springer, Berlin 1977.

Greenwood, N. N., Boron, Pergamon, Oxford 1975.

10.1 Phase Diagram

Differential thermal analysis, metallographic analysis and X-ray analysis have been used to establish phase data of the binary Pd–B system over the temperature range 500 to 1600°C. Isothermal reaction data are collected in the table below. Compounds formed are Pd_4B, Pd_3B, Pd_5B_2 and Pd_2B. Each melts to form a liquid phase which extends to approximately $PdB_{0.57}$ at 1237 K whereupon solid boron forms. Solid palladium dissolves in boron to give a saturated composition near $PdB_{0.25}$ [1].

Isothermal reactions in the Pd–B system:

t in °C	reaction	at% B	type of reaction
1138 ± 2	$L + (Pd) \rightleftharpoons Pd_{\sim 4}B$	22.0	peritectic
933 ± 3	$Pd_{\sim 4}B \rightleftharpoons (Pd) + Pd_3B$	21.5	eutectoid
1107 ± 2	$L \rightleftharpoons Pd_{\sim 4}B + Pd_3B$	24.0	eutectic
1125 ± 5	$L \rightleftharpoons Pd_3B$	25.0	congruent melting
1062 ± 2	$L \rightleftharpoons Pd_3B + Pd_5B_2$	29.0	eutectic
1077 ± 3	$L \rightleftharpoons Pd_5B_2$	30.0	congruent melting
944 ± 3	$L + Pd_5B_2 \rightleftharpoons Pd_2B$	31.0	peritectic
964 ± 3	$L \rightleftharpoons Pd_2B + (B)$	36.5	eutectic

A partial phase diagram, see **Fig. 65**, is reproduced in a later paper together with thermodynamic data for vaporisation of Pd–B (liquid) and Pd–B–C (liquid) obtained by Knudsen effusion technique. Thermodynamic data for Pd/B compounds at their melting points are given in the table below [2].

Thermodynamic values for the compounds at their melting points:

T in K	B atom fraction	B activity	Pd activity	ΔG_f kcal/mol	kJ/mol
eutectic 1237	0.426	1.0	0.012	−6.23	−26.1
Pd_2B 1267[a]	0.310[a]	0.034	0.102	−6.61	−27.7
Pd_5B_2 1350[a]	0.300[a]	0.014	0.171	−6.73	−28.2
Pd_3B 1398[a]	0.250[a]	0.0038	0.297	−6.40	−26.8

T in K	B atom fraction	B activity	Pd activity	ΔG_f kcal/mol	kJ/mol
Pd$_4$B					
1411[a]	0.220[a]	0.0021	0.354	−6.02	−25.2

[a] Ref. [1].

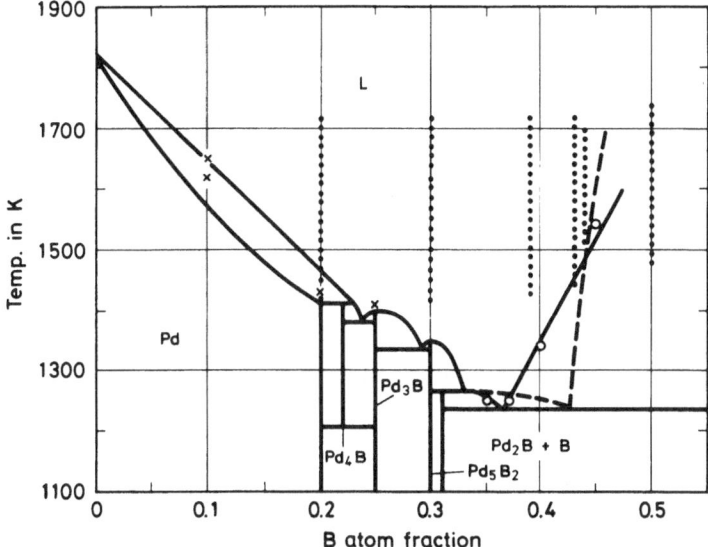

Fig. 65. Partial phase diagram for the Pd–B system. The composition and temperature range are shown by dotted lines, the calculated phase boundary is indicated by a dashed line.

References:

[1] Ipser, H.; Rogl, P. (J. Less-Common Metals **82** [1981] 363).
[2] Storms, E. K.; Szklarz, E. G. (J. Less-Common Metals **135** [1987] 217/28).

10.2 Binary Palladium Borides

PdB$_{0.021}$ to PdB$_{0.114}$. Solid solutions of boron in palladium display a miscibility gap below a critical temperature of 410°C which has been shown by X-ray analysis to extend from PdB$_{0.021}$ to PdB$_{0.114}$ at 312°C. The miscibility gap is obtained in good approximation from an evaluation of activity measurements at 1000°C using excess potentials of boron [1].

Pd$_{16}$B$_3$. An ordered phase of this composition exists in Pd–B alloys containing ≦30 at% B below a critical temperature of 397°C [2].

Pd$_{76}$B$_{24}$. The local atomic arrangement of glassy Pd$_{76}$B$_{24}$ has been determined by X-ray analysis. The experimental correlation function is in fairly good agreement with that calculated from the Percus-Yevick equation for a hard-sphere binary mixture [3].

Pd$_3$B is prepared by heating mixtures of amorphous palladium and boron powders (atom ratio 3:1) at 700 to 900°C for 5 to 6 h [4, 5], at 800 to 1100°C for an unspecified period [6] or at 700 to 1400°C for ≦80 h [7]. Preparation of Pd$_3$B in good yield by reduction of PdCl$_2$(CH$_3$CN)$_2$ in toluene with dispersions of boron on potassium has recently been described [8]. The melting point is given as 1110 to 1120°C [4, 5]. Pd$_3$B has the cementite [Fe$_3$C] structure, space group Pnma-D$_{2h}^{16}$ with a = 5.463, b = 7.567, c = 4.852 Å and Z = 4. A projection of the structure on the a-c-plane and a diagram of the B environment are given in the paper. Shortest Pd–B distances average 2.17 Å giving a value of ~0.80 Å for the boron radius [6]. Microhardness is given as 470 ± 40 kg/mm^2 [5]. The behaviour of Pd$_3$B towards boiling H$_2$SO$_4$, HNO$_3$, HCl and aqua regia has been investigated [7]. Pd$_3$B, along with various metal carbides, has been evaluated as an indicator electrode material for potentiometric methods based on simple redox systems [9].

Pd$_{76}$B$_{14}$–Pd$_{63}$B$_{37}$. Enthalpies of mixing of solid boron and solid palladium to form liquid alloys at 1400 ± 2 K have been tabulated for mole fractions of B between 0.24 and 0.37; the minimum value of ΔH_{mix}^{s-s} is about −12.9 kJ/mol and occurs near $X_B = 0.27$ [10].

Pd$_5$B$_2$ is prepared by heating together amorphous powders of palladium and boron at 800 to 950°C for 60 to 70 h [4, 5], at 800 to 1100°C for an unspecified period [6] or at 700 to 1400°C for ≦80 h [7]. The melting point is given as 1190 ± 20°C [4, 5], a later reference gives a congruent melting point of 1388 K [11]. Pd$_5$B$_2$ is monoclinic, space group C2/c-C$_{2h}^6$, with a = 12.786, b = 4.955, c = 5.472 Å, β = 97°2', Z = 4; it is isomorphous with Mn$_5$C$_2$. Shortest Pd–B distances average 2.18 Å. A projection of the structure on the a-c-plane, a schematic picture along the b axis and a diagram of the B environment are all reproduced [6]. The standard enthalpy of formation of Pd$_5$B$_2$ is given as −179.3 ± 6.3 kJ/mol [10] or −147 kJ/mol [12]. Decomposition of Pd$_5$B$_2$ by acids (H$_2$SO$_4$, HNO$_3$, HCl and aqua regia) has been described [7].

Pd$_{68.5}$B$_{31.5}$. The transition from the amorphous to the crystalline state has been studied for a metallic glass of this composition using isothermal and nonisothermal differential scanning calorimetry (DSC) tests [13, 14]. The crystallisation kinetics follow the Avrami rate equation over a wide range of transformed fractions (5 to 95%). Plots of DSC data are reproduced in the paper [13].

Pd$_2$B was first detected as the third component in sintered alloys prepared from Pd and B but was not characterised [6]. It was later prepared by arc melting stoichiometric mixtures of Pd and B powder, and shown to probably crystallise with the anti-CaCl$_2$ type structure (C35), space group Pnnm-D$_{2h}^{12}$ (No. 58) with a = 4.6918(4), b = 5.1271(4), c = 3.1096(3) Å, Z = 2; D$_{calc}$ = 9.93 g/cm^3 [15, 16]. A stereoscopic view [15] and a projection [16] of the structure are reproduced and powder diffraction data are tabulated [15]. Pd$_2$B has also been obtained by mixing 5% PdCl$_2$ with 25% K (or Na) BH$_4$ in water at room temperature and forms a voluminous nonpyrophoric black precipitate. Catalytic properties were investigated [17]. The predicted heat of formation is −21 kJ/mol [12].

"Pd$_3$B$_2$". A product of this apparent composition, made by heating Pd and B together in a silica tube [18], is now thought to be Pd$_2$Si formed by a reaction between palladium and the silica tube [6].

PdB has been detected in the vapours effusing from a Knudsen cell at high temperatures. The dissociation energy D$_0^o$ (third law), determined by mass spectrometry, is 325.3 ± 21 kJ/mol [19]. A value of −29 kJ/mol has been predicted for the heat of formation [12]. The structure of a PdB catalyst, obtained from palladium salts and NaBH$_4$, has been investigated by X-ray diffraction and charge curves plotted. The specific surface area and bond energy of PdB are 36 m^2/g and 271.5 kJ/mol, respectively [20]. Features of the electronic structure of PdB including constituent band limits and calculated compound state densities are reproduced in a paper on theory of bonding of transition metals to nontransition metals [21].

PdB$_2$. The predicted heat of formation is −25 kJ/mol [12].

References:

[1] Brodowsky, H.; Schaller, H. J. (Ber. Bunsenges. Physik. Chem. **80** [1976] 656/61).

[2] Alqasmi, R. A.; Brodowsky, H.; Schaller, H. J. (Z. Metallk. **73** [1982] 331/4).

[3] Cocco, G.; Schiffini, L.; Lucci, A.; Battezzati, L.; Sampoli, M. (Phys. Status Solidi A **75** [1983] K53/K56).

[4] Samsonov, G. V.; Kosenko, V. A.; Rud, B. M.; Sidorova, V. G. (Izv. Vysshik. Uchebn. Zavedenii Fiz. **15** No. 6 [1972] 146/7; Soviet Phys. J. **15** [1972] 912/3; C.A. **77** [1972] No. 119528).

[5] Samsonov, G. V.; Kosenko, V. A.; Ivanchenko, V. G. (Zh. Prikl. Khim. [Leningrad] **44** [1971] 2304/6; J. Appl. Chem. [USSR] **44** [1971] 2352/3).

[6] Stenberg, E. (Acta Chem. Scand. **15** [1961] 861/70).

[7] Kosenko, V. A. (Mater. Izdeliya Poluch. Metodom Poroshk. Metall. **1974** 27/35; C.A. **85** [1976] No. 40210).

[8] Carturan, G.; Cocco, G.; Semenzato, D. (React. Solids **1** [1985] 31/42).

[9] Pungor, E.; Weser, A. (Anal. Chim. Acta **47** [1969] 145/8).

[10] Kleppa, O. J.; Topor, L. (J. Less-Common Metals **106** [1985] 269/76).

[11] Moffatt, W. G. (Handbook of Binary Phase Diagrams, General Electric Company, Schenectady, N.Y., 1978/84).

[12] Niessen, A. K.; de Boer, F. R. (J. Less-Common Metals **82** [1981] 75/80).

[13] Lucci, A.; Battezzati, L. (Thermochim. Acta **54** [1982] 343/8).

[14] Lucci, A.; Battezzati, L. (Proc. Eur. Symp. Therm. Anal. **2** [1981] 567/71; C.A. **96** [1982] No. 43925).

[15] Tergenius, L. E.; Lundström, T. (J. Solid State Chem. **31** [1980] 361/7).

[16] Hassler, E.; Lundström, T.; Tergenius, L. E. (J. Less-Common Metals **67** [1979] 567/72).

[17] Polkovnikov, B. D.; Balandin, A. A.; Taber, A. M. (Katalit. Reakts. Zhidk. Faze Tr. Vses. Konf., Alma-Ata 1962 [1963], pp. 25/9; C.A. **60** [1964] 15190).

[18] Buddery, J. H.; Welch, A. J. E. (Nature **167** [1951] 362).

[19] Auwera-Mahieu, A. V.; Peeters, R.; McIntyre, N. S.; Drowart, J. (Trans. Faraday Soc. **66** [1970] 809/16).

[20] Varushchenko, V. M.; Polkovnikov, B. D.; Bogdanovskii, G. A. (Katalit. Reakts. Zhidk. Faze Tr. 3rd Vses. Konf., Alma-Ata 1971 [1972], pp. 162/6; C.A. **79** [1963] No. 97416).

[21] Gelatt, C. D.; Williams, A. R.; Moruzzi, V. L. (Phys. Rev. [3] B **27** [1983] 2005/13).

10.3 Pd Borides with Other Metals

LiPdB is readily obtained by heating a stoichiometric mixture of the elements contained in a tantalum crucible under argon in a steel bomb. It is a moisture-sensitive grey microcrystalline powder. The crystals are cubic, lattice parameters a = 6.73_3 Å, D_{calc} = 6.02, D_{exp} = 5.95 g/cm³ [1].

EuPd$_6$B$_4$ has been shown by ^{151}Eu Mössbauer and magnetic susceptibility measurements to contain the Eu ion in a mixed valence state. Experimental results are consistent with an almost temperature independent interconfiguration excitation energy (Eu^{2+} above Eu^{3+}) of 500 ± 50 K with an inhomogeneous width of 120 ± 30 K. The Mössbauer spectrum and magnetic susceptibility curve (~20 to 250 K) are reproduced in the paper [2].

References:

[1] Jung, W.; Schmidt, B. (Naturwissenschaften **63** [1976] 583/4).

[2] Felner, I.; Nowik, I. (Solid State Commun. **39** [1981] 61/3).

10.4 Palladoboranes

Cs$_2$[Pd(B$_{10}$H$_{12}$)$_2$] is prepared from Cs$_2$[B$_{10}$H$_{14}$] and PdCl$_2$ (cyclo-octa-1,5-diene) in aqueous acetone; no other details are given [1].

[N(CH$_3$)$_4$]$_2$[Pd(B$_{10}$H$_{12}$)$_2$]. The complex anion is prepared from PdCl$_2$ (cyclo-octa-1,5-diene) and Na[B$_{10}$H$_{13}$] (from NaH and B$_{10}$H$_{14}$) in tetrahydrofuran/MeCN solution and is precipitated as the tetramethylammonium salt by treatment with aqueous [N(CH$_3$)$_4$]Cl. The salt forms air-sensitive yellow crystals, m.p. 240°C (with decomposition), from ethanol/acetonitrile. Electron-ic spectrum λ_{max} = 365, 306 and 212 mμ, infrared spectrum ν(B–H) = 2505, ν(BHB) = 1950 cm^{-1}, ^{11}B NMR spectrum (reproduced in paper) δ in ppm (J, Hz) = −10.2 (119), 1.03 (127), 4.60 (115) and 24.5 (128) relative to BF$_3\cdot$OEt$_2$ [2].

[PH(C$_4$H$_9$)$_3$]$_2$[Pd(B$_{10}$H$_{12}$)$_2$] is obtained by stirring a mixture of PdCl$_2$\{P(C$_4$H$_9$)$_3$\}$_2$ and Na[B$_{10}$H$_{13}$] in diethylether for 30 min, and forms yellow crystals of poorly defined habit (m.p. 158 to 161°C, decomposition) from tetrahydrofuran. Infrared spectrum ν(B–H) = 2500, ν(P–H) = 2300, ν(B–H–B) = 1950 cm^{-1} [2].

References:

[1] Siedle, A. R.; Hill, T. A. (J. Inorg. Nucl. Chem. **31** [1979] 3874/5).
[2] Klanberg, F.; Wegner, P. A.; Parshall, G. W.; Muetterties, E. L. (Inorg. Chem. **7** [1968] 2072/7).

10.5 Miscellaneous Compounds

Pd(BR$_4$)$_2$ R=

This complex is prepared by adding a solution of K[BR$_4$] in dimethylformamide (DMF) to a cold aqueous solution of PdCl$_2$, and deposits as a light brown solid, decomposition >300°C. Other data recorded are molar conductance (DMF solution) = 10.9 cm$^2\cdot\Omega^{-1}\cdot$mol^{-1}, electronic spectrum λ_{max} = 40.8 (C.T.), 25.1 (^{1}E$_{1g} \leftarrow ^1$A$_{1g}$), 22.9 (^{1}B$_{1g} \leftarrow ^1$A$_{1g}$), 17.8 (^{1}A$_{2g} \leftarrow ^1$A$_{1g}$) kK, and infrared spectrum 4000 to 400 cm^{-1} [1].

Pd(HBR$_3$)$_2$ is similarly prepared using a solution of K[HBR$_3$]; it deposits as a light brown solid, decomposition >300°C. Other data recorded are molar conductance (DMF solution) = 7.6 cm$^2\cdot\Omega^{-1}\cdot$mol^{-1}, electronic spectrum λ_{max} = 40.8 (C.T.), 24.6 (^{1}E$_{1g} \leftarrow ^1$A$_{1g}$), 22.3 (^{1}B$_{1g} \leftarrow ^1$A$_{1g}$), 17.5 (^{1}A$_{2g} \leftarrow ^1$A$_{1g}$) kK, and infrared spectrum (4000 to 400 cm^{-1}) [1].

Pd(H$_2$BR$_2$)$_2$ is similarly prepared using a solution of K[H$_2$BR$_2$] and deposits as an orange-brown solid, decomposition at 268°C. Other data recorded are molar conductance (DMF solution) = 1.98 cm$^2\cdot\Omega^{-1}\cdot$mol^{-1}, electronic spectrum 39.22 (C.T.), 31.25 (^{1}E$_{1g} \leftarrow ^1$A$_{1g}$), 22.73 (^{1}B$_{1g} \leftarrow ^1$A$_{1g}$) kK, and infrared spectrum 4000 to 400 cm^{-1} [2].

PdB$_2$O$_4$ is prepared by heating a 1:1 mixture of Pd(NO$_3$)$_2\cdot$2H$_2$O and B$_2$O$_3$ at 800°C and 40 kbar pressure. The dark brown to black crystals up to 500 μm are idiomorphic and show tetragonal scalenohedron combined with tetragonal prisms. The crystals belong to the space group I$\bar{4}$2d–D$_{2h}^{12}$ with a = 11.675(5) and c = 5.703(5) Å, Z = 12; D$_{calc}$ = 4.92 g/cm^3. A list of d values and intensities is given. PdB$_2$O$_4$ is isotypic with CuB$_2$O$_4$ [3].

References:

[1] Siddiqi, Z. A.; Khan, S.; Zaidi, S. A. A. (Syn. React. Inorg. Metal-Org. Chem. **13** [1983] 425/47).

[2] Siddiqi, Z. A.; Khan, S.; Zaidi, S. A. A. (Syn. React. Inorg. Metal-Org. Chem. **12** [1982] 433/53).

[3] Depmeier, W.; Schmid, H. (Naturwissenschaften **67** [1980] 456).

11 Palladium and Carbon

Although the palladium-carbon interaction is important in catalysis no authentic palladium carbides appear to have been experimentally characterised. However, some unsubstantiated syntheses have been described and a number of theoretical studies have been reported. The use of palladium on carbon as a catalyst for the reduction of organic compounds is well documented in the literature of organic chemistry. A review is given in "Platinum" Suppl. Vol. A1, 1985, pp. 225/7.

11.1 Solid Solution PdC_x (x = 0.10 to 0.15)

Penetration of carbon atoms into the palladium lattice to form solid solutions in this composition range occurs when metallic palladium is exposed to a flow of carbon containing gases (C_2H_4, C_2H_2, CO) at moderate temperatures [1 to 4]. Exposure of $PdC_{0.15}$ to hydrogen does not yield β-PdH [1, 2, 3]. However, a continuous series of ternary solid solutions PdC_xH_y (0 < x < 0.15; 0 < y < 0.65) can be obtained for samples containing lesser amounts of carbon [3, 5, 6]. X-ray diffraction patterns for Pd-black treated with acetylene [1] or ethylene [3, 4] and H_2/He under various conditions are reproduced. Dilatation of the palladium lattice by dissolved carbon has been measured, the lattice parameter a ranges from 3.8900 (0.062 at% C) to 3.9029 (2.020 at% C) [7]. The solubility of carbon in palladium has been measured over the temperature range 900 to 1197°C (data tabulated in the paper) and shown to display Henryan behaviour, indicating minimal interaction between Pd and C atoms [8]. The Pd–C phase is metastable and decomposes above 600°C (inert atmosphere) or ~150°C (H_2 or O_2 atmosphere) [1].

References:

[1] Ziemecki, S. B.; Jones, G. A.; Swartzfager, D. G.; Harlow, R. L.; Faber, J. (J. Am. Chem. Soc. **107** [1985] 4547/8).
[2] Stachurski, J.; Frakiewicz, A. (J. Less-Common Metals **108** [1985] 249/56).
[3] Ziemecki, S. B.; Jones, G. A.; Swartzfager, D. G. (J. Less-Common Metals **131** [1987] 157/72).
[4] Ziemecki, S. B.; Jones, G. A. (J. Catalysis **95** [1985] 621/2).
[5] Ziemecki, S. B. (React. Solids **1** [1986] 195).
[6] Stachurski, J. (J. Chem. Soc. Faraday Trans. I **81** [1985] 2813/9).
[7] Siller, R. H.; McLellan, R. B.; Rudee, M. L. (J. Less-Common Metals **18** [1969] 432/3).
[8] Siller, R. H.; Oates, W. A.; McLellan, R. B. (J. Less-Common Metals **16** [1968] 71/3).

11.2 Palladium Carbides (see "Palladium" 1942, p. 298)

General Reference:

Frad, W. A., Metal Carbides, Advan. Inorganic Chem. Radiochem. **11** [1968] 153/247.

"Pd_5C_2". Formation of this compound, reported in 1934 (see Palladium 1942, p. 299), does not appear to have been confirmed.

"PdC". Formation of palladium carbide at 1500°C and its decomposition to graphite and palladium on slow cooling has been described [1]. A product approximating to the stoichiometry "PdC" has been obtained by subjecting mixtures of palladium metal and hydrocarbon decomposition products to impulse electric spark discharge, and is reported to have a face-centred cubic lattice with a = 3.956 Å. The X-ray diffraction pattern of this product is reproduced in the paper. Annealing at 800°C in argon, followed by slow cooling, gave palladium metal and graphite [2]. Although it is widely held that PdC has not been experimentally characterised [3, 4] its electronic structure has been probed by theoretical procedures including the pseudopotential multi-reference double excitations configuration interaction method (PP. MRD. Cl) [5] and all-electron, ab initio, Hartree-Fock and C. I. calculations [3, 4]. Mulliken population analysis data [3, 4], spectroscopic constants for the ground state [4] and a deformation density map for "PdC" at the internuclear distance 3.60 a. u. are reproduced [3]. The deviation of the "PdC" compound volume from Vegard's law is reported to be −6% [6]. Knudsen cell mass spectrometry has been used to confirm that "PdC"(g) is considerably less stable than other monocarbides of the platinum-group metals. Measurements of a Pd–Y–Cu–Ag–graphite system led to an upper value of 430 ± 20 kJ/mol for the dissociation energy of "PdC" [3]. Equations have been derived and tabulated for the estimation of carbon solubility in palladium as a function of temperature [7]. The minimum eutectic temperature of the Pd/C system has been given as 1504 ± 16°C [8].

Studies on the catalytic activity of Pd/C systems include kinetic and mechanistic investigations of hydrogen oxidation over lamellar graphite/palladium catalysts [9] and measurements of heats of reaction for catalysed hydrogenation of alkyl fluorides [10]. Bond energies of palladium catalyst surfaces with carbon have been measured [11].

PdC₂. A red-brown nonexplosive "acetylide" of this composition has been examined as a possible precipitate for gravimetric determination of palladium. A thermolysis curve is reproduced [12].

Pd–Fe–C. γ-Fe–Pd alloys are stabilised by additional elements including carbon to give materials with large linear magnetostrictions and good work abilities [13].

References:

[1] Hansen, M.; Anderko, K. (Constitution of Binary Alloys, McGraw-Hill, New York 1958, p. 377).
[2] Lisichkin, G. V.; Pisarenko, O. I.; Khinchagashvili, V. Yu.; Lunina, M. A. (Zh. Neorgan. Khim. **22** [1977] 1131/2; Russ. J. Inorg. Chem. **22** [1977] 619/20).
[3] Shim, I.; Gingerich, K. A. (J. Chem. Phys. **76** [1982] 3833/4).
[4] Shim, I.; Gingerich, K. A. (Surf. Sci. **156** [1985] 623/34).
[5] Pacchioni, G.; Koutecky, J.; Fantucci, P. (Chem. Phys. Letters **92** [1982] 486/92).
[6] Gelatt, C. D.; Williams, A. R.; Moruzzi, V. L. (Phys. Rev. [3] B **27** [1983] 2005/13).
[7] Burylev, B. P. (Izv. Vysshikh Uchebn. Zavedenii Chernaya Met. **1963** No. 6, pp. 17/21).
[8] Nadler, M. R.; Kempter, C. P. (J. Phys. Chem. **64** [1960] 1468/71).
[9] Il'Chenko, N. I.; Maksimovich, P. N.; Novikov, Yu. N.; Golodets, G. I.; Vol'pin, M. E. (Kinetika Kataliz **21** [1980]1463/8; Kinet. Catal. [USSR] **21** [1980] 1045/9).
[10] Lacher, J. R.; Kianpour, A.; Park, J. D. (J. Phys. Chem. **60** [1956] 1454/5).

[11] Kiperman, S. L.; Balandin, A. A. (Zh. Fiz. Khim. **33** [1959] 2045/52; Russ. J. Phys. Chem. **33** [1959] 277/81; C.A. **1959** 1920).
[12] Champ, P.; Fauconnier, P.; Duval, C. (Anal. Chim. Acta **6** [1952] 250/8).
[13] Nakayama, T.; Kikuchi, M.; Fukamichi, K. (IEEE Trans. Magn. MAG-16 [1980] 1071/3; C.A. **93** [1980] No. 214419).

11.3 Cyanides and Cyano Complexes

Palladium dicyanide, $Pd(CN)_2$, is well established. There is an extensive chemistry of $[Pd(CN)_4]^{2-}$ and there are a few studies on $[Pd(CN)_6]^{2-}$. The lower oxidation states are in general less well characterised.

General References:

Sharpe, A. G., Chemistry of Cyano Complexes of the Transition Metals, Academic, London 1976, p. 243.

Chadwick, B. M.; Sharpe, A. G., Transition Metal Cyanides and Their Complexes, Advan. Inorg. Chem. Radiochem. **8** [1966] 83/176, 149.

11.3.1 Palladium Dicyanide Pd(CN)₂ (see "Palladium" 1942, pp. 300/1)

This is made by reaction of $(NH_4)_2[PdCl_4]$ or of $PdCl_2$ in warm water with a little HCl and an aqueous solution of KCN [1]; from $PdCl_2$ and $Hg(CN)_2$ [2, 3]; $[Pd(CN)_4]^{2-}$ and dilute acids [1]. It can also be made by reaction of palladium(II) salts with acrolein [4], and is said to be formed during titration of $PdCl_2$ with NaCN in a LiCl–NaCl eutectic at 450°C [5].

It is cream-coloured, insoluble in water [3].

Spectroscopic and Miscellaneous Data. The X-ray photoelectron spectrum shows binding energies of 339.6(2) and 344.9(2) eV for $3d_{5/2}$ and $3d_{3/2}$, respectively [6]. Ab initio calculations on PdX_2 complexes, including $Pd(CN)_2$, have been made using effective core potentials [7].

Reactions. With ligands L (L = quinoline, 8-hydroxyquinoline, β-naphthoquinoline), $Pd(CN)_2L_2$ was formed; with L' (L' = ethylenediamine, 2,2'-bipyridyl, 1,10-phenanthroline), $Pd(CN)_2L'$ was formed. The compound will also form "lakes" with a variety of dyes [3].

Amongst the three cyanides $Ni(CN)_2$, $Pd(CN)_2$ and $Cu(CN)$, $Pd(CN)_2$ was found to be the only effective reagent for the cyanation of ethylene under pressure to give acrylonitrile and propionitrile [8]. The manufacture of unsaturated aliphatic nitriles from ethylene and propylene using $Pd(CN)_2$ in aprotic solvents under pressure has been described [9]. With 4% aqueous HCN under pressure, $HPd(CN)_3$ is said to be produced [10]. With molten KCN, $K_2[Pd(CN)_4]$ and a red polymeric material are formed [11].

Analytical methods have been applied to several samples of $Pd(CN)_2$ obtained by different procedures [2]. The compound catalyses the reaction of cyanotrimethylsilane with oxiranes to give 2-trimethylsiloxy-isocyanides [12].

References:

[1] Bigelow, J. H. (Inorg. Syn. **2** [1946] 245/6).
[2] Latimer, G. W.; Payne, L. R.; Smith, M. (Anal. Chem. **46** [1974] 311/2).
[3] Feigl, F.; Heisig, G. B. (J. Am. Chem. Soc. **73** [1951] 5630/1).
[4] Tsutsumi, M.; Shoda, S. M. (Japan. 74-27519 [1974]; C. A. **82** [1975] No. 142244).
[5] Inman, D.; Jones, B.; White, S. H. (J. Inorg. Nucl. Chem. **32** [1970] 927/36).
[6] Kumar, G.; Blackburn, J. R.; Albridge, R. G.; Moddeman, W. E.; Jones, M. M. (Inorg. Chem. **11** [1972] 296/300).
[7] Backväll, J. E.; Björkman, E. J.; Peterson, L.; Siegbahn, P. (J. Am. Chem. Soc. **107** [1985] 7265/7).
[8] Odaira, Y.; Oishi, T.; Yukawa, T.; Tsutsumi, S. (J. Am. Chem. Soc. **88** [1966] 4105/6).

[9] Tsutsumi, S. (Japan. 69-19849 [1969]; C.A. **73** [1970] No. 44941).

[10] Nuzaki, K. (U.S. 3835123 [1974]; C.A. **83** [1975] No. 132273).

[11] Magnuson, W. L.; Griswold, E.; Kleinberg, J. (Inorg. Chem. **3** [1964] 88/93).

[12] Imi, K.; Yanagihara, N.; Utimoto, K. (J. Org. Chem. **52** [1987] 1013).

11.3.2 Cyano Complexes

See "Palladium" 1942, pp. 301, 309, 317/8, 324, 336/9, 341/2.

11.3.2.1 Palladium(0) Complexes

$R_2[Pd(CN)_2]$ ($R = Ph_4P$, $MePh_3P$). These red salts are made from the appropriate cations and $K_2[Pd(CN)_2]$ in liquid ammonia. In the infrared they show ν_{CN} bands at 2035 cm^{-1} (Ph_4P) and at 2028 and 2018 cm^{-1} ($MePh_3P$) [1].

$K_2[Pd(CN)_2]$. Reaction of $K_2[Pd(CN)_4]$ and potassium in liquid ammonia at $-40°C$ gives the deep violet, diamagnetic salt. It is insoluble in polar solvents such as ether, tetrahydrofuran and dimethyl sulphoxide, and is decomposed by water, ethanol and sulphuric acid. It is soluble in liquid ammonia [1].

The infrared spectrum shows a ν_{CN} band at 2055 cm^{-1} [1]. A Cotton-Kraihanzel force-constant treatment for $[Pd(CN)_2]^{2-}$ has been briefly mentioned [8].

With NH_4CN in liquid ammonia, $K_2[Pd(CN)_4]$ is formed [2]. With $[MePh_3P]^+$, $[PdMe(CN)_3]^{2-}$ is formed, while $[Ph_4As]^+$ gives $[PdPh(CN)_3]^{2-}$ salts [2]. With CO, salts containing $[Pd(CN)(CO)]^-$ can be made [7].

$Ba[Pd(CN)_2]\cdot 0.6\,NH_3$. This black material is obtained by reaction of a barium salt with $K_2[Pd(CN)_2]$ in liquid ammonia. The infrared spectrum shows a band at 2060 cm^{-1} [1].

$K_4[Pd(CN)_4]$. This has been made by reaction of $K_2[Pd(CN)_4]$ with potassium in liquid ammonia at $-33°C$ as a yellowish white material. It is stable at $0°C$ for some 6 h in liquid ammonia, 12 h in liquid ammonia at $-33°C$, and 2 h in an ammonia atmosphere at room temperature [3]. The salt cannot be obtained from $Pd(CN)_2$ and molten KCN [6].

There is some evidence that $[Pd(CN)_4]^{4-}$ may be present in melts of $PdCl_2$ in molten KCN [4, 5], though it has also been suggested that polymeric palladium(II) cyano species are present in such mixtures [6]. Infrared bands at 2038 cm^{1-} (melt) and at 2022 cm^{1-} (quenched melt) have been associated with such a species, prepared by adding sodium cyanide to melts containing $PdCl_2$ and KCN. Electronic absorption spectra were measured (180 to 600 nm, reproduced in paper) [4]; see also [5].

Polarographic evidence for the existence of $[Pd(CN)_4]^{4-}$ has been presented [9].

References:

[1] Nast, R.; Bülck, J.; Kramolowsky, R. (Chem. Ber. **108** [1975] 3461/8).

[2] Dodsworth, E. S.; Eaton, J. P.; Ellerby, M. P.; Nicholls, D. (Inorg. Chim. Acta **89** [1984] 143/5).

[3] Burbage, J. J.; Fernelius, W. C. (J. Am. Chem. Soc. **65** [1943] 1484/6).

[4] de Haas, K. S.; Fouché, K. F. (Inorg. Chim. Acta **24** [1977] 269/76).

[5] de Haas, K. S.; Fouché, K. F. (Inorg. Chim. Acta **26** [1979] 213/20).

[6] Magnuson, W. L.; Griswold, E.; Kleinberg, J. (Inorg. Chem. **3** [1964] 88/93).

[7] Nast, R.; Bülck, J. (Z. Anorg. Allgem. Chem. **416** [1975] 285/8).
[8] Alvarez, S.; Lopez, C.; Berjemo, M. J. (Transition Metal Chem. [Weinheim] **9** [1984] 123/6; C.A. **100** [1984] No. 182475).
[9] Hirota, M.; Fujiwara, S. (J. Inorg. Nucl. Chem. **35** [1973] 3883/9).

11.3.2.2 Palladium(I) Complexes

No definite evidence for such species is available. In early work a palladium(I) cyano species was said to be formed on reduction of $K_2[Pd(CN)_4]$ in alkaline solution by sodium amalgam [1]. Some evidence for the presence of unspecified palladium(I) cyanide complexes in $PdCl_2$–KCN melts has been given; an infrared band at 2130 cm^{-1} in a quenched melt may arise from this species [2]. There is some polarographic evidence for the existence of $[Pd(CN)_4]^{3-}$ [3].

References:

[1] Manchot, W.; Schmid, H. (Ber. Deut. Chem. Ges. **63** [1930] 2782/6).
[2] de Haas, R. S.; Fouché, K. F. (Inorg. Chim. Acta **24** [1977] 269/76).
[3] Hirota, M.; Fujiwara, S. (J. Inorg. Nucl. Chem. **35** [1973] 3883/9).

11.3.2.3 Palladium(II) Complexes

11.3.2.3.1 Tetracyanopalladates $[Pd(CN)_4]^{2-}$

These have been extensively studied, particularly the electronic absorption and emission spectra of the alkaline earth salts. The structural and electronic spectral aspects are the subject of a review.

The general method of preparation is to treat $PdCl_2$ or other palladium(II) salts with excess of a cyanide of the appropriate cation and recrystallising the product from water.

Reference:

Moreau-Colin, M. L. (Struct. Bonding [Berlin] **10** [1972] 167/90).

Vibrational Spectra

Data on these have been reviewed [1], and in the table below frequencies for the ions are listed using the numbering system of this reference. The most up-to-date Raman spectra (and infrared data for v_{13}) concern $[Pd(^{12}C^{14}N)_4]^{2-}$, $[Pd(^{12}C^{15}N)_4]^{2-}$ and $[Pd(^{13}C^{14}N)_4]^{2-}$ (250 to 2200 cm^{-1}) [2] but some older data are sometimes more extensive (80 to 2200 cm^{-1}) [3, 4]. There are also values for the solutions in the infrared for v_{13} (E_u) [2,5]. The integrated intensity of the v_{13} (E_u) band has been measured [12].

Force constant data using a general quadratic force field were calculated using data from [2] and from data (see p. 273) on solid $K_2[Pd(CN)_4]$ [2]; a general discussion of such force constant treatments has been given [6]. A simple valence force field using the data of [3] was also used [7], and force constants using the data of [4] calculated [8,9]. The general valence and Cotton-Kraihanzel force fields (for k_{CN} only) for $[Pd(CN)_4]^{2-}$ have been compared [10], and metal-ligand stretching force constants for a number of species, including $[Pd(CN)_4]^{2-}$, compared [11].

For further data see under individual salts, pp. 270/80.

Vibrational spectra of $[Pd(CN)_4]^{2-}$ in solution:

	(a)	(b)	(b)	(c)	(d)	(e)
ν_1 (ν_{CN}; A_{1g})	2159p	2160p	2160.5	2130.7	2113.1	2150p
ν_2 (ν_{MC}; A_{1g})	439p	437p	428.0	420.0	423.7	—
ν_3 (ν_{MC}; A_{2g})	—	—	—	—	—	—
ν_4 (ν_{CN}; B_{1g})	2147dp	2145dp	2146.4	2116.8	2099.7	2138dp
ν_5 (ν_{MC}; B_{2g})	—	—	427.6	416.8	419.0	—
ν_6 (δ_{MCN}; B_{2g})	290dp	290dp	—	—	—	—
ν_7 (δ_{CMC}; B_{2g})	94dp	92	—	—	—	—
ν_8 (δ_{MCN}; E_g)	—	390dp	294.0	290.4	285.2	—
ν_{13} (ν_{CN}; E_u)*)	2139 [4, 5]	2135.8	2104.8	2089.7	—	—
Ref.	[3]	[4]	[2]	[2]	[2]	[13]

(a) Raman spectrum of aqueous solution of $Na_2[Pd(CN)_4]$. – (b) Raman spectrum of aqueous solution of $K_2[Pd(CN)_4]$. – (c) Raman spectrum of aqueous solution of $K_2[Pd(^{12}C^{15}N)_4]$. – (d) Raman spectrum of aqueous solution of $K_2[Pd^{13}C^{14}N]_4$. – (e) Raman spectrum of $K_2[Pd(CN)_4]$ in liquid ammonia [13]. – *) Infrared spectrum of aqueous solution of $K_2[Pd(CN)_4]$ [5].

References:

[1] Jerome-Lerutte, S. (Struct. Bonding [Berlin] 10 [1972] 153/66).
[2] Kubas, G. J.; Jones, L. H. (Inorg. Chem 13 [1974] 2816/9).
[3] Mathieu, J.-P.; Cornevin, S. (J. Chim. Phys. 36 [1939] 271/9).
[4] Bonino, G. B.; Chiorboli, P.; Fabbri, G. (Atti Accad. Nazl. Lincei Classe Sci. Fis. Mat. Nat. Rend. [8] 26 [1959] 137/46; C.A. 1960 14943).
[5] Mathieu, J.-P.; Poulet, H. (Compt. Rend. 248 [1959] 2315/7).
[6] Jones, L. H.; Swanson, B. L. (Accounts Chem. Res. 9 [1976] 128/34).
[7] Pistorius, C. W. F. T. (Z. Physik. Chem. [Frankfurt] 23 [1960] 206/9).
[8] Salvetti, O. (Atti Accad. Nazl. Lincei Classe Sci. Fis. Mat. Nat. Rend. [8] 26 [1959] 225/30).
[9] Salvetti, O. (Atti Accad. Nazl. Lincei Classe Sci. Fis. Mat. Nat. Rend. [8] 26 [1959] 383/5).
[10] Alvarez, S. (Transition Metal Chem. [Weinheim] 7 [1982] 116/8).

[11] Hancock, R. D.; Evers, A. (J. Inorg. Nucl. Chem. 35 [1973] 2558/61).
[12] Memering, M. N.; Jones, L. H.; Bailar, J. C. (Inorg. Chem. 12 [1973] 2793/801).
[13] Gans, P.; Gill, J. B.; MacIntosh, D. (Polyhedron 6 [1987] 79/84).

Electronic Spectra

The topic has been reviewed for $[Pd(CN)_4]^{2-}$ [1]. Spectra of $[Pd(CN)_4]^{2-}$ in aqueous solution, with assignments and with reproduction of the spectra, have been given: 200 to 265 nm [1]; 200 to 400 nm [2]; 180 to 260 nm [3]; 180 to 300 nm [4]; 200 to 250 nm [5]; 200 to 500 nm [12]. Bands were listed and assigned (200 to 700 nm) [6] and 200 to 250 nm [7,8] without reproduction of the spectrum. The spectra of $(Bu_4^nN)_2[Pd(CN)_4]$ at room temperature in H_2O, CH_3CN and an ether-isopentane-ethanol mixture at room temperature and at 77 K were listed (200 to 300 nm) and assignments proposed [9]. Electronic spectra of $PdCl_2$ in molten cyanide have also been measured (300 to 500 nm; reproduced in paper) [10].

The magnetic circular dichroism (MCD) spectrum was measured (200 to 250 nm, reproduced in paper) [5]. The luminescence spectrum of $[Pd(CN)_4]^{2-}$ (200 to 700 nm, reproduced in paper) has been measured [6].

Dichroism of $[Pd(CN)_4]^{2-}$ salts has been discussed [11].

References:

[1] Moreau-Colin, M. L. (Struct. Bonding [Berlin] **10** [1972] 167/90, 172).
[2] Moreau-Colin, M. L. (J. Chim. Phys. **67** [1970] 498/506, 499).
[3] Macadré, A.; Moncuit, C. (Compt. Rend. **261** [1965] 2339/42).
[4] Moncuit, C. (J. Chim. Phys. **64** [1967] 494/8).
[5] Piepho, S. B.; Schatz, P. N.; McCaffery, A. J. (J. Am. Chem. Soc. **91** [1969] 5994/6001, 5995).
[6] Rossiello, L. A.; Furlani, C. (Atti Accad. Nazl. Lincei Classe Sci. Fis. Mat. Nat. Rend. [8] **38** [1965] 207/13).
[7] Gray, H. B.; Ballhausen, C. J. (J. Am. Chem. Soc. **85** [1963] 260/5).
[8] Perumareddi, J. R.; Liehr, A. D.; Adamson, A. W. (J. Am. Chem. Soc. **85** [1963] 249/59, 259).
[9] Mason, W. R.; Gray, H. B. (J. Am. Chem. Soc. **90** [1968] 5721/9).
[10] de Haas, K. S.; Fouché, K. F. (Inorg. Chim. Acta **26** [1978] 213/20).

[11] Brasseur, H. (Bull. Classe Sci. Acad. Roy. Belg. [5] **49** [1963] 1028/9).
[12] Kiss, A.; Bán, M. L. (Acta Chim. [Budapest] **40** [1964] 397/417, 410).

Photochemistry

Irradiation of $[Pd(CN)_4]^{2-}$ with light at 254 nm causes no apparent reaction [1]. Quenching of the triplet state of a number of organic molecules by d^8 square-planar complexes has been studied: $[PdCl_4]^{2-}$ exhibits such quenching but $[Pd(CN)_4]^{2-}$ does not, having a first excited state at much higher energy [2].

References:

[1] Moggi, L.; Bolletta, F.; Balzani, V.; Scandola, F. (J. Inorg. Nucl. Chem. **28** [1966] 2589/97).
[2] Marshall, K. C.; Wilkinson, F. (Z. Physik. Chem. [N. F.] **101** [1976] 67/78, 74).

Miscellaneous Spectroscopic Techniques

The ^{13}C NMR spectrum of $[Pd(CN)_4]^{2-}$ in D_2O shows a sharp line, $\delta = 131.9$ ppm relative to Me_4Si; spin-lattice relaxation for $[Pd(CN)_4]^{2-}$ was also recorded from 25 to 75°C [1]. There are earlier chemical shift data for $[Pd(CN)_4]^{2-}$ [2]. The ^{15}N chemical shift for the ion was also observed [3] and also the ^{14}N shift [4]. Energies of relaxation on ionisation of valence levels in $[Pd(CN)_4]^{2-}$ were calculated using second-order perturbation theory [5].

References:

[1] Pesek, J. J.; Mason, W. R. (Inorg. Chem. **18** [1979] 924/8).
[2] Hirota, M.; Koike, Y.; Ishizuka, H.; Yamasaki, A.; Fujiwara, S. (Chem. Letters **1973** 853/4).
[3] Wasylishen, R. E. (Can. J. Chem. **60** [1982] 2194/7).
[4] Griffith, W. P.; Mockford, M. J.; Skapski, A. C. (Inorg. Chim. Acta **328** [1987] 179/86).
[5] Voitzuk, A. A.; Mazalov, L. N. (Zh. Strukt. Khim. **23** [1982] 61/6; J. Struct. Chem. USSR **23** [1982] 50/4).

Electrochemical Data

The E_0 for $[Pd(CN)_4]^{2-}$ has been found to be -0.564 V at 10°C and -0.553 V at 60°C for solutions in 0.5 M CN^- [1].

Polarographic reduction of $[Pd(CN)_4]^{2-}$ in aqueous solution at the dropping mercury electrode was studied with a variety of supporting electrolytes, and a half-wave potential near −1.70 V vs. standard calomel electrode noted. Evidence for reduction to $[Pd(CN)_4]^{3-}$ and $[Pd(CN)_4]^{4-}$ was presented [2].

The effect of added CN^- on the electrolytic deposition of Pd has been studied [3], and the EMF changes consequent on addition of NaCN to $PdCl_2$ in a LiCl–KCl eutectic at 450°C plotted [4].

References:

[1] Fasman, A. B.; Kutyukov, G. G.; Sokol'skii, D. V. (Zh. Neorgan. Khim. **10** [1965] 1338/44; Russ. J. Inorg. Chem. **10** [1965] 727/30).
[2] Hirota, M.; Fujiwara, S. (J. Inorg. Nucl. Chem. **35** [1973] 3883/9).
[3] Visomirskis, R.; Morgenshtern, Ya. L. (Lietuvos TSR Mokslu Akad. Darbai B **1966** No. 4, pp. 29/37; C.A. **67** [1967] No. 60262).
[4] Inman, D.; Jones, B.; White, S. H. (J. Inorg. Nucl. Chem. **32** [1970] 927/36, 932).

Stability Constants

Preliminary spectrophotometric data suggest that log β_4 is very high, between 65 and 75, while other potentiometric data suggested a value close to 60 [1]. Earlier potentiometric data gave log $\beta_4 = 42.4$ [2, 3], 51.6 [4] and 49.9 [5].

For the reaction $Pd^{2+} + 4CN^- \rightarrow [Pd(CN)_4]^{2-}$ $\Delta H°$ and $\Delta G°$ were measured as -386 ± 4 and -242 ± 2 kJ/mol with $\Delta S°$ as -485 ± 12 J·deg^{-1}·mol^{-1} [2], and $\Delta H° = -385 \pm 2$, $\Delta G° = -242$ J/mol, $\Delta S°$ -476 J·deg^{-1}·mol^{-1} also given [3].

References:

[1] Hancock, R. D.; Evers, E. (Inorg. Chem. **15** [1976] 995/6).
[2] Izatt, R. M.; Watt, G. D.; Christensen, J. J. (J. Chem. Soc. A **1967** 1304/8).
[3] Watt, G. D.; Eataugh, D.; Izatt, R. M.; Christensen, J. J. (Proc. Utah Acad. Sci. Arts Letters **42** No. 2 [1965] 298/302; C.A. **66** [1967] No. 59526).
[4] Fasman, A. B.; Kutyukov, G. G.; Sokol'skii, D. V. (Zh. Neorgan. Khim. **10** [1965] 1338/44; Russ. J. Inorg. Chem. **10** [1965] 727/30).
[5] Visomirskis, R.; Morgenshtern, Ya. L. (Lietuvos TSR Mokslu Akad. Darbai B **1966** No. 4, pp. 29/37; C.A. **67** [1967] No. 60262).

Reactions of $[Pd(CN)_4]^{2-}$

Exchange of $^{14}CN^-$ with aqueous $[Pd(CN)_4]^{2-}$ is complete within 3 min at 25°C at pH = 10 [1]. Later data using ^{13}C NMR methods gave a rate constant of 120 mol^{-1}·s^{-1} at 24°C; $\Delta H^* = 17 \pm 2$ kJ/mol, $\Delta S^* = 143 \pm 8$ J·K^{-1}·mol^{-1} [2]. Rates of reaction of $[Pd(CN)_4]^{2-}$ with hydrated electrons have been determined [3]. The reaction of $[Pd(CN)_4]^{2-}$ with a variety of cations was studied and the pH values for precipitation determined [4]. Unlike its nickel analogue, $[Pd(C_2S_2O_3)_2]^{2-}$ does not react with $[Pd(CN)_4]^{2-}$ [5]. Titrimetric methods for determination of complex cyanides, including $[Pd(CN)_4]^{2-}$, have been given [6]. The adsorption of $[Pd(CN)_4]^{2-}$ on ion-exchange resins has been compared with that for other anions [7].

Although I_2 reacts rapidly with CN^- it does not readily react with cyano complexes. It has been used as a scavenger to measure the dissociation rates of a number of cyano complexes including $[Pd(CN)_4]^{2-}$; the rate constant for the dissociation of the first CN^- ligand from $[Pd(CN)_4]^{2-}$ is $(4.3 \pm 0.2) \times 10^{-5}$ s^{-1} at 25°C [8].

References:

[1] Adamson, A. W.; Welker, J. P.; Volpe, M. (J. Am. Chem. Soc. **72** [1950] 4030/6).

[2] Pešek, J. J.; Mason, W. R. (Inorg. Chem. **22** [1983] 2958/9).

[3] Anbar, M.; Hart, E. J. (Advan. Chem. Ser. **81** [1968] 79/94, 82).

[4] Mendez, J. H.; Medina, J. E.; Gimeno, J. V. G. (Quim. Anal. **29** [1975] 32/5).

[5] Kida, S. (Bull. Chem. Soc. Japan **33** [1960] 1204/6).

[6] Selig, W. S. (Microchem. J. **32** [1985] 18/23).

[7] Aveston, J.; Everest, D. A.; Wells, R. A. (J. Chem. Soc. **1958** 231/9).

[8] Crouse, W. C.; Bennett, D. A.; Margerum, D. W. (React. Kinet. Catal. Letters **1** [1974] 135/40).

Applications

Electroplating of Pd from $[Pd(CN)_4]^{2-}$ solutions has been reported.

Reference:

Valsiuniene, J.; Prokopchik, A. J.; Kaskelis, A. (Lietuvos TSR Mokslu Akad. Darbai B **1976** No. 4, pp. 25/32; C.A. **86** [1977] No. 125406).

Cyanopalladates of Nonmetals

$H_2[Pd(CN)_4]$. This is made by reaction of conc. HCl saturated with diethyl ether and an aqueous solution of $K_2[Pd(CN)_4]$. It is white [1]. It can also be made in aqueous solution from $K_2[Pd(CN)_4]$ and HCl; such a solution should be freshly made [2].

The infrared spectrum was measured from 400 to 4000 cm^{-1}; ν_{CN} is at 2202 cm^{-1} and ν_{MC} at 488 cm^{-1}. It was concluded that the acid contains symmetric $N \cdots H \cdots N$ hydrogen bonds [1].

$(Ph_2I)_2[Pd(CN)_4]$. The thermal decomposition of this has been measured; PhI and PhCN are evolved below 400°C and $(CN)_2$ above 400°C [3].

$(EtNH_3)_2[Pd(CN)_4]$. This is made from $Ba[Pd(CN)_4]$ with $(EtNH_3)_2SO_4$; the colourless material is tetragonal, a=17.15, c=18.90 Å, space group $I4_1/acd-D_{4h}^{20}$ [4]. The specular electronic reflectance spectrum was measured (180 to 500 nm, reproduced in paper) [11].

$(Et_2NH_2)_2[Pd(CN)_4]$. This colourless salt is triclinic, a=15.79(1), b=9.10(1), c=6.35(1) Å, α=84°11(1)′, β=92°42(1)′, γ=94°5(1)′, space group $P\bar{1}-C_i^1$, Z=2, density 1.305(10) g/cm³. Coordination about the palladium is square planar, mean Pd–C=1.98(2) Å, mean C–N=1.16(2) Å. There is hydrogen-bonding between the protons of the cation and the nitrogen atoms ($N \cdots H$ 1.69 to 1.90 Å) of the cyanide giving chains as shown schematically in **Fig. 66** [5]; see also [4, 6].

The electronic absorption spectrum of the solid was measured (200 to 400 nm), reproduced in paper [7].

$(Et_3NH)_2[Pd(CN)_4] \cdot 2H_2O$. This colourless material is monoclinic, a=16.54, b=9.14, c=8.22 Å, β=93°32′, space group either C2/m, C2 or Cm (C_{2h}^3, C_2^3, or C_s^3). The coordination about the palladium is square planar, Pd–C=2.05(4) Å, C–N=1.13(5) Å [4].

$(Et_4N)_2[Pd(CN)_4]$. The infrared spectrum (200 to 2200 cm^{-1}) has been measured and assigned; Urey-Bradley and simple valence force field data were calculated for the anion on the basis of these data [8].

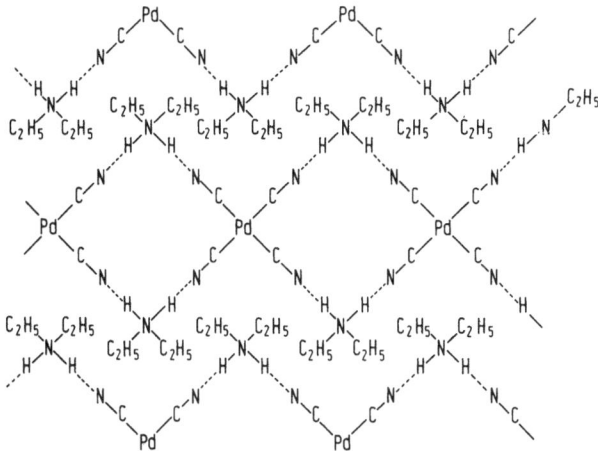

Fig. 66. Schematic representation of chains
in $(Et_2NH_2)_2[Pd(CN)_4]$ [5].

$(Bu_4^nN)_2[Pd(CN)_4]$. This greenish-grey salt is made from $K_2[Pd(CN)_4]$ in water with a slight deficiency of $(Bu_4^nN)Cl$, extracted with CH_2Cl_2 and the organic phase, then evaporated.

Electronic spectra of the salt in water, acetonitrile, and at room temperature and at 77 K in an ether-isopentane-ethanol mixture (200 to 300 nm, data listed and assignments proposed) [9].

$(C_{12}H_{12}N_2H_2)[Pd(CN)_4]$. This is a white crystalline material obtained by recrystallising the product of reaction of $K_2[Pd(CN)_4]$ in HCl with benzidine [10].

$(C_{10}H_6O_2H)_2[Pd(CN)_4]$. This white crystalline salt is obtained by recrystallisation of "$H_2[Pd(CN)_4]$" in HCl with 1,2-naphthoquinone [10].

$(C_{19}H_{16}N_4H_2)[Pd(CN)_4]$. This is a solid material obtained from nitron in HCl with "$H_2[Pd(CN)_4]$" [10].

References:

[1] Evans, D. F.; Jones, D.; Wilkinson, G. (J. Chem. Soc. **1964** 3164/7).
[2] Feigl, F.; Heisig, G. B. (J. Am. Chem. Soc. **73** [1951] 5630/1).
[3] Gyoryova, K.; Mohai, B. (Zb. Celostatnej 8th Konf. Term. Anal.; Vysoke Tatry, Czech., 1979, pp. 235/8; C.A. **92** [1980] No. 121003).
[4] Moreau-Colin, M. L. (Struct. Bonding [Berlin] **10** [1972] 167/90, 169).
[5] Jérôme-Lerutte, S. (Acta Cryst. B **27** [1971] 1624/30).
[6] Jérôme-Lerutte, S. (Bull. Soc. Roy. Sci. Liège **36** [1968] 206/11).
[7] Moreau-Colin, M. L. (J. Chim. Phys. **67** [1970] 498/506).
[8] Jérôme-Lerutte, S. (Struct. Bonding [Berlin] **10** [1972] 153/66).
[9] Mason, W. R.; Gray, H. B. (J. Am. Chem. Soc. **90** [1968] 5721/9).
[10] Feigl, F.; Heisig, G. B. (J. Am. Chem. Soc. **73** [1951] 5631/2).

[11] Musselman, R. C.; Anex, B. G. (J. Phys. Chem. **91** [1987] 4460/3).

Alkali Metal and Ammonium Cyanopalladates

Li$_2$[Pd(CN)$_4$]. The salt is diamagnetic at room temperature, with $\chi_A = -87.1 \times 10^{-6}$ c.g.s.u. [6]. The absorption spectrum and birefringence of Li$_2$[Pd(CN)$_4$]·3H$_2$O were briefly mentioned [13].

Na$_2$[Pd(CN)$_4$]·3H$_2$O. This is made by dissolving Pd(CN)$_2$ in aqueous NaCN solution [1].

X-ray investigation of the trihydrate [2, 3] shows the crystals to be triclinic, Z = 4; a = 15.45, b = 9.05, c = 7.42 Å, measured density 2.025 g/cm^3 at 20°C [3]. There are two very similar environments for the palladium atoms with Pd–C = 2.000 Å and C–N = 1.145 Å in one and Pd–C = 1.986 Å, C–N = 1.153 Å in the other [2]. Details of the structure, shown in the papers [4, 5, 14].

The anhydrous salt is diamagnetic, $\chi_A = -86.6 \times 10^{-6}$ c.g.s.u. [6], and the diamagnetic anisotropy of the trihydrate was measured [7].

Vibrational and Electronic Spectra

The infrared spectrum of the trihydrate has been measured from 400 to 4000 cm^{-1} (reproduced diagramatically in paper) [8] and the Raman spectrum in aqueous solution (80 to 3000 cm^{-1}) [1]; using these data force constants were calculated [9, 10]. The infrared spectrum of solid Na$_2$[Pd(CN)$_4$] has also been recorded from 250 to 700 and 2000 to 2250 cm^{-1} (reproduced in paper) [11].

The polarised electronic absorption spectra of the trihydrate were measured (200 to 400 nm, reproduced in paper) [12], and the absorption spectrum of the solid (200 to 700 nm, reproduced in paper) [13]. The ^{14}N nuclear quadrupole resonance of the salt has been measured at 77 K [15].

References:

[1] Mathieu, J.-P.; Cornevin, S. (J. Chim. Phys. **36** [1939] 271/9).
[2] Ledent, J. (Bull. Soc. Roy. Sci. Liège **43** [1974] 172/189).
[3] Ledent, J. (Bull. Soc. Roy. Sci. Liège **36** [1967] 295/301).
[4] Ledent, J. (Bull. Soc. Roy. Sci. Liège **41** [1972] 537/41).
[5] Ledent, J. (Bull. Soc. Roy. Sci. Liège **38** [1969] 14/26).
[6] Simon, J. (Bull. Soc. Roy. Sci. Liège **36** [1967] 480/4).
[7] Simon, J. (Z. Krist. **124** [1967] 369/74).
[8] Bonino, G. B.; Chiorboli, P.; Fabbri, G. (Atti Accad. Nazl. Lincei Classe Sci. Fis. Mat. Nat. Rend. [8] **26** [1959] 137/46).
[9] Salvetti, O. (Atti Accad. Nazl. Lincei Classe Sci. Fis. Mat. Nat. Rend. [8] **26** [1959] 383/5).
[10] Salvetti, O. (Atti Accad. Nazl. Lincei Classe Sci. Fis. Mat. Nat. Rend. [8] **26** [1959] 225/30).

[11] Hidalgo, A.; Mathieu, J.-P. (Anales Real Soc. Espan. Fis. Quim. [Madrid] A **56** [1960] 9/18).
[12] Moreau-Colin, M. L. (J. Chim. Phys. **67** [1970] 498/506).
[13] Moreau-Colin, M. L. (Bull. Soc. Roy. Sci. Liège **34** [1965]˙763/71, 764).
[14] Moreau-Colin, M. L. (Struct. Bonding [Berlin] **10** [1972] 167/90, 170).
[15] Murgich, J.; Oda, T. (J. Chem. Soc. Dalton Trans. **1987** 1637/40).

Potassium Tetracyanopalladate K$_2$[Pd(CN)$_4$]·3H$_2$O

This compound is made by reaction of Pd(CN)$_2$ in aqueous suspension with excess KCN solution [1, 9], from Pd(DMG)$_2$ with KCN [2], from K$_2$[PdCl$_4$] with excess KCN [3], and from [Pd(NO)Cl]$_2$ and KCN [4]. Isotopically enriched (^{13}C, ^{15}N) K$_2$[Pd(CN)$_4$] was made from PdCl$_2$ in excess labelled KCN solution [5].

In aqueous solution it is made from $Pd(CN)_2$ and sufficient KCN to dissolve the dicyanide [6]. Although it is said that $[Pd(CN)_4]^{2-}$ is present in melts of $PdCl_2$ in molten KCN [10] it is claimed that lower oxidation state species are actually present in such melts [10, 11].

It forms white crystals [1] which are triclinic [3]; the trihydrate is efflorescent in air, losing two molecules of water of crystallisation at 100°C and a third at 200°C. It is readily soluble in water and liquid ammonia and sparingly soluble in alcohol. With dilute acids it gives $Pd(CN)_2$. At high temperatures it decomposes to Pd, C_2N_2 and KCN [1]. In liquid ammonia it is reduced by potassium to $K_2[Pd(CN)_2]$ [8].

It is diamagnetic, $\chi_A = -86.8 \times 10^{-6}$ c.g.s.u. [7].

References:

[1] Bigelow, J. H. (Inorg. Syn. **2** [1946] 245/6).
[2] Oi, N. (Nippon Kagaku Zasshi **71** [1950] 570/1; C.A. **1951** 7463).
[3] Yamada, S. (J. Am. Chem. Soc. **73** [1951] 1182/4).
[4] Smidt, J.; Jira, J. (Chem. Ber. **93** [1960] 162/5).
[5] Kubas, G. I.; Jones, L. H. (Inorg. Chem. **13** [1974] 2816/8).
[6] Feigl, F.; Heisig, G. B. (J. Am. Chem. Soc. **73** [1951] 5630/1).
[7] Simon, J. (Bull. Soc. Roy. Sci. Liège **36** [1967] 480/4).
[8] Nast, R.; Bülck, J.; Kramolowsky, R. (Chem. Ber. **108** [1975] 3461/8).
[9] Adamson, A. W.; Welker, J. P.; Volpe, M. (J. Am. Chem. Soc. **72** [1950] 4030/6).
[10] Magnuson, W. L.; Griswold, E.; Kleinberg, J. (Inorg. Chem. **3** [1964] 88/93).

[11] de Haas, R. S.; Fouché, K. F. (Inorg. Chim. Acta **24** [1977] 269/76).

Vibrational Spectra

Data on the vibrational spectra of the $(Et_4N)^+$ and potassium salts are given in the table below using the numbering scheme from [1]; see also table below. The most recent data concern $K_2[Pd(^{12}C^{14}N)_4]$, $K_2[Pd(^{12}C^{15}N)_4]$, and $K_2[Pd(^{13}C^{14}N)_4]$ for which solid state data are given in the table below (for solution data and assignments see table on p. 267); spectra were measured from 250 to 4000 cm^{-1} of both the Raman and infrared spectra of the solids [2]. The infrared spectrum of $K_2[Pd(CN)_4] \cdot H_2O$ has been measured (400 to 2200 cm^{-1}, spectrum reproduced diagramatically in paper) [3], also of solid $K_2[Pd(CN)_4]$ (200 to 2200 cm^{-1}, band maxima listed) [4]. The far-infrared spectrum of $K_2[Pd(CN)_4] \cdot 3H_2O$ (60 to 300 cm^{-1}, reproduced in paper) was measured [5]. Data for low-frequency modes of "$[Pd(CN)_4]^{2-}$" with assignments have been given; they may refer to $K_2[Pd(CN)_4]$ though this is not clear [6]. Infrared ν_{CN} values for a number of cyano complexes, including $K_2[Pd(CN)_4]$, have been compared [7]. The ν_{CN} frequency for $K_2[Pd(CN)_4]$ in a NaCN–KCN eutectic melt is 2140 cm^{-1} [8].

Force constant treatments for the anion in $K_2[Pd(CN)_4]$ have been given based on some of the solution data in the table below and data from [3] in [9, 10]; using data from [2] and a general valence force field [2]; using data from [6] for low frequency modes [6].

Vibrational spectra of solid salts $M_2^I[Pd(CN)_4]$:

M^I	Et_4N	$K^{2)}$	$K^{2)}$	$K^{3)}$	$K^{4)}$
ν_1 (ν_{CN}; A_{1g})	$2175^{1)}$	$2160^{1)}$	$2159.0^{1)}$		
ν_2 (ν_{MC}; A_{1g})	$447^{1)}$	$440^{1)}$	$424.5^{1)}$		
ν_3 (ν_{MC}; A_{2g})					

(continued)

M^I	Et_4N	$K^{2)}$	$K^{2)}$	$K^{3)}$	$K^{4)}$
ν_4 (ν_{CN}; B_{1g})	2158[1)	2146[1)	2146.8[1)		
ν_5 (ν_{MC}; B_{2g})	423[1)	430[1)			
ν_6 (δ_{MCN}; B_{2g})	399[1)	290[1)			
ν_7 (δ_{CMC}; B_{2g})					
ν_8 (δ_{MCN}; E_g)	297				
ν_9 (δ_{MCN}; A_{2u})	410	415			
ν_{10} (δ_{CMC}; A_{2u})					
ν_{11} (ν_{MC}; B_{2u})		140 [11]			
ν_{12} (δ_{CMC}; B_{2u})		180 [11]			
ν_{13} (ν_{CN}; E_u)	2138	2140	2134.0	2103.0	2088.0
ν_{14} (ν_{MC}; E_u)		494	485.1	482.7	474.1
ν_{15} (δ_{MCN}; E_u)	388, 380		383.0	379.5	376.2
ν_{16} (δ_{CMC}; E_u)			296.0		
Ref.	[1]	[3]+	[2]	[2]	[2]

Infrared or [1)] Raman data on solid, [2)] $K_2[Pd(^{12}C^{14}N)_4]$, [3)] $K_2[Pd(^{12}C^{15}N)_4]$, [4)] $K_2[Pd(^{13}C^{14}N)_4]$, [+] data from [3], assigned in [1].

Electronic Spectra

The dichroic absorption spectrum of the solid trihydrate has been recorded (200 to 750 nm, reproduced in paper) [11]. The absorption and luminescence spectra of the solid monohydrate were also measured (190 to 450 nm; luminescence spectrum from 200 to 600 nm reproduced in paper) [12].

References:

[1] Jerome-Lerutte, S. (Struct. Bonding [Berlin] 10 [1972] 153/66).
[2] Kubas, G. I.; Jones, L. H. (Inorg. Chem. 13 [1974] 2816/8).
[3] Bonino, G. B.; Chiorboli, P.; Fabbri, G. F. (Atti. Accad. Nazl. Lincei Classe Sci. Fis. Mat. Nat. Rend. [8] 26 [1959] 137/46; C.A. 54 [1960] 14943).
[4] Hidalgo, A.; Mathieu, J.-P. (Compt. Rend. 249 [1959] 233/5).
[5] Lorenzelli, V.; Delorme, P. (Spectrochim. Acta 19 [1963] 2033/45).
[6] Mizushima, S.; Nakagawa, I. (Nippon Kagaku Zasshi 80 [1959] 124/8; C.A. 1959 6757).
[7] El-Sayed, M. F. A.; Sheline, R. K. (J. Inorg. Nucl. Chem. 6 [1958] 187/93).
[8] de Haas, K. S.; Fouché, K. F. (Inorg. Chim. Acta 24 [1977] 269/76).
[9] Salvetti, O. (Atti. Accad. Nazl. Lincei Classe Sci. Fis. Mat. Nat. Rend. [8] 26 [1959] 225/30; C.A. 54 [1960] 14943).
[10] Salvetti, O. (Atti. Accad. Nazl. Lincei Classe Sci. Fis. Mat. Nat. Rend. [8] 26 [1959] 383/5).

[11] Yamada, S. (J. Am. Chem. Soc. 73 [1951] 1182/4).
[12] Rossiello, L. A.; Furlani, C. (Atti. Accad. Nazl. Lincei Classe Sci. Fis. Mat. Nat. Rend. [8] 38 [1965] 207/13).

Miscellaneous Data

Photoelectronic spectroscopy (ESCA) of $K_2[Pd(CN)_4]$ showed binding energies of 339.2(2) eV for $3d_{5/2}$ and 344.3(3) eV for $3d_{3/2}$ [1]; also 344.9 eV for $3d_{5/2}$ and 350.2 eV for $3d_{3/2}$ [2].

The indices of refraction of $K_2[Pd(CN)_4]\cdot 3H_2O$ are $N_g=1.576$, $N_m=1.550$, $N_p=1.468$, the corresponding values for the monohydrate being 1.577, 1.531 and 1.428 [3].

The ionic character of the bonds in $K_2[Pd(CN)_4]$ has been calculated by the electronegativity method [4]. The nuclear quadrupole resonance of the salt has been measured at 77 K [5].

References:

[1] Kumar, G.; Blackburn, J. R.; Moddemann, W. E.; Jones, M. M. (Inorg. Chem. **11** [1972] 296/300).

[2] Jørgensen, C. K.; Berthou, H. (Mat. Fys. Medd. Kgl. Danske Videnskab Selsk. **38** [1972] 1/93; C.A. **79** [1973] No. 11782).

[3] Batsanov, S. S.; Aleksandrova, O. P. (Zh. Neorgan. Khim. **3** [1958] 2666/70; Russ. J. Inorg. Chem. **3** No. 12 [1958] 94/9).

[4] Batsanov, S. S. (Zh. Neorgan. Khim. **9** [1964] 1323/7; Russ. J. Inorg. Chem. **9** [1964] 722/5).

[5] Murgich, J.; Oda, T. (J. Chem. Soc. Dalton Trans. **1987** 1637/40).

Compounds with NH$_4$, Rb and Cs

$(NH_4)_2[Pd(CN)_4]$. The salt is diamagnetic, $\chi_A = -86.8\times 10^{-6}$ c.g.s.u [4]. The birefringence of the trihydrate was briefly mentioned [6].

$Rb_2[Pd(CN)_4]\cdot H_2O$. This is made from $Ba[Pd(CN)_4]$ and Rb_2SO_4 [9]. The polarised specular reflectance spectrum (180 to 500 nm) was reproduced in [9]. The X-ray photoelectronic spectrum gave binding energies of 345.65 eV for Pd $3d_{5/2}$ and of 350.9 eV for Pd $3d_{3/2}$ [8]. The X-ray crystal structure of the orthorhombic salt shows the space group to be Pncn-D_{2h}^6, $Z=4$; $a=10.01(2)$, $b=13.74(2)$, $c=7.44(2)$ Å, measured density 2.56(4) g/cm^3. The unit cell is shown in **Fig. 67**; the $[Pd(CN)_4]^{2-}$ anions are planar with mean Pd–C and C–N distances of 2.01 and 1.13 Å, respectively [1]; see also [2]. Earlier X-ray data gave $a=9.88(3)$, $b=13.74(7)$, $c=7.39(4)$ Å, with mean Pd–C$=1.94$ and C–N$=16$ Å [3].

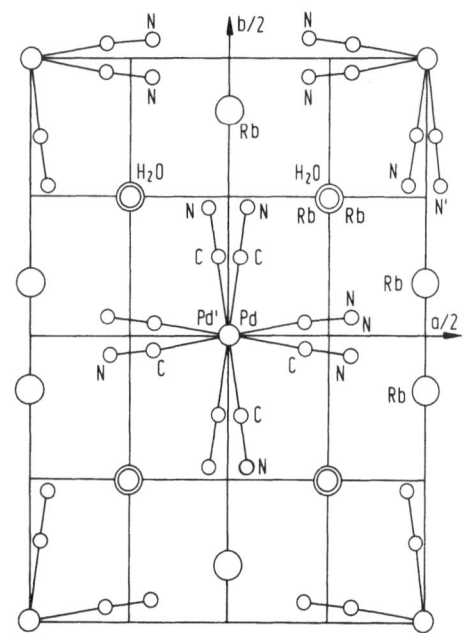

Fig. 67.   Unit cell of $Rb_2[Pd(CN)_4]\cdot H_2O$ [1].

The salt is diamagnetic, $\chi_A = -87.2 \times 10^{-6}$ c.g.s.u. [4].

The electronic absorption spectrum of the solid trihydrate was recorded (200 to 450 nm, reproduced in paper) [5].

$Cs_2[Pd(CN)_4] \cdot H_2O$. This is made from $Ba[Pd(CN)_4]$ and Cs_2SO_4 in aqueous solution.

The space group is $P6_1\text{-}C_6^2$ or $P6_5\text{-}C_6^3$, $Z = 6$, for the hexagonal crystals; $a = 9.704(2)$, $c = 19.388(4)$ Å; measured density 3.12(4) g/cm^3, calculated density 3.115 g/cm^3. The crystals are isotypic with $Cs_2[M(CN)_4] \cdot H_2O$ (M = Ni, Pt) with a helical arrangement of $[Pd(CN)_4]^{2-}$ groups about the z axis [7].

References:

[1] Dupont, L. (Acta Cryst. B **26** [1970] 964/71).
[2] Moreau-Colin, M. L. (Struct. Bonding [Berlin] **10** [1972] 167/90, 170).
[3] Dupont, L. (Bull. Soc. Roy. Sci. Liège **36** [1967] 471/5).
[4] Simon, J. (Bull. Soc. Roy. Sci. Liège **36** [1967] 480/4).
[5] Moreau-Colin, M. L. (J. Chim. Phys. **67** [1970] 498/506).
[6] Moreau-Colin, M. L. (Bull. Soc. Roy. Sci. Liège **34** [1965] 763/71).
[7] Holzapfel, W.; Otto, H. H.; Yersin, H.; Gliemann, G. (J. Appl. Cryst. **12** [1979] 241/2).
[8] Jørgensen, C. K.; Berthold, H. (Mat. Fys. Medd. Kgl. Danske Videnskab Selsk. **38** [1972] 1/93, 73; C.A. **79** [1973] No. 11782).
[9] Musselman, R. L.; Anex, B. G. (J. Phys. Chem. **91** [1987] 4460/3).

Heavy-Metal Cyanopalladates

Magnesium Tetracyanopalladate Mg[Pd(CN)_4]

This is made from $Ba[Pd(CN)_4]$ and $MgSO_4$ in aqueous solution; the $BaSO_4$ is filtered off and the solution evaporated. The polarised specular reflectance spectrum was measured.

Reference:

Musselman, R. L.; Anex, B. G. (J. Phys. Chem. **91** [1987] 4460/3).

Calcium Tetracyanopalladate Ca[Pd(CN)_4]·5H_2O

This is made by reaction of $K_2[Pd(CN)_4]$ to give $Cu[Pd(CN)_4]$ which is then reacted with $Ca(OH)_2$ to yield $Ca[Pd(CN)_4] \cdot 5H_2O$ [1]. It is colourless [5], and is attacked by atmospheric CO_2 to give $Pd(CN)_2$ and HCN [1].

X-ray investigations of the pentahydrate show the crystals to be orthorhombic with space group $Pcab\text{-}D_{2h}^{15}$, $Z = 8$; $a = 17.33$, $b = 19.29$, $c = 6.84$ Å. The unit cell is shown in **Fig. 68** [2]; for earlier X-ray data see [1, 3], for a general summary see [5]. The Pd–C distances vary between 1.92 and 2.00 Å and C–N between 1.11 and 1.16 Å [5].

The salt is diamagnetic, $\chi_A = -86.4 \times 10^{-6}$ c.g.s.u. [6], and the anisotropy of its diamagnetism measured [4].

There are a number of studies on the electronic spectrum [5]. The polarised electronic absorption spectrum of a single crystal has been measured from 200 to 450 nm (reproduced in paper) [7, 8, 9]. The dichroism of the salt has been measured (300 to 800 nm, reproduced in paper) [10], and the photographic birefringence spectrum reproduced [11]; see also [15]. Polarised electronic absorption spectral studies on $Ba[Pd_nPt_{1-n}(CN)_4] \cdot 4H_2O$ (200 to 500 nm, reproduced in paper) were made [12]. The polarised specular reflectance spectrum was recorded (180 to 500 nm, reproduced in paper) [16].

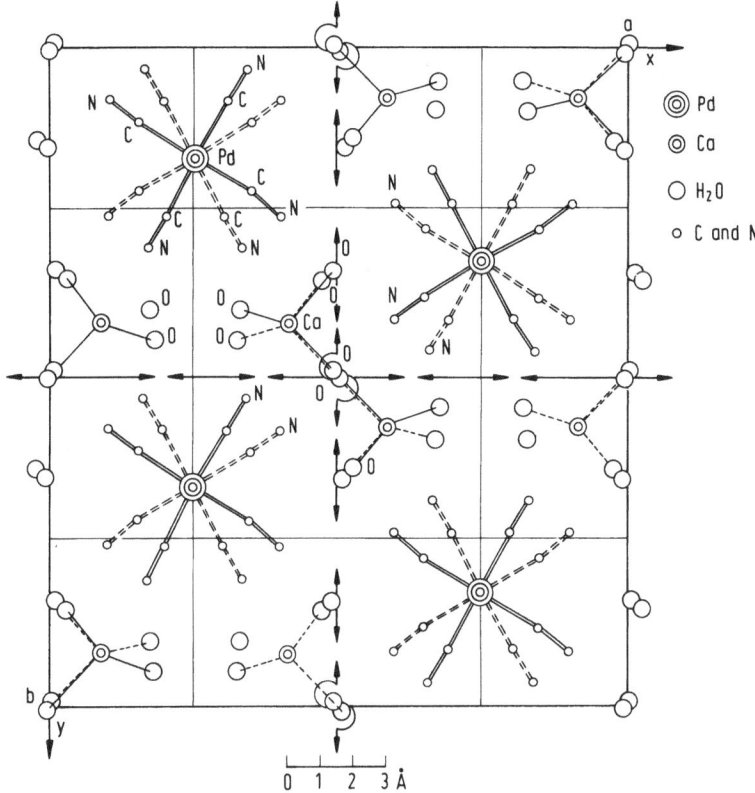

Fig. 68.   Unit cell of Ca[Pd(CN)$_4$]·5H$_2$O [1].

The salt has been briefly mentioned in connection with quasi-one-dimensional solids [13]. Differential thermal analysis (DTA) from −140 to −20°C suggested a specific heat change at −70°C, and this indication of a phase change was confirmed by Brillouin scattering studies [14].

References:

[1] Fontaine, F. (Bull. Soc. Roy. Sci. Liège **33** [1964] 178/89).
[2] Fontaine, F. (Bull. Soc. Roy. Sci. Liège **36** [1967] 437/44).
[3] Fontaine, F. (Bull. Soc. Roy. Sci. Liège **34** [1965] 750/62, 761).
[4] Simon, J.; Toussaint, J. (Bull. Soc. Roy. Sci. Liège **32** [1963] 881/5).
[5] Moreau-Colin, M. L. (Struct. Bonding [Berlin] **10** [1972] 167/90).
[6] Simon, J. (Bull. Soc. Roy. Sci. Liège **36** [1967] 480/4).
[7] Le Bras, R.; Moncuit, C. (Compt. Rend. B **267** [1968] 1032/5).
[8] Macadré, A.; Moncuit, C. (Compt. Rend. **261** [1965] 2339/42).
[9] Moreau-Colin, M. L. (J. Chim. Phys. **67** [1970] 498/506).
[10] Moreau-Colin, M. L. (Bull. Classe Sci. Acad. Roy. Belg. [5] **49** [1963] 973/89).

[11] Moreau-Colin, M. L. (Bull. Classe Sci. Acad. Roy. Belg. [5] **51** [1965] 916/31).
[12] Holzapfel, W.; Gliemann, G. (Ber. Bunsenges. Physik. Chem. **89** [1985] 935/9).
[13] Stevens, G. C. (Platinum Metals Rev. **23** [1979] 23/8).

[14] Luspin, Y.; Chabin, M.; Moncuit, C. (Solid State Commun. **44** [1982] 71/3).
[15] Moreau-Colin, M. L. (Bull. Soc. Roy. Sci. Liège **34** [1965] 763/71).
[16] Musselman, R. L.; Anex, B. G. (J. Phys. Chem. **91** [1987] 4460/3).

Strontium Tetracyanopalladate Sr[Pd(CN)₄]·5H₂O

The salt is colourless [1] and is monoclinic; space group C2/n, Z = 4; a = 10.64, b = 15.51, c = 7.26 Å, $\beta = 97°17'$ [1]. The andydrous salt is diamagnetic, $\chi_A = -86.9 \times 10^{-6}$ c.g.s.u. [5].

The electronic absorption spectrum of microcrystalline Sr[Pd(CN)₄]·5H₂O has been measured (200 to 400 nm, reproduced in paper) [1,2,3] and assigned [2,3]. The birefringence of Sr[Pd(CN)₄]·5H₂O has been listed (356 to 600 nm) [4].

References:

[1] Moreau-Colin, M. L. (Struct. Bonding [Berlin] **10** [1972] 167/90).
[2] Moreau-Colin, M. L. (J. Chim. Phys. **67** [1970] 498/506).
[3] Macadré, A.; Moncuit, C. (Compt. Rend. **261** [1965] 2339/42).
[4] Moreau-Colin, M. L. (Bull. Soc. Roy. Sci. Liège **34** [1965] 763/71).
[5] Simon, J. (Bull. Soc. Roy. Sci. Liège **36** [1967] 480/4).

Barium Tetracyanopalladates Ba[Pd(CN)₄]·4H₂O

The tetrahydrate is obtained from an equimolar mixture of Pd(CN)₂ and Ba(CN)₂ in aqueous solution after evaporation and crystallisation. It forms needle-shaped crystals [1] which are colourless [5]. It can also be made from K₂[Pd(CN)₄] and BaCl₂ [2].

The tetrahydrate is monoclinic, space group C2/c-C_{2h}^6, Z = 4; a = 11.98, b = 13.83, c = 6.73 Å, $\beta = 104°28'$ [5].

The salt is diamagnetic, $\chi_A = -87.4 \times 10^{-6}$ c.g.s.u. [3], and the diamagnetic anisotropy shows that the crystal axes of the pentahydrate are monoclinic with the b axis as the principle axis of the ellipsoid of susceptibility [4].

Electronic spectra of the salt have been reviewed [5]. Polarised electronic spectra of single crystals of Ba[Pd(CN)₄]·4H₂O have been measured (200 to 400 nm, reproduced in papers) [2, 6, 7, 16] and 200 to 600 nm, reproduced in paper [1] and also of Ba[Pd,Pt(CN)₄]·4H₂O [1, 2]. The dichroism of the solid was measured (300 to 800 nm, reproduced in paper) [8] and the birefringence of the salt measured (400 to 800 nm, photograph reproduced in paper) [9]. The emission spectrum of Ba[Pd₀.₅Pt₀.₅(CN)₄]·4H₂O (400 to 700 nm, reproduced in paper) was recorded [10], the luminescence and absorption spectra of Ba[Pd,Pt(CN)₄]·4H₂O [11], and the fluorescence spectrum of Ba[Pd(CN)₄]·4H₂O (400 to 600 nm, reproduced in paper) [12]. For birefringence of the salt see [15].

The salt has been briefly mentioned in the context of one-dimensional solids [13].

With H₂S at 870 K, Ba[Pd(CN)₄]·4H₂O gives Ba[PdS₂] [14].

References:

[1] Anex, B. G.; Musselman, R. L. (J. Phys. Chem. **84** [1960] 883/7).
[2] Holzapfel, W.; Gliemann, G. (Ber. Bunsenges. Physik. Chem. **89** [1985] 935/9).
[3] Simon, J. (Bull. Soc. Roy. Sci. Liège **36** [1967] 480/4).
[4] Simon, J. Toussaint, J. (Bull. Soc. Roy. Sci. Liège **32** [1963] 881/5).
[5] Moreau-Colin, M. L. (Struct. Bonding [Berlin] **10** [1972] 167/90).
[6] Macadré, A.; Moncuit, C. (Compt. Rend. **261** [1965] 2339/42).

[7] Moreau-Colin, M. L. (J. Chim. Phys. **67** [1970] 498/506).

[8] Moreau-Colin, M. L. (Bull. Classe Sci. Acad. Roy. Belg. [5] **49** [1963] 973/80).

[9] Moreau-Colin, M. L. (Bull. Classe Sci. Acad. Roy. Belg. [5] **51** [1965] 916/31, 921).

[10] Moreau-Colin, M. L. (Bull. Soc. Roy. Sci. Liège **43** [1974] 653/8).

[11] Viswanath, A. K.; Vetuskey, J.; Leighton, R.; Krogh-Jespersen, M.-B.; Patterson, H. H. (Mol. Phys. **48** [1983] 567/79).

[12] Niki, B.; Shirai, H. (Kogyo Kagaku Zasshi **56** [1953] 406/10, Chem. Abstr. **1954** 11195).

[13] Stevens, G. C. (Platinum Metals Rev. **23** [1979] 23/8).

[14] Huster, J.; Bronger, W. (J. Less-Common Metals **119** [1986] 159/65).

[15] Moreau-Colin, M. L. (Bull. Soc. Roy. Sci. Liège **34** [1965] 763/71).

[16] Musselman, R. L.; Anex, B. G. (J. Phys. Chem. **91** [1987] 4460/3).

Other Heavy-Metal Cyanopalladates

$Tl_2[Pd(CN)_4]$. This is made from equimolar amounts of TlCN and $Pd(CN)_2$ in aqueous solution followed by evaporation to give colourless plates [9].

The crystals belong to the $P2_1/c$-C_{2h}^5 space group; they are monoclinic, $a = 6.31(5)$, $b = 7.39(2)$, $c = 10.03(1)$ Å, $\beta = 113.12°$; $Z = 2$ [10]; measured density 4.37 g/cm³ [9, 10].

$Zn(NH_3)_2[Pd(CN)_4] \cdot 2L$ (L = benzene or aniline). This clathrate is made from $ZnCl_2$ and $[Pd(CN)_4]^{2-}$ in the presence of the organic molecule and NH_4OH [1, 8]; the benzene clathrate is white and has a tetragonal structure, $a = 7.58$, $c = 8.32$ Å [1].

$Cd(NH_3)_2[Pd(CN)_4] \cdot 2L$ (L = benzene or aniline). This white material is made in similar fashion to the zinc analogue using $CdCl_2$ [1, 8]; the benzene clathrate is also tetragonal, $a = 7.77$, $c = 8.38$ Å [1].

trans-$[CrF(H_2O)dn_2][Pd(CN)_4]$. This is made by reaction of $K_2[Pd(CN)_4]$ in water with trans-$[CrF(H_2O)dn_2](ClO_4)_2$ (dn = 1,3-diaminopropane). The thermal decomposition curve was plotted (298 to 438 K, thermogravimetric curve reproduced in paper) and thermochemical parameters derived for the salt [2].

cis-$[Cren_2(H_2O)Cl][Pd(CN)_4] \cdot H_2O$. This is made from cis-$[Cren_2(H_2O)Cl]Br_2$ and $[Pd(CN)_4]^{2-}$ in acid. It is pink; in the infrared ν_{CN} lies at 2130 cm⁻¹. On heating it gives $(NC)_3Pd–C–N–Cren_2Cl$ [7].

trans-$[CrF(chxn)_2H_2O][Pd(CN)_4]$ is made from $[Pd(CN)_4]^{2-}$ and trans-$[CrF(chxn)_2H_2O]^{2+}$. On heating it gives $(NC)_3Pd–C–N–cis-CrF(chxn)_2$ (chxn = diaminocyclohexane) [12].

cis-$[Cren_2(H_2O)Br][Pd(CN)_4] \cdot H_2O$. This is similarly made from cis-$[Cren_2(H_2O)Br]Br_2$ and $[Pd(CN)_4]^{2-}$ in water with acid, or from $K_2[Pd(CN)_4]$ and cis-$[Cren_2(H_2O)Br]Br_2$ in methanol. It is red, and in the infrared ν_{CN} lies at 2140 cm⁻¹. On dehydration it yields $(NC)_3Pd–C–N–Cren_2Br$ [7].

$[Ni(NH_3)_2][Pd(CN)_4] \cdot 2L$ (L = benzene or aniline). This pale violet material is made from $NiCl_2$, ammonia $K_2[Pd(CN)_4]$ and the organic solvent [1, 8]. The benzene clathrate is tetragonal, $a = 7.44$, $c = 8.39$ Å [1].

$[Ni en_3][Pd(CN)_4]$ and $[Pd en_2][Pd(CN)_4]$. The X-ray photoelectronic (ESCA) spectra of these salts gave binding energies of 349.5 eV for Pd $3d_{3/2}$ and 344.2 eV for Pd $3d_{5/2}$, respectively, in both salts [11].

$[Co(NH_3)_5(H_2O)]_2[Pd(CN)_4]_3$. This is made from $K_2[Pd(CN)_4]$ and $[Co(NH_3)_5(H_2O)](ClO_4)_3$ in water. On heating it gives $[(NC)_3Pd–C–N–Co(NH_3)_5]_2[Pd(CN)_4]$; the de-aquation kinetics were measured [5].

cis-[Co en$_2$(NH$_3$)(H$_2$O)]$_2$[Pd(CN)$_4$]$_3$. This is made from K$_2$[Pd(CN)$_4$] and cis-[Co en$_2$(NH$_3$)(H$_2$O)]-(ClO)$_3$ in water. It is pink; in the infrared ν_{CN} is at 2135 cm^{-1}. On heating to 70°C it gives [(NC)$_3$Pd–C–N–Co en$_2$(NH$_3$)]$_2$[Pd(CN)$_4$] [6].

Co[Pd(CN)$_4$]. The X-ray photoelectronic (ESCA) spectrum gave binding energies of 345.4 eV for Pd 3d$_{5/2}$ and 350.7 eV for Pd 3d$_{3/2}$ [11].

[Cu(NH$_3$)$_2$][Pd(CN)$_4$]·2L (L = benzene or aniline). This sky-blue clathrate is made from aqueous CuSO$_4$, ammonia, K$_2$[Pd(CN)$_4$] and the organic molecule [1, 8]. The benzene clathrate is tetragonal, a = 7.58, c = 8.29 Å [1].

Cu[Pd(CN)$_4$]. This is made by addition of crystals of CuSO$_4$·5H$_2$O to an aqueous solution of K$_2$[Pd(CN)$_4$] [3].

[Cu(NH$_3$)$_2$]$_2$[Pd(CN)$_4$]. This blue salt is made from K$_2$[Pd(CN)$_4$] in solution with [Cu(NH$_3$)$_4$]NO$_3$ [3].

Ag$_2$[Pd(CN)$_4$]. This material is made from a solution of AgNO$_3$ in excess to K$_2$[Pd(CN)$_4$] [3].

[Ag(NH$_3$)$_2$]$_2$[Pd(CN)$_4$]. This white, crystalline, but impure material is made from an ammoniacal solution of Ag(NO$_3$) with K$_2$[Pd(CN)$_4$]. On heating it gives Ag$_2$[Pd(CN)$_4$] [3].

[Rh(NH$_3$)$_5$(H$_2$O)]$_2$[Pd(CN)$_4$]$_3$ is made from [Rh(NH$_3$)$_5$(H$_2$O)](ClO$_4$)$_2$ and K$_2$[Pd(CN)$_4$]; it is white. Thermogravimetric analysis was carried out (20 to 200°C, curves reproduced in paper). The kinetics of de-aquation, electronic and IR spectra were measured [11].

[Pd(NH$_3$)$_4$][Pd(CN)$_4$]. Attempts to make this resulted in formation of cis-Pd(CN)$_2$(NH$_3$)$_2$ (see p. 285) [4].

References:

[1] Morita, M.; Miyoshi, T.; Miyamoto, T.; Iwamoto, T.; Sasaki, Y. (Bull. Chem. Soc. Japan **40** [1967] 1556).

[2] Serra, M.; Escuer, A.; Ribas, J.; Baro, M. D. (Thermochim. Acta **64** [1983] 237/46).

[3] Feigl, F.; Heisig, G. B. (J. Am. Chem. Soc. **73** [1951] 5630/1).

[4] Penavic, M. (Acta Cryst. C **42** [1986] 1283/4).

[5] Ribas, J.; Serra, M.; Escuer, A. (Inorg. Chem. **23** [1984] 2236/42).

[6] Ribas, J.; Escuer, A. (Transition Metal Chem. [Weinheim] **10** [1985] 466/9).

[7] Ribas, J.; Serra, M.; Escuer, A. (Transition Metal Chem. [Weinheim] **9** [1984] 287/90).

[8] Iwamoto, T.; Miyoshi, T.; Miyamoto, T.; Sasaki, Y.; Fujiwara, S. (Bull. Chem. Soc. Japan **40** [1967] 1174/8).

[9] Bok, L. D. C.; Leipoldt, J. G. (Z. Anorg. Allgem. Chem. **344** [1966] 86/91).

[10] Bok, L. D. C.; Leipoldt, J. G.; Basson, S. S. (J. South African Chem. Inst. **19** [1966] 62/72, 66).

[11] Jørgensen, C. K.; Berthou, H. (Mat. Fys. Medd. Kgl. Danske Videnskab Selsk. **38** [1972] 1/93, 73; C.A. **79** [1973] No. 11782).

[12] Corbella, M.; Ribas, J. (Ann. Qim. B **82** [1986] 273/8).

11.3.2.3.2 Palladium(II) Complexes of Coordination Number >4

[Pd(CN)$_5$]$^{3-}$. No evidence could be found for the existence of this from spectrophotometric studies of [Pd(CN)$_4$]$^{2-}$ with CN$^-$ [1]. A reported log K$_5$ value of 2.9 for the Pd^{II}–CN$^-$ reaction [2] is thought to be due to non-Nernstian behaviour of redox potentials [1]; another quoted value is 2.1 ± 0.2 for log K$_5$ [3]. Infrared spectra also give no evidence for existence of the species [1].

Values of -0.8 ± 1.6 and -16 ± 3 kJ/mol were given for $\Delta H°$ and $\Delta G°$ for $[Pd(CN)_4]^{2-} + CN^- \rightarrow [Pd(CN)_5]^{3-}$ with $\Delta S° = 33 \pm 8$ J·deg^{-1}mol^{-1} [2].

References:

[1] Hancock, R. D.; Evers, E. (Inorg. Chem. **15** [1976] 995/6).
[2] Izatt, R. M.; Watt, G. D.; Eastaugh, D.; Christensen, J. J. (J. Chem. Soc. A **1967** 1304/8).
[3] Watt, G. D.; Eastaugh, D.; Izatt, R. M.; Christensen, J. J. (Proc. Utah Acad. Sci. Art Letters **42** No. 2 [1965] 298/302; C.A. **66** [1967] No. 59526).

11.3.2.4 Hexacyanopalladates(IV) [Pd(CN)₆]²⁻

Salts of $[PdCl_6]^{2-}$ are made by reaction of $[PdCl_4]^{2-}$ with cyanide and persulphate ion [1, 2].

Vibrational and Electronic Spectra. Infrared and Raman data for the solutions of the ion and a number of salts have been measured (see following table). Force constants for the ion were calculated. The electronic spectrum of an aqueous solution was also measured and assigned (200 to 700 nm) [1].

Reactions. The rates of reaction of hydrated electrons with $[Pd(CN)_6]^{2-}$ have been studied [2].

Vibrational spectra of $[Pd(CN)_6]^{2-}$ and its salts [1]:

	(a)*⁾	(b)	(c)	(d)	(e)	(f)	(g)
ν_1 (ν_{CN}; A_{1g})	2199p	2236, 2234²⁾		2202			
ν_2 (ν_{PdC}; A_{1g})	438p			438			
ν_3 (ν_{CN}; E_g)	2189dp			2192			
ν_4 (ν_{PdC}; E_g)	435dp						
ν_5 (δ_{PdCN}; T_{2g})	403dp			402			
ν_6 (δ_{CPdC}; T_{2g})	104dp			125			
ν_7 (ν_{CN}; T_{1u})		2160 [2177]¹⁾	2206	2185	2171, 2127	2230	2212, 2204, 2195
ν_8 (ν_{PdC}; T_{1u})		400 [388]¹⁾	436, 406	398	408, 392, 373	414	425, 413, 408
ν_9 (δ_{PdCN}; T_{1u})		497 [488]¹⁾	505	488	483	511	499

(a) = $[Pd(CN)_6]^{2-}$; (b) = $(H_3O)_2[Pd(CN)_6]$; (c) = $Li_2[Pd(CN)_6] \cdot 2H_2O$; (d) = $K_2[Pd(CN)_6]$; (e) = $Cs_2[Pd(CN)_6]$; (f) = $Zn[Pd(CN)_6]$; (g) = $Ag_2[Pd(CN)_6]$; *⁾ $Na_2[Pd(CN)_6]$ in aqueous solution. – ¹⁾ IR of $(H_3O)_2[Pd(CN)_6]$ in aqueous solution, all other data for solids. A_{1g}, E_g and T_{2g} bands observed in Raman; T_{1u} in infrared. ²⁾ Raman bands (unassigned) also at 602, 423, 474 and 116 cm^{-1} observed.

References:

[1] Siebert, H.; Siebert, A. (Z. Anorg. Allgem. Chem. **378** [1970] 160/7).
[2] Anbar, M.; Hart, F. J. (Advan. Chem. Ser. **81** [1968] 79/94, 88).

11.3.2.4.1 Salts of [Pd(CN)₆]²⁻

(H₃O)₂[Pd(CN)₆]. This is made from an aqueous solution of $K_2[Pd(CN)_6]$ on a strongly acid cation exchange column, and was obtained as a solid by evaporation in vacuum. The Raman and infrared spectra of the solid were measured (100 to 3000 cm⁻¹ Raman, 350 to 3000 cm⁻¹ infrared; bands and assignments listed in paper) and also the Raman spectrum of the solution (see p. 281).

Reference:

Siebert, H.; Siebert, A. (Z. Anorg. Allgem. Chem. **378** [1970] 160/7).

Salts of Alkali Metals and Ammonium

Li₂[Pd(CN)₆]·2H₂O; Na₂[Pd(CN)₆]. These were made from the acid and the appropriately loaded cation-exchange resin [1].

K₂[Pd(CN)₆]. Attempts to prepare this by direct reaction of $K_2[PdX_6]$ (X = Cl, Br) with CN⁻ lead to reduction to Pd and $[Pd(CN)_4]^{2-}$. However, in the presence of $S_2O_8^{2-}$ and KCN in aqueous solution. $K_2[PdCl_6]$ yields a mixture of $K_2[Pd(CN)_6]$ and $K_2[Pd(CN)_4]$, and the former is separated by fractional crystallisation.

The salt is colourless and forms domed prisms; it is fairly soluble in water and slightly soluble in methanol. It is stable at room temperature in water, dilute acids, concentrated HNO_3 and Br_2–H_2O [2].

X-ray Debye diagram data were accumulated for $K_2[Pd(CN)_6]$; it is hexagonal with a = 7.42 and c = 6.65 Å. The measured density is 1.78 g/cm. The X-ray diffraction lines were listed. The Raman spectrum of solid $K_2[Pd(CN)_6]$ from 40 to 3000 cm⁻¹ was measured and assigned, as was the infrared spectrum (400 to 2500 cm⁻¹); force constant data were obtained. The electronic spectrum of the aqueous solution of the potassium salt was also measured [1].

The ESCA spectrum shows binding energies at 344.4(2) (Pd $3d_{3/2}$), 400.9(1) (N1s), and 287.3(2) (C1s) eV [3]. The ¹³C NMR shift has been measured in aqueous solution [4].

(NH₄)₂[Pd(CN)₆]; Rb₂[Pd(CN)₆]; Cs₂[Pd(CN)₆]. The ammonium salt is made from the acid and 2M NH₄OH; the rubidium salt from the ammonium salt and RbOH and the caesium salt from $K_2[Pd(CN)_6]$ and CsCl. The infrared spectrum of $Cs_2[Pd(CN)_6]$ was measured from 200 to 2500 cm⁻¹ and assigned [1].

References:

[1] Siebert, H.; Siebert, A. (Z. Anorg. Allgem. Chem. **378** [1970] 160/7).
[2] Siebert, H.; Siebert, A. (Angew. Chem. Intern. Ed. Engl. **8** [1969] 600/1).
[3] Inoue, H.; Fluck, E. (Z. Naturforsch. **38b** [1983] 687/91).
[4] Pesek, J. J.; Mason, W. R. (Inorg. Chem. **18** [1979] 924/8).

Heavy-Metal Salts of [Pd(CN)₆]²⁻

A number of these is known, mostly of the Prussian Blue or Berlin Green type, and strictly therefore are polymeric species, but for convenience we consider them here with other $[Pd(CN)_6]^{2-}$ salts.

Zn[Pd(CN)₆]. This is made from $Mn[Pd(CN)_6]$ and $ZnSO_4$ [2]. For X-ray and infrared data see table below [8].

Cd[Pd(CN)₆]. This is made from $(H_3O)_2[Pd(CN)_6]$ and $Cd(NO_3)_2$, and forms white crystals [3].

X-ray measurements show it to have a cubic face-centered structure with the space group Fm3m-O_h^5, Z = 4; a = 10.911(4); measured density (by flotation) 1.92(1) g/cm³, calculated density 1.92 g/cm³ [1, 2]; for earlier data see [3]. The unit cell is shown in **Fig. 69**. It consists of an uninterrupted Pd–C–N–Cd framework; Pd–C = 2.07(2), Cd–N = 2.27(3), C–N = 1.11(4) Å [4]. The structure is briefly discussed also in [5].

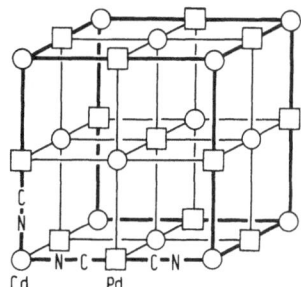

Fig. 69. Unit cell of Cd[Pd(CN)₆], with most of the C and N atoms omitted for clarity [4].

The polymeric nature is also suggested by the infrared spectrum, where the C–N stretch of 2223 cm⁻¹ is substantially higher [3] than in K₂[Pd(CN)₆] (2185 cm⁻¹) [1].

X-ray photoelectronic (ESCA) data show binding energies of 355.3 eV for Pd 3d$_{3/2}$ and 350.5 eV for Pd 3d$_{5/2}$ [9].

Mn[Pd(CN)₆]. This is made from (H₃O)₂[Pd(CN)₆] and MnSO₄ solutions [2]. For X-ray and infrared data see table below [8].

Ni[Pd(CN)₆]·2½H₂O. This blue-violet finely crystalline powder is made from NiCl₂ and Na₂[Pd(CN)₆] in water. For X-ray and infrared data see table below [8].

Co[Pd(CN)₆]·2H₂O. This forms roseate crystals and is prepared from solutions of CoCl₂ and Na₂[Pd(CN)₆] [8]. For X-ray and infrared data see table below [8].

Fe[Pd(CN)₆]. This is made from FeSO₄ and K₂[Pd(CN)₆] [6], or, as a yellow monohydrate, from FeCl₂ in HCl with Na₂[Pd(CN)₆] [8].

The anhydrous salt is face-centered cubic with space group Fm3m−O_h^5, Z = 4, a ≈ 10.6 Å. The structure is of the Prussian Blue type, see **Fig. 70** [6]. For X-ray and infrared data on the monohydrate see table below [8].

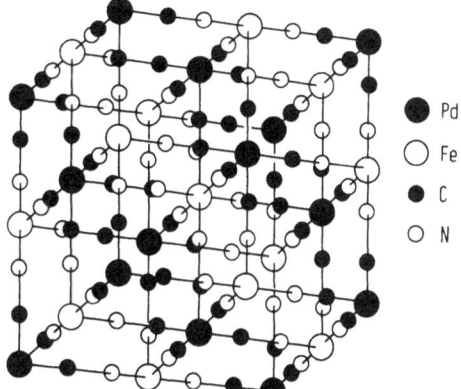

● Pd
○ Fe
• C
○ N

Fig. 70. Unit cell of Fe[Pd(CN)₆] [6].

The magnetic susceptibility was measured from 78 to 300 K and compared with that of Fe[Pt(CN)$_6$], see **Fig. 71**. The ^{57}Fe Mössbauer spectrum was measured (4.2 to 298 K) and also the ESCA spectrum, Pd 3d$_{5/2}$ = 344.7(1), Fe 2p$_{3/2}$ = 712.9(1), N1s = 401.1(1), C1s = 287.1(1) eV binding energies, respectively [6]. At 140 K the Mössbauer spectrum showed a single absorption with a relatively large quadrupole splitting suggesting the presence of PdC$_6$ and FeN$_6$ sites, with high-spin iron(II) [7].

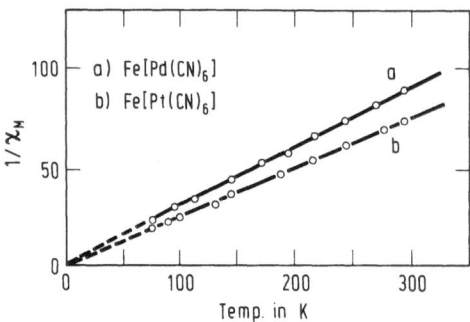

Fig. 71. Variation of reciprocal of magnetic susceptibility of Fe[Pd(CN)$_6$] with temperature [6].

X-ray and infrared data for M^{II}[Pd(CN)$_6$] [8]:

	a (Å)	ν_{CN} (cm^{-1})	δ_{PdCN} (cm^{-1})	ν_{PdC} (cm^{-1})
Zn[Pd(CN)$_6$]	10.58	2230	511	414
Mn[Pd(CN)$_6$]	10.75	2221	507	418
Ni[Pd(CN)$_6$]·2½H$_2$O	10.43	2234	514	429
Co[Pd(CN)$_6$]·2H$_2$O	10.53	2230	514	425
Fe[Pd(CN)$_6$]·H$_2$O	10.61	2223	511	421
Cu[Pd(CN)$_6$]	—	2248	526	438, 406

Cu[Pd(CN)$_6$]. This is made from aqueous solutions of CuCl$_2$ and Na$_2$[Pd(CN)$_6$] and forms blue-green crystals.

Unlike the other salts in the table above it has a tetragonal structure, a = 7.27, c = 11.01 Å with space group I4/mmm-D$_{4h}^{17}$, Z = 2 [8]. For infrared data see table above [8].

Ag$_2$[Pd(CN)$_6$]. This is made from K$_2$[Pd(CN)$_6$] and AgNO$_3$ in aqueous solution; the infrared spectrum was measured and assigned (300 to 2500 cm^{-1}); ν_{CN} = 2212, 2204 and 2195 cm^{-1}.

With CH$_3$I under pressure it gives Pd(NCMe)$_2$I$_2$ [2].

References:

[1] Siebert, H.; Siebert, A. (Angew. Chem. Int. Ed. **8** [1969] 600/1).
[2] Siebert, H.; Siebert, A. (Z. Anorg. Allgem. Chem. **378** [1970] 160/7).
[3] Ludi, A.; Ron, G. (Chimia [Switz.] **25** [1971] 333/4).
[4] Buser, H.-J.; Ron, G.; Ludi, A.; Engel, P. (J. Chem. Soc. Dalton Trans. **1974** 2473/4).
[5] Ludi, A.; Güdel, H. U. (Struct. Bonding [Berlin] **14** [1973] 1/21, 10).
[6] Inoue, H.; Fluck, E. (Z. Naturforsch. **38b** [1983] 687/91).

[7] Inoue, H.; Fluck, E.; Shirai, T.; Yanagisawa, S. (J. Phys. Colloq. [Paris] **40** [1979] C2-361/C2-362).

[8] Siebert, H.; Weise, M. (Z. Naturforsch. **27b** [1972] 865/6).

11.3.2.4.2 Substituted Complexes

Cyanohydride Complexes

An unspecified cyanide hydride complex of palladium has been made by reduction of $[Pd(CN)_4]^{2-}$ with sodium amalgam. The complex could not be isolated in a pure state, but a ^1H NMR shift (-912 Hz relative to H_2O) and infrared bands at 1980 and 1952 cm^{-1} (reproduced in paper) may be due to Pd–H modes, and a band at 2095 cm^{-1} was assigned to a C–N stretch [1].

HPd(CN)$_3$. This has been described in the patent literature as a white material, made from $Pd(CN)_2$ and 4% aqueous HCN under pressure or from $K_2[Pd(CN)_4]$ in water treated with an acid cation exchange column. It catalyses the co-polymerisation of CO and C_2H_4 [2].

References:

[1] de Haas, K. S.; Fouché, K. F. (Inorg. Chim. Acta **24** [1977] 269/76).

[2] Nozaki, K. (U.S.P. 3835123 [1974]; C.A. **83** [1975] No. 132273).

Other Substituted Complexes

[Pd(CN)$_3$(NH$_3$)]$^-$. This species is observed in solutions of $[Pd(NH_3)_4](NO_3)_2$ with KCN in liquid ammonia; it gives rise to polarised Raman bands at 2148 and 2134 cm^{-1}, and a depolarised one at 2136 cm^{-1}. In the infrared bands at 2146 and 2134 cm^{-1} are observed [6].

Pd(CN)$_2$(NH$_3$). This compound has been briefly mentioned; it is made by treating $Pd(CN)_2$-$(NH_3)_2$ with dilute acetic acid [3].

cis-Pd(CN)$_2$(NH$_3$)$_2$. This is made from $Pd(CN)_2$ with 6M aqueous NH_4OH, and is white [1].

A single crystal X-ray study showed the crystals to be monoclinic, space group $P2_1/n$; $Z=4$; $a=6.825(2)$, $b=12.733(3)$, $c=6.779(2)$ Å; $\beta=111.82(2)°$; measured density 2.41 g/cm^3, calculated 2.34 g/cm^3. The structure shown in **Fig. 72**, p. 286, is square planar with cis cyano groups; mean Pd–C $=1.958(6)$, mean C–N $=1.138(7)$, mean Pd–N $=2.081(5)$ Å; CPdC angle 87.8(2)°, NPdN angle 90.5(2)° [2]. With dilute acetic acid it yields $Pd(CN)_2(NH_3)$ [3].

cis- and trans-Pd(CN)$_2$(NH$_3$)$_2$. These have been detected by Raman spectroscopy in solutions of $[Pd(NH_3)_4](NO_3)_2$ in liquid ammonia with KCN; for the cis form ν_{CN} bands at 2145 and 2141 cm^{-1} were observed, and for the trans at 2142 cm^{-1}. The cis form gave an infrared band at 2141 cm^{-1} [6].

[Pd(CN)(NH$_3$)$_3$]$^+$. This species was detected in solutions of $[Pd(NH_3)_4](NO_3)_2$ in liquid ammonia with KCN by a polarised Raman band at 2134 cm^{-1}, observed also in the infrared [6].

Pd(CN)$_2$py$_2$. This is made by the exothermic reaction of $Pd(CN)_2$ with pyridine. The product is white [3].

286

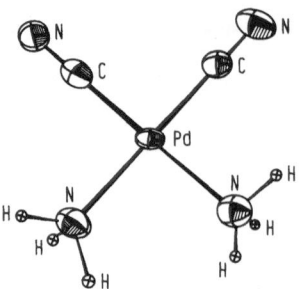

Fig. 72.   Structure of cis-$Pd(CN)_2(NH_3)_2$ [2].

$Pd(CN)_2$bipy. Reaction of a suspension of $Pd(CN)_2$ in ethanol containing excess 2,2′-bipyridyl gives the complex after 2 d of heating on a water bath [3].

$Pd(CN)_2$phen. Reaction of $Pd(CN)_2$ with an almost saturated solution of 1,10-phenanthroline gives the white crystalline product [3].

$Pd(CN)_2$en. This is made by addition of excess ethylenediamine to $Pd(CN)_2$ in water, and forms white needles [3]. The species has also been detected in solution by conductimetric studies on ethylenediamine-$[Pd(CN)_4]^{2-}$ mixtures [4].

$Pd(CN)_2hq_2$. This is made from $Pd(CN)_2$ and a saturated aqueous solution of 8-hydroxyquinoline (hq). It forms an amorphous yellow powder. Although formulated as $Pd(CN)_2hq_2$ it is presumably $Pd(CN)_2hq$ [3].

$Pd(CN)_2quin_2$. Reaction of $Pd(CN)_2$ in excess quinoline (quin) yields the fine crystalline complex [3].

$[Pd(CN)_2naphth]_2$. This is prepared as white crystals from a melt of β-naphthoquinoline (naphth) and $Pd(CN)_2$ [3].

$K_2[Pd(CN)_2L]$. This is made from $K_2[Pd(CN)_4]$ and 4,4′-, 5,5′-tetracyano-2,2′-biimidazole (H_2L) in boiling water [5].

Chloro-cyano Complexes. Preliminary spectrophotometric data on reaction of $[PdCl_4]^{2-}$ with CN^- suggested the existence of $[Pd(CN)_xCl_{4-x}]^{2-}$; the pseudo-first-order rate constant for formation of $[Pd(CN)_2Cl_2]^{2-}$ was some 100 times higher than that for formation of $[Pd(CN)_3Cl]^{2-}$ [7].

Iodo-cyano Complexes. There is evidence, from rate of dissociation studies of $[Pd(CN)_4]^{2-}$ with I_2, that kinetically stable iodo-cyano complexes exist [8].

References:

[1] Jonassen, H. B.; Sistrunk, T. O. (J. Phys. Chem. **59** [1955] 290/3).
[2] Penavić, M.; Šoptrajanova, L.; Jovanovski, G.; Šoptrajanov, B. (Acta Cryst. C **42** [1986] 1283/4).
[3] Feigl, F.; Heinig, G. B. (J. Am. Chem. Soc. **73** [1951] 5631/5).
[4] Kida, S. (Bull. Chem. Soc. Japan **33** [1960] 1365/7).
[5] Rasmussen, P. G.; Bailey, O. H.; Bayon, J. C. (Inorg. Chim. Acta **86** [1984] 107/11).
[6] Gans, P.; Gill, J. B.; McIntosh, D. (Polyhedron **6** [1987] 79/84).
[7] Hancock, R. D.; Evers, A. (Inorg. Chem. **15** [1976] 995/6).
[8] Crouse, W. C.; Bennett, D. A.; Margerum, D. W. (React. Kinet. Catal. Letters **1** [1974] 135/40).

11.3.2.5 Polynuclear Species

Palladium(I). The $[Pd_2(CN)_6]^{4-}$ species has been mentioned in connection with KCN melts containing Pd or $PdCl_2$ [1].

Palladium(II). It is possible that the red material formed when $Pd(CN)_2$ is heated with molten KCN contains $K_2[Pd_2(CN)_6]$ [2].

[(NC)$_3$Pd–CN–Co(NH$_3$)$_4$(H$_2$O)]$_2$[Pd(CN)$_4$]·2H$_2$O. This is made by reaction of $Na_2[Pd(CN)_4]$ and cis-$[Co(NH_3)_4(H_2O)_2](ClO_4)_3$ in a 3:2 molar ratio. The complex is orange and has ν_{CN} bands at 2200 and 2125 cm^{-1}; electronic maxima were observed at 491, 348 and 260 nm [3].

[(NC)$_3$Pd–CN–Coen$_2$(H$_2$O)]$_2$[Pd(CN)$_4$]·3H$_2$O. This red-orange material is made from $Na_2[Pd(CN)_4]$ and cis-$[Coen_2(H_2O)_2](ClO_4)_3$ in aqueous solution in a 3:2 molar ratio. The infrared spectrum has ν_{CN} bands at 2180 and 2135 cm^{-1}, and electronic spectra maxima at 482, 350 and 240 nm [3].

[(NC)$_3$Pd–CN–Co(NH$_3$)$_5$]$_2$[Pd(CN)$_4$]. This was made by heating $[Co(NH_3)_5(H_2O)]_2[Pd(CN)_4]_3$ at 110°C for 30 min. The kinetics of the solid-state de-aquation reaction were studied.

The infrared spectrum has ν_{CN} at 2200 and 2135 cm^{-1}; the electronic spectrum was also measured (300 to 600 nm) [4].

[(NC)$_3$Pd–CN–Coen$_2$(NH$_3$)]$_2$[Pd(CN)$_4$]. This is made by heating cis-$[Coen_2(NH_3)(H_2O)]_2$ $[Pd(CN)_4]_3$ to 70°C for 45 min. In the infrared ν_{CN} is observed at 2190 and 2140 cm^{-1}; electronic maxima at 540, 468 and 345 nm were recorded [5].

[(NC)$_3$Pd–CN–Cren$_2$Cl]·nH$_2$O. The dihydrate and the anhydrous complex can be made from cis-$[Cren_2(H_2O)Cl]Br_2$ and $K_2[Pd(CN)_4]$ in water, or by dehydration of cis-$[Cr en_2(H_2O)Cl]$-$[Pd(CN)_4]$.

The red salts have ν_{CN} at 2175 and 2130 cm^{-1}; electronic spectra were also recorded (200 to 600 nm) [6].

(NC)$_3$Pd–CN–Cren$_2$Br. This is made from $K_2[Pd(CN)_4]·3H_2O$ and cis-$[Cren_2Br(H_2O)]Br_2$ in refluxing methanol, or by dehydration of cis-$[Cren_2Br(H_2O)][Pd(CN)_4]$.

The infrared spectrum has ν_{CN} at 2180 and 2135 cm^{-1}; electronic spectra from 200 to 600 nm were recorded [6].

(NC)$_3$Pd–CN–cis-CrF(chxn)$_2$ (chxn = 1,2-diaminocyclohexane) is made by heating $[Pd(CN)_4]$-$[trans-CrF–(chxn)_2–H_2O]$ [7].

References:

[1] de Haas, K. S.; Fouché, K. F. (Inorg. Chim. Acta **24** [1977] 269/76).
[2] Magnuson, W. L.; Griswold, E.; Kleinberg, J. (Inorg. Chem. **3** [1964] 88/93).
[3] Escuer, A.; Ribas, J. (Inorg. Chim. Acta **104** [1985] 143/7).
[4] Ribas, J.; Serra, M.; Escuer, A. (Inorg. Chem. **23** [1984] 2236/42).
[5] Ribas, J.; Escuer, A. (Transition Metal Chem. [Weinheim] **10** [1985] 466/9).
[6] Ribas, J.; Serra, M.; Escuer, A. (Transition Metal Chem. [Weinheim] **9** [1984] 287/90).
[7] Corbella, M.; Ribas, J. (Anales Quim. **82b** [1986] 273/8).

For species of the Prussian Blue or Berlin Green type $M^{II}[Pd(CN)_6]$ see above, p. 282.

288

11.4 Palladium Cyanate(?) and Cyanato Complexes

Palladium cyanate does not seem to have been reported. An infrared band at 2264 cm^{-1} has been assigned to the ν_{CN} of adsorbed NCO$^-$ on Pd, formed when reaction mixtures of CO, NO and N$_2$ were passed over Pd (5% on Al$_2$O$_3$) at 400°C [1].

The [Pd(NCO)$_4$]$^{2-}$ Ion. The pK$_1$ of [Pd(NCO)$_4$]$^{2-}$ has been given as 1.29 [2]. Extraction of Pd from aqueous solutions containing Pd by KNCO into organic solvents has been described [3].

(PPh$_4$)$_2$[Pd(NCO)$_4$]. The X-ray photoelectronic (ESCA) spectrum of this shows the Pd 3d$_{5/2}$ binding energy to lie at 338.4 eV, P2p at 134.0 eV and N1s at 398.8 eV [4].

Pd(NCO)$_2$py$_2$. This is made from PdCl$_2$, Ag(NCO) and pyridine for 4 h at 60°C [5].

Pd(NCO)$_2$(TMEN). This is made from PdCl$_2$, Ag(NCO) and tetramethylethylenediamine (TMEN) [5].

Pd(NCO)$_2$(PPh$_3$)$_2$. This is made from PdCl$_2$, Ag(NCO) and PPh$_3$ [5].

References:

[1] Unland, M. L. (Science **179** [1973] 567/9).
[2] Lodzinska, A. (Roc. Chem. **41** [1967] 1007/14).
[3] Ziegler, M. (Ger. P. 1057078 [1959]; C.A. **55** [1961] 4900).
[4] Nefedov, V. I.; Koehler, H.; Fiedler, R. (Koord. Khim. **9** [1983] 1561/2; C.A. **100** [1984] No. 42324).
[5] Ozaki, S.; Tamaoki, A.; Mitsui Toatsu Chemicals Co. Ltd. (Jap. P. 72-19526 [1972]; C.A. **77** [1972] No. 90754).

11.5 Thiocyanates

See "Palladium" 1942, pp. 301, 319/320, 337, 339, 342, 386, 398.

There is a general review on thiocyanates and thiocyanate complexes [1].

Palladium thiocyanate, Pd(NCS)$_2$, is well known, but the commonest palladium thiocyanate complex is K$_2$[Pd(SCN)$_4$], made by reaction of palladium(II) salts with excess KSCN. In addition there is a large number of substituted thiocyanate complexes, mostly of the form Pd(SCN)$_2$L$_2$ or Pd(NCS)$_2$L$_2$.

Since thiocyanate is an ambidentate ligand there are a number of instances where –S bonding, denoted here by Pd(SCN), or –N bonding, denoted by Pd(NCS), to the palladium occurs. In a few cases both –S and –N bonding may occur within the same molecule. Only in relatively rare cases has the actual donor atom been identified; where in doubt the Pd(SCN) form is written here. A number of reviews on this aspect are available [2 to 5].

References:

[1] Chemistry of Pseudohalides, Elsevier, Amsterdam 1986, p. 275.
[2] Norbury, A. H. (Advan. Inorg. Chem. Radiochem. **17** [1974/75] 231/386, 315).
[3] Burmeister, J. L. (Coord. Chem. Rev. **3** [1968] 225/45).
[4] Burmeister, J. L. (Coord. Chem. Rev. **1** [1966] 205/21).
[5] Norbury, A. H.; Sinha, A. I. P. (Quart. Rev. [London] **24** [1970] 69/94, 82).

11.5.1 Palladium Thiocyanate Pd(SCN)$_2$

This is made by reaction of a palladium(II) nitrate solution with SCN$^-$ [1], from PdCl$_2$ and NH$_4$NCS [2], or by heating Pd(NCS)$_2$(PBu$_3^n$)$_2$ to 290°C [6].

It is an orange compound, sparingly soluble in water and insoluble in acetone, ethanol and ether [1, 2].

The solubility product has been calculated as 1.5×10^{-18}, and its solubility in aqueous SCN$^-$ solutions was measured [1]. A reference seemingly to the formation constant of "Pd(SCN)$_2$" actually refers to that of [Pd(SCN)$_2$Cl$_2$]$^{2-}$ [3], and one to extraction of Pd(SCN)$_2$ probably refers to [Pd(SCN)$_2$(H$_2$O)$_2$] [4].

With SCN$^-$, [Pd(SCN)$_4$]$^{2-}$ is formed [1, 2], while pyridine yields Pd(SCN)$_2$py$_2$ [1], and 4-azido-pyridine (azpy) gives Pd(SCN)$_2$(azpy)$_2$ [5]. With isoxazole (L), Pd(SCN)$_2$L is formed [6], and with 3-nicotinamide (L'), 2-pyrazinecarboxamide (L') and 2-aminopyrimidine (L"), Pd(SCN)$_2$L'$_2$ and Pd(SCN)$_2$L" are formed [7].

The use of Pd(SCN)$_2$ in analysis of Pd has been investigated [2].

Pd(SCN)OH. This is made by boiling the material extracted with (NH$_4$)SCN from palladium(II) nitrate solutions [8].

References:

[1] Golub, A. M.; Pomerants, G. B. (Zh. Neorgan. Khim. **9** [1964] 1624/9; Russ. J. Inorg. Chem. **9** [1964] 879/82).

[2] Przheval'skii, E. S.; Shlenskaya, V. I. (Vestn. Mosk. Univ. **7** No. 5 [1952] Ser. Fiz. Mat. Estestv. Nauk. No. **3** 61/7; C.A. **1953** 4789).

[3] Joshi, S. B.; Pundalik M. D.; Mattoo, B. N. (Indian. J. Chem. **11** [1973] 1297/9).

[4] Al-Bazi, S. J.; Chow, A. (Talanta **30** [1983] 487/92).

[5] El'tsov, A. V.; Khanchmann, A.; Rtishchev, N. I. (Zh. Org. Khim. **13** [1977] 465; J. Org. Chem. [USSR] **13** [1977] 425/6).

[6] Pinna, R.; Ponticelli, G.; Preti, C. (J. Inorg. Nucl. Chem. **37** [1975] 1681/4).

[7] Singh, P. P.; Seth, J. N.; Khan, S. A. (Inorg. Nucl. Chem. Letters **11** [1975] 525/8).

[8] Prihoda, J.; Kyrs, M.; Purkyne, J. E. (Jad. Energ. **28** [Prague] [1982] 127/9; C.A. **97** [1982] No. 48 506).

11.5.2 Palladium Thiocyanato Complexes

11.5.2.1 The [Pd(SCN)$_4$]$^{2-}$ Ion

This is made by reaction of palladium(II) salts with excess thiocyanate. There is a review on thiocyanato complexes giving a little data on [Pd(SCN)$_4$]$^{2-}$ [1]; see also [2].

References:

[1] Singh, F. P. (Coord. Chem. Rev. **32** [1980] 33/65, 39).

[2] Chemistry of Pseudohalides, Elsevier, Amsterdam 1986.

Vibrational Spectra

For a review on vibrational spectra of thiocyanato complexes see [1]. Most of the data are for specific salts in the solid state and are referred to in the appropriate sections.

A solution of $K_2[Pd(SCN)_4]$ in CH_3CN shows ν_{CN} at 2100 and 2045 cm^{-1} [2], or at 2110 cm^{-1} and at 2110 cm^{-1} in $CH_3COC_2H_5$, with ν_{CS} at 730 cm^{-1} [3]; in $(CH_3)_2SO$ the bands of free SCN$^-$ at 2045 and 740 cm^{-1} are observed [2]. The predominant presence of the S-bonded anion in CH_3CN was deduced [2]. Intensity measurements have been made on the ν_{CN} peak at 2110 cm^{-1} of $[Pd(SCN)_4]^{2-}$ in CH_3CN and in $CH_3COC_2H_5$ solutions [3] and compared with intensity data for other thiocyanato complexes [4].

References:

[1] Bailey, R. A.; Kozak, S. L.; Michelsen, T. W.; Mills, W. N. (Coord. Chem. Rev. **6** [1971] 407/45).

[2] Sodnomov, B. G.; Polovnyak, V. K.; Troitskaya, A. D.; Rusetskii, O. I. (Koord. Khim. **6** [1980] 1061/3; Soviet J. Coord. Chem. **6** [1980] 534/5).

[3] Pecile, C. (Inorg. Chem. **5** [1966] 210/14).

[4] Bailey, R. A.; Michelsen, T. W.; Mills, W. N. (J. Inorg. Nucl. Chem. **33** [1971] 3206/10).

Electronic Spectra

Electronic spectra have been reported as follows (spectra reproduced in papers over the cited ranges): in aqueous SCN$^-$ (250 to 700 nm) [1]; of $Pd(ClO_4)_2$ in excess aqueous SCN$^-$ (280 to 500 nm) [2] and 260 to 400 nm [3]; of $[PdCl(SCN)_3]^{2-}$ in excess SCN$^-$ (200 to 400 nm) [4]; of $Pd(SCN)_2$ in excess SCN$^-$ (240 to 370 nm) [5].

The electronic spectrum of $((n-C_4H_9)_4N)_2[Pd(SCN)_4]$ in molten KSCN–NaSCN has been measured (250 to 500 nm, reproduced in paper) [1]; the spectrum of $(Ph_4As)_2[Pd(SCN)_4]$ in $CHCl_3$ (200 to 340 nm, reproduced in paper) [6]; the spectra of $K_2[Pd(SCN)_4]$ at low temperatures in glassy 4:1 C_2H_5OH–CH_3OH [7]; the spectrum of the rhodamine-B salt of $[Pd(SCN)_4]^{2-}$ in butyl acetate (480 to 680 nm, reproduced in paper) [13].

The molar absorption coefficients for electronic spectra of $(CV)_2[Pd(SCN)_4]$ (CV = crystal violet, $C_{15}H_{30}N_3$) have been listed [9].

Data were listed for the aqueous solution of $K_2[Pd(SCN)_4]$ (250 to 600 nm, with assignments) [12] and, misprinted in paper as $K[Pd(SCN)_4]$, from 250 to 550 nm [8].

The validity of Beer's law to solutions of Pd salts in SCN$^-$ was tested and the use of the absorption spectrum of $[Pd(SCN)_4]^{2-}$ as an analytical procedure for Pd shown to be useful [10]. Formation of a red species, thought to be $[Pd(SCN)_4]^{2-}$, has been used for the spectrophotometric determination of Pd in plating baths containing the metal [11].

References:

[1] de Haas, K. S. (J. Inorg. Nucl. Chem. **35** [1973] 3231/40).

[2] Golub, A. M.; Pomerants, G. B. (Zh. Neorgan. Khim. **9** [1964] 1624/9; Russ. J. Inorg. Chem. **9** [1964] 879/82).

[3] Biryukov, A. A.; Shlenskaya, V. I.; Alimarin, I. P. (Izv. Akad. Nauk SSSR Ser. Khim. **1966** 3/8; Bull. Acad. Sci. USSR Div. Chem. Sci. [1966] 1/5).

[4] Biryukov, A. A.; Shlenskaya, V. I. (Zh. Neorgan. Khim. **12** [1967] 2579/82; Russ. J. Inorg. Chem. **12** [1967] 1362/3).

[5] Shlenskaya, V. I.; Khvostova, V. P.; Peshkova, V. M. (Zh. Analit. Khim. **17** [1962] 598/603; J. Anal. Chem. [USSR] **17** [1962] 596/601).

[6] Magee, R. J.; Khattak, M. A. (Microchem. J. **8** [1964] 285/94).

[7] Francke, E.; Moncuit, C. (Mol. Spectrosc. Dense Phases Proc. 12th Eur. Congr. Mol. Spectrosc., Strasbourg 1975 [1976], pp. 429/32; C.A. **86** [1977] No. 24032).

[8] Babaeva, A. V.; Rudyl, R. I. (Zh. Neorgan. Khim. **1** [1956] 921/9; J. Inorg. Chem. [USSR] **1** [1956] 42/51).

[9] Pilipenko, A. T.; Ol'Khovich, I. F. (Izv. Sibirsk. Otd. Akad. Nauk SSSR Ser. Khim. Nauk **1970** 87/91; Sib. Chem. J. **1970** 528/31).

[10] Peshkova, V. M.; Shlenskaya, V. I.; Khvostova, V. P. (Vopr. Anal. Blagorodn. Metal. Tr. 5th Vses. Soveshch., Novosibirsk 1960 [1963], pp. 57/63; C.A. **61** [1964] 4955).

[11] Erdos, E.; Hasko, F. (Galvanotechnik **56** [1965] 726).

[12] Jørgensen, C. K. (Absorption Spectra and Chemical Bonding in Complexes, Pergamon, Oxford 1962, p. 287).

[13] Lopez-Garcia, I.; Martinez Aviles, J.; Hernandez Cordoba, M. (Talanta **33** [1986] 411/4).

^{13}C, ^{14}N and ^{15}N NMR Data

The ^{13}C chemical shift for $[Pd(SCN)_4]^{2-}$ in aqueous solution is $\delta = 128.4$ (as compared with $\delta = 134.0$ for SCN^-) vs. $(CH_3)_4Si$, suggesting that S-bonding of the thiocyanate group is retained in solution [1, 2].

The ^{14}N chemical shift for an aqueous solution of $Na_2[Pd(SCN)_4]$ was found to be at $\delta = -18$ ppm compared with free SCN^-, and at $\delta = +148$ ppm compared with the NO_3^- ion [3]. The ^{15}N NMR spectrum of $(Bu_4N)_2[Pd(SC^{15}N)_4]$ in CH_2Cl_2 showed a shift relative to $^{15}NH_4Cl$ of 225.3 ppm, thought to be suggestive of thiocyanate being S-bonded to the metal [4].

References:

[1] Kargol, J. A.; Crecely, R. W.; Burmeister, J. L. (Inorg. Chem. **18** [1979] 2532/5).

[2] Kargol, J. A.; Crecely, R. W.; Burmeister, J. L. (Inorg. Chim. Acta **25** [1979] L109/L110).

[3] Howarth, O. W.; Richards, R. E.; Venanzi, L. M. (J. Chem. Soc. **1964** 3335/7).

[4] Pregosin, P. S.; Streit, H.; Venanzi, L. M. (Inorg. Chim. Acta **38** [1980] 237/42).

Stability Constants

The equilibrium constant for $[Pd(SCN)_4]^{2-}$ has been calculated by spectrophotometric methods to be 42.2 [1]; the dissociation constant K_4 of $[Pd(SCN)_4]^{2-}$, from spectrophotometric and potentiometric data, has been calculated as 3.5×10^{-20} [1] and, from spectrophotometric data, as 1×10^{-8} [6]. From spectrophotometric data on the $[Pd(SCN)_4]^{2-}-Br^-$ and $[PdBr_4]^{2-}-$ SCN$^-$ system the log K_4 value for $[Pd(SCN)_4]^{2-}$ at 25°C and total ionic strength log β_4 was found to be 28.22 [2]; similar data on $[PdCl_4]^{2-}-SCN^-$ mixtures gave log β_4 as 28.67 ± 0.36 at 25°C and 1M ionic strength [3]. From potentiometric data on $[Pd(SCN)_4]^{2-}-SCN^-$ solutions at 20°C a value of log β_4 of 27.6 ± 0.2 was measured [4], and from spectrophotometric data on the $PdCl_2-SCN^-$ reaction a log β_4 value of 25.6 at 25°C and total ionic strength 0.1M was determined [5]. Other spectrophotometric determinations of β_4 values have been carried out [6]. A pK value (instability constant) of 1.69 for $[Pd(SCN)_4]^{2-}$ has been determined [7].

References:

[1] Golub, A. M.; Pomerants, G. B. (Zh. Neorgan. Khim. **9** [1964] 1624/9; Russ. J. Inorg. Chem. **9** [1964] 879/82).

[2] Biryukov, A. A.; Shlenskaya, V. I.; Alimarin, I. P. (Izv. Akad. Nauk SSSR Ser. Khim. **1966** 3/8; Bull. Acad. Sci. USSR Div. Chem. Sci. **1966** 1/5).

[3] Biryukov, A. A.; Shlenskaya, V. I. (Zh. Neorgan. Khim. **12** [1967] 2579/82; Russ. J. Inorg. Chem. **12** [1967] 1362/3).

[4] Grinberg, A. A.; Kiseleva, N. V.; Gel'fman, M. I. (Dokl. Akad. Nauk SSSR **153** [1963] 1327/9; Dokl. Chem. Proc. Acad. Sci. USSR **148/153** [1963] 1025/7).

[5] Joshi, N. B.; Pundalit, M. D.; Mattoo, B. N. (Indian J. Chem. **11** [1973] 1297/9).

[6] Shlenskaya, V. I.; Khvostova, V. P.; Peshkova, V. M. (Zh. Analit. Khim. **17** [1962] 598/603; J. Anal. Chem. [USSR] **17** [1962] 596/601).

[7] Lodzinska, A. (Roczniki Chem. **41** [1967] 1007/14).

Electrochemical Behaviour

The E_0 for $[Pd(SCN)_4]^{2-}/Pd$ in aqueous solution is given as 0.240 V at 10°C and 0.328 V at 60°C vs. the standard calomel electrode.

Reference:

Gasman, A. B.; Kutyukov, G. G.; Sokol'skii, D. V. (Zh. Neorgan. Khim. **10** [1965] 1338/43; Russ. J. Inorg. Chem. **10** [1965] 727/30).

Extraction and Ion Exchange

Extraction of $[Pd(SCN)_4]^{2-}$ in benzene or toluene using diantipyrylmethane has been studied and optimum conditions for the formation of $(CV)_2[Pd(SCN)_4]$ (CV = crystal violet) were determined [1]. The extraction of $[Pd(SCN)_4]^{2-}$ (or of mixed thiocyanato complexes) with a chloroform sulution of 1-carbethoxypentadecyl-trimethylammonium bromide (KPTB) has been reported [2], and also the extraction of Pd^{II} complexes with KSCN from slightly acidic solutions into polyethylene glycol with CH_3Cl, CH_3Br or CH_3I [3]. The separation by ion exchange of Pd and Pt by nucleophilic substitution with SCN^- or thiourea from weak base anion-exchange resins has been described [4]. Distribution coefficients for $[PdX_4]^{2-}$ (X = SCN, Br, I) for various concentrations of alcohols and cyclohexanone in benzene have been determined; the thiocyanato complex can be completely extracted at pH < 4 [5].

Extraction from palladium(II) nitrate solution with NH_4SCN has been studied and "Pd(SCN)OH" obtained from the extracted material [6]. The use of tetraethyldiamide heptyl-phosphate in xylene for extracting $[Pd(SCN)_4]^{2-}$ and $[Pd(SCN)_2Cl_2]^{2-}$ has been described [7].

Extraction of $[Pd(SCN)_4]^{2-}$ by polyether-type polyurethane foam has been investigated [8, 9, 10]; the rate of such extraction is higher from aqueous HCl than from aqueous KCl solutions [8], and "$Pd(SCN)_2$" (presumably $[Pd(SCN)_2(H_2O)_2]$ is extracted at low SCN^- concentrations [9]. Extraction of $[Pd(SCN)_4]^{2-}$ as the rhodamine-B ion association pair into organic solvents has been used to analyse the metal spectrophotometrically [11]. Palladium(II) has been extracted as an unspecified species, presumably as $[Pd(SCN)_4]^{2-}$, from aqueous solutions containing palladium(II) salts with NH_4NCS into ethylacetate [12].

Extraction of palladium from $PdCl_2$ and SCN^- with dibenzo-18-crown-6 and its derivatives has been studied [13].

References:

[1] Pilipenko, A. T.; Ol'Khovich, I. F. (Izv. Sibirsk. Otd. Akad. Nauk SSSR Ser. Khim. Nauk **1970** 87/91; Sib. Chem. J. **1970** 528/31).

[2] Malat, M. (Z. Anal. Chem. **297** [1979] 417).

[3] Ziegler, M. (Z. Anal. Chem. **171** [1959] 111/4).

[4] Warshawsky, A. (Sepn. Purif. Methods **12** [1983] 37/48).

[5] Golub, A. M.; Pomerants, G. V. (Ukr. Khim. Zh. **31** [1965] 104/12).

[6] Prihoda, J.; Kyrs, M. (Jad. Energ. [Prague] **28** [1982] 127/9; C. A. **97** [1982] No. 48506).

[7] Fadeeva, V. I.; Nasonovskii, I. S.; Minochkina, L. N.; Volynskii, A. B.; Zorov, N. B. (Vestn. Mosk. Univ. II **21** No. 1 [1980] 98; C.A. **92** [1980] No. 170088).
[8] Al-Bazi, S. J.; Chow, A. (Talanta **29** [1982] 507/10).
[9] Al-Bazi, S. J.; Chow, A. (Talanta **30** [1983] 487/92).
[10] Chow, A.; Ginsberg, S. L. (Talanta **30** [1983] 620/2).

[11] Lopez-Garcia, I.; Martinez Aviles, J.; Hernandez Cordoba, M. (Talanta **33** [1986] 411/4).
[12] Paria, P. K.; Majumdar, S. K. (Indian J. Chem. A **14** [1976] 820/1).
[13] Poladyan, V. E.; Burtnenko, L. M.; Avlasovich, L. M.; Andrianov, A. M. (Zh. Neorgan. Khim. **32** [1987] 737/40; Russ. J. Inorg. Chem. **32** [1987] 413/5).

Chemical Reactions

With N Donors. Salts containing $[Pd(SCN)_4]^{2-}$ react with pyridine (py) to give $Pd(SCN)_2py_2$ [1] and possibly $Pd(NCS)_2py_2$ (see p. 302) [2]. There are also complexes with a wide range of substituted pyridines R-py, $Pd(SCN)_2(R-py)_2$, made in a similar fashion [1, 2, 3]. Reaction of $[Pd(SCN)_4]^{2-}$ in ethanol with 2,2'-bipyridyl (bipy) or 1,10-phenanthroline (phen) yields $Pd(SCN)_2bipy$ [2, 4] and $Pd(SCN)_2phen$ [1, 2]; there are also substituted bipyridyl and phenanthroline species made in similar fashion, e.g., $Pd(NCS)(SCN)(4,4-Mebipy)$ using 4,4'-dimethyl-2,2'-bipyridyl and $[Pd(SCN)_4]^{2-}$ [3] and both $Pd(SCN)_2(5-nitrophen)$ and $Pd(NCS)_2$-(5-nitrophen) can be got from $[Pd(SCN)_4]^{2-}$ and 5-nitro-1,10-phenanthroline [3, 5]. A wide range of complexes $Pd(SCN)_2(R-phen)$ with substituted phenanthrolines R-phen was similarly made [3, 5]. With 2,2',6',2''-terpyridyl (terpy), $[Pd(NCS)(terpy)]NCS$ is formed [2]. With hexamethylenetetramine (HMTA), $Pd(SCN)_2HMTA_2$ is formed [6].

References:

[1] Miezis, A. (Acta Chem. Scand. **27** [1973] 3746/60).
[2] Burmeister, J. L.; Basolo, F. (Inorg. Chem. **3** [1964] 1587/93).
[3] Bertini, I.; Sabatini, A. (Inorg. Chem. **5** [1966] 1025/8).
[4] Basolo, F.; Burmeister, J. L.; Poë, A. J. (J. Am. Chem. Soc. **85** [1963] 1700/1).
[5] Fultz, W. C.; Burmeister, J. L. (Inorg. Chim. Acta **45** [1980] L271/L273).
[6] Sinitsyn, N. M.; Buslaeva, T. M.; Efanov, V. I. (Zh. Neorgan. Khim. **29** [1984] 2986/7; Russ. J. Inorg. Chem. **29** [1984] 1708/9).

With Halides. Reaction of $[PdCl_4]^{2-}$ with SCN^- gives $[Pd(SCN)_nCl_{4-n}]^{2-}$; stability constants for the mixed species were given [1]. Similar data for $[Pd(SCN)_nBr_{4-n}]^{2-}$ have been given [2].

References:

[1] Biryukov, A. A.; Shlenskaya, V. I. (Zh. Neorgan. Khim. **12** [1967] 2579/82; Russ. J. Inorg. Chem. **12** [1967] 1362/3).
[2] Biryukov, A. A.; Shlenskaya, V. I.; Alimarin, I. P. (Izv. Akad. Nauk SSSR Ser. Khim. **1966** 3/8; Bull. Acad. Sci. USSR Div. Chem. Sci. **1966** 1/5).

With S Donors. With N-methyl O-ethylthiocarbamate (MTC) $[Pd(SCN)_4]^{2-}$ yields trans-$Pd(SCN)_2MTC_2$ [1], while thiourea gives $Pd(SCN)_2(thiourea)_2$ [2]. Ethylenethiourea (etu) and $[Pd(SCN)_4]^{2-}$ give $Pd(SCN)_2(etu)_2$ [2, 3].

References:

[1] Bardi, R.; Del Pra, A.; Piazzessi, A. M.; Sindellari, L.; Zarli, B. (Inorg. Chim. Acta **47** [1981] 231/4).

[2] Burmeister, J. L.; Basolo, F. (Inorg. Chem. **3** [1964] 1587/93).

[3] Miezis, A. (Acta Chem. Scand. **27** [1973] 3746/60).

With C Donors. The reactivity of $[PdX_4]^{2-}$ (X = SCN, Cl, Br, I, NO_2, NH_3, CN, thiourea) in aqueous solution towards CO have been compared.

Reference:

Fasman, A. B.; Kutyukov, G. G.; Sokol'skii, D. V. (Zh. Neorgan. Khim. **10** [1965] 1338/43; Russ. J. Inorg. Chem. **10** [1965] 727/30).

With P Donors. With $P(OMe)_3$, $[Pd(SCN)_4]^{2-}$ yields $Pd(SCN)_2[P(OMe)_3]_2$ [1]. The kinetics of reaction of $[Pd(SCN)_4]^{2-}$ with $PPh_2(OEt)$ and $PEt_2(OPh)$ were studied; kinetic and activation parameters of the replacement reaction were studied [2], and similar thermodynamic stability and kinetic observations made on the substitution reactions of $[Pd(SCN)_4]^{2-}$ with PR_3 (R = Et, Ph, OEt, OPh) using stopped-flow spectrophotometry [3].

Reaction of $[Pd(SCN)_4]^{2-}$ with PBu_3^n gives $Pd(SCN)_2(PBu_3^n)_2$; likewise PPh_3 yields $Pd(NCS)_2$-$(PPh_3)_2$ [4, 5].

References:

[1] Troitskaya, A. D.; Orlova, I. A. (Zh. Obshch. Khim. **45** [1975] 951; J. Gen. Chem. [USSR] **45** [1975] 938).

[2] Sodnomov, B. D.; Polovnyak, V. K.; Troitskaya, A. D. (Zh. Obshch. Khim. **50** [1980] 1459/61; J. Gen. Chem. [USSR] **50** [1980] 942/4).

[3] Sodnomov, B. G.; Polovnyak, V. K.; Troitskaya, A. D.; Kormachev, V. V. (Koord. Khim. **6** [1980] 1737/40; Soviet J. Coord. Chem. **6** [1980] 872/5).

[4] Burmeister, J. L.; Basolo, F. (Inorg. Chem. **3** [1964] 1587/93).

[5] Miezis, A. (Acta Chem. Scand. **27** [1973] 3746/60).

With As and Sb Donors

With $K_2[Pd(SCN)_4]$, $AsBu_3^n$ in ethanol gives $Pd(SCN)_2(AsBu_3^n)_2$ [1]. The N-bonded isomer $Pd(NCS)_2(AsBu_3^n)_2$ may be made in an analogous fashion [2]. Reaction of $[Pd(SCN)_4]^{2-}$ with $AsPh_3$ yields $Pd(SCN)_2(AsPh_3)_2$ [3, 4]. Reaction of $[Pd(SCN)_4]^{2-}$ with $SbPh_3$ yields $Pd(SCN)_2$-$(SbPh_3)_2$ [2, 4]; the chelating ligand $o\text{-}C_6H_4PPh_2SbPh_2$ reacts with $[Pd(SCN)_4]^{2-}$ to give $Pd(NCS)(SCN)(o\text{-}C_6H_4PPh_2SbPh_2)$ [5].

References:

[1] Sabatini, A.; Bertini, I. (Inorg. Chem. **4** [1965] 1665/7).

[2] Burmeister, J. L.; Basolo, F. (Inorg. Chem. **3** [1964] 1587/93).

[3] Basolo, F.; Burmeister, J. L.; Poë, A. J. (J. Am. Chem. Soc. **85** [1963] 1700/1).

[4] Tsytsyktueva, L. A.; Polovnyak, V. K.; Gazizov, K. K.; Gazizova, D. M. (Zh. Neorgan. Khim. **31** [1986] 2146/8; Russ. J. Inorg. Chem. **31** [1986] 1237/8).

[5] Levason, W.; McAuliffe, C. A. (Inorg. Chim. Acta **16** [1976] 167/72).

Miscellaneous Reactions of $[Pd(SCN)_4]^{2-}$

Reaction of solutions of urea-formaldehyde polymers with palladium(II) salts gives a bright red thiocyanato palladate precipitate [1]. The $[Pd(SCN)_4]^{2-}$ ion, unlike $[PdX_4]^{2-}$ (X = Cl, Br) does not catalytically decompose H_2O_2 [2].

References:

[1] Ziegler, M. (Z. Anal. Chem. **168** [1959] 29/30).
[2] Grinberg, A. A.; Kukushkin, Yu. N.; Vlasova, R. A. (Zh. Neorgan. Khim. **13** [1968] 2177/83; Russ. J. Inorg. Chem. **13** [1968] 1126/9).

11.5.2.2 Free Acid and Salts with Organic Cations (see "Palladium" 1942, p. 301)

[(CH₃)₄N]₂[Pd(SCN)₄]. This is made from $K_2[Pd(SCN)_4]$ and $[(CH_3)_4N]NCS$ in water. The infrared spectrum shows the Pd–S stretches to lie at 298, 289 and 207 cm^{-1} [1].

[(C₂H₅)₄N]₂[Pd(SCN)₄]. The salt is made by reaction of "$H_2[PdCl_4]$", $(C_2H_5)_4NCl$ and KSCN for 1 h in alcohol [2].

The infrared spectrum showed 2112 and 2100 cm^{-1}, ν_{CS} as 698 and 694 cm^{-1}, δ_{NCS} at 465, 433, 429 and 418 cm^{-1} and ν_{PdS} as 294 cm^{-1} [2].

[C₂H₈N₄Se₂][Pd(SCN)₄]. The existence of this species was revealed by a potentiometric study of $[Pd(SCN)_4]^{2-}$ with the selenourea-formamidine diselenide system [3].

(n-C₄H₉)₂[Pd(SCN)₄]. This is made by reaction of an aqueous solution of $K_2[Pd(SCN)_4]$ in excess KSCN with $n-C_4H_9Br$. The salt is soluble in chloroform [4]. It is dark red [5].

The ^{15}N enriched form was made from $Pd(NO_3)_2 \cdot 2H_2O$ and $NaSC^{15}N$ in water with Bu_4N-(NO_3). The ^{15}N NMR spectrum was measured (see p. 244) [5].

The electronic absorption spectrum of the material in molten NaSCN–KSCN was measured (250 to 700 nm, reproduced in paper) [4].

(Ph₄P)₂[Pd(SCN)₄]. This is made from an unnamed Pd salt, SCN^- ion and $(PPh_4)Cl$ in aqueous solution [6].

The X-ray photoelectronic (ESCA) spectrum showed binding energies for Pd $3d_{5/2}$ of 337.9, N1s = 398.4, S 2p = 163.7, C1s = 285.5 and P 2p = 133.5 eV. Effective changes on the atoms were computed and compared with data for other thiocyanato complexes.

Binding energies for $Pd_{5/2}$ of 337.9, P 2p of 133.1 and N1s of 398.5 eV were also reported for the salt [7].

(Ph₄As)₂[Pd(SCN)₄]. This is made by reaction of $PdCl_2$ with KSCN and $(Ph_4As)Cl$ to give the red salt [8].

The electronic absorption spectrum in $CHCl_3$ was recorded (700 to 500 nm, reproduced in paper) [8].

(C₉H₁₀NH)₂[Pd(SCN)₄]. This quinaldinium salt ist made from $[Pd(SCN)_4]^{2-}$ and $NH_4(SCN)$ with quinaldinium chloride, and is precipitated at pH = I [9].

(C₁₅H₃₀N₃)₂[Pd(SCN)₄]. It is not clear whether this crystal violet salt was isolated, but its molar absorption coefficient was measured [10].

References:

[1] Forster, D.; Goodgame, D. M. L. (Inorg. Chem. **4** [1965] 715/8).
[2] Sabatini, A.; Bertini, I. (Inorg. Chem. **4** [1965] 959/61).
[3] Shul'man, V. M.; Tyuleneva, L. I. (Izv. Sibirsk. Otd. Akad. Nauk SSSR Ser. Khim. Nauk **1972** No. 6, pp. 124/6; C.A. **78** [1973] No. 76484).

[4] de Haas, K. S. (J. Inorg. Nucl. Chem. **35** [1973] 3231/40).

[5] Pregosin, P. S.; Streit, H.; Venanzi, L. M. (Inorg. Chim. Acta **38** [1980] 237/42).

[6] Folkesson, B.; Larsson, R. (J. Electron Spectrosc. Relat. Phenom. **26** [1982] 157/66).

[7] Nefedov, V. I.; Koehler, H.; Fiedler, R. (Koord. Khim. **9** [1983] 1561/2; C.A. **100** [1984] No. 42324).

[8] Magee, R. I.; Khattak, M. A. (Microchem. J. **8** [1964] 285/94).

[9] Gagliardi, E.; Höhn, P. (Mikrochim. Ichnoanal. Acta **1965** 852/4).

[10] Pilipenko, A. T.; Ol'Khovich, I. F. (Izv. Sibirsk. Otd. Akad. Nauk SSSR Ser. Khim. Nauk **1970** No. 4, pp. 87/91; Sib. Chem. J. **1970** 528/31).

(C$_{23}$C$_{24}$O$_2$N$_4$H)$_2$[Pd(SCN)$_4$]. It is not clear whether this diantipyrylmethane salt was isolated but its instability content was given as 6.3×10^{-26} and its formation content as 1.7×10^7 [2].

(C$_{26}$H$_{30}$O$_2$N$_4$H)$_2$[Pd(SCN)$_4$]. This diantipyrylpropylmethane salt was made from Pd^{II} solutions in 0.5 M H$_2$SO$_4$ with diantipyrylmethane and excess thiocyanate. It is orange, insoluble in water but soluble in acetone, chloroform, acetonitrile, dichloroethane and dimethylformamide. The electronic spectra of organic solutions were measured. The use of the reagent in analysis of Pd was discussed [1]. Infrared data on the salt were given [3].

(PyrazoloneH)$_2$[Pd(SCN)$_4$]. In addition to the diantipyrylmethane salt (C$_{26}$H$_{30}$O$_2$N$_4$H)$_2$[Pd(SCN)$_4$] described above similar orange salts (RH)$_2$[Pd(SCN)$_4$] (R = diantipyrylmethylmethane, diantipyrylpropylmethane, diantipyrylphenylmethane) have been prepared from Pd^{II} solutions in 0.5 M H$_2$SO$_4$ with the pyrazolone in the presence of excess thiocyanate. All are insoluble in water but soluble in acetone, chloroform, acetonitrile, dichloroethane and dimethylformamide. The electronic spectra of the organic solutions were measured, and the possible application of these species to analysis of Pd discussed [1]. Infrared data on the complexes were given [3].

(AsPC$_{30}$H$_{24}$)$_2$[Pd(SCN)$_4$]. This orange material is made by reaction of Na$_2$[Pd(SCN)$_4$] in 1-butanol with diphenyl(o-diphenylarsinophenyl)phosphine. It is insoluble in CH$_3$CN, CHCl$_3$ and CH$_2$Cl$_2$ [4].

The infrared spectrum shows ν_{CN} at 2128 and 2102 cm^{-1}, similar to K$_2$[Pd(SCN)$_4$] (spectrum of the solid from 2000 to 2200 cm^{-1} reproduced in paper). The reflectance electronic spectrum was measured; this and the infrared spectrum suggests it to be a Magnus-type salt [4].

On heating it turns yellow at 220°C and finally melts at 253.5 to 254.5°C to a red-orange liquid. Differential thermal analysis showed a strong exotherm reaction at 216 ±1°C to give the yellow Pd(SCN)(NCS)(AsPC$_{30}$H$_{24}$), and this is also formed by heating a solution of the complex in dimethylformamide [4].

[Pd(Ph$_2$P(CH$_2$)$_2$PPh$_2$)$_2$][Pd(SCN)$_4$]. Reaction of 1,2-bis(diphenylphosphino)ethane with [Pd(SCN)$_4$]$^{2-}$ in ethanol gives the light orange Magnus-type salt.

On recrystallising from dimethylformamide, Pd(SCN)(NCS)(Ph$_2$P(CH$_2$)$_2$PPh$_2$) is formed [5].

References:

[1] Dzhishkariani, G. I.; Akimov, V. K.; Emel'yanova, I. A.; Busev, A. I. (Soobshch. Akad. Nauk Gruz. SSSR **73** [1974] 61/4; C.A. **80** [1974] No. 140825).

[2] Pilipenko, A. T.; Ol'Khovich I. F. (Izv. Sibirsk. Otd. Akad. Nauk SSSR Ser. Khim. Nauk **1976** No. 4, pp. 87/91; Sib. Chem. J. **1970** 528/31).

[3] Dzhishkariani, G. I.; Zaitsev, B. E.; Akimov, V. K.; Emel'yanova, I. A. (Soobshch. Akad. Nauk Gruz. SSR **73** [1974] 597/9; C.A. **81** [1974] No. 56089).

[4] Nicpon, P.; Meek, D. W. (Inorg. Chem. **6** [1967] 145/9).

[5] Meek, D. W.; Nicpon, P. E.; Meek, V. I. (J. Am. Chem. Soc. **92** [1970] 5351/9).

11.5.2.3 Salts with Alkali Metal Cations

Na₂[Pd(SCN)₄]. The blood-red solution can be prepared from aqueous solutions of $Na_2[PdCl_4]$ and NaSCN [1].

Reaction of $Na_2[Pd(SCN)_4]$ with Amberlite A21 resin gave an orange product, the infrared spectrum of which was recorded ($v_{CN} = 2090$ and 2040 cm^{-1}).

Reference:

Allen, N. P.; Bamiro, F. O.; Burns, R. P.; Dwyer, J.; McAuliffe, C. A. (Inorg. Chim. Acta **28** [1978] 231/5).

K₂[Pd(SCN)₄] (see "Palladium" 1942, pp. 319/20). This is made by reaction of a saturated aqueous solution of $K_2[PdCl_4]$ and KSCN [1, 2]; after evaporation to dryness the salt is crystallised from acetone [1].

The crystals are dichroic on the prismatic face [2]. The X-ray crystal structure of $K_2[Pd(SCN)_4]$ show the brownish red needle-shaped crystals to be monoclinic, space group $P2_1/a-C_{2h}^5$, $Z = 2$; $a = 11.11(2)$, $b = 12.90(2)$, $c = 4.28(3)$ Å, $\beta = 98.2(2)°$; calculated density 2.39 g/cm³, density by flotation 2.37 g/cm³ [3]. Earlier data gave $a = 11.11$, $b = 12.90$, $c = 4.28$ Å, $\beta = 98.2°$ [4]. Bond lengths are given in **Fig. 73**. Standard deviations are 0.009 Å for Pd–S, 0.05 Å for S–C and C–N; the S–C–N angle has a standard deviation of 4° and that for S(1)PdS(2) is 0.3°. The complex ions are stacked in colums such that the S(2) of one ligand is in the correct position to interact with the Pd atom below as shown in **Fig. 74**, p. 298, to give distorted octahedra about the Pd atoms; the Pd–S distance is 3.657 Å for such axial positions and this relatively long distance is thought to result from repulsion between the S atoms and the filled d_{z^2} orbital on the metal atoms [3].

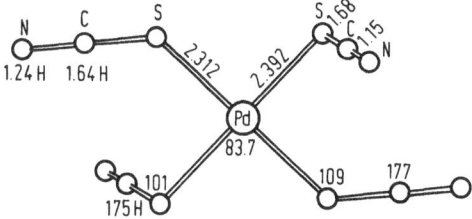

Fig. 73. Structure of the anion in $[Pd(SCN)_4]^{2-}$ [3].

References:

[1] de Stefano, N. J.; Burmeister, J. L. (Syn. Inorg. Metal-Org. Chem. **3** [1973] 313/5).

[2] Yamada, S.; Tsuchida, R. (Bull. Chem. Soc. Japan **26** [1953] 489/93).

[3] Mawby, A.; Pringle, G. E. (J. Inorg. Nucl. Chem. **34** [1972] 2213/17).

[4] Mawby, A.; Pringle, G. E. (Chem. Commun. **1970** 385).

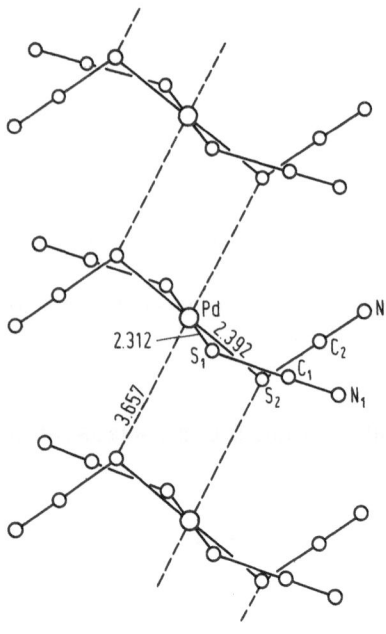

Fig. 74. Arrangement of complex ions in $[Pd(SCN)_4]^{2-}$ [3].

Vibrational Spectra. Most of the data concern the infrared spectrum of the solid, and are listed in the table below.

Infrared spectra of $K_2[Pd(SCN)_4]$:

ν_{CN}	$2\delta_{NCS}$	ν_{CS} in cm^{-1}	δ_{NCS}	ν_{M-S}	Ref.
2125, 2095	942, 936, 929, 885, 876, 867	703, 697	474, 467, 442, 432		[1]
2118, 2086		707, 703, 696			[2]
2124, 2094		704, 698			[3]
2122, 2093		709, 703, 697	474, 468, 442, 432	300, 286	[4]
2122, 2116, 2098, 2092, 2088, 2047	940, 927, 883, 873, 865	708, 703, 694	473, 465, 439, 430		[5]
2115, 2090					[6]
2122, 2091, 2044	940, 928, 883, 875, 868	705, 699, 693	470, 463, 438, 427		[7]
2124, 2093					[8]

References:

[1] Bailey, R. A.; Kozak, S. L.; Michelsen, T. W.; Mills, W. N. (Coord. Chem. Rev. **6** [1971] 407/45, 414).
[2] Turco, A.; Pecile, C. (Nature **191** [1961] 66/7).
[3] Miezis, A. (Acta Chem. Scand. **27** [1973] 3746/60, 3751).
[4] Sabatini, A.; Bertini, I. (Inorg. Chem. **4** [1965] 959/61).
[5] Lewis, J.; Nyholm, R. S.; Smith, P. W. (J. Chem. Soc. **1961** 4590/9).
[6] Chiswell, B.; Livingstone, S. E. (J. Chem. Soc. **1959** 2931/6).
[7] de Stefano, M. J.; Burmeister, J. L. (Syn. Inorg. Metal-Org. Chem. **3** [1973] 313/5).
[8] Bailey, R. A.; Michelsen, T. W.; Mills, W. N. (J. Inorg. Nucl. Chem. **33** [1971] 3206/10).

Electronic Spectra

For a brief review of early work on single-crystal spectra including those for $K_2[Pd(SCN)_4]$ see [1].

Single-crystal absorption spectra of $K_2[Pd(SCN)_4]$ were measured at 10 and 295 K (250 to 600 nm, reproduced in paper) and assignments proposed; 10 Dq was calculated as 22600 cm^{-1}. The spectra and data calculated from them were compared with those for $K_2[Pt(SCN)_4]$ and $K_2[MX_4]$ (M = Pd, Pt; X = Cl, Br) [2]. The dichroic electronic absorption spectra for $K_2[Pd(SCN)_4]$ and $K_2[Pt(SCN)_4]$ were measured (250 to 800 nm, reproduced in paper) [3]. The low temperature polarised crystal spectra of oriented and polished single-crystal plates of $K_2[Pd(SCN)_4]$ were reported and assigned [4].

The electronic absorption maxima of crystalline $K_2[Pd(SCN)_4]$ (misprinted in original paper as $K[Pd(SCN)_4]$) were listed (300 to 550 nm) [5], and those of a Nujol mull of the salt (300 to 550 nm) [6].

References:

[1] Hush, N. S.; Hobbs, R. J. M. (Progr. Inorg. Chem. **10** [1968] 259/486, 433).
[2] Tuszynski, W.; Gliemann, G. (Z. Naturforsch. **34a** [1979] 211/9).
[3] Yamada, S.; Tsuchida, R. (Bull. Chem. Soc. Japan **26** [1953] 489/93).
[4] Francke, E.; Moncuit, C. (Mol. Spectrosc. Dense Phases Proc. 12th Eur. Congr. Mol. Spectrosc., Strasbourg 1975 [1976], pp. 429/32; C.A. **86** [1977] No. 24032).
[5] Babaeva, A. V.; Rudyl, R. I. (Zh. Neorgan. Khim. **1** [1956] 921/9; J. Inorg. Chem. [USSR] **1** [1956] 42/51).
[6] Burmeister, J. L.; Basolo, F. (Inorg. Chem. **3** [1964] 1587/93).

ESCA Data

The X-ray photoelectronic spectrum (ESCA) of $K_2[Pd(SCN)_4]$ shows binding energies of Pd $3d_{5/2} = 337.0$, N1s $= 397.8$, S 2p $= 162.5$ and C1s $= 284.7$ eV. Data were compared with those for other thiocyanato complexes (see also $(PPh_4)_2[Pd(SCN)_4]$, p. 295) [1].

The X-ray SK$_\beta$ spectrum of sulphur in $K_2[Pd(SCN)_4]$ was measured (reproduced in paper) [2].

References:

[1] Folkesson, B.; Larsson, R. (J. Electron Spectrosc. Relat. Phenom. **26** [1982] 157/66).
[2] Mazalov, L. N.; Voityuk, A. A.; Parygina, G. K. (Zh. Strukt. Khim. **23** No. 3 [1982] 48/53; J. Struct. Chem. [USSR] **23** [1982] 364/8).

Heavy-Metal Salts of [Pd(SCN)$_4$]$^{2-}$

Ba[Pd(SCN)$_4$] (see "Palladium" 1942, p. 337).

[Pd(SCN)$_4$][Zn bipy$_3$]. This yellow salt, melting at 240°C, is made from Zn(NCS)$_2$, Pd(SCN)$_2$ and 2,2'-bipyridyl (bipy). The molar conductance in CH$_3$CN is 110.8 $\Omega^{-1} \cdot$ cm^2 [1].

[Pd(SCN)$_4$][Cd py$_4$]. This yellow salt melts at 195°C and is made from Cd(NCS)$_2$, Pd(SCN)$_2$ and py. The molar conductance in CH$_3$CN is 116.8 $\Omega^{-1} \cdot$ cm^2.

The infrared spectrum shows ν_{CN} at 2098 cm^{-1}, ν_{CS} at 760 and 755 cm^{-1} and δ_{SCN} at 470 and 462 cm^{-1} [1].

[Pd(SCN)$_4$][Cd bipy$_3$]. This red salt is obtained from Pd(SCN)$_2$, Cd(NCS)$_2$ and bipy, and melts at 190°C. Its molar conductance in CH$_3$CN is 115.3 $\Omega^{-1} \cdot$ cm^2.

The infrared spectrum shows ν_{CN} at 2085 and 2045 cm^{-1}, ν_{CS} at 840 and 754 cm^{-1} and δ_{NCS} at 482 and 476 cm^{-1} [1].

Tl$_2$[Pd(SCN)$_4$] (see "Palladium" 1942, p. 339).

[Pd(SCN)$_4$][Ni bipy$_3$]. This orange material is prepared from Ni(NCS)$_2$, Pd(SCN)$_2$ and bipy. It melts at 230°C and has a molar conductance in CH$_3$CN of 116.6 $\Omega^{-1} \cdot$ cm^2 [1].

[Pd(SCN)$_4$][Ni phen$_3$]. Reaction of Pd(SCN)$_2$ with Ni(NCS)$_2$ and 1,10-phenanthroline gives this orange material, melting at 228°C. The molar conductance in CH$_3$CN is 118.8 $\Omega^{-1} \cdot$ cm^2.

The infrared spectrum shows ν_{CN} at 2100 and 2075 cm^{-1}, ν_{CS} at 850 and 750 cm^{-1} and δ_{SCN} at 485 and 476 cm^{-1} [1].

[Pd(SCN)$_4$][Co bipy$_3$]. This is pink, melting at 170°C, made by reaction of Pd(SCN)$_2$, Co(NCS)$_2$ and bipy. The molar conductance in CH$_3$CN is 110.8 $\Omega^{-1} \cdot$ cm^2.

The infrared spectrum shows ν_{CN} at 2100 and 2070 cm^{-1}, ν_{CS} at 760 and 750 cm^{-1} and δ_{NCS} at 474 and 458 cm^{-1} [1].

[Pd(SCN)$_4$][Co phen$_3$]. This red material is obtained from Pd(SCN)$_2$, Co(NCS)$_2$ and phen. It melts at 180°C. The molar conductance in CH$_3$CN is 114.3 $\Omega^{-1} \cdot$ cm^2 [1].

Ag$_2$[Pd(SCN)$_4$] (see "Palladium" 1942, p. 342).

[Pd(SCN)$_4$][Pd en$_2$]. This is made from ethylenediamine (en) and K$_2$[Pd(SCN)$_4$] [2].

The infrared spectrum shows ν_{CN} at 2116 cm^{-1}, ν_{C-S} at 756 and δ_{SCN} at 458 and 428 cm^{-1} [3].

[Pd(SCN)$_4$][Pd bipy$_2$]. This pink salt is made from K$_2$[Pd(SCN)$_4$] in acetone and [Pd bipy$_2$]-(BPh$_4$)$_2$.

The electronic spectrum of a liquid paraffin mull of the solid was measured (300 to 550 nm) [4].

[Pd(SCN)$_4$][Pd(piperidine)$_4$]. The infrared spectrum shows ν_{CN} at 2129 and 2106 cm^{-1}, ν_{CS} at 760 and 700 cm^{-1} and δ_{CS} at 490 and 455 cm^{-1} [3].

[Pd(SCN)$_4$][Pd(thiourea)$_4$]. The infrared spectrum shows ν_{CN} at 2115 cm^{-1}, ν_{CS} between 787 and 708 cm^{-1} and δ_{SC} at 472 and 435 cm^{-1} [3].

Thermal decomposition of the salt was measured (curves reproduced in paper from 100 to 600°C) [5].

[Pd(SCN)₄][PdR₂]. This, the [Pd(SCN)₄]²⁻ salt of [PdR₂]²⁺ (R = phenothiazene) is made from PdCl₂ and phenothiazine in a 1:1 ratio with KSCN (Pd:SCN⁻ =1:2). It is cerise coloured, insoluble in ether, partly soluble in alcohol and soluble in acetone. It melts at 236 to 238°C [6].

References:

[1] Singh, P. P.; Yadav, S. P.; Pal, R. B. (J. Inorg. Nucl. Chem. **40** [1978] 247/51).
[2] Spacu, P.; Camboli, D. (Rev. Chim. Acad. Rep. Populaire Roumaine **7** [1962] 1311/6).
[3] Spacu, P.; Camboli, D. (Rev. Roumaine Chim. **11** [1966] 157/9).
[4] Burmeister, J. L.; Basolo, F. (Inorg. Chem. **3** [1964] 1587/93).
[5] Fatu, D.; Cimpu, V.; Camboli, D. (Rev. Roumaine Chim. **22** [1977] 133/6).
[6] Cimpu, V.; Camboli, D. (Rev. Roumaine Chim. **21** [1976] 1327/30).

11.5.2.4 Substituted Complexes

There has been much interest in complexes of the type Pd(SCN)₂L₂ where L is a monodentate or bidentate ligand, since in many cases the –N and/or –S bonded linkage isomers of thiocyanate may be separated. There are a number of reviews on the subject [1 to 6].

The factors affecting such bonding with the ambidentate thiocyanate ligand have been discussed for palladium complexes [7, 8].

In general "Pd(SCN)" in the text implies the presence of a Pd–S bond and "Pd(NCS)" of a Pd–N bond, though in many cases the evidence for this is tenuous.

References:

[1] Burmeister, J. L. (Coord. Chem. Rev. **3** [1968] 225/45).
[2] Burmeister, J. L. (Coord. Chem. Rev. **1** [1966] 205/21).
[3] Norbury, A. H.; Sinha, A. I. P. (Quart. Rev. [London] **24** [1970] 69/94, 82).
[4] Chemistry of Pseudohalides, Elsevier, Amsterdam 1986.
[5] Bailey, R. A.; Kozak, S. L.; Michelsen, T. W.; Mills, W. N. (Coord. Chem. Rev. **6** [1971] 407/45, 420).
[6] Norbury, A. H. (Advan. Inorg. Chem. Radiochem. **17** [1974/75] 231/386, 250).
[7] Turco, A.; Pecile, C. (Nature **191** [1961] 66/7).
[8] Burmeister, J. L.; Hassel, R. L.; Phelan, R. J. (Chem. Commun. **1970** 679/80).

11.5.2.4.1 Complexes with Oxygen Donors

[Pd(SCN)₃(H₂O)]⁻. This has been mentioned (as "[Pd(SCN)₃]⁻") as a likely intermediate in the Pd(SCN)₂–SCN⁻ reaction [1].

Pd(SCN)₂(H₂O)₂ referred to in the original paper as "Pd(SCN)₂" is presumably the species extracted at low SCN⁻ concentrations by polyether-type polyurethane foam from aqueous solutions of PdCl₂ in HCl with SCN⁻ [2].

[Pd(SCN)₃H₂O]⁻. Extraction of PdCl₂ in H₂SO₄ treated with SCN⁻ ion and then extracted with crystal violet gives solutions of this species. The electronic spectrum of the extract by crystal violet in benzene has been measured (420 to 650 nm, reproduced in paper) [3].

Pd(NCS)₂(OAsPh₃)₂. This is made by heating Pd(SCN)₂(AsPh₃)₂ in air to 200°C [4].

References:

[1] Shlenskaya, V. I.; Khvostova, V. P.; Peshkova, V. M. (Zh. Analit. Khim. **17** [1962] 598/603; J. Anal. Chem. [USSR] **17** [1962] 596/601).

[2] Al-Bazi, S. J.; Chow, A. (Talanta **30** [1983] 487/92).

[3] Khvatova, Z. M.; Golovina, V. V. (Zh. Analit. Khim. **34** [1979] 2035/9; J. Anal. Chem. [USSR] **34** [1979] 1577/80).

[4] Sedova, G. N.; Kirillova, M. A. (Zh. Neorgan. Khim. **32** [1987] 1510/2; Russ. J. Inorg. Chem. **32** [1987] 905/6).

11.5.2.4.2 Complexes with Nitrogen Donors

Complexes with Pyridine and Substituted Pyridines

Pd(SCN)$_2$py$_2$. This is made by reaction of [Pd(SCN)$_4$]$^{2-}$ in alcohol with pyridine [1, 2, 3, 8].

The infrared spectrum shows ν_{CN} at 2114 and 2102 [2], 2115 [1], 2118 [4], 2112 cm^{-1} [3, 5], ν_{CS} at 697 [2] or 767, 748 and 691 cm^{-1} [4] or at 865 cm^{-1} [1], and δ_{SCN} at 426 [2], 470 and 433 [4] or 424 cm^{-1} [3, 5].

The Pd–S stretch was assigned to a band at 305 cm^{-1} [6]. The integrated intensities of the ν_{CN} and ν_{CS} bands were measured [2]. Although it was deduced on the basis of these infrared data that the complex contains S-bonded thiocyanate ligands [2, 4, 5, 6] it has been suggested that it contains –N rather than –S bonded thiocyanate ligands; i.e., it should be formulated as Pd(NCS)$_2$py$_2$ [1]. This question seems to have remained unresolved.

Pd(NCS)$_2$py$_2$. This is made by reaction of K$_2$[Pd(SCN)$_4$] in ethanol with an aqueous solution of pyridine [1].

The infrared spectrum showed ν_{CN} to be at 2115 cm^{-1} and ν_{CS} at 865 cm^{-1}. On this basis it was suggested that Pd–N bonds were present in the thiocyanate linkages [1]. Similar infrared data (see above) have been obtained for the complex formulated as Pd(SCN)$_2$py$_2$ [2, 4, 5, 6], and it is not clear whether the complex contains –N or –S bonded thiocyanate.

Pd(SCN)$_2$(Rpy)$_2$. These substituted pyridine complexes are made from K$_2$[Pd(SCN)$_4$] and the ligand (Rpy) in ethanol (Rpy = 4-cyanopyridine (4-cyano); 4-nitropyridine (4-nitropy), iso-nicotinamide (iso-nic)) [3]; γ-picoline (4-methylpyridine; γ-pic, 4-n-amylpyridine (4-ampy)) [1]; α-picoline (2-methylpyridine, α-pic); β-picoline (3-methylpyridine, β-pic); 2,4-lutidine (2,4-dimethylpyridine, 2,4-lut); 2,6-lutidine (2,6-dimethylpyridine, 2,6-lut); γ-collidine (2,4,6-trimethylpyridine, γ-coll) [4]; α-aminopyridine (a-py) [8]. Reaction of Na$_2$[Pd(SCN)$_4$] and the appropriate ligand in ethanol was used to make the complexes with 4-nitropyridine (4-nitro); 4-benzoylpyridine (4-benzoylpy); pyridine-4-aldehyde (4-CHOpy) isonicotinamide; 4-benzylpyridine (4-benzylpy); 4-aminopyridine (4-aminopy); 4-acetylpyridine (4-acpy); 4-cyanopyridine (4-cyanopy), 4-methylpyridine (γ-pic) and iso-nicotinic acid (iso-nic) [2]. The 3-nicotinamide (nic) complex was prepared from 3-nicotinamide and Pd(SCN)$_2$; it melts at 260°C with decomposition [9]. The 4-azidopyridine complex is similarly made from the ligand and Pd(SCN)$_2$ and is thought to have a cis arrangement of thiocyanato ligands [7].

Infrared data for these are listed in the table below.

The integrated intensities of the ν_{CN} and ν_{CS} bands were measured (4-nitropy, 4-cyanopy, 4-benzoylpy, γ-pic, 4-acpy, 4-CHOpy, 4-benzylpy, 4-aminopy; isonicotinamide) and the Raman spectrum of the 4-cyanopy complex was recorded [2].

Infrared spectra of $Pd(SCN)_2(Rpy)_2$ complexes:

	ν_{CN}	ν_{CS} in cm^{-1}	δ_{SCN}	Ref.
4-cyanopy	2114	702, 696	420	[3]
	2113	701, 696	419	[2]
	2119*)	702, 699*)	—	[2]
4-nitropy	2115	obsc.	427	[3]
4-n-ampy	2111	707	—	[1]
4-benzoylpy	2119	obsc.	—	[2]
4-acpy	2120	706	—	[2]
4-CHO py	2117	697	—	[2]
4-benzylpy	2117	702	—	[2]
4-aminopy	2116	716	—	[2]
α-pic	2120	776, 724, 700	467, 452, 432	[4]
β-pic	2118	750, 696	462, 428	[4]
γ-pic	2116	723, 705	467, 432	[4]
	2109	702	—	[1]
	2110	698	—	[2]
2,4-lut	2120	758, 702	465, 454, 430	[4]
2,6-lut	2120	705	466, 424	[4]
γ-coll	2118	772, 728, 700	463, 425	[4]
iso-nic	2117	obsc.	424	[2, 3]
nia	2115	718	430	[9]
α-py	2118	770, 741, 708	458, 435	[4]

*) Raman data, other data IR

[Pd(SCN)$_3$(Rpy)]$^-$. Evidence for the existence of these species (Rpy = lutidine and picoline) was obtained from conductimetric measurements on the $PdCl_2$–SCN^-–Rpy system. The corresponding $Pd(SCN)_2(Rpy)_2$ species were also detected [10].

Pd(NCS)$_2$(4-nitropy)$_2$. The compound was made by reaction of $Na_2[Pd(SCN)_4]$ with the ligand. IR data were measured [2].

References:

[1] Burmeister, J. L.; Basolo, F. (Inorg. Chem. **3** [1964] 1587/93).
[2] Miezis, A. (Acta Chem. Scand. **27** [1973] 3746/60).
[3] Bertini, I.; Sabatini, A. (Inorg. Chem. **5** [1966] 1025/8).
[4] Spacu, P.; Camboli, D. (Rev. Roumaine Chim. **11** [1966] 157/9).
[5] Sabatini, A.; Bertini, I. (Inorg. Chem. **4** [1965] 1665/7).
[6] Keller, R. N.; Johnson, N. B.; Westmoreland, L. L. (J. Am. Chem. Soc. **90** [1968] 2729/30).
[7] El'tsov, A. V.; Khanchmann, A.; Rtishchev, N. I. (Zh. Org. Khim. **13** [1977] 465; J. Org. Chem. [USSR] **13** [1977] 425/6).
[8] Spacu, P.; Camboli, D. (Rev. Chim. Acad. Rep. Populaire Roumaine **7** [1962] 1311/6).
[9] Singh, P. P.; Seth, J. N.; Khan, S. A. (Inorg. Nucl. Chem. Letters **11** [1975] 525/8).
[10] Ambrus, C.; Demian, N.; Buteceanu, E.; Camboli, D. (Bul. Inst. Politeh. Gheorghe Gheorghiu-Dej Bucuresti **38** [1976] 43/6; C. A. **87** [1977] No. 173530).

Complexes with 2,2′-Bipyridyl (bipy) and 1,10-Phenanthroline (phen)

Pd(SCN)₂bipy. This light orange-yellow complex is made by reaction of $K_2[Pd(SCN)_4]$ in C_2H_5OH with 2,2′-bipyridyl (bipy) in $C_2H_5OH–(C_2H_5)_2O$ [1] or from $Na_2[Pd(SCN)_4]$ in C_2H_5OH with 2,2′-bipyridyl [2].

It melts with decomposition at 270°C, being bright yellow at 130°C [1]. On heating to 156°C for 30 min it gave the light yellow Pd(NCS)₂bipy. The molar conductance in dimethylformamide is 20.3 $\Omega^{-1} \cdot cm^2$ [1, 9].

The infrared spectrum of the solid showed ν_{CN} at 2117 and 2108 [4, 9] or at 2100 cm⁻¹ [11] with ν_{CS} at 700 cm⁻¹ [1, 2, 4]. The Pd–S stretches are at 316 and 304 [10] or at 319 cm⁻¹ [6]. On the basis of these spectral data it was thought to contain S-bonded thiocyanate ligands [1, 9], though N-bonding of the thiocyanate groups has also been inferred [2, 5].

The electronic spectrum in dimethylformamide, $(CH_3)_2SO$, CH_3CN and acetone were measured [8].

Pd(NCS)₂bipy. This is made by heating Pd(SCN)₂bipy to 156°C for 30 min; it is light yellow and melts at 270°C with decomposition [1].

The molar conductance in dimethylformamide is 20.8 $\Omega^{-1} \cdot cm^2$ [1, 9].

The infrared spectrum of the solid shows ν_{CN} at 2100 [4, 5, 9] or 2099 [2] cm⁻¹ with ν_{CS} at 845 [4, 5] or 849 and 843 cm⁻¹ [2], and δ_{NCS} at 458 cm⁻¹ [2, 4, 5]. The Pd–N stretch is at 261 cm⁻¹ [6] or at 345 and 332 cm⁻¹ [10]. Band intensities were measured from ν_{CN} and ν_{CS} [2]. An N-bonded mode for the thiocyanate ligands was deduced on the basis of these data [2, 4, 5, 6].

The electronic spectra in CCl_4 and $CHCl_3$ were measured [8].

Pd(NCS)(SCN)(4,4-Me₂bipy). This is made by reaction of 4,4′-dimethyl-2,2′-bipyridyl (4,4-Me bipy) with $K_2[Pd(SCN)_4]$ in ethanol [5] or by a similar method but using $Na_2[Pd(SCN)_4]$ [2].

The infrared spectrum of the solid shows ν_{CN} at 2120 and 2090 cm⁻¹ (spectrum from 2500 to 1700 cm⁻¹ reproduced in paper) [5] or at 2121 and 2099 cm⁻¹ [2]; with ν_{CS} at 839 [2] and δ_{SCN} at 455 [2] or at 458 and 452 cm⁻¹ [5]. Integrated absorption intensities for ν_{CN} and ν_{CS} were given [2]. The splitting of ν_{CN} has lead to the suggestion that the complex contains both –N and –S bonded thiocyanate [2, 5].

Pd(SCN)₂phen. This is made by reaction of $K_2[PdCl_4]$ in excess KSCN with 1,10-phenanthroline (phen) [3] or from $K_2[Pd(SCN)_4]$ and the ligand in C_2H_5OH [1]; $Na_2[Pd(SCN)_4]$ can also be used [2].

The infrared spectrum of the solid shows ν_{CN} at 2114 [1] or 2116 [2, 4, 5] with ν_{CS} at 696 [1], 695 [2] or 697 cm⁻¹ [4, 5] and δ_{SCN} at 418 [2] or 460 and 418 cm⁻¹ [4, 5]; the Pd–S stretch lies at 305 cm⁻¹ [6]. Band intensities were measured for ν_{CN}, ν_{CS} and δ_{SCN} [2]. The infrared spectrum of Pd(SCN)₂phen from 400 to 900 cm⁻¹ was reproduced in [5]. On the basis of these data S-bonding for the thiocyanate ligand to the metal was proposed [1, 2, 4, 5].

Electronic spectra in dimethylformamide, $(CH_3)_2SO$, CH_3CN, CCl_4, $CHCl_3$ and acetone were measured [8].

Pd(SCN)₂(5-nitrophen). This is made from $K_2[Pd(SCN)_4]$ in ethanol with 5-nitro-1,10-phenanthroline (5-nitrophen) at 60°C, or by heating Pd(NCS)₂(5-nitrophen). It is orange and decomposes at 248 to 250°C [7].

The infrared spectrum shows ν_{CN} at 2122 cm⁻¹ (spectrum from 2040 to 2140 cm⁻¹ reproduced in paper); on the basis of this and the position of the ¹⁴N NQR resonance it was suggested that it contained two S-bonded thiocyanate ligands [7].

Pd(NCS)$_2$(5-nitrophen). Reaction of K$_2$[Pd(SCN)$_4$] in dimethylformamide with 5-nitro-1,10-phenanthroline gives the yellow complex which decomposes at 248 to 250°C [7]. An earlier report of the complex, also made from K$_2$[Pd(SCN)$_4$] and the ligand [5] could not be fully substantiated [7].

The infrared spectrum shows ν_{CN} at 2120 and 2093 cm^{-1} [7], at 2120 and 2095 cm^{-1} [5] or at 2090 cm^{-1} [4, 5], δ_{NCS} at 458 cm^{-1} [4, 5], though the latter data have been questioned [7]. The spectra from 2060 to 2140 cm^{-1} at different temperatures (29 to 101°C) were reproduced in [7].

On heating from 95 to 100°C it gave Pd(SCN)$_2$(5-nitrophen) [7].

Pd(SCN)(NCS)(5-nitrophen). This is made by reaction of Na$_2$[Pd(SCN)$_4$] with 5-nitro-1,10-phenanthroline in ethanol [2], though the purity and identity of the product has been questioned [7].

The infrared spectrum shows ν_{CN} at 2120 and 2090 cm^{-1}, ν_{CS} at 692 cm^{-1} and δ_{SCN} at 424 and 419 cm^{-1}; band intensities were measured. On the basis of these data it has been suggested that one –N and one –S bonded thiocyanate group are present in the complex [2]; however, later work suggested that the product is probably a mixture [7].

Pd(SCN)$_2$(4,7-Ph$_2$-phen). Reaction of 4,7-diphenyl-1,10-phenanthroline in absolute ethanol with K$_2$[Pd(SCN)$_4$] at room temperature gives this complex [5].

The infrared spectrum shows ν_{CN} at 2120 and 2113 cm^{-1} with ν_{CS} obscured and δ_{SCN} at 419 cm^{-1}. It was concluded from these data that it contains S-bonded thiocyanate [5].

Pd(NCS)$_2$(4,7-Ph$_2$-phen). Reaction of K$_2$[Pd(SCN)$_4$] in absolute ethanol at 0°C with 4,7-diphenyl-1,10-phenanthroline gives a material which has an infrared spectrum quite different from that of Pd(SCN)$_2$(4,7-Ph$_2$-phen); ν_{CN} is at 2110 cm^{-1}, but ν_{CS} and δ_{SCN} are obscured. The suggestion that this contains –N rather than –S bonded thiocyanate was made [5].

Pd(SCN)$_2$(R-phen). These complexes are made by reaction of K$_2$[Pd(SCN)$_4$] and the ligand in absolute ethanol (R-phen = 5-methyl-1,10-phenanthroline (5-Mephen); 5,6-dimethyl-1,10-phenanthroline (5,6-Me$_2$phen); 4,7-dimethyl-1,10-phenanthroline (4,7-Me$_2$phen); 5-methyl-6-nitro-1,10-phenanthroline (5Me6NO$_2$phen); 5-chloro-1,10-phenanthroline (5Clphen)) [5].

The infrared spectra are listed in the following table. On the basis of these data it was proposed that S-bonded thiocyanate was present [5]. The intensities of the ν_{CN} and ν_{CS} bands for the 5,6-dimethyl-1,10-phenanthroline complex were recorded [2].

Infrared spectra of Pd(SCN)$_2$(R-phen) complexes:

	ν_{CN}	ν_{CS}	δ_{SCN}	Ref.
			in cm^{-1}	
5-Mephen	2115, 2110		453, 418	[5]
5,6-Me$_2$phen	2125, 2117, 2111	700	420	[5]
	2112, 2106	698	420	[2]
4,7-Me$_2$phen	2115, 2102		426, 419	[5]
5Me6NO$_2$phen	2114		455, 421	[5]
5Clphen	2116, 2105	700	465, 415	[5]

For 5-nitro-1,10-phenanthroline and 4,7-diphenyl-1,10-phenanthroline see above.

Pd(NCS)$_2$terpy. Reaction of K$_2$[Pd(SCN)$_4$] with 2,2',6',2''-terpyridyl (terpy) in ethanol gives the complex [1].

The infrared spectrum gave ν_{CN} at 2088 cm^{-1} with ν_{CS} at 848 cm^{-1}. N-bonding of the thiocyanate ligand to the metal was suggested on the basis of the infrared data, with the terpy ligand functioning in a bidentate mode [1].

References:

[1] Burmeister, J. L.; Basolo, F. (Inorg. Chem. **3** [1964] 1587/93).
[2] Miezis, A. (Acta Chem. Scand. **27** [1973] 3746/60).
[3] Livingstone, S. E. (J. Proc. Roy. Soc. N. South Wales **85** [1952] 151/6).
[4] Sabatini, A.; Bertini, I. (Inorg. Chem. **4** [1965] 1665/7).
[5] Bertini, I.; Sabatini, A. (Inorg. Chem. **5** [1966] 1025/8).
[6] Keller, R. N.; Johnson, N. B.; Westmoreland, L. L. (J. Am. Chem. Soc. **90** [1968] 2729/30).
[7] Fultz, W. C.; Burmeister, J. L. (Inorg. Chim. Acta **45** [1980] L271/L273).
[8] Burmeister, J. L.; Hassel, R. L.; Phelan, R. J. (Inorg. Chem. **10** [1971] 2032/8).
[9] Basolo, F.; Burmeister, J. L.; Poë, A. J. (J. Am. Chem. Soc. **85** [1963] 1700/1).
[10] Goodgame, D. M. L.; Malerbi, B. W. (Spectrochim. Acta A **24** [1968] 1254/5).

[11] Chatt, O.; Duncanson, L. A. (Nature **178** [1956] 997/8).

Complexes with Other Heterocyclic Systems

Pd(SCN)$_2$(2-aminopyrimidine). This is made by reaction of Pd(SCN)$_2$ with the ligand. It melts at 260°C with decomposition [1].

The infrared spectrum shows ν_{CN} at 2105 and 2065 cm^{-1}, ν_{CS} at 780 and 710 cm^{-1} and δ_{SCN} at 445 and 430 cm^{-1}; S-bonding of the thiocyanate ligand was suggested [1].

Pd(SCN)$_2$(2-pyrazinecarboxamide)$_2$. This is made from Pd(SCN)$_2$ and the ligand, and melts at 195°C [1].

On the basis of the infrared spectrum ($\nu_{CN} = 2105$, $\nu_{CS} = 720$, $\delta_{SCN} = 435$ cm^{-1}) S-bonding of the thiocyanate was proposed [1].

Pd(SCN)$_2$(phthalazine)$_2$. The infrared spectrum shows ν_{CN} at 2120, ν_{SC} at 762 and δ_{SCN} at 427 cm^{-1}, suggesting –S bonding of thiocyanate to palladium [2].

Pd(SCN)$_2$(acridine)$_2$. The infrared spectrum shows ν_{CN} at 2120, ν_{CS} at 738 and 672 cm^{-1} and δ_{SCN} at 468 and 425 cm^{-1}, consistent with the presence of Pd–S bonds [2].

Pd(SCN)$_2$(morpholine)$_2$. The infrared spectrum shows ν_{CN} to be at 2108, ν_{CS} at 780 and 675 cm^{-1}, and δ_{SCN} at 460 and 435 cm^{-1} [2].

Pd(SCN)$_2$(piperidine)$_2$. This compound melts at 126 to 127°C [3]. The infrared spectrum has ν_{CN} at 2108, ν_{CS} at 652 and δ_{SCN} at 490 and 453 cm^{-1}, consistent with the presence of Pd–S bonding [2].

References:

[1] Singh, P. P.; Seth, J. N.; Khan, S. A. (Inorg. Nucl. Chem. Letters **11** [1975] 525/8).
[2] Spacu, P.; Camboli, D. (Rev. Roumaine Chim. **11** [1966] 157/9).
[3] Spacu, P.; Camboli, D. (Rev. Chim. Acad. Rep. Populaire Roumaine **7** [1962] 1311/6).

Complexes with Miscellaneous Donors

trans-Pd(SCN)$_2$(NH$_3$)$_2$. The infrared spectrum has bands at 2116 and 2100 cm^{-1} (ν_{CN}) and 701 cm^{-1} (ν_{CS}) and, on the basis of this, –S bonded thiocyanate is thought to be involved [1].

The MO-SCF method in the INDO approximation has been used to calculate the electronic structure in cis-Pd(SCN)$_2$(NH$_3$)$_2$ [2].

cis-Pd(SCN)$_2$(NH$_3$)$_2$. The MO-SCF method in the INDO approximation was used to calculate the electronic structure in this molecule [2].

[Pd(SCN)(Et$_4$dien)]Y. The light yellow **tetraphenylborate** is made by reaction of NaSCN with a suspension of PdCl$_2$ in acetone to which was then added 1,1,7,7-tetraethyldiethylenetriamine and NaBPh$_4$ [3]. The **thiocyanate** is made from K$_2$[Pd(SCN)$_4$] and the ligand in acetone, and the **hexafluorophosphate** by a similar method but with addition of NH$_4$(PF$_6$) [4].

The infrared spectrum of the hexafluorophosphate shows ν_{CN} at 2125, ν_{CS} at 710, δ_{SCN} at 435 and ν_{PdS} at 295 cm^{-1}, suggesting it to be an S-bonded isomer [5].

Kinetics of the isomerisation reactions of [Pd(SCN)(Et$_4$dien)]$^+$ to [Pd(NCS)(Et$_4$dien)]$^+$ were studied [4] as were substitution reactions with Br$^-$ of [Pd(SCN)(Et$_4$dien)]$^+$ [6].

[Pd(NCS)(Et$_4$dien)](BPh$_4$). This is made by stirring a solution of [Pd(SCN)(Et$_4$dien)](BPh$_4$) in dimethylformamide for 3 h at 25°C [3].

The infrared spectrum of [Pd(NCS)(Et$_4$dien)](NCS) shows ν_{CN} at 2080, ν_{CS} at 832, δ_{NCS} at 472 and ν_{PdN} at 268 cm^{-1}, suggesting that it contains an N-bonded thiocyanato ligand [5].

The rates of isomerisation of [Pd(SCN)(Et$_4$dien)]$^+$ to [Pd(NCS)(Et$_4$dien)]$^+$ were studied [3, 6].

Pd(SCN)$_2$(HMTA)$_2$. This yellow material is made from K$_2$[Pd(SCN)$_4$] and hexamethylene-tetramine (HMTA) in aqueous solution at 70°C. It is insoluble in alcohol, ether, CCl$_4$ and hexane [7].

The spectroscopic and antitumour properties have been studied [8].

Pd(SCN)$_2$L (L = (NH(R^1)C(R^{2}R^3)C(R^{4}R^5)HN(R^6)). These complexes are made from the bidentate N,N-donor ligand (R groups are H or C$_{1\,to\,6}$ alkyl or C$_{2\,to\,6}$ alkylene); their antitumour properties were studied [9].

Pd(SCN)$_2$(δNC$_{12}$H$_{19}$)$_2$. This is made from 1:1 PdCl$_2$ and phenothiazine to which is added KSCN (Pd:SCN$^-$ 1:2). It is soluble in acetone, slightly soluble in alcohol and insoluble in ether, and melts at 208 to 210°C [10]. In the infrared δ_{CN} lies at 2160 and 2114 cm^{-1} [10].

Pd(SCN)$_2$(1,2-diaminocyclohexane). This is prepared from palladium(II) salts, NCS$^-$ and the ligand. Its antitumour properties were studied [9].

Pd(SCN)$_2$(isox). This is made from Pd(SCN)$_2$ and isoxazole (O–N=CH–CH=CH) in acetone. It is red-brown, decomposing at 255 to 260°C [11].

The infrared spectrum showed the Pd–S stretch at 287 cm^{-1} and the Pd–O stretch at 310 and 230 cm^{-1}; the electronic absorption spectrum was also measured [11].

Pd(SCN)$_2$(CNCH$_3$)$_2$. This is made from SCN$^-$ and PdCl$_2$(CNCH$_3$)$_2$. The infrared spectrum shows ν_{CN} at 2111 and ν_{CS} at 700 cm^{-1} for the solid, with ν_{CN} at 2124 cm^{-1} in CHCl$_3$ solution. The integrated intensity of the 2124 cm^{-1} band was measured [12].

Pd(SCN)$_2$(CN^{i}C$_3$H$_7$)$_2$. This is made from SCN$^-$ and PdCl$_2$(CN^{i}C$_3$H$_7$)$_2$. The infrared spectrum of the solid showed ν_{CN} at 2113 and ν_{CS} at 700 cm^{-1}. In CHCl$_3$ and C$_6$H$_6$ solutions ν_{CN} appeared at 2122 and 2125 cm^{-1}, respectively, and their integrated intensities were measured [12].

References:

[1] Turco, A.; Pecile, C. (Nature **191** [1961] 66/7).

[2] Voityuk, A. A.; Vadash, P. I.; Mazalov, L. N. (Zh. Strukt. Khim. **21** No. 3 [1980] 28/33; J. Struct Chem. [USSR] **21** [1980] 276/80).

[3] Burmeister, J. L.; Gysling, H. J.; Lim, J. C. (J. Am. Chem. Soc. **91** [1969] 44/7).

[4] Basolo, F.; Baddley, W. H.; Weidenbaum, K. J. (J. Am. Chem. Soc. **88** [1966] 1576/8).

[5] Keller, R. N.; Johnson, N. B.; Westmoreland, L. L. (J. Am. Chem. Soc. **90** [1968] 2719/30).

[6] Johnson, K. A.; Lim, J. C.; Burmeister, J. L. (Inorg. Chem. **12** [1973] 124/8).

[7] Sinitsyn, N. M.; Buslaeva, T. M.; Efanov, V. I. (Zh. Neorgan. Khim. **29** [1984] 2986/7; Russ. J. Inorg. Chem. **29** [1984] 1708/9).

[8] Umreiko, D. S.; Kachurina, D. N.; Chernikova, I. E.; Novitskii, G. G.; Sinitsyn, N. M.; Buslaeva, T. M.; Efanov, V. I. (Zdravookhr. Beloruss. **12** [1986] 22/4; C.A. **107** [1987] No. 17330).

[9] Amundsen, A. R.; Stern, E. W.; Engelhard Corp. (Eur. Appl. 98133 [1984]; C.A. **100** [1984] No. 161787).

[10] Cimpu, V.; Camboli, D. (Rev. Roumaine Chim. **21** [1976] 1327/30).

[11] Pinna, R.; Ponticelli, G.; Preti, C. (J. Inorg. Nucl. Chem. **37** [1975] 1681/4).

[12] Pecile, C. (Inorg. Chem. **5** [1966] 210/4).

11.5.2.4.3 Complexes with Halogen Donors

The $[Pd(SCN)_nCl_{4-n}]^{2-}$ System. From electronic spectra of solutions of $[PdCl_4]^{2-}$ with SCN^- (200 to 400 nm, reproduced in paper) the existence of three intermediate species was deduced and stability constants calculated at 25°C and ionic strength 1.1 [1]:

	log k_i	log β		log k_i	log β
$[Pd(SCN)_4]^{2-}$	3.03 ± 0.05	28.67 ± 0.36	$[Pd(SCN)Cl_3]^{2-}$	6.03 ± 0.04	17.15 ± 0.20
$[Pd(SCN)_3Cl]^{2-}$	3.59 ± 0.06	25.64 ± 0.31	$[PdCl_4]^{2-}$		11.12 ± 0.16
$[Pd(SCN)_2Cl_2]^{2-}$	4.90 ± 0.04	22.05 ± 0.24			

For $[Pd(SCN)_3Cl]^{2-} + SCN^- \rightleftharpoons [Pd(SCN)_4]^{2-} + Cl^-$ a log K_4 value of 3.03 was given, assuming log β_4 for $[Pd(SCN)_4]^{2-}$ to be 28.22 [2, 3]. A log β_2 value for formation of $[Pd(SCN)_2Cl_2]^{2-}$ from $PdCl_2$ and SCN^- of 16.85 has been determined spectrophotometrically [3].

$(RH)_2[Pd(SCN)_2Cl_2]$. This species has been briefly mentioned as resulting from extraction of Pd^{II} species in SCN^- into xylene with tetraethyldiamide heptylphosphate [4].

The $[Pd(SCN)_nBr_{4-n}]^{2-}$ System. From electronic spectra of solutions of $[PdBr_4]^{2-}$ with SCN^- and of $[Pd(SCN)_4]^{2-}$ with Br^- (spectra from 250 to 400 nm reproduced in paper) the existence of the three species was deduced.

From these data the following stability constants were determined for $[Pd(SCN)_nBr_{4-n}]^{2-}$ at 25°C and total ionic strength 1.1 [2]:

	log K_i	log β_{ni}		log K_i	log β_{ni}
$[Pd(SCN)_4]^{2-}$	log K_4 2.37	log β_{04} 28.22	$[Pd(SCN)_2Br_2]^{2-}$	log K_2 4.10	log β_{22} 22.25
$[Pd(SCN)_3Br]^{2-}$	log K_3 3.60	log β_{13} 25.85	$[Pd(SCN)Br_3]^{2-}$	log K_1 5.10	log β_{31} 18.15

[Pd(SCN)$_2$I$_2$]$^{2-}$. From changes of solubility of PdI$_2$ in aqueous solutions of NaSCN, the existence of [Pd(SCN)$_2$I$_2$]$^{2-}$ in solution was deduced [5]. The equilibrium constant for its formation was found to be 0.34 [5].

[Pd$_2$I$_4$(SCN)]$^-$. From solubility studies of PdI$_2$ in solutions of NaSCN in acetone and dimethylformamide values of K = 2.0 ± 0.1 (acetone) and K = 1.7 ± 0.01 (dimethylformamide) were calculated from the equilibrium constant [Pd$_2$I$_4$(SCN)]$^-$/[SCN$^-$] [6].

References:

[1] Biryukov, A. A.; Shlenskaya, V. I. (Zh. Neorgan. Khim. **12** [1967] 2579/82; Russ. J. Inorg. Chem. **12** [1967] 1362/3).

[2] Biryukov, A. A.; Shlenskaya, V. I.; Alimarin, I. P. (Izv. Akad. Nauk SSSR **1966** 3/8; Bull. Acad. Sci. USSR Div. Chem. Sci. **1966** 1/5).

[3] Joshi, S. B.; Pundalik, M. D.; Mattoo, B. N. (Indian J. Chem. **11** [1973] 1297/9).

[4] Fadeeva, V. I.; Nasonovskii, I. S.; Minochkina, L. N.; Volynskii, A. B.; Zorov, N. B. (Vestn. Mosk. Univ. II **21** [1980] 98; C. A. **92** [1980] No. 170088).

[5] Golub, A. M.; Pomerants, G. B. (Zh. Neorgan. Khim. **9** [1964] 1624/9; Russ. J. Inorg. Chem. **9** [1964] 879/82).

[6] Golub, A. M.; Pomerants, G. B.; Ivanova, S. A. (Zh. Neorgan. Khim. **14** [1969] 2826/31; Russ. J. Inorg. Chem. **14** [1969] 1488/91).

11.5.2.4.4 Complexes with Sulphur Donors

Pd(SCN)$_2$(SC(NH$_2$)$_2$)$_2$. This is made from thiourea and K$_2$[Pd(SCN)$_4$] in ethanol [1, 2] or similarly from Na$_2$[Pd(SCN)$_4$] [2].

The infrared spectrum shows ν_{CN} at 2107 [1], 2110 [2] or 2128 cm^{-1} [3], with ν_{CS} at 703 [1], 714 and 704 [2] or 712 cm^{-1} [3], and δ_{SCN} at 474 cm^{-1} [3]. The Raman spectrum showed ν_{CN} at 2110 and ν_{CS} at 716 and 704 cm^{-1}. The integrated intensities of the ν_{CN} and ν_{CS} modes were measured [2]. On the basis of these infrared data, S-bonding of the thiocyanate groups to the metal was inferred [1, 2, 3].

Thermal decomposition studies on the complex were carried out from 100 to 600°C [4].

Pd(SCN)$_2$(etu)$_2$. This ethylenethiourea complex is made from the ligand and K$_2$[Pd(SCN)$_4$] in ethanol [1] or by an analogous method but using Na$_2$[Pd(SCN)$_4$] [2].

The infrared spectrum shows ν_{CN} at 2101 [1] or 2103 cm^{-1} [2] and ν_{CS} at 701 [1] or 703 cm^{-1} [2]; S-bonding of the thiocyanate ligands was inferred [1, 2]. Integrated intensities of the ν_{CN} and ν_{CS} bonds were measured [2].

trans-Pd(SCN)$_2$(SC(NCH$_3$)OC$_2$H$_5$)$_2$. This is made by reaction of N-methyl O-ethylthiocarbamate with [Pd(SCN)$_4$]$^{2-}$ in ethanol. It forms orange crystals melting at 146°C with decomposition.

The X-ray crystal structure showed the crystals to be triclinic, space group P$\bar{1}$-C$_i^1$; Z = 1, a = 9.522(7), b = 8.352(7), c = 5.658(6) Å, α = 93.1(1)°, β = 96.5(1)°, γ = 88.1°. The observed density is 1.72 g/cm^3 and the calculated density 1.717 g/cm^3. The palladiums has square-planar coordination, the two S-bonded SCN-groups (Pd–S = 2.331(2) Å, Pd–S–C angle = 109.7(2)°) are trans, and the organic ligand is S-bonded, Pd–S = 2.319(1) Å; see **Fig. 75**, p. 310 [5].

The ν_{CN} frequency is at 2110 cm^{-1}, ν_{CS} = 707 cm^{-1}, δ_{SCN} 432 cm^{-1} with the ν_{CN} for the ligand at 1578 cm^{-1} [5].

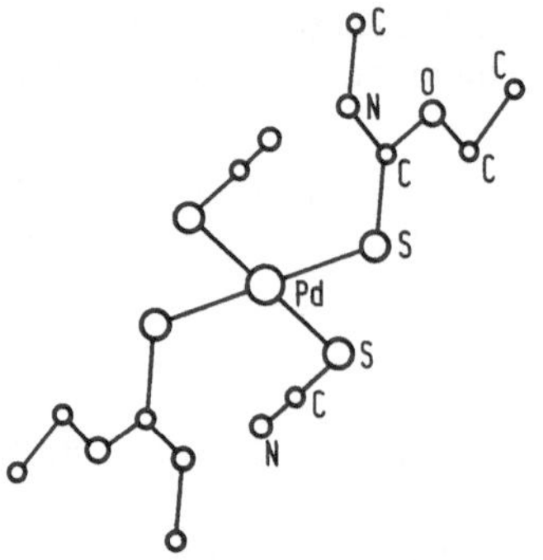

Fig. 75.   Crystal structure of trans-Pd(SCN)$_2$(SC(NCH$_3$)OC$_2$H$_5$)$_2$ [5].

References:

[1] Burmeister, J. L.; Basolo, F. (Inorg. Chem. **3** [1964] 1587/93).
[2] Miezis, A. (Acta Chem. Scand. **27** [1973] 3746/60).
[3] Spacu, P.; Camboli, D. (Rev. Roumaine Chim. **11** [1966] 157/9).
[4] Fatu, D.; Cimpu, V.; Camboli, D. (Rev. Roumaine Chim. **22** [1977] 133/6).
[5] Bardi, R.; del Pra, A.; Piazzesi, A. M.; Sindellari, L.; Zarli, B. (Inorg. Chim. Acta **47** [1981] 231/4).

11.5.2.4.5   Miscellaneous Substituted Complexes

Reaction of [Pd(SCN)$_4$]$^{2-}$ with Amberlite A21 resin with excess SCN$^-$ gave an orange product, the infrared spectrum of which was recorded. It was suggested, that on the basis of the infrared spectrum, the structure involved [Pd(NCS)L$_3$]$^+$ species, with –NMe$_2$ groups from the resin and the thiocyanate group being N-bonded to the palladium [1].

PdCl$_2$[(SCN)$_2$SiMe$_2$]. This is formed by reaction of PdCl$_2$ with Me$_2$Si(NCS)$_2$; the ligand is thought to be bonded through the S atoms to the palladium [2].

References:

[1] Allen, N. P.; Bamiro, F. O.; Burns, R. P.; Dwyer, J.; McAuliffe, C. A. (Inorg. Chim. Acta **28** [1978] 231/5).
[2] Zhakova, I. V.; Garnovskii, A. D.; Kolodyazhnyi, Yu. V.; Ukhin, L. Yu. (Koord. Khim. **9** [1983] 1143; C.A. **99** [1983] No. 132729).

11.5.2.4.6   Polymeric Heavy-Metal Thiocyanato Complexes

Pd(SCN)$_2$M(NCS)$_2$(CH$_3$OH)$_2$ (M = Zn, Cd, Hg). These materials are made by reaction of M(NCS)$_2$ and Pd(SCN)$_2$ in dry methanol. Their properties are summarised in the table below. A polymeric structure involving thiocyanato bridges was proposed.

Pd(SCN)$_2$Zn(NCS)$_2$py$_4$. This is prepared from Pd(SCN)$_2$, Zn(NCS)$_2$ and pyridine; its properties are summarised in the table below.

A binuclear structure involving two thiocyanato bridges between octahedral zinc and square-planar palladium was proposed.

Pd(SCN)$_2$M(NCS)$_2$L$_4$ with L = pyridine, nicotinamide (nia). These are made from Pd(SCN)$_2$-M(NCS)$_2$(CH$_3$OH)$_2$ and the ligand; see table below.

It was suggested that these are binuclear, octahedral nickel and square-planar palladium being bridged by two thiocyanato ligands.

Pd(SCN)$_2$Ni(NCS)$_2$(C$_2$H$_5$OH)$_2$. This yellow material is made from Pd(SCN)$_2$, Co(NCS)$_2$ and ethanol; see table below. A polymeric structure with thiocyanato bridges was proposed.

Pd(SCN)$_2$Co(NCS)$_2$(C$_2$H$_5$OH)$_2$. This pink material is obtained from Pd(SCN)$_2$, Co(NCS)$_2$ and ethanol. Its properties are given in the table below.

Pd(SCN)$_2$Co(NCS)$_2$L$_4$ with L = pyridine, nicotinamide (nia). These materials are made from Pd(SCN)$_2$, Co(NCS)$_2$ and the ligand. For properties see table below. It was suggested that these species were binuclear with two thiocyanato groups bridging octahedral cobalt and square-planar palladium.

Physical properties of polymeric thiocyanato complexes:

complex	colour	m.p. in °C	molar conduct- ance[a]	infrared data[b]		
				ν_{CN}	ν_{CS}	δ_{NCS}
Pd(SCN)$_2$Zn(NCS)$_2$(CH$_3$OH)$_2$	yellow	130	—	2160, 2140	780, 725	480, 440
Pd(SCN)$_2$Zn(NCS)$_2$py$_4$	orange	182	10.3	2140, 2100, 2060	865, 826, 762, 723	470, 435
Pd(SCN)$_2$Cd(NCS)$_2$(CH$_3$OH)$_2$	brown	188	8.4	—	—	—
Pd(SCN)$_2$Hg(NCS)$_2$(CH$_3$OH)$_2$	red	110	8.3	2175, 2160, 2138	790, 730, 720	460, 450, 435, 420
Pd(SCN)$_2$Ni(NCS)$_2$(C$_2$H$_5$OH)$_2$	yellow	105	—	—	—	—
Pd(SCN)$_2$Ni(NCS)$_2$py$_4$	yellow	165	10.8	—	—	—
Pd(SCN)$_2$Ni(NCS)$_2$nia$_4$	yellow	175	10.8	2120, 2068, 2050	740, 730,	485, 445
Pd(SCN)$_2$Co(NCS)$_2$(C$_2$H$_5$OH)$_2$	pink	110	—	2130	780	472, 431
Pd(SCN)$_2$Co(NCS)$_2$py$_4$	pink	182	8.1	2175, 2128, 2104, 2060	805, 768, 758, 715	480, 443
Pd(SCN)$_2$Co(NCS)$_2$nia$_4$	pink	202	8.2	—	—	—

The magnetic moments μ_{eff} of the last six compounds were 2.93, 3.10, 3.05, 5.00, 4.93 and 5.03 BM, respectively.

m.p. = melting points (all with decomposition); nia = nicotinamide;
[a] in $\Omega^{-1} \cdot cm^2$, in CH$_3$CN; [b] frequencies in cm^{-1}.

Reference:

Singh, P. P.; Yadav, S. P.; Pal, R. B. (J. Inorg. Nucl. Chem. **40** [1978] 247/51).

12 Palladium and Silicon

General References:

Langer, H.; Wachtel, E., Constitution of the Binary Palladium–Silicon System in the Concentration Range from 0 to 33.3 at% Silicon, Z. Metallk. **74** [1983] 535/44.

Mason, K. N., Growth and Characterization of Transition Metal Silicides, Progr. Cryst. Growth Charact. **2** [1979] 269/307.

Ottaviani, G., Review of Binary Alloy Formation by Thin Film Interactions, J. Vac. Sci. Technol. **16** [1979] 1112/9.

Remarks. Crystalline and amorphous phases of the Pd–Si system have been prepared and investigated mainly for technological reasons, for instance construction of diodes, IR detectors, contacts, or catalysts resistant to high temperatures.

As these uses are beyond the scope of this volume the following chapter is limited to the description of the binary system and to the treatment of formation, physical properties and chemical behaviour of defined silicides, especially Pd_2Si and $PdSi$.

12.1 The Pd–Si System

12.1.1 Diffusion

The kinetics of transformation of roller-quenched amorphous $Pd_{83}Si_{17}$ and $Pd_{80}Si_{20}$ after linear and isothermal heating were investigated by using electron microscopy, electron diffraction and resistivity measurements. Crystallisation of amorphous $Pd_{83}S_{17}$ began with a metastable ordered face-centred solid solution which transformed to an ordered orthorhombic metastable phase. The crystallisation of amorphous $Pd_{80}Si_{20}$ began with formation of orthorhombic Pd_3Si spheroids and Pd-rich silicide. Activation energies determined by resistivity measurements were 28.5 kcal/mol for $Pd_{83}Si_{17}$ and 80 kcal/mol for $Pd_{80}Si_{20}$ [1]. The lateral diffusion of atoms in a device geometry structure prepared by the deposition of Pd films onto patterned Si substrates was studied using SEM and scanning Auger microanalyses. It was found that Pd silicides grow laterally under thermally activated diffusion with the very low energy of ~0.3 eV [2].

As shown by electron-transmission observations, the dominant phase in Pd–Si diffusion couples is Pd_2Si at 450°C. Small amounts of PdSi were also detected. The Pd_2Si–PdSi transformation is determined by the kinetics of its mechanism, not by thermodynamic stability [3].

The influence of substrate orientation and doping on the growth kinetics of Pd_2Si was studied using ⟨100⟩ and ⟨111⟩ oriented Si wafers with P concentrations of 5×10^{13} to 4.8×10^{18} atoms/cm³ and As concentrations of 2.8×10^{19} to 5×10^{20} atoms/cm³. Doping concentrations $<10^{19}$ do not affect the growth kinetics, but a significant influence is found for As doping levels $>10^{19}$ cm⁻³. For 5×10^{20} As atoms/cm³ the activation energy is lowered by ~0.35 eV, and the pre-exponential factor is lowered by 3 orders of magnitude (see also figures in the paper) [4].

The following table shows doping effects on the activation energies E_a and the diffusion coefficients D_0 on Si ⟨100⟩ and Si⟨111⟩ [4]:

substrate	atoms/cm³	E_a in eV	D_0 in cm²/s
Si⟨100⟩	P: ~5×10¹³	1.40±0.10	0.48
	As: 5×10²⁰	1.05±0.10	1.2×10⁻⁴
Si⟨111⟩	P: 4.6×10¹⁴	1.35±0.10	0.18
	As: 5.0×10²⁰	1.05±0.10	4.6×10⁻⁴

References:

[1] Duhaj, P.; Barancok, D.; Ondrejka, A. (J. Non-Cryst. Solids **21** [1976] 411/28).
[2] Majni, G.; Panini, F.; Sodo, G.; Cantoni, P. (Solid Films **125** [1985] 313/20, 319).
[3] Drobek, J.; Sun, R. C.; Tisone, T. C. (Phys. Status Solidi A **8** [1971] 243/8).
[4] Wittmer, M.; Tu, K. N. (Phys. Rev. B [3] **27** [1983] 1173/9).

12.1.2 Phase Diagram

In the literature there are discrepancies as to the composition and stability of some phases in the Pd–Si system. This may be due to impurities in the components and to a strong tendency to form metastable phases.

The Pd-rich part of the equilibrium diagram is shown by **Fig. 76** [1] and the Si-rich part by **Fig. 77**, p. 314 [2].

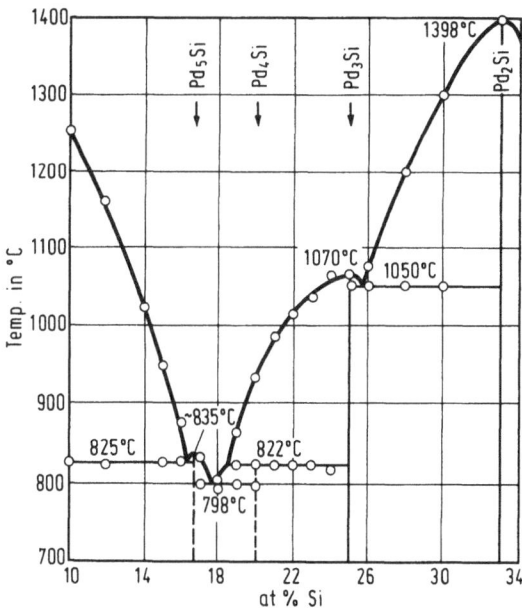

Fig. 76. Pd-rich part of the Pd-Si equilibrium diagram [1].

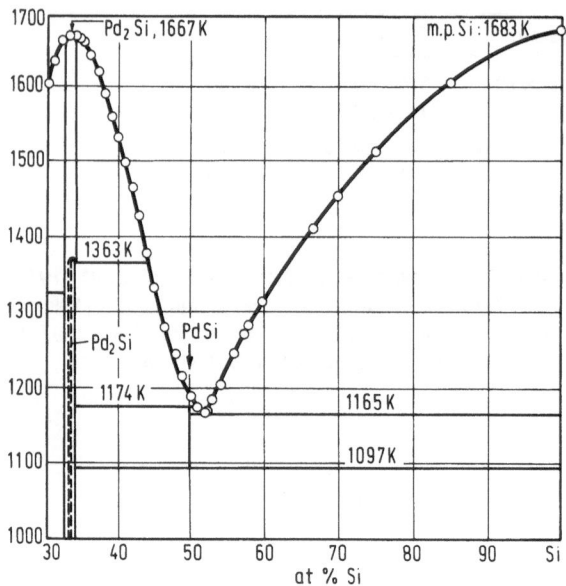

Fig. 77. Si-rich part of the Pd–Si equilibrium diagram [2].

Stable phases in the Pd–Si system are listed in the following table (c = congruent, p = peritectic).

composition	melting point in °C	Ref.	composition	melting point in °C	Ref.
Pd$_5$Si	~835c	1	Pd$_2$Si	1398c	1
	~810 to 823p	3		1394c	2,4
	~837c	4		1330	5
Pd$_{4.5}$Si(?) (Pd$_9$Si$_2$)	823p	3	PdSi	901p	2,4
Pd$_4$Si(?)	822p	1		900c	6
Pd$_3$Si	1070c	1,4			
	1045p	3			

Eutectics containing ~16, ~18 and ~27 at% Si were found melting at 825, 798 and 1050°C, respectively [1].

In the concentration range 0 to 33.3 at% Si two metastable systems appear the constituents of which are formed from melt due to well defined cooling rates. At cooling rates R > 10^4 K/s in the concentration range 15 to 23 at% Si, amorphous phases and metastable Pd$_{5.25}$Si and Pd$_{4.5}$Si are formed. At 10^2 K/s < R < 10^4 K/s the formation of Pd$_{4.5}$Si is suppressed, and Pd$_2$Si and Pd$_4$Si are stable (for combined stable and metastable phase diagrams see figures in paper) [4]. No evidence of the existence of Pd$_4$Si is found by [3]. During the crystallisation of amorphous Pd$_{80}$Si$_{20}$ at 645 K, the formation of Pd$_3$Si, Pd$_2$Si, Pd$_4$Si and other metastable phases was observed. Metastable Pd$_4$Si transforms to stable Pd$_4$Si after annealing at 923 K for 25 h [7].

References:

[1] Röschel, E.; Raub, C. J. (Z. Metallk. **62** [1971] 840/2).
[2] Langer, H.; Wachtel, E. (Z. Metallk. **72** [1981] 769/75).
[3] Wysocki, J. A.; Duwez, P. E. (Met. Trans. A **12** [1981] 1455/60).
[4] Langer, H.; Wachtel, E. (Z. Metallk. **74** [1983] 535/44, 536/8).
[5] Saitoh, S.; Ishiwara, H.; Asano, T.; Furukawa, S. (Japan. J. Appl. Phys. **20** [1981] 1649/56).
[6] Pfisterer, H.; Schubert, K. (Z. Metallk. **41** [1950] 358/67, 360).
[7] Xiao, K.; Dong, Y.; He, Y. (Phys. Status Solidi A **102** [1987] 221/9).

12.1.3 Formation and Preparation

All phases in the Pd–Si system can be prepared by melting together the components in an arc furnace under Ar, then annealing to reach equilibrium. Experiments with 99.999% Pd and Si are described in [1, 2].

Thin Pd and Pd_2Si films deposited on Si were reacted by using electron beam pulses of 60 ns duration in the energy range of 1.2 to 2.4 J/cm^2. **Fig. 78** shows the sequence of phases after electron beam annealing and postannealing in a conventional furnace [3, 4].

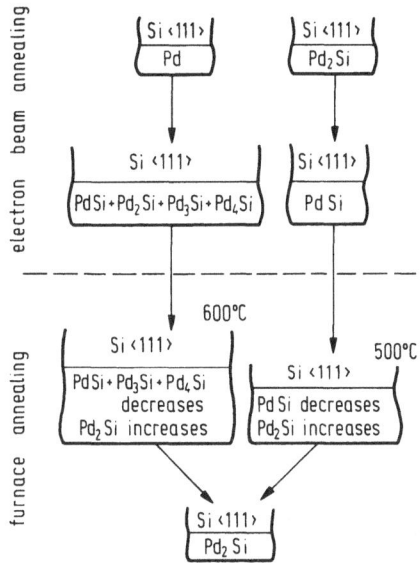

Fig. 78.   Pd silicides formed in the electron
beam and by furnace annealing [3].

Thin Pd films on Si single crystals were laser-irradiated for 18 ns. Except for compositions close to the congruently melting phases and the pure elements, amorphous films are formed. Thermal decomposition of Si-rich films results in the formation of equilibrium PdSi. At Pd-rich compositions a number of metastable phases are formed [5, 6].

The **enthalpies** of mixing in the Pd–Si system as determined by calorimetric measurements at 1600 K are strongly negative with a maximum of -72.2 kJ/g-atom for Pd_2Si [7]. The standard enthalpy of Pd_2Si formation at 1400 K was found to be -193.5 ± 7.5 kJ/mol [8].

316

References:

[1] Langer, H.; Wachtel, E. (Z. Metallk. **72** [1981] 769/75).
[2] Langer, H.; Wachtel, E. (Z. Metallk. **74** [1983] 535/44, 536).
[3] Majni, G.; Nava, F.; Ottaviani, G.; Danna, E.; Leggieri, G.; Luches, A.; Celotti, G. (J. Appl. Phys. **52** [1981] 4055/61).
[4] Majni, G.; Nava, F.; Ottaviani, G.; Luches, A.; Nassisi, U.; Galli, E. (Vide Couches Minces No. 201 [1980] 715/9).
[5] v. Allmen, M.; Lau, S. S.; Mäenpää, M.; Tsaur, B. Y. (Appl. Phys. Letters **37** [1980] 84/6).
[6] v. Allmen, M.; Wittmer, M. (Appl. Phys. Letters **34** [1979] 68/70).
[7] Arpshofen, I.; Pool, M. J.; Gerling, U.; Sommer, F.; Schultheiss, E.; Predel, B. (Z. Metallk. **72** [1981] 776/81).
[8] Topor, L.; Kleppa, O. J. (Z. Metallk. **77** [1986] 65/71).

12.1.4 Reactions of Pd–Silicides

At room temperature and a constant current density of 8.9 mA/cm^2 no anodic oxidation of Pd_2Si was detected in an N-methylacetamide (2% H_2O, 1% KNO_3) electrolyte [1].

On a clean surface of Pd_2Si grown on (111) Si, a thin layer of elementary Si is accumulated by heat treatment at 250 to 600°C (temperature dependence of the Si-layer thickness shown by figure in paper) [2].

PdSi, when in contact with crystalline Si, transformed into Pd_2Si + Si at 500 to 700°C, a process contrary to the equilibrium phase diagram. The rate of transformation depended on the structure and orientation of the Si. Above 750°C, Pd_2Si transformed back to PdSi. However, PdSi was stable against annealing when in contact with Pd_2Si or SiO_2. It was proposed that the decomposition of PdSi in the presence of Si is due to a lower energy of the Pd_2Si/Si interface compared to that of the PdSi/Si interface [3]. Thin films of Pd_2Si on (111) Si wafers transformed to PdSi at 810±10°C, and PdSi transformed back to Pd_2Si when annealed below that temperature. Contrary to the equilibrium diagram, Pd_2Si rather than PdSi was believed to be the thermodynamically stable phase on Si at low temperatures. To confirm this conclusion, bulk Pd–Si diffusion couples were annealed at 750, then at 850°C and again at 750°C [4].

The sensitivity of 7 transition-metal silicides to Ar$^+$ bombardment increased in the following order: Pd_2Si, Pt_2Si, PdSi, Ni_2Si, PtSi, NiSi, $NiSi_2$. This was shown by sheet resistance measurements on polycrystalline films [5].

At room temperature only Si atoms were oxidised in Pd_4Si, Pd_2Si and PdSi. The oxidation of Pd_4Si is enhanced compared with that of the two other silicides, and the growth of the surface oxide layer leaves behind a phase enriched with Pd [6, 7]. The oxidation of 50 nm Pd_2Si films on (100) Si substrates in oxygen and air at room temperature was investigated by [8]. After exposure of a Pd_2Si surface to an O_2 pressure of 5×10^{-6} Torr for 2000s, only Si was found to react at room temperature. The reaction rate was increased with respect to that of pure Si. Obviously the main effect of Pd is to by-pass the dissociation of O_2 at the silicide/gas interface [9].

In wet O_2 the energy for the oxidation of Pd_xSi on amorphous Si is ~1.09 eV at 900 to 1000°C. Pd_xSi dissociates on the surface, and both Pd and Si diffuse through the silicide to the substrate. The only species oxidised is Si which forms SiO_2 [10]. The SiO_2 growth through a Pd_2Si or PdSi layer on a Si substrate at ~800°C was described as a process limited by the diffusion of the oxidant from the ambient gas to the silicide/oxide interface [11, 12].

An investigation of $Al/Pd_2Si/Si$ junctions revealed the formation of a compound Al_3Pd_4Si by thermal annealing [13]. A quasi-binary system of limited mutual solubility was detected after interaction of Pd_2Si with $MoSi_2$ or Mo_5Si_3 at 700°C [14].

References:

[1] Strydom, W. J.; Lombaard, J. C.; Pretorius, R. (Solid-State Electron. **30** [1987] 947/51).
[2] Oura, K.; Okada, S.; Hanawa, T. (Appl. Phys. Letters **35** [1979] 705/6).
[3] Tsaur, B.-Y.; Nicolet, M.-A. (Appl. Phys. Letters **37** [1980] 708/11).
[4] Tu, K. N. (J. Appl. Phys. **53** [1982] 428/32).
[5] Tsaur, B.-Y.; Anderson, C. H. (J. Appl. Phys. **53** [1982] 940/2).
[6] Cros, A.; Pollak, R. A.; Tu, K. N. (Thin Solid Films **104** [1983] 221/5).
[7] Cros, A.; Pollak, R. A.; Tu, K. N. (J. Appl. Phys. **57** [1985] 2253/7).
[8] Brunner, A. J.; Oelhafen, P.; Güntherodt, H. J. (Surf. Sci. **189/190** [1987] 1122/8).
[9] Valeri, S.; del Pennino, U.; Lomellini, P.; Ottaviani, G. (Surf. Sci. **161** [1985] 1/11, 3).
[10] Lue, J. T.; Yang, C. S. (J. Mater. Sci. Letters **4** [1985] 463/6).

[11] Bartur, M.; Nicolet, M.-A. (J. Appl. Phys. **54** [1983] 5404/15).
[12] Bartur, M. (Thin Solid Films **107** [1983] 55/65).
[13] Köster, U.; Ho, P. S.; Lewis, J. E. (J. Appl. Phys. **53** [1982] 7436/44).
[14] Lashuk, E. P.; Raevskaya, M. V.; Kazakova, E. F. (Vestn. Mosk. Univ. Ser. II Khim. **25** [1984] 413/4; C.A. **101** [1984] No. 196460).

12.2 Dipalladium Monosilicide Pd_2Si

12.2.1 Formation and Preparation

Following Pd deposition on (111) Si, Pd_2Si formation was observed up to a Pd coverage of ~ 10 Å at room temperature [1]. Pd_2Si forms below ~ 700°C in a wide temperature range, stable in contact with (111) Si [2]. It was formed at room temperature and became epitaxial above 200°C. Surface segregation occurred above 400°C; annealing between 200 and 400°C produced epitaxial Pd_2Si islands grown on the Si substrate [3]. Pd_2Si can be formed at 300 to 720°C when Pd films are deposited on clean (111) Si surfaces and annealed in vacuum [4]; at 300°C a Pd film on (111) Si was annealed for 40 min in an oil-free vacuum system under pressures $< 10^{-6}$ Torr [5]. The growth and transformation of Pd_2Si on (111), (110) and (100) Si was elaborately investigated by [6]. Formation of Pd_2Si induced by laser irradiation was observed by [7, 8]. Of course the silicide can also be obtained by the methods outlined in Section 12.1.3, p. 315.

The rate-determining step for the formation of Pd_2Si is lattice diffusion of Si along the grain boundaries of Pd and/or Pd_2Si and subsequent diffusion and transformation in the Pd grains. The kinetics is independent of substrate orientation [9]. Between 160 and 222°C, parabolic growth of a Pd layer on (111) Si was observed. Rate constants showed Arrhenius behaviour with an activation energy of 1.06 eV and prefactor $k_0 = 7 \times 10^{-4}$ cm^2/s [10].

The phase transformation from PdSi to Pd_2Si was described by [11]; see also Sections 12.1.1 and 12.1.3, pp. 312 and 315.

Further investigations on the growth of Pd_2Si layers were performed by [12 to 20].

The activation energies for Pd_2Si formation on hydrogenated amorphous Si films and on (111) Si substrates are both 0.9 ± 0.1 eV, but the rate constants differ by a pre-exponential factor of 2; the growth rate on hydrogenated Si is faster [21, 22].

Both N and O impurities were found to retard Pd_2Si growth when initially present in Si. When the impurities were initially in Pd there was no significant retardation [23]. No effect of implanted O on the growth kinetics was found by [24]. The implanted O was said to be incorporated into Pd_2Si in a deep-lying band near the Pd_2Si/Si interface; additional O was also found at the surface [24]. Pd_2Si was obtained by reaction of Pd films with (100) Si covered with interfacial SiO_2 films (10 to 50 Å thick). The minimum temperature necessary for the reaction was ~400°C for ~24 Å of SiO_2 [25].

Reaction of Pd with SiC at high temperatures produced first Pd_3Si, then Pd_2Si; C compounds were not detected [26]. The reaction between Pd and SiC began at ~1000°C [27]. Reacting thin layers of Pd with As-implanted Si by rapid thermal annealing, an enhanced growth rate for Pd_2Si was measured which did not obey the diffusion-limited growth kinetics as reported for furnace-reacted Pd_2Si [28]; see also Section 12.1.1, p. 312.

References:

[1] Butz, R.; Wagner, H. (J. Vac. Sci. Technol. [2] B **1** [1983] 816/8).
[2] Ottaviani, G.; Tu, K. N.; Mayer, J. W. (Phys. Rev. B [3] **24** [1981] 3354/9).
[3] Oustry, A.; Berty, J.; Caumont, M.; David, M. J. (J. Microsc. Spectrosc. Electron. **9** [1984] 49/56).
[4] Saitoh, S.; Ishiwara, H.; Asano, T.; Furukawa, S. (Japan. J. Appl. Phys. **20** [1981] 1649/56).
[5] Lue, J. T.; Chen, H.-W.; Lew, S.-I. (Phys. Rev. B [3] **34** [1986] 5438/42).
[6] Hutchins, G. A.; Shepela, A. (Thin Solid Films **18** [1973] 343/63).
[7] Geiler, H. D.; Thrum, F.; Goetz, G. (Phys. Status Solidi A **70** [1982] K 159/K 162).
[8] Shibata, T.; Sigmon, T. W.; Gibbons, J. F. (Proc. Electrochem. Soc. **80** [1980] 458/68).
[9] Lau, S. S.; Sigurd, D. (J. Electrochem. Soc. **121** [1974] 1538/40).
[10] Coulman, B.; Chen, H. (J. Appl. Phys. **59** [1986] 3467/74).

[11] Chikyow, T.; Ohdomari, I.; Suzuki, S. (Phys. Rev. B [3] **34** [1986] 4807/11).
[12] Yang, C.-S.; Lue, J.-T. (Phys. Status Solidi A **101** [1987] 151/7).
[13] Zhang, J.; Liu, A.; Wu, Z.; Guo, K. (Wuli Xuebao **35** [1986] 965/8; C.A. **105** [1986] No. 162483).
[14] Ho, K. T.; Lien, C.-D.; Shreter, U.; Nicolet, M.-A. (J. Appl. Phys. **57** [1985] 227/31).
[15] Hung, L. S.; Mayer, J. W.; Pai, C. S.; Lau, S. S. (J. Appl. Phys. **58** [1985] 1527/36, 1528).
[16] Levy, D.; Ponpon, J. P.; Grob, A.; Grob, J. J.; Siffert, P. (Physica B + C **129** [1985] 205/9).
[17] Levy, D.; Grob, A.; Grob, J. J.; Ponpon, J. P. (Appl. Phys. A **35** [1984] 141/4).
[18] Cheung, N. W.; Nicolet, M.-A., Wittmer, M.; Evans, C. A.; Sheng, T. T. (Thin Solid Films **79** [1981] 51/60).
[19] Cheung, N. W.; Lau, S. S.; Nicolet, M.-A.; Mayer, J. W.; Sheng, T. T. (Proc. Electrochem. Soc. **80** [1980] 494/505).
[20] Pretorius, R.; Liau, Z. L.; Lau, S. S.; Nicolet, M.-A. (J. Appl. Phys. **48** [1977] 2886/90).

[21] Hung, L. S.; Kennedy, E. F.; Palmstrøm, C. J.; Olowolafe, J. O.; Mayer, J. W.; Rhodes, H. (Appl. Phys. Letters **47** [1985] 236/8).
[22] Paccagnella, A.; Majni, G.; Ottaviani, G.; della Mea, G. (Appl. Phys. Letters **47** [1985] 806/8).
[23] Ho, K. T.; Lien, C.-D.; Nicolet, M.-A. (J. Appl. Phys. **57** [1985] 232/6).
[24] Scott, D. M.; Nicolet, M.-A. (Nucl. Instrum. Methods Phys. Res. **209/210** [1983] 297/301).
[25] Scott, D. M.; Lau, S. S.; Pfeffer, R. L.; Lux, R. A.; Mikkelson, J.; Wielunski, L.; Nicolet, M.-A. (Thin Solid Films **104** [1983] 227/33).
[26] Pai, C. S.; Hanson, C. M.; Lau, S. S. (J. Appl. Phys. **57** [1985] 618/9).
[27] Suzuki, H.; Iseki, T.; Imanaka, T. (J. Nucl. Sci. Technol. [Tokyo] **14** [1977] 438/42).

[28] Alvi, N. S.; Kwong, D. L.; Hopkins, C. G.; Bauman, S. G. (Appl. Phys. Letters **48** [1986] 1433/5).

12.2.2 Physical Properties

Structure

Pd$_2$Si belongs to the hexagonal C 22 structure [1, 3]; Fe$_2$P type, space group P$\bar{6}$2m-D$_{3h}^3$, Z = 9 [3]; lattice parameters of the Pd-rich silicide were a = 6.497, c = 3.432 Å [1] or a = 6.496, c = 3.433 Å [3], and a = 6.528, c = 3.437 Å for the Si-rich compound [1]. Later a hexagonal superstructure was assumed with a = 13.05$_5$, c = 27.49$_0$ Å [2, 3].

Pd$_2$Si became epitaxial after annealing at 300 to 700°C when formed on (111) Si but not on (100) Si. A sequence of superstructures was observed on (111) Si as the annealing time was increased [4]. Domain structure in Pd$_2$Si films at Pd/Si interfaces was studied by [5].

In Pd$_2$Si the d-states of the Pd interact strongly with the p-states of the Si. The resulting p-d hybrid complex is composed of Pd d-states and two groups of Si states separated by ~5 eV. The lower-lying group of Si states forms the Si(3p)–Pd(4d) bonding levels while the higher-lying group forms the corresponding antibonding states [6]. Binding energies of 336.8 and 99.72 eV were determined for Pd 3d$_{5/2}$ and Si 2p$_{3/2}$, respectively. The strength of the metal (d)–Si(p) interaction increases in the order Ni$_2$Si < Pd$_2$Si < Pt$_2$Si [7]. Stoichiometry and structural disorder effects of Pd$_2$Si layers (thickness 1000 to 2000 Å) were investigated by [8]. A theoretical investigation of the electronic structure was performed by [9].

References:

[1] Aronsson, B.; Nylund, A. (Acta Chem. Scand. **14** [1960] 1011/8, 1012).
[2] Nylund, A. (Acta Chem. Scand. **20** [1966] 2381/6).
[3] Langer, H.; Wachtel, E. (Z. Metallk. **72** [1981] 769/75).
[4] Okada, S.; Oura, K.; Hanawa, T.; Satoh, K. (Surf. Sci. **97** [1980] 88/100).
[5] Chen, H.; White, G. E.; Stock, S. R. (Mater. Res. Soc. Symp. Proc. **12** [1982] 165/73).
[6] Ho, P. S.; Rubloff, G. W.; Lewis, J. E.; Moruzzi, V. L.; Williams, A. R. (Phys. Rev. B [3] **22** [1980] 4784/90).
[7] Grunthaner, P. J.; Grunthaner, F. J.; Madhukar, A. (J. Vac. Sci. Technol. **20** [1982] 680/3).
[8] Chabal, Y. J.; Rowe, J. E.; Poate, J. M.; Franciosi, A.; Weaver, J. H. (Phys. Rev. B [3] **26** [1982] 2748/58).
[9] Bisi, O.; Calandra, C. (J. Phys. C **14** [1981] 5479/94).

Density. Thermal Expansion

The density was calculated as 9.46 g/cm^3 [1]. The thermal expansion coefficient of Pd$_2$Si was determined as 23 × 10^{-6} (K^{-1}). Stress relaxation was observed after prolonged annealing at temperatures > 200°C [2].

References:

[1] Wysocki, J. A.; Duwez, P. E. (Metallurg. Trans. A **12** [1981] 1455/60).
[2] Chen, H.; White, G. E.; Stock, S. R. (Mater. Letters **4** [1986] 61/4).

Electrical and Optical Properties

Electrical resistivity and Hall mobility measurements were performed on Pd_2Si films grown on (100) Si and (111) Si substrates as a function of temperature and film thickness. The Debye temperature is 120 ± 20 K, and the concentration of charge carriers (electrons) is 4×10^{21} cm^{-3}. The bulk value of the resistivity at room temperature is 25 to 30 $\mu\Omega \cdot cm$, and the Hall mobility is 50 to 60 $cm^2 \cdot V^{-1} \cdot s^{-1}$, both depending on the structure which is epitaxial on (111) Si but not on (100) Si [1]. The resistivities of oriented Pd_2Si films on (111) Si were closely related to the film crystallinity and decreased by 30% on increasing the formation temperature from 300 to 700°C. Under most suitable conditions, $\varrho = 20$ $\mu\Omega \cdot cm$ was obtained. For crystalline thin films grown on (100) Si the resistivities were greatly affected by grain size. A rapid increase was observed for grain sizes $\leqq 1000$ Å; for grain sizes $\geqq 1000$ Å the resistivities were approximately equal to those of the oriented films [2]. Other measurements on Pd_2Si films were performed by [3 to 6].

The optical absorption and dispersion of as-deposited Pd_2Si were measured with an ellipsometer at wavelengths of 400 to 700 nm [7]; for further investigations on optical properties see [6].

References:

[1] Wittmer, M.; Smith, D. L.; Lew, P. W.; Nicolet, M.-A. (Solid State Electron. **21** [1978] 573/80).

[2] Sorimachi, Y.; Asano, T.; Saitoh, S.; Ishiwara, H.; Furukawa, S. (Denshi Tsushin Gakkai Ronbunshi C J 65-C [1982] 73/9 from C.A. **96** [1982] No. 227060).

[3] Saitoh, S.; Ishiwara, H.; Asano, T.; Furukawa, S. (Japan. J. Appl. Phys. **20** [1981] 1649/56).

[4] Nylandsted-Larsen, A.; Chevallier, J.; Pedersen, A. S. (Mater. Letters **3** [1985] 242/6).

[5] Wei, C. S.; van der Spiegel, J.; Santiago, J. (J. Appl. Phys. **58** [1985] 4200/6).

[6] Bisi, O.; Betti, M. G.; Nava, F.; Borghesi, A.; Guizetti, G.; Nosenzo, L.; Piaggi, A. (Vide Couches Minces **42** [1987] 215/8).

[7] Lue, J. T.; Chen, H.-W.; Lew, S.-I. (Phys. Rev. B [3] **34** [1986] 5438/42).

12.3 Monopalladium Monosilicide PdSi

12.3.1 Formation and Preparation

PdSi was formed by annealing a Pd film on (111) Si at 840°C [1], or by annealing at 800°C for 40 min in an oil-free vacuum system under pressures $<10^{-6}$ Torr [2]. It was also obtained by laser irradiation of thin Pd films on (111) Si [3, 4].

Quartz-halogen lamps with power densities of 10, 15 and 25 W/cm^2 were used to obtain Pd–Si films. Metal films of 83 to 200 nm thickness were evaporated on (111) oriented Si single-crystals and subsequently processed in vacuum for 5 to 60 s. A nonuniform PdSi film with dendrite-like surface topography was formed at 25 W/cm^2, and a discontinuous low-resistance film of predominantly PdSi at 15 W/cm^2 [5].

PdSi was obtained by implanting energetic Xe ions through a thin Pd or Pd_2Si film on a Si substrate. The PdSi phase formed near the Pd_2Si/Si interface and grew by subsequent thermal annealing at 300 to 400°C [6]. The influence of Pt atoms on the low-temperature formation of epitaxial PdSi was investigated by [7].

PdSi was detected in the vapours effusing from a Knudsen cell at high temperatures. For the reaction $Pd(gas) + Si_2(gas) \rightleftharpoons PdSi(gas) + Si(gas)$ at 2025 to 2185 K, a dissociation energy of 309.6 ± 14 kJ/mol has been estimated [8].

See also Section 12.1.3, p. 315.

References:

[1] Ottaviani, G.; Tu, K. N.; Mayer, J. W. (Phys. Rev. B [3] **24** [1981] 3354/9).

[2] Lue, J. T.; Chen, H.-W.; Lew, S.-I. (Phys. Rev. B [3] **34** [1986] 5438/42).

[3] Geiler, H. D.; Thrum, F.; Goetz, G. (Phys. Status Solidi A **70** [1982] K 151/K 162).

[4] Shibata, T.; Sigmon, T. W.; Gibbons, J. F. (Proc. Electrochem. Soc. **80** [1980] 458/68).

[5] Wei, C. S.; van der Spiegel, J.; Santiago, J. (J. Appl. Phys. **58** [1985] 4200/6).

[6] Tsaur, B.-Y.; Lau, S. S.; Mayer, J. W. (Appl. Phys. Letters **35** [1979] 225/7).

[7] Kawarada, H.; Mizugaki, K.; Ohdomari, I. (J. Appl. Phys. **57** [1985] 244/8).

[8] van der Auwera-Mahieu, A.; Peeters, R.; McIntire, N. S.; Drowart, J. (Trans. Faraday Soc. **66** [1970] 809/16, 813).

12.3.2 Physical Properties

PdSi is orthorhombic (B 31 type) [1, 2, 3], space group Pnma–D_{2h}^{16}, Z = 4 molecules [3, 4]; lattice parameters: a = 6.121, b = 5.588, c = 3.374 kX [1]; a = 5.599, b = 3.381, c = 6.133 Å [2]; a = 5.6173(10), b = 3.3909(6), c = 6.1534(12) Å [3]; a = 6.133, b = 5.599, c = 3.381 Å [4].

For an investigation of the electronic structure see [4].

A uniform PdSi layer exhibited a sheet resistivity of 18 $\mu\Omega \cdot$cm [5].

The optical absorption and dispersion of as-deposited PdSi were measured with an ellipsometer at wavelengths of 400 to 700 nm. The binding energy of the Pd $3d_{5/2}$ core electrons shifted by 1.8 ± 0.2 eV as studied by X-ray photoelectron spectroscopy [6].

Reactions see Section 12.1.4, pp. 316/7.

References:

[1] Pfisterer, H.; Schubert, K. (Z. Metallk. **41** [1950] 358/67, 360).

[2] Aronsson, B.; Nylund, A. (Acta Chem. Scand. **14** [1960] 1011/8, 1012).

[3] Engström, I. (Acta Chem. Scand. **24** [1970] 1466/8).

[4] Bisi, O.; Calandra, C. (J. Phys. C **14** [1981] 5479/94, 5484).

[5] Tsaur, B.-Y.; Lau, S. S.; Mayer, J. W. (Appl. Phys. Letters **35** [1979] 225/7).

[6] Lue, J. T.; Chen, H.-W.; Lew, S.-I. (Phys. Rev. B **34** [1986] 5438/42).

12.4 Other Palladium Silicides

Pd₅Si is said to belong to the monoclinic space group $P2_1$–C_2^2 with a = 8.465, b = 7.485, c = 5.555 Å, β = 100.7°, Z = 24 atoms; calculated density 10.76 g/cm³ [9].

Pd₄.₅Si was claimed to crystallise in the orthorhombic system with a = 7.418, b = 9.936, c = 9.048 Å [1]; space group Pnma–D_{2h}^{16}, a = 9.414, b = 7.4188, c = 9.0548 Å, Z = 44 atoms; calculated density 10.65 g/cm³ [9].

Pd₄Si was found to be stable up to 650°C in the presence of unreacted Pd. The crystals are claimed to be triclinic, space group P$\bar{1}$–C_i^1, with a = 4.402, b = 7.700, c = 8.990 Å, α = 111.02°, β = 96.52°, γ = 89.15°, Z = 4; calculated density 10.67 g/cm³ [2].

Pd₃Si. This silicide can be prepared using the methods outlined in Section 12.1.3, p. 315. It can also be obtained by annealing an amorphous Pd–Si alloy with 20 at% Si, after quenching from the melt, under vacuum at 800°C for 40 h. X-ray diffraction diagrams showed the existence of a face-centred cubic solid solution of Si in Pd and an orthorhombic phase Pd_3Si [3, 4]. The amorphous structure transformed to stable Pd_3Si on heating at temperatures ≥ 300°C [5].

Pd_3Si was formed (besides Pd_2Si) by reaction of Pd with SiC at temperatures ≥ 1000°C [6].

As derived from single-crystal X-ray data. Pd_3Si belongs to the orthorhombic space group Pnma–D_{2h}^{16} with a = 5.735, b = 7.555, c = 5.260 Å, Z = 4 formula units. Thus, Pd_3Si crystallises in the cementite (DO_{11}) structure (interatomic distances in paper) [7], see also [8]. According to [9] the calculated density is 10.12 g/cm³.

References:

[1] Nylund, A. (Acta Chem. Scand. **20** [1966] 2381/6).
[2] Canali, C.; Silvestri, L.; Celotti, G. (J. Appl. Phys. **50** [1979] 5768/72).
[3] Lesueur, D. (CEA-CONF-1693 [1970] 1/8 from N.S.A. **25** [1971] No. 29901).
[4] Lesueru, D. (Fizika [Zagreb] Suppl. **2** No. 2 [1970] 13.1/13.8; C.A. **75** [1971] No. 83349).
[5] Masumoto, T.; Maddin, R. (Acta Met. **19** [1971] 725/41).
[6] Suzuki, H.; Iseki, T.; Imanaka, T. (J. Nucl. Sci. Technol. [Tokyo] **14** [1977] 438/42).
[7] Aronsson, B.; Nylund, A. (Acta Chem. Scand. **14** [1960] 1011/8).
[8] Röschel, E.; Raub, C. J. (Z. Metallk. **62** [1971] 840/2).
[9] Wysocki, J. A.; Duwez, P. E. (Met. Trans. A **12** [1981] 1455/60).

13 Palladium and Phosphorus

General References:

Ward, R., Structural Chemistry of Condensed Systems of Transition Metals, MTP (Med. Tech. Publ. Co.) Int. Rev. Sci. Inorg. Chem. Ser. One **5** [1972] 93/174.

Suchet, J. P., Crystal Chemistry and Semiconduction in Transition-Metal Binary Compounds, Academic Press, New York 1971.

Hulliger, F., Crystal Chemistry of the Chalcogenides and Pnictides of the Transition Elements, Structure and Bonding [Berlin] **4** [1968] 83/229.

Rundqvist, S., Binary Transition Metal Phosphides and Crystal Chemical Relations between them and Transition Metal Compounds with other Nonmetals of Small Atomic Radius, Arkiv Kemi **20** [1962] 67/113.

13.1 Pd–P Phase Diagram

An equilibrium diagram for the Pd/P system has been constructed using data from DTA and X-ray analyses [1]. A portion of the same diagram (25 to 35 at% P) is reproduced in [2].

The following intermediate phases have been found: Pd_8P, Pd_6P, $Pd_{4.8}P$, Pd_3P, Pd_5P_2, Pd_7P_3, and PdP_2. $Pd_{4.8}P$, Pd_3P, and probably PdP_2 melt congruently; the other phases decompose peritectically, see **Fig. 79** according to [1].

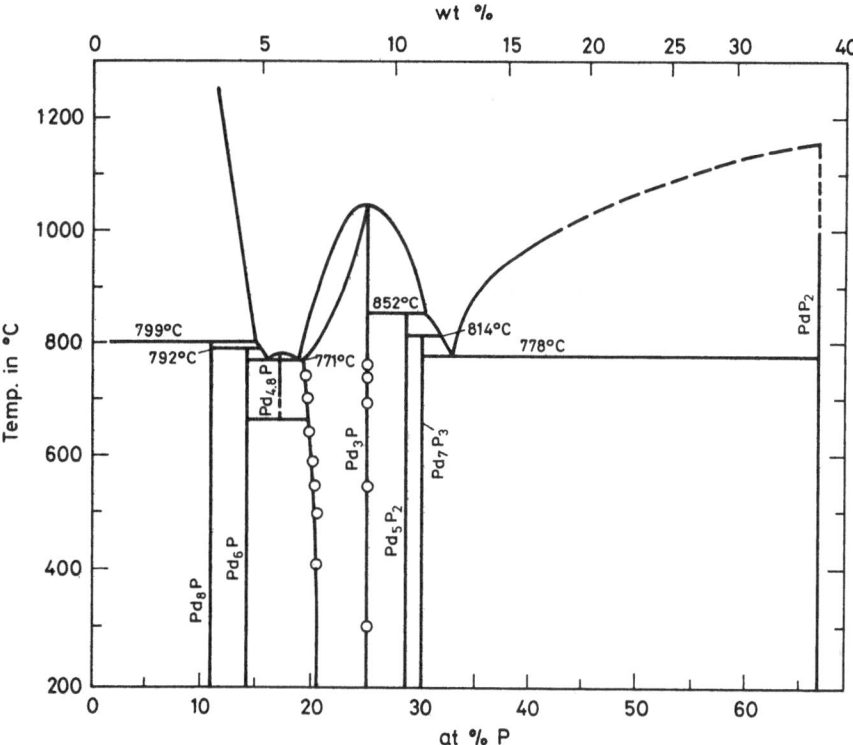

Fig. 79. Pd-rich part of the Pd–P phase diagram [1].

$Pd_{4.8}P$ melts at 780°C. On cooling, it decomposes eutectoidally into Pd_6P and Pd_3P [1].

The P content of the Pd_3P phase varies on the Pd-rich side from 19.3 at% at 741 ± 3°C to 20.4 at% at 498 ± 3°C; it remains constant 25.0 at% between 762 ± 3°C and 305 ± 3°C as derived from X-ray determinations of the unit cell parameters (for lattice constants see p. 325) [1].

Palladium/phosphorus alloys in the composition region 20 to 25 at% P, originally thought to yield a single Pd_5P_2 phase [3] were subsequently shown to form two phases [4]. These were first characterised as polymorphic modifications of Pd_7P_3 [4] but were later identified as Pd_7P_3 and authentic Pd_5P_2 [1]. Thermal data for the Pd/P system show a horizontal at 852°C corresponding to the peritectic formation of Pd_5P_2 from Pd_3P and melt. X-ray data show that Pd_5P_2 is stable down to room temperature and confirm the narrow homogeneity range [1].

References:

[1] Gullman, L.-O. (J. Less-Common Metals **11** [1966] 157/67).
[2] Matković, T.; Schubert, K. (J. Less-Common Metals **55** [1977] 177/84).
[3] Wiehage, G.; Weibke, F.; Biltz, W.; Meisel, K.; Wiechmann, F. (Z. Anorg. Allgem. Chem. **228** [1936] 357/71).
[4] Raub, C. J.; Zachariasen, W. H.; Geballe, T. H.; Matthias, B. T. (J. Phys. Chem. Solids **24** [1963] 1093/100).

13.2 Binary Compounds

"Pd_8P". This compound, which has subsequently been reformulated as $Pd_{15}P_2$ [3], is prepared by heating a stoichiometric mixture of pure palladium metal and phosphorus in a sealed silica tube at 700 to 750°C for 2 d, then annealing the crushed material for several days. X-ray intensity data and d-values have been tabulated for a sample quenched from 591°C [1].

$Pd_{15}P_2(Pd_{7.5}P)$ was originally formulated as Pd_8P [1] but subsequently shown to possess the ideal crystallographic formula $Pd_{15}P_2$ [3]. The compound is prepared in polycrystalline form as described above for "Pd_8P". Single crystals for X-ray work were obtained by heating well-mixed pellets of pure palladium and phosphorus powder at 830°C for 2 d and 795°C for 5 months. The irregular needle-shaped crystals are hexagonal, space group $R\bar{3}$-C_{3i}^2 with a = 7.1067(2), c = 17.0867(6) Å, Z = 3. The structure consists of Pd icosahedra arranged in similar fashion to the spheres in cubic close packing with pairs of Pd atoms in the octahedral sites and phosphorus atoms in the tetrahedral sites, and is related geometrically to the α-boron structure [3]. A neutron powder diffraction investigation of $Pd_{15}P_2$ and the deuterated product $Pd_{15}P_2D_{0.46}$ has been reported. Data for $Pd_{15}P_2$ are identical with those given above, the deuterated product belongs to the same space group with a = 7.1166(3) and c = 17.1281(20) Å. The deuterated compound which was prepared at 296 K and 500 kPa deuterium pressure, has D atoms in distorted tetrahedral interstices between the palladium atoms [4].

Pd_6P is obtained by heating a stoichiometric mixture of palladium and phosphorus at 700 to 750°C for 2 d in an evacuated silica tube, then annealing the crushed product for 1 to 2 d [1] or by annealing a Pd_6/P alloy at 1030 K for 7 d [5]. X-ray intensities and d-spacings have been tabulated for a sample quenched from 758°C [1]. A later investigation of Pd_6P by powder diffraction methods gave a monoclinic cell with a = 2.8370, b = 9.4409, c = 7.6945 Å, β = 90.20°, Z = 2. The space group symmetry is uncertain but a refinement based on C2/c symmetry converged satisfactorily to a final R_w of 0.113. Powder diffraction data and a projection of the structure on (100) are reproduced in the paper [6]. A neutron diffraction study, which revealed new low-angle data, established a larger unit cell, space group $P2_1/c$-C_{2h}^5 with a = 5.6740(4), b = 9.4409(6), c = 8.2100(6) Å, β = 110.414(4)°, Z = 4 or a = 5.6740(4), b = 9.4409(6), c = 8.1916(6) Å, β = 110.065(4)° [5].

$Pd_{4.8}P$. This phase which was originally formulated as Pd_5P [9] has now been reformulated as $Pd_{4.8}P$ [1]. It is obtained by heating pressed pellets of the mixed elements at 750°C for several days, then quenching in water [10]. X-ray powder data (tabulated in the paper) gave unit cell dimensions a = 5.0059 ± 0.0003, b = 7.6082 ± 0.0005, c = 8.4199 ± 0.0005 Å, β = 95.640° ± 0.005° [1]. A single crystal X-ray structure determination gave the space group $P2_1$-C_2^2 with a = 5.004, b = 7.606, c = 8.416 (all ± 0.0005)Å, β = 95.63° ± 0.005°. Cell parameters for samples of nominal composition Pd_xP (x = 4.75, 4.65, 4.5) are tabulated. A projection of the $Pd_{4.8}P$ structure on (010) is reproduced in the paper [10].

Pd_3P to $Pd_3P_{0.75}$ (see "Palladium" 1942, p. 306). The existence of a Pd-P phase with the composition Pd_3P [9] has been confirmed and the homogeneity range, determined by quenching alloys from 740°C, has been shown to extend from Pd_3P to ~$Pd_3P_{0.75}$ [11]; see also phase diagram on p. 323.

Preparation from the elements at 990 K in an evacuated silica tube has been reported, crushing followed by annealing for 4 to 6 d at 1020 to 1040 K gives complete homogeneity [14]. Samples in black powder form have been obtained by dropwise addition of $PdCl_2(CH_3CN)_2$ in toluene solution to molten potassium and phosphorus at 200°C [18].

The structure is of the cementite (DO_{11}) type. Lattice parameters are a = 5.980, b = 7.440, c = 5.164 Å (P-rich limit) and a = 5.645, b = 7.558, c = 5.071 Å (Pd-rich limit). Structural data for $Pd_3P_{0.95}$ (later corrected to $Pd_3P_{0.99}$ by [1] are space group $Pnma$-D_{2h}^{16}, a = 5.947, b = 7.451, c = 5.170 Å, Z = 4, interatomic distances are tabulated in the paper [11]. Lattice parameters for a sample prepared by heating palladium metal and red phosphorus are a = 5.166(1), b = 5.971(1), c = 7.445(1) Å [12]. Lattice parameters change anisotropically with composition, as the phosphorus content increases from 7 to 11%, a and b and the cell volume U increase but c decreases. Data are tabulated and plotted in the paper. Changes are attributed to gaps in the space lattice due to lack of P when P < 8.83% and to interstitial insertion of P atoms when P > 8.83% [13]. Lattice parameters are a = 5.680, b = 7.546, c = 5.101 Å (410°C); a = 5.637, b = 7.561, c = 5.065 Å (741°C) on the Pd-rich phase limit, and a = 5.979, b = 7.441, c = 5.164 Å (305 to 762°C) on the P-rich phase limit [1]. Data for lattice parameters and density as a function of composition are also tabulated [1] and have been plotted in [14]. The crystal structure of $Pd_3P_{0.80}$ has been analysed by Rietveld-type profile refinements of neutron diffraction intensity data. A cementite-type structure, space group $Pnma$-D_{2h}^{16}, with random phosphorus vacancies has been confirmed. Lattice parameters are a = 5.7019(2), b = 7.5366(2), c = 5.1191(2) Å [15]. Values for stoichiometric Pd_3P are a = 5.983, b = 7.442, c = 5.166 Å [14]. Pd_3P is a superconductor below 0.75 K [16].

Valence-band and core-level photoemission spectra together with photon induced Auger spectra provide evidence for formation of a palladium phosphide, probably Pd_3P, at Pd/InP (110) interfaces [17].

Pd_5P_2 (see "Palladium" 1942, p. 306). Intensities and d-spacings for a sample quenched from 735°C are tabulated in [1]. The phase subsequently identified as Pd_5P_2 is reported to be a superconductor (T_c = 0.70 K) [16].

Pd_7P_3. High and low temperature polymorphic modifications have been claimed [16], however the latter was later characterised as Pd_5P_2 [1]. Diffraction lines for authentic Pd_7P_3 were indexed on the basis of a hexagonal cell, a = 11.94, c = 7.04 Å or, alternatively, a rhombohedral cell with a = 7.28 Å, α = 110.12° [16]. A later determination found a = 11.952, c = 7.037 Å at 736°C [1]. The crystals are rhombohedral, space group C_3^4-R3, with a = 11.976(2), c = 7.055(2) Å and

326

Z = 2. The crystal structure is related to the W-structure family and has fully occupied CuZn binding [2].

Pd_7P_3 is a superconductor below 1.0 K [16].

PdP₂ (see "Palladium" 1942, p. 306). Preparation by heating stoichiometric mixtures of palladium powder and red phophorus in evacuated silica tubes at 800 to 1100°C has been described. X-ray powder data are tabulated and the monoclinic space group C2/c or Cc (C_{2h}^6 or C_s^4) has been tentatively assigned, unit cell parameters are a = 6.777(1), b = 5.856(3), c = 6.206(3) Å, β = 126.42(7)° [7]. Another source gives a = 6.207, b = 5.857, c = 5.874 Å, β = 111.80°, and Z = 4 [16]. A single crystal X-ray diffraction study showed PdP_2 to be monoclinic, space group I2/a-C_{2h}^6 (or possibly Ia) with a = 6.207(1), b = 5.857(1), c = 5.874(1) Å, β = 111.80(1)°, Z = 4, and D_{calc} = 5.631 g/cm³. The structure contains square planar palladium with Pd–P distances of 2.335 and 2.341 Å, the phosphorus atoms are linked in endless zig-zag chains with P–P distances of 2.201 and 2.224 Å and bond angles of 102.0°. Elements of the structure are reproduced in the paper [8]. PdP_2 is a semiconductor with an energy gap, derived from resistivity measurement, of 0.6 to 0.7 eV [21]. The photoelectrochemical properties have been investigated [22].

PdP₃. A phase of this composition has been obtained by heating palladium powder and red phosphorus together in an evacuated silica tube at 500 to 1100°C. X-ray powder data established a D2 (skutterudite) type structure, space group Im3-T_h^5 with a = 7.705 Å [19]. An electron-valence diagram for skutterudite type semiconductors including PdP_3 has been described [20].

References:

[1] Gullman, L.-O. (J. Less-Common Metals **11** [1966] 157/67).
[2] Matković, T.; Schubert, K. (J. Less-Common Metals **55** [1977] 177/84).
[3] Andersson, Y. (Acta Chem. Scand. A **31** [1977] 354/8).
[4] Andersson, Y.; Rundqvist, S.; Tellgren, R.; Flanagan, T. B. (Z. Physik. Chem. [N.F.] **145** [1985] 43/9).
[5] Andersson, Y.; Rundqvist, S.; Tellgren, R.; Thomas, J. O.; Flanagan, T. B. (Acta Cryst. B **37** [1981] 1965/72).
[6] Andersson, Y.; Kaewchansilp, V.; del Rosario Casteleiro Soto, M.; Rundqvist, S. (Acta Chem. Scand. A **28** [1974] 797/802).
[7] Rundqvist, S. (Acta Chem. Scand. **15** [1961] 451/3).
[8] Zachariasen, W. H. (Acta Cryst. **16** [1963] 1253/5).
[9] Wiehage, G.; Weibke, F.; Biltz, W.; Meisel, K.; Wiechmann, F. (Z. Anorg. Allgem. Chem. **228** [1936] 357/71).
[10] Sellberg, B. (Acta Chem. Scand. **20** [1966] 2179/86).

[11] Rundqvist, S.; Gullman, L.-O. (Acta Chem. Scand. **14** [1960] 2246/7).
[12] Fruchart-Triquet, E.; Fruchart, R.; Michel, A. (Compt. Rend. **252** [1961] 1323/4).
[13] Fruchart, E.; Fruchart, R.; Michel, A. (Compt. Rend. **252** [1961] 3263/5).
[14] Flanagan, T. B.; Biehl, G. E.; Clewley, J. D.; Rundqvist, S.; Andersson, Y. (J. Chem. Soc. Faraday Trans. I **76** [1980] 196/208).
[15] Andersson, Y.; Rundqvist, S.; Tellgren, R.; Thomas, J. O.; Flanagan, T. B. (J. Solid State Chem. **32** [1980] 321/7).
[16] Raub, C. J.; Zachariasen, W. H.; Geballe, T. H.; Matthias, B. T. (J. Phys. Chem. Solids **24** [1963] 1093/1100).
[17] Kendelewicz, T.; Petro, W. G.; Lindau, I.; Spicer, W. E. (Phys. Rev. [3] B **28** [1986] 3618/21).
[18] Carturan, G.; Cocco, G.; Semenzato, D. (React. Solids **1** [1985] 31/42).

[19] Rundqvist, S. (Nature **185** [1960] 31/2).
[20] Kuz'min, R. N. (Khim. Svyaz Poluprov Tverd. Telakh **1965** 335/46; C.A. **64** [1966] 5894).

[21] Hulliger, F. (Nature **200** [1963] 1064/5).
[22] Folmer, J. C. W.; Turner, J. A.; Parkinson, B. A. (J. Solid State Chem. **68** [1987] 28/37).

13.3 Polynary Compounds

Pd_6PH_x (x = 0.15) is prepared by dissolving hydrogen in Pd_6P at 1 atm and 298 K to form an interstitial solution. Obedience to Sieverts' law of ideal solubility at low values of x indicates that the molecular hydrogen dissociates on entering the Pd_6P lattice. Solubility data for hydrogen in Pd_6P at 273, 292, 303, and 323 K are plotted in the paper and thermodynamic parameters for solution of hydrogen in Pd_6P are calculated. A diagram of the Pd_6P unit cell showing interstices occupied by hydrogen is reproduced [1].

Pd_6PH_x (x = 0.39). According to neutron diffraction data, unit cell dimensions are a = 5.6846(2), b = 9.4638(4), c = 8.2097(5) Å, β = 10.431(7)°. Structural parameters and a projection of the structure are given in the paper [2].

Pd_6PD_x (x = 0.15 to 0.26). The solubilities of H and D in Pd_6P at 323 K have been compared (plot reproduced in paper) and thermodynamic data for the solution of deuterium in Pd_6P have been tabulated [1]. Crystallographic data for x = 0.15, 0.22, and 0.26 are tabulated, and the Pd_6PD_x structure projected on (010) and on a plane perpendicular to the a axis is reproduced [3].

$Pd_3P_xH_y$ (x = 1 to 0.8, y = 0 to 0.2). The solution of molecular hydrogen in compounds of the type Pd_3P_x has been investigated, solubility of H_2 increases as the nonstoichiometry of the Pd_3P_x phase increases and reaches a maximum corresponding to the composition $Pd_3P_{0.8}H_{0.15}$ at 296 K and 93.3 kPa. There is no indication of hydride phase formation. Lattice expansion accompanying H_2 solution is anisotropic, the b axis expands at less than half the rate of the a and c axes. Plots of unit cell volume expansion and equilibrium H_2 pressure against hydrogen content H/Pd_3P_x at different temperatures and values of x are reproduced in the paper. A model based on H occupation of one of the quadrilateral faces of each trigonal prism not occupied by a central phosphorus atom has been advanced to account for H-pressure/H-content data. For partial thermodynamic parameters for H_2 solution in Pd_3P_x compounds see table below [4]. Neutron diffraction data have been reported for $Pd_3P_{0.80}H_{0.17}$, lattice parameters are a = 5.7144(3), b = 7.5364(4), c = 5.1271(3) Å [2].

Relative partial thermodynamic parameters at infinite dilution of H for the solution of ½ H_2 in Pd_3P_{1-x} compounds at 323 K [4]:

Pd_3P_{1-x}	$\Delta H_H^\circ/J(mol\ H)^{-1}$	$\Delta S_{H \to 0}/J(mol\ H)^{-1} \cdot K^{-1}$	$\Delta S_H^\circ/J(mol\ H)^{-1} K^{-1}$
x = 1	− 9.800(2)	− 56.2(2)	− 56.5(2)
0.12	− 10.900	− 58.3	− 46.4
0.17	− 15.250	− 56.7	− 47.7
0.20	− 16.300	− 56.3	− 48.7

$Pd_3P_xD_y$ (x = 0.80 to 1, y = 0 to 0.20). Studies on the uptake of D_2 by Pd_3P_x phases give results similar to those reported for the $Pd_3P_x–H_2$ systems. Plots of equilibrium deuterium pressures against D/Pd_3P_x ratios at 323 K for x = 0.80 and 0.83 are reproduced in [4]. A neutron diffraction study has been reported for the deuterated phase $Pd_3P_{0.80}D_{0.15}$ (equilibrium composition at 296 K and 500 kPa deuterium pressure). The D atoms occupy the same sites as the H atoms in $Pd_3P_{0.80}H_x$, lattice parameters are a = 5.7182(4), b = 7.5448(6), c = 5.1304(4) Å. The crystal structure projected on (010) is reproduced and interatomic distances <3.6 Å are tabulated in the paper. The D atoms have 5 Pd neighbours (average Pd-D distance 2.14 Å) in distorted square pyramidal coordination [5].

PdPS. Well crystallised samples can be obtained by heating the elements together at 1000 to 1200°C and 65 kbar pressure in a sealed silica tube [6, 7]. The graphite-like crystals have orthorhombic symmetry, space group Pbcn-D_{2h}^{14} with lattice constants a = 13.3045(37), b = 5.6777(5), c = 5.6932(5) Å, Z = 8; D_{exp} = 5.13 g/cm³, D_{calc} = 5.23 g/cm³ [8]. An earlier paper reports space group $P2_1ca$-C_{2v}^5 or Pmca-D_{2h}^{11} with a = 5.693, b = 13.305, c = 5.678 Å [6]. The structure shown in **Fig. 80** contains square planar Pd coordinated to 2 S and 2 P atoms, tetrahedral P coordinated to 1 S, 1 P and 2 Pd atoms, and tetrahedral S coordinated to 1 P, 2 Pd and a lone pair of electrons. The bonding has been described in terms of an ionic structure 2Pd²⁺ [S–P–P–S]⁴⁻. PdPS is a diamagnetic semiconductor [8]. The photoelectrochemical properties of PdPS have been extensively reported [10, 11]. Resistivities are $\varrho_{298\,K}$ = 9 × 10⁷ and $\varrho_{425\,K}$ = 3 × 10⁴ Ω·cm, the activation energy of resistivity, E_a, is given as 0.7 eV [6].

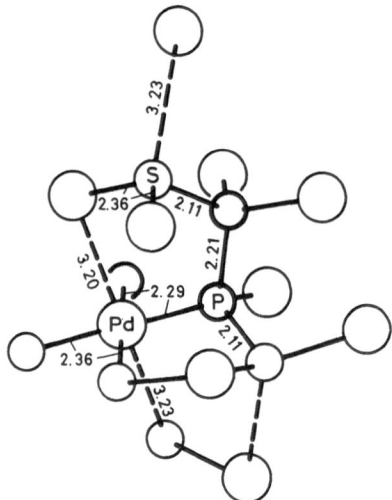

Fig. 80.   Near-neighbour environments in PdPS [8].

PdPSe. Well crystallised examples can be obtained by heating the elements together at 1000 to 1200°C and 25 to 65 kbar pressure in a sealed silica tube [6, 7]. The graphite-like crystals have orthorhombic symmetry, space group Pbcn-D_{2h}^{14} with lattice constants a = 13.569(4), b = 5.824(1), c = 5.856(1) Å, Z = 8 [6, 8]. PdPSe is isostructural with PdPS and is a diamagnetic semiconductor [8]. More recently single crystals have been shown to be n-type semiconductors, displaying weak Pauli paramagnetic behaviour consistent with the presence of delocalised electrons. Room temperature resistivity, ϱ = 70 Ω·cm, activation energy of resistivity, E_a = 0.32 eV, Hall mobility, μ = 34 cm²·V⁻¹·s⁻¹, have been reported [12]. Resistivities are $\varrho_{60\,K}$ = 4 × 10³, $\varrho_{300\,K}$ = 30, $\varrho_{400\,K}$ = 1 Ω·cm [6].

Photoelectronic measurements in aqueous solutions of I⁻/I₃⁻ indicate that PdPSe has high quantum efficiencies below 800 nm. Plots of resistivity and magnetic susceptibility as a function of temperature are reproduced [12].

PdP$_y$Se$_{2-y}$ ($0 < y < 1$). A range of pyrite type phases with variable P/Se ratio. Unit cell dimension a = 6.05 to 6.13 Å [6].

PdPS$_x$Se$_{1-x}$ (x = 0 to 1). These products are obtained by heating the requisite elements or binary compounds at ~900 to 1000°C and 20 to 65 kbar pressure. X-ray diffraction data have been indexed on the basis of orthorhombic cells. The compounds are semiconductors and have uses in electronic solid state devices [9]. The X-ray powder diffraction pattern for PdPS$_{0.67}$Se$_{0.33}$ is tabulated in [6].

References:

[1] Flanagan, T. B.; Bowerman, B. S.; Rundqvist, S.; Andersson, Y. (J. Chem. Soc. Faraday Trans. I **79** [1983] 1605/16).
[2] Andersson, Y.; Rundqvist, S.; Tellgren, R. (J. Solid State Chem. **52** [1984] 327/9).
[3] Andersson, Y.; Rundqvist, S.; Tellgren, R.; Thomas, J. O.; Flanagan, T. B. (Acta Cryst. B **37** [1981] 1965/72).
[4] Flanagan, T. B.; Biehl, G. E.; Clewley, J. D.; Rundqvist, S.; Andersson, Y. (J. Chem. Soc. Faraday Trans. I **76** [1980] 196/208).
[5] Andersson, Y.; Rundqvist, S.; Tellgren, R.; Thomas, J. O.; Flanagan, T. B. (J. Solid. State Chem. **32** [1980] 321/7).
[6] Bither, T. A.; Donohue, P. C.; Young, H. S. (J. Solid State Chem. **3** [1971] 300/7).
[7] Bither, T. A. (U.S. 3655348 [1972]; C.A. **77** [1972] No. 26335).
[8] Jeitschko, W. (Acta Cryst. B **30** [1974] 2565/72).
[9] Bither, T. A. (U.S. 3761572[1973]; C.A. **80** [1974] No. 8017).
[10] Folmer, J. C. W.; Turner, J. A.; Parkinson, B. A. (J. Solid State Chem. **68** [1987] 28/37).

[11] Folmer, J. C. W.; Turner, J. A.; Parkinson, B. A. (Inorg. Chem. **24** [1985] 4028/30).
[12] Marzik, J. V.; Kershaw, R.; Dwight, K.; Wold, A. (J. Solid State Chem. **44** [1982] 382/7).

K$_2$PdP$_2$ is obtained by heating the elements at 900°C over a period of 1 to 3 d. The crystals are orthorhombic, space group Cmcm-D$_{2h}^{17}$ with a = 6.355(1), b = 13.898(1), c = 5.900(1) Å, Z = 4, D$_{x-ray}$ = 3.142 g/cm^3 and D$_{calc}$ = 3.24 g/cm^3. The structure, which is isotypic with K$_2$PdAs$_2$, has infinite Pd–P–P zig-zag chains.

Reference:

Rózsa, S.; Schuster, H.-U. (Z. Naturforsch. **36b** [1981] 1666/7).

Eu$_{0.05}$Y$_{0.95}$Pd$_2$P$_2$. [151]Eu Mössbauer isomer shifts in the dilute compound suggest that Eu^{2+} is present.

Reference:

Sampathkumaran, E. V.; Wortmann, G.; Kaindl, G. (J. Magn. Magn. Mater. **54/57** [1986] 347/8).

Compounds APd$_2$P$_2$ (A = Ca, Sr, Y, La, Ce, Nd, Sm, Eu, Gd, Tb, Dy, Ho, Er, Yb)

These compounds are prepared by heating the respective elements in silica tubes for 3 d at 1050 K or 2 d at 1150 K, and form shiny black microcrystals after annealing at 1050 K for 10 d. Their Guinier powder patterns indicate ThCr$_2$Si$_2$ structures, cell dimensions are given in the table below [1].

Cell dimensions of ternary palladium phosphides with tetragonal $ThCr_2Si_2$-type structures [1]:

compound	a in Å	c in Å	c/a
$CaPd_2P_2$	4.147(1)	9.656(3)	2.328
	4.145(1)*	9.659(2)*	
$SrPd_2P_2$	4.241(1)	9.719(3)	2.292
	4.241(1)*	9.724(2)*	
$BaPd_2P_2$	4.256*	5.647*	
YPd_2P_2	4.053(2)	9.850(3)	2.430
$LaPd_2P_2$	4.188(1)	9.857(4)	2.354
$CePd_2P_2$	4.156(1)	9.887(3)	2.379
$PrPd_2P_2$	4.134(2)	9.879(5)	2.390
$NdPd_2P_2$	4.124(2)	9.878(5)	2.395
$SmPd_2P_2$	4.100(1)	9.866(4)	2.406
$EuPd_2P_2$	4.181(1)	9.742(4)	2.330
$GdPd_2P_2$	4.081(1)	9.858(4)	2.416
$TbPd_2P_2$	4.056(1)	9.845(3)	2.427
$DyPd_2P_2$	4.042(2)	9.844(4)	2.435
$HoPd_2P_2$	4.022(1)	9.833(3)	2.445
$ErPd_2P_2$	4.022(1)	9.841(3)	2.447
$YbPd_2P_2$	4.088(1)	9.721(2)	2.378

* Values from [2].

The standard deviations for the least significant digits are listed in parentheses. They do not reflect the possible homogeneity ranges.

APd_2P_2 (A = Ca, Sr, Ba). These are obtained as air and moisture stable steel-grey powders by heating together stoichiometric amounts of the elements under argon at 1000°C for 10 h, then twice tempering at 900°C for 15 h (A = Ca, Sr) or at 800°C for 10 h, then twice tempering at 1000°C for 15 h (A = Ba). They adopt the $ThCr_2Si_2$-type structure (A = Ca, Sr) or the $CeMg_2Si_2$-type structure (A = Ba) [2]. For cell dimensions see table above.

No crystals of APd_2P_2 compounds large enough for structure determination could be found. However, comparison with the isotypic nickel compound $EuNi_2P_2$ suggests the oxidation state formula $A^{3+} Pd^{0.5+} Pd^{0.5+}(P_2)^{4-}$ (A = lanthanide) with considerable Pd–Pd bonding [1].

$B_6Pd_6P_{17}$ (B = La or Ce). These compounds are prepared by heating the respective elements for 12 d at 1020 K, and form shiny black microcrystals. They adopt $La_6Ni_6P_{17}$-type structures with a = 10.411(1) Å (B = La) or 10.339(1) Å (B = Ce) [1].

$PdP_{0.67}S_{1.33}$. Prepared from the appropriate stoichiometric mixture of Pd, P and S at 1000 to 1200°C and 65 kbar. Forms silvery crystals of pyrite type, a = 5.844 Å, D = 5.65 g/cm³. X-ray diffraction powder data are tabulated [3].

$PdP_{0.33}S_{1.67}$. Obtained from a 1:1:3 mixture of Pd, P and S at 1000 to 1200°C and 65 kbar, silvery crystals, tetragonal cells with a = 5.63 to 5.67, c = 6.48 to 6.43 Å. Powder X-ray data are tabulated [3].

PdP_2S_7. A blue-black glassy material formed on heating Pd, P and S together in the ratio 1:1:3 at 1200°C and 65 kbar. It has D = 2.90 g/cm³ and is amorphous to X-rays. Susceptible to slow hydrolytic decomposition [3].

Pd$_3$P$_2$S$_8$[Pd$_3$(PS$_4$)$_2$]. Red purple crystals obtained by heating Pd, P and S (ratio 1:1:2.5 to 3.5) in silica tubes at 900°C and autoclave pressure or 1200°C and 3 kbar. Single crystal X-ray data indicate trigonal symmetry in space group P321, P3m1 or P$\bar{3}$m1, hexagonal cell dimensions are a = 6.836, c = 7.239 Å. Powder diffraction data, refined parameters, bond distances and bond angles are tabulated in the paper, the structure is illustrated. The compound is a true thiophosphate (3Pd^{2+} 2PS$_4^{3-}$). It is stable under argon to 730°C but oxidises in oxygen above 450°C [3]. Photochemical properties have been described [4].

References:

[1] Jeitschko, W.; Hofmann, W. K. (J. Less-Common Metals **95** [1983] 317/22).

[2] Mewis, A. (Z. Naturforsch. **39b** [1984] 713/20).

[3] Bither, T. A.; Donohue, P. C.; Young, H. S. (J. Solid State Chem. **3** [1971] 300/7).

[4] Folmer, J. C. W.; Turner, J. A.; Parkinson, B. A. (J. Solid State Chem. **68** [1987] 28/37).

13.4 Palladium Phosphite Pd(PO$_3$)$_2$

This is made by reaction of 1 mol PdCl$_2$ in 10 mol of 85% H$_3$PO$_4$ at 380°C.

The crystals are orthorhombic, space group Pcnm-D$_{2h}^7$, Z = 2; a = 4.233(3), b = 4.571(1), c = 12.472(3) Å; density 3.55 g/cm^3. The Pd has square-planar coordination with Pd–O = 1.993(3) Å, and there are chains of linked PO$_4$ units: two sets of P–O distances, two at 1.485(4) Å and two at 1.587(2) Å (see figure in paper).

Palkina, K. K.; Maksimova, S. I.; Lavrov, A. V.; Chalisova, N. A. (Dokl. Akad. Nauk SSSR **242** [1978] 829/31; C.A. **90** [1979] No. 64805).

13.5 Palladium Phosphate and Phosphato Complexes

As with palladium sulphates and sulphato complexes there is little structural information on these ill-defined species.

13.5.1 Palladium Phosphate(?)

Oxidation of alkanes by palladium(II) in H$_3$PO$_4$ with BF$_3$ has been studied.

Rudakov, E. S.; Lutsyk, A. I.; Rudakova, R. I. (Kinetika Kataliz **18** [1977] 525; Kinet. Catal. [USSR] **18** [1977] 441).

13.5.2 Phosphato Complexes

NaPdPO$_4$. This is made by reaction of PdCl$_2$ with Na$_3$PO$_4$ in melts [1].

KPdPO$_4$. This is made by reaction of PdCl$_2$ with K$_3$PO$_4$ in a melt [1], from PdCl$_2$ with K$_3$PO$_4$ at 790 to 820°C [2], or from Pd(HPO$_4$)$_2$ (assumed formula; the paper mentions it as "palladium acid phosphate" without formulation or further details) and K$_3$PO$_4$ at 500 to 600°C for 2 to 3 h [2].

The compound is cubic, a = 6.998 Å; Z = 3, measured density 3.49 g/cm^3, calculated density 3.50 g/cm^3 [2].

Thermal gravimetric analysis was carried out at the compound (curves from 40 to 1000°C reproduced in paper). The solubility of KPdPO$_4$ in a NaCl–KCl melt at 700, 750 and 800°C was determined [2].

H$_7$[Pd$_4$(PO$_4$)$_5$(NH$_3$)$_6$]. Reaction of trans-Pd(NH$_3$)$_2$(NO$_2$)$_2$ in 84% H$_3$PO$_4$ gives the green material. At higher temperatures a rose coloured material apparently having the same formula was obtained.

Infrared spectra were measured (400 to 4000 cm^{-1}, reproduced in paper) and electronic spectra (350 to 700 nm, also reproduced in paper) [3].

Na$_3$[Pd$_6$(PO$_4$)$_5$]·3H$_2$O. This is made from PdCl$_2$ in HCl and H$_3$PO$_4$.

The infrared and electronic spectra were measured, as well as thermal gravimetric analytical data [4].

K$_2$(H$_3$O)$_2$[Pd(HPO$_4$)$_3$]. This species is formed as an intermediate during the reaction of K$_2$[Pd(NO$_2$)$_4$] and H$_3$PO$_4$ at 130 to 150°C [4].

K$_2$[Pd$_6$(PO$_4$)$_4$(HPO$_4$)]·3H$_2$O. This is said to be the formula of the material obtained by reaction of [PdCl$_4$]$^{2-}$ or [Pd(NO$_2$)$_4$]$^{2-}$ with H$_3$PO$_4$. It is sparingly soluble in water and dilute acids [5].

References:

[1] Kryukova, A. I.; Korshunov, I. A.; Egorova, L. Yu. (Fiz. Khim. Elektrokhim. Rasplavl. Tverd. Elektrolitov Tezisy Dokl. 7th Vses. Konf. Fiz. Khim. Ionnykh Rasplavov Tverd. Eletrolitov, Sverdlovsk 1979, pp. 131/2; C.A. **93** [1980] No. 160357).

[2] Kryukova, A. I.; Korshunov, I. A.; Egorova, L. Yu. (Radiokhimiya **24** [1982] 374/7; Soviet Radiochem. **24** [1982] 317/20).

[3] Muraveiskaya, G. S.; Abashkin, V. E.; Estaf'eva, O. N.; Golovaneva, I. F. (Koord. Khim. **6** [1980] 284/90; C.A. **92** [1980] No. 173730).

[4] Chalisova, N. N.; Leonova, O. G.; Kochubei, D. I.; Yuz'ko, M. I. (Zh. Neorgan. Khim. **33** [1988] 409/14; C.A. **108** [1988] No. 215210).

[5] Volynets, M. P.; Ermakov, A. N.; Ginzburg, S. I.; Chalisova, N. N.; Gur'eva, R. F.; Dubrova, T. V.; Yuz'ko, M. I.; Fomina, T. A. (Zh. Analit. Khim. **32** [1977] 914/21; J. Anal. Chem. [USSR] **32** [1977] 914/21).

13.6 Palladium Pyrophosphate and Pyrophosphato Complexes

13.6.1 Palladium Pyrophosphate Pd$_2$P$_2$O$_7$(?)

This can be made from PdCl$_2$ and polyphosphoric acid reacting at 300°C for 5 h (though this formula was not ascribed to the material in the reference).

With an H$_2$–O$_2$ mixture, H$_2$O$_2$ is produced [1].

13.6.2 Pyrophosphato Complexes

Li$_2$[PdP$_2$O$_7$]. This is made by reaction of LiH$_2$PO$_4$ and PdCl$_2$ at 650°C for 50 h. It is orange [2].

The X-ray powder diffraction diagram was measured (diagrammatically reproduced in paper) and the infrared spectrum was measured (200 to 1300 cm^{-1}, reproduced in paper; ν_{PO} = 1200, 1010 and 910 cm^{-1}) [2].

Na$_2$[PdP$_2$O$_7$]. This is yellow and is made from NaH$_2$PO$_4$ and PdCl$_2$ at 650°C for 50 h [2].

The X-ray powder diffraction diagram was measured (diagrammatically reproduced in paper) and the infrared spectra (200 to 1300 cm^{-1}, reproduced in paper; v_{PO}=1180, 1010 and 920 cm^{-1}) [2].

K$_3$(H$_3$O)$_3$[Pd(P$_2$O$_7$)$_2$]·2H$_2$O. This is made by reaction of K$_2$[PdCl$_4$] or K$_2$[Pd(NO$_2$)$_4$] with H$_3$PO$_4$ and is readily soluble in water.

The electronic spectrum was measured (300 to 700 nm, reproduced in paper) as a function of H$_3$PO$_4$ concentration.

Thin layer chromatography (TLC) studies were made on the complex [3].

K$_4$(H$_3$O)[Pd$_2$(HP$_2$O$_7$)$_3$]·3H$_2$O. This is made by reaction of [PdCl$_4$]$^{2-}$ or [Pd(NO$_2$)$_4$]$^{2-}$ with H$_3$PO$_4$.

The infrared and electronic spectra were measured, as were thermal gravimetric analytical data [4].

References:

[1] Campbell, J. S.; Imperial Chemical Industries Ltd. (Brit. 1056125 [1967]; C.A. **66** [1967] No. 80021).

[2] Sokolova, I. D.; Markina, I. B.; Shaplygin, I. S. (Zh. Neorgan. Khim. **27** [1982] 302/4; Russ. J. Inorg. Chem. **27** [1982] 172/3).

[3] Volynets, M. P.; Ermakov, A. N.; Ginzburg, S. I.; Chalisova, N. N.; Gur'eva, R. F.; Dubrova, T. V.; Yuz'ko, M. I.; Fomina, T. A. (Zh. Analit. Khim. **32** [1977] 914/21; J. Anal. Chem. [USSR] **32** [1977] 914/21).

[4] Chalisova, N. N.; Leonova, O. G.; Kochubei, D. I.; Yuz'ko, M. I. (Zh. Neorgan. Khim. **33** [1988] 409/14; C.A. **108** [1988] No. 215210).

13.7 Miscellaneous Phosphato Compound PdP$_5$Cr$_{1.5}$O$_{15.8}$

This material is made from Pd(NO$_3$)$_2$, 85% H$_3$PO$_4$ and Cr(NO$_3$)$_3$, heated to 130°C for 8 h and then heated to 450°C for 4 h. It is a catalyst for the oxidation of methacrolein to methacrylic acid.

Reference:

Ogawa, M., Nippon Kayaku Co. Ltd. (Japan. Kokai 78-02416 [1978]; C.A. **88** [1978] No. 153235).

13.8 Complexes Containing Thiocyanate and Tertiary Phosphines

Pd(NCS)$_2$(PEt$_3$)$_2$. The infrared spectrum shows v_{CN} at 2089 and v_{CS} at 846 and 842 cm^{-1}, suggestive of N-bonding of the thiocyanate ligand [1].

Pd(NCS)$_2$(PPr$_3^i$)$_2$. This is made from K$_2$[Pd(SCN)$_4$] and the ligand in water-acetone [12] or from Na$_2$[Pd(SCN)$_4$] and the ligand in ethanol [2].

The infrared spectrum shows v_{CN} at 2094 [1, 13] or 2106 cm^{-1} [2], v_{CS} at 852 and 844 cm^{-1} [1, 12], 845 [13] or 848 cm^{-1} [2]; band intensities were measured [2]. In CCl$_4$ solution v_{CN} appears at 2115 cm^{-1}; its intensity was measured [12]. On the basis of these data N-bonding of the thiocyanate to the metal was proposed [1, 2].

Pd(NCS)$_2$(PBu$_3^n$)$_2$. This is made by reaction of K$_2$[Pd(SCN)$_4$] and PBu$_3^n$ in alcohol at room temperature [3, 8] or by a similar method using Na$_2$[Pd(SCN)$_4$] [2].

The infrared spectrum of the solid showed ν_{CN} at 2102 [3], 2105 [2], 2100 cm^{-1} [4] or 2110 cm^{-1} [9], with ν_{CS} at 853 [3], 847 [2, 3, 8], 846 cm^{-1} [4], and δ_{SCN} at 448 cm^{-1} [2]. A band at 352 cm^{-1} may arise from a Pd–N stretch [8], though in another report the expected band for this near 270 cm^{-1} was obscured [5]. Integrated absorption intensity data for ν_{CN} and ν_{CS} were given [2]. On the basis of the infrared results Pd–N binding to thiocyanate in the complex has been proposed [2, 3, 4, 8].

Diffuse reflectance spectra of the complex before and after heating were recorded (200 to 800 nm, reproduced in paper) [8].

The complex decomposes at 290°C to Pd(SCN)$_2$; differential thermal analysis was carried out [8].

Pd$_2$(SCN)$_4$(PBu$_3^n$)$_2$. The infrared spectrum of this shows ν_{CN} of the terminal SCN$^-$ ligand at 2120 cm^{-1} and that of the bridging SCN$^-$ ligand at 2162 cm^{-1} [9].

Pd$_2$(SCN)$_2$Cl$_2$(PBu$_3^n$)$_2$. The infrared spectrum of this shows the ν_{CN} of the bridging SCN$^-$ ligand at 2150 cm^{-1} [9].

Pd(NCS)$_2$(PPh$_3$)$_2$. This is made by reaction of K$_2$[Pd(SCN)$_4$] [3] or Na$_2$[Pd(SCN)$_4$] [2] in ethanol with a stoichiometric quantity of PPh$_3$, from K$_2$[Pd(SCN)$_4$] and PPh$_3$ in water-acetone [12].

The infrared spectra of the solid are consistent with the presence of N-bonded thiocyanate: ν_{CN} lies at 2093 [3], 2092 [12] or 2095 cm^{-1} [2, 4], ν_{CS} at 853 [2, 3, 4, 7], 852 [12] or 851 and 846 cm^{-1} [2]. Band intensities were recorded [2]. The Pd–N stretch was assigned to a band at 270 cm^{-1} [5]. The infrared spectrum in benzene showed ν_{CN} at 2087 to 2075 cm^{-1}, suggesting retention of the N-bonded coordination mode in solution [7].

Electronic spectra were measured in (CH$_3$)$_2$SO, CH$_3$CN, CHCl$_3$, CCl$_4$ and C$_6$H$_6$ [7]. The diffuse reflectance spectrum was measured (200 to 800 nm, reproduced in paper) [9].

Pd(SCN)$_2$(PPh$_3$)$_2$. This is made by reaction of [Pd(SCN)$_4$]$^{2-}$ in CH$_3$CN with PPh$_3$ at −10°C [6] or from K$_2$[Pd(SCN)$_4$] and PPh$_3$ in C$_2$H$_5$OH [9]. It is yellow [8].

The infrared spectrum of the solid before and after heating was measured from 150 to 400 cm^{-1} (spectra reproduced in paper) [8]. The infrared spectrum in (CH$_3$)$_2$SO and in CH$_3$CN gave bands at 2040 and 2050 cm^{-1}, respectively, for ν_{CN} and at 740 and 730 cm^{-1} for ν_{CS}, suggesting that displacement of coordinated SCN$^-$ had occured [6]. The Pd–S stretch was assigned to a band at 300 cm^{-1} [9]. The diffuse electronic spectrum of the complex before and after heating the solid was recorded (200 to 800 nm, reproduced in paper) [8].

Differential thermal analysis was carried out on the complex; it was said to isomerise to Pd(NCS)$_2$(PPh$_3$)$_2$ at 220°C [8].

The E$_{1/2}$ for the complex in dimethylformamide is 0.36 V; dissociation constants for the thiocyanate ligand in CH$_3$CN and CHCl$_3$ were calculated [10].

Pd(NCS)$_2$(P(C$_6$H$_{11}$)$_3$)$_2$. This is made from K$_2$[Pd(SCN)$_4$] and the ligand in water-acetone [12].

The infrared spectrum shows ν_{CN} at 2102 cm^{-1} and ν_{CS} at 845 cm^{-1}, suggesting N-binding of the thiocyanate ligands [12].

Pd(SCN)$_2$(p-CH$_3$C$_6$H$_4$PPh$_2$)$_2$. This is made from K$_2$[Pd(SCN)$_4$] and the ligand in ethanol. The E$_{1/2}$ in dimethylformamide is 0.35 V; dissociation constants for thiocyanate ion from the complex in CH$_3$CN and CHCl$_3$ were calculated [10].

Pd(SCN)₂(p-CH₃OC₆H₄PPh₂)₂. This is made from K₂[Pd(SCN)₄] and the ligand in ethanol. The E½ in dimethylformamide is 0.345 V; dissociation constants for the SCN⁻ ligands from the complex in CH₃CN and CHCl₃ were calculated [10].

Pd(SCN)₂(p-XC₆H₄Ph₂P)₂. These (X = F, Cl) are made from K₂[Pd(SCN)₄] and the ligand in ethanol [10].

The E½ values in dimethylformamide are 0.340 for both fluoro and chloro complexes; dissociation constants of the thiocyanate ligands from the complex in CH₃CN and CHCl₃ were calculated [10].

Pd(SCN)₂(Ph₂PCH₂COOH)₂. This is made by reaction of Na₂[Pd(SCN)₄] and diphenylphosphine acetic acid with excess Na(SCN). It is yellow, melting from 179 to 181°C.

The infrared spectrum shows ν_{CN} at 2140 cm⁻¹ and ν_{CS} at 665 cm⁻¹, suggesting S-bonding of the thiocyanate ligands; ν_{PdS} is at 318 cm⁻¹ and ν_{PdP} at 219 cm⁻¹. The electronic absorption spectrum (300 to 600 nm) was measured and assigned [11].

With Na₂[Pd(SCN)₄] in ethanol it gives Pd₂(SCN)₄(PPh₂PCH₂COOH)₂·2C₂H₅OH [11].

Pd₂(SCN)₄(Ph₂PCH₂COOH)₂·2C₂H₅OH. This is prepared by reaction of Pd(SCN)₂-(Ph₂PCH₂COOH)₂ with Na₂[Pd(SCN)₄] in ethanol. It is red-orange, melting at 194 to 196°C.

The infrared spectrum shows ν_{CN} at 2173 and 2138 cm⁻¹, ν_{PdS} at 299 cm⁻¹, ν_{PdN} at 260 cm⁻¹ and ν_{PdP} at 210 cm⁻¹. A structure involving terminal S-bonded thiocyanato bridges and bridging SCN ligands was proposed. The electronic absorption spectrum was measured (300 to 600 nm) and assigned [11].

trans-Pd(SCN)₂(Ph₂PC≡CBuᵗ)₂. This is made by reaction of K₂[Pd(SCN)₄] in ethanol with Ph₂PC≡CBuᵗ (3,3-dimethylbutynyl)diphenylphosphine followed by recrystallisation from cyclohexane [13].

The crystals are monoclinic, space group P2₁/c-C²ₕ⁵, Z = 2; a = 12.563(4), b = 10.460(3), c = 14.781(4) Å; β = 97.8°; measured density 1.309, calculated density 1.303 g/cm³. The structure is shown in **Fig. 81** with bond lengths and angles; both thiocyanate groups are S-bonded [13].

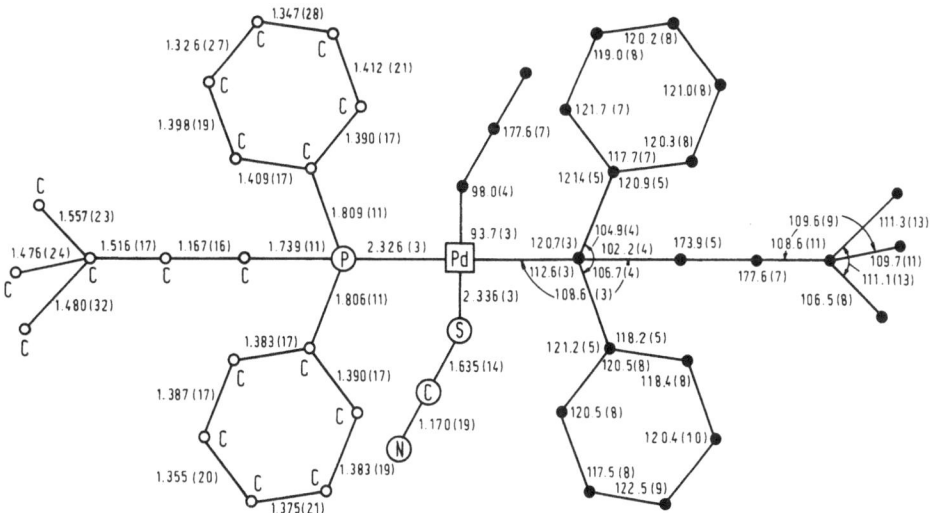

Fig. 81.   The crystal structure of trans-Pd(SCN)₂(Ph₂PC≡CBuᵗ)₂ [13].

The infrared C–N stretch is at 2116 cm^{-1}; far-infrared data were also listed [13].

cis-Pd(SCN)(NCS)(Ph$_2$PC≡CBu^t)$_2$. This is made by reaction of K$_2$[Pd(SCN)$_4$] with (3,3-dimethylbutynyl)diphenylphosphine in ethanol and exists in two isomeric forms, one orange-yellow and the other lemon-yellow. The latter form is made from the former by recrystallisation from C$_2$H$_5$OH–CH$_2$Cl$_2$ or from cis-PdCl$_2$(Ph$_2$PC≡CBu^t)$_2$ and KSCN in acetone followed by recrystallisation from C$_2$H$_5$OH–CH$_2$Cl$_2$ [13].

A brief report of the X-ray crystal structure shows the orange-yellow isomer to be disordered so that one thiocyanate group is 50% N- and 50% S-bonded; the Pd–P distances are 2.279(5) and 2.270(4) Å, Pd–S = 2.367(5) and 2.428(10) Å, Pd–N = 1.951(28) Å. The Pd–S–C angles are 105.9(0.7)° and 96.5(1.5)° [13].

The orange isomer shows ν_{CN} at 2124, 2120 and 2089 cm^{-1} for the solid, at 2122 and 2088 cm^{-1} in C$_6$H$_6$ or CH$_2$Cl$_2$, 2123 and 2094 cm^{-1} in CH$_3$CN, 2116 and 2098 cm^{-1} in CHCl$_3$, 2122, 2094 and 2086 cm^{-1} in acetone, 2121 and 2094 cm^{-1} in dimethylformamide and 2120 and 2090 cm^{-1} in CCl$_4$ solution. The yellow isomer showed ν_{CN} at 2125, 2119 and 2090 cm^{-1} for the solid; far-infrared data for both solids (150 to 500 cm^{-1}) were also listed [13].

References:

[1] Turco, A.; Pecile, C. (Nature **191** [1961] 66/7).
[2] Miezis, A. (Acta Chem. Scand. **27** [1973] 3746/60).
[3] Burmeister, J. L.; Basolo, F. (Inorg. Chem. **3** [1964] 1587/93).
[4] Sabatini, A.; Bertini, I. (Inorg. Chem. **4** [1965] 1665/7).
[5] Keller, R. N.; Johnson, N. B.; Westmoreland, L. L. (J. Am. Chem. Soc. **90** [1968] 2729/30).
[6] Sodnomov, B. G.; Polovnyak, V. K.; Troitskaya, A. D.; Rusetskii, O. I. (Koord. Khim. **6** [1980] 1061/3; Soviet J. Coord. Chem. **6** [1980] 534/5).
[7] Burmeister, J. L.; Hassel, R. L.; Phelan, R. J. (Inorg. Chem. **10** [1971] 2032/8).
[8] Sedova, G. N.; Vlasova, R. A.; Pak, V. N.; Kirillova, M. A. (Zh. Neorgan. Khim. **31** [1986] 960/3; Russ. J. Inorg. Chem. **31** [1986] 546/8).
[9] Chatt, J.; Duncanson, L. A. (Nature **178** [1956] 997/8).
[10] Tsytskytueva, L. A.; Polovnyak, V. K.; Gazizov, K. K.; Gazizova, D. M. (Zh. Neorgan. Khim. **31** [1986] 2146/8; Russ. J. Inorg. Chem. **31** [1986] 1237/8).

[11] Ruzickova, J.; Podlahova, J. (Collection Czech Chem. Commun. **43** [1978] 2853/61).
[12] Pecile, C. (Inorg. Chem. **5** [1966] 210/4).
[13] Beran, G.; Carty, A. J.; Chieh, P. C.; Patel, H. A. (J. Chem. Soc. Dalton Trans. **1973** 488/94).

13.9 Complexes Containing Thiocyanate and Chelating Phosphines

Pd(SCN)$_2$(PPh$_2$CH$_2$PPh$_2$). The X-ray crystal structure shows the crystals to be monoclinic, space group P2$_1$/n, Z = 4; a = 10.426(8), b = 29.353(10), c = 9.884(6) Å, β = 119.86(4)°; measured density 1.54, calculated density 1.536 g/cm^3. The structure is shown in **Fig. 82**. The metal has square-planar coordination with bond parameters as shown in the figure, with two S-bonded thiocyanato ligands [1].

Pd(SCN)(NCS)(PPh$_2$(CH$_2$)$_2$PPh$_2$). This is made by reaction of Pd(NO$_3$)$_2$(Ph$_2$P(CH$_2$)$_2$PPh$_2$) and NaSCN or from the ligand and [Pd(SCN)$_4$]$^{2-}$ in ethanol. It is light yellow [2].

The X-ray crystal structure showed the orthorhombic crystals to belong to the P2$_1$2$_1$2$_1$-D$_2^4$ space group; Z = 4; a = 13.773(6), b = 23.212(15), c = 8.502(4) Å; measured density 1.511, calculated density 1.517 g/cm³. The structure is shown in **Fig. 83**. There is approximate square-planar coordination about the palladium, with dimensions as shown in the figure [1, 3].

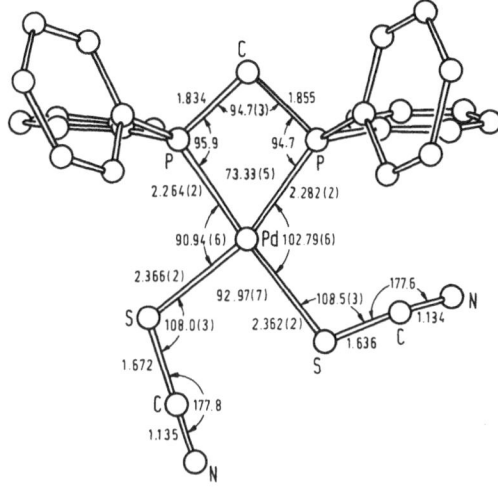

Fig. 82.   X-ray crystal structure of Pd(SCN)$_2$(Ph$_2$PCH$_2$PPh$_2$) [1].

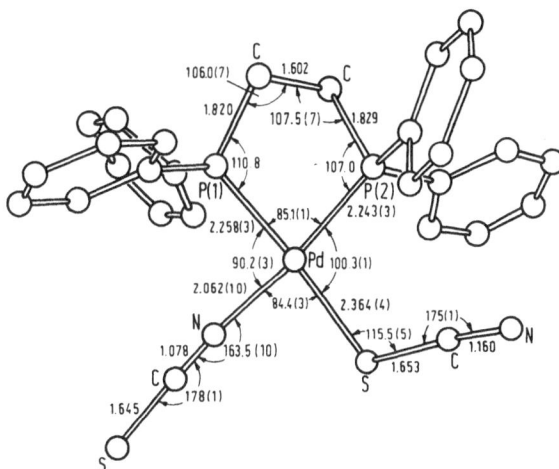

Fig. 83.   X-ray crystal structure of Pd(SCN)(NCS)(Ph$_2$P(CH$_2$)$_2$PPh$_2$) [3].

The infrared spectrum of the solid shows ν_{CN} at 2118 and 2095 cm^{-1}; in CH$_2$Cl$_2$ solution at 2120 and 2086 cm^{-1} (spectrum 2000 to 2200 cm^{-1}, reproduced in paper). Band intensities were measured for the solution data [1].

Pd(NCS)$_2$(Ph$_2$P(CH$_2$)$_3$PPh$_2$). The X-ray crystal structure shows this to be monoclinic, space group I2/a; Z = 4; a = 14.774(6), b = 9.181(5), c = 21.182(10) Å; β = 95.48(2)°; measured density 1.48, calculated density 1.475 g/cm³. The bond parameters are shown in **Fig. 84**, p. 338; both thiocyanato groups are N-bonded [1].

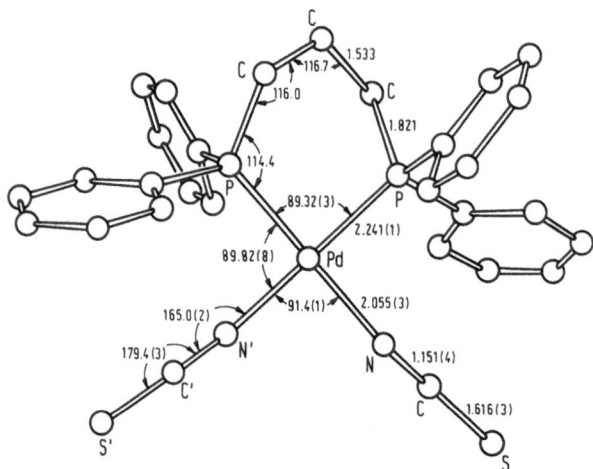

Fig. 84. X-ray crystal structure of Pd(NCS)$_2$(Ph$_2$P(CH$_2$)$_3$PPh$_2$) [1].

Pd(SCN)$_2$(Ph$_2$PCH$_2$(CF$_2$)CHPPh$_2$). This is made by reaction of [Pd(SCN)$_4$]$^{2-}$ with Ph$_2$PCCCF$_3$.

The X-ray crystal structure shows the yellow crystal to be monoclinic, space group P2$_1$/c-C$_{2h}^5$; Z = 4; a = 11.019(1), b = 17.180(2), c = 17.800(2) Å; β = 102.46(1)°, density by flotation 1.58 g/cm^{-3}, calculated density 1.580 g/cm^{-3}. The bond parameters are shown in **Fig. 85** [4].

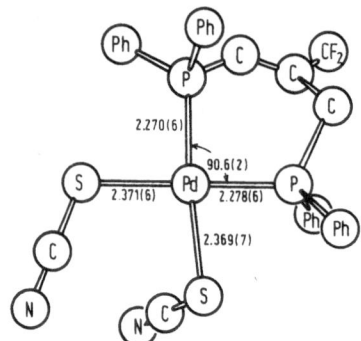

Fig. 85. X-ray crystal structure of Pd(SCN)$_2$(Ph$_2$PCH$_2$C(CF$_2$)CHPPh$_2$) [4].

Pd$_2$(SCN)$_4$(Ph$_2$PC≡CPPh$_2$)$_2$. This complex of bis(diphenylphosphino)acetylene is made by reaction of Pd$_2$Cl$_4$(Ph$_2$PC≡CPPh$_2$)$_2$ with KSCN, and forms orange crystals, melting above 300°C [5].

The infrared spectrum showed ν$_{CN}$ at 2118 and 2072 cm^{-1}; other frequencies from 200 to 2200 cm^{-1} were also listed. The infrared spectrum from 2000 to 2200 cm^{-1} was reproduced in the paper. A possible structure involving terminal S-bonded thiocyanato ligands with bridging phosphine ligands was suggested [5].

Pd(SCN)$_2$(Ph$_2$PCH=CHPPh$_2$). This is made from [Pd(SCN)$_4$]$^{2-}$ and cis-1,2-bis-diphenyl-phosphinoethylene in n-butanol at room temperature [6].

The infrared spectrum showed ν_{CN} at 2105 cm^{-1} and δ_{SCN} at 410 cm^{-1}, suggestive of S-bonding of the thiocyanato ligands. The electronic absorption spectrum of the complex was measured [6].

Pd(NCS)(SCN)(o-C$_6$H$_4$(PPh$_2$)$_2$). This is made from Na$_2$[Pd(SCN)$_4$] and the ligand o-phenylene-bis-diphenylphosphine in refluxing dimethylformamide. It is pale yellow and very insoluble. It melts at 278°C and the molar conductance in C$_6$H$_5$NO$_2$ is <0.1 Ω^{-1}·cm^2 [7].

The infrared spectrum shows ν_{CN} at 2120 and 2080 cm^{-1} for the solid and at 2115 and 2080 cm^{-1} in CH$_2$Cl$_2$ and integrated intensities were measured. The electronic absorption spectrum was reported. On the basis of the infrared data it was suggested that the complex contains an –N and an –S bonded thiocyanate ligand [7].

Pd(SCN)(NCS)(P$_2$C$_{30}$H$_{24}$). This is made by reaction of o-Ph$_2$PC$_6$H$_4$PPh$_2$ with [Pd(SCN)$_4$]$^{2-}$ in ethanol. It is yellow. It can also be made from Pd(P$_2$C$_{30}$H$_{24}$)(NO$_3$)$_2$ and NaSCN in hot dimethylformamide [1].

The infrared spectrum of the solid shows ν_{CN} at 2118 and 2095 cm^{-1}; in CH$_2$Cl$_2$ these bands are at 2121 and 2086 cm^{-1} and in CH$_3$NO$_2$ at 2121 and 2091 cm^{-1}. Band intensities were measured for the solutions. The infrared spectrum of the CH$_2$Cl$_2$ solution was reproduced in the paper (2000 to 2200 cm^{-1}), and also those of the solid over the same range [1].

trans-Pd(SCN)$_2$(C$_6$F$_5$PMe$_2$)$_2$. Brief mention has been made of this complex. The Pd–S distance is 2.351(1) Å [1].

References:

[1] Palenik, G. J.; Mathew, M.; Steffen, W. L.; Beran, G. (J. Am. Chem. Soc. **97** [1975] 1059/66).
[2] Meek, D. W.; Nicpon, P. E.; Meek, V. I. (J. Am. Chem. Soc. **92** [1970] 5351/9).
[3] Beran, G.; Palenik, G. J. (Chem. Commun. **1970** 1354/5).
[4] Simpson, A. T.; Jacobson, G.; Carty, A. J.; Mathew, M.; Palenik, G. J. (J. Chem. Soc. Chem. Commun. **1973** 388/9).
[5] Carty, A. J.; Efraty, A. (Can. J. Chem. **47** [1969] 2573/8).
[6] Chow, K. K.; McAuliffe, C. A. (Inorg. Nucl. Chem. Letters **8** [1972] 1031/3).
[7] Levason, W.; McAuliffe, C. A. (Inorg. Chim. Acta **16** [1976] 167/72).

13.10 Complexes with Thiocyanate and Phosphorus-Containing Chelates

Phosphorus-Nitrogen Donors

Pd(SCN)(NCS)(Ph$_2$P(CH$_2$)$_3$N(CH$_3$)$_2$). The X-ray crystal structure of this orange, monoclinic complex of 1-diphenylphosphine-3-dimethylaminopropane has been determined; the space group is P2$_1$/c-C$_{2h}^5$, Z = 4; a = 11.684(3), b = 12.961(4), c = 14.641(3) Å, β = 110.04(1)°; observed density 1.567, calculated density 1.578 g/cm^3 [1, 2]. The structure is shown in **Fig. 86**, p. 340. The palladium is square-planar and contains an S-bonded thiocyanate with Pd–S = 2.295(2) Å, Pd–S–CN angle = 107.3(3)°, and an N-bonded one [1, 2]; Pd–N = 2.063(2) Å, Pd–N–CS angle = 177.7(6)° [1].

The complex is a nonelectrolyte in nitromethane (molar conductance 4.2 Ω^{-1}·cm^2). The infrared spectrum of the solid has ν_{CN} at 2120 and 2080 cm^{-1} which in CH$_2$Cl$_2$ solution lie at 2125 and 2087 cm^{-1}; this suggests that the mixed mode of coordination is retained in solution [2].

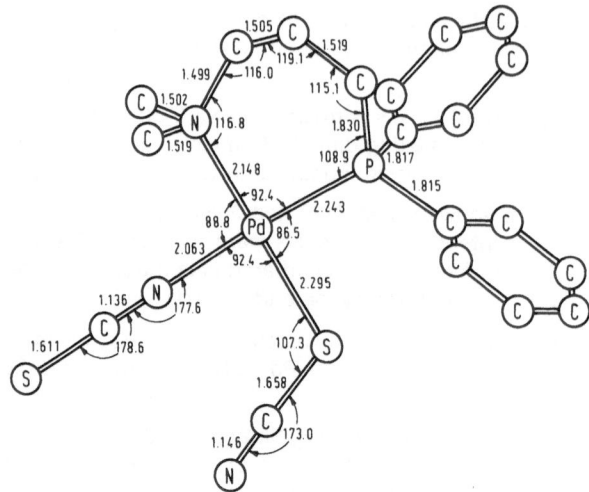

Fig. 86.   X-ray crystal structure of Pd(SCN)(NCS)(Ph$_2$P(CH$_2$)$_3$N(CH$_3$)$_2$) [1].

Pd(SCN)(NCS)(Ph$_2$P(CH$_2$)$_2$N(CH$_3$)$_2$) is made from the ligand [PdCl$_4$]$^{2-}$ and a fourfold excess of SCN$^-$ in ethanol [3].

The infrared spectrum shows ν_{CN} for the solid at 2126 and 2108 cm^{-1}; in CH$_2$Cl$_2$ at 2126 and 2085 cm^{-1}, and in CH$_3$NO$_2$ at 2125 and 2089 cm^{-1}. The intensities and half-widths of these bands were recorded. The spectra from 2000 to 2200 and 700 to 900 cm^{-1} were reproduced in the paper. A structure involving –N and –S bonded thiocyanate was proposed [3].

Phosphorus-Sulphur Donors

Pd(SCN)$_2$(PSC$_{19}$H$_{17}$). This yellow complex is made from o-Ph$_2$PC$_6$H$_4$SCH$_2$ and [Pd(SCN)$_4$]$^{2-}$ in ethanol [3].

The infrared spectrum of the solid shows ν_{CN} at 2122 and 2111 cm^{-1} (2122 and 2112 cm^{-1} in CH$_2$Cl$_2$); band intensities for the solution were measured. S-bonding of the thiocyanate ligands was deduced [3].

Pd(SCN)$_2$(Ph$_2$PC$_6$F$_4$SMe). Reaction of o-Ph$_2$PC$_6$F$_4$SMe with [Pd(SCN)$_4$]$^{2-}$ in ethanol gives this yellow complex [3].

The infrared spectrum of the solid shows ν_{CN} at 2123 and 2110 cm^{-1} (2124 and 2113 cm^{-1} in CH$_2$Cl$_2$); band intensities for these latter modes were measured [3].

Pd(SCN)$_2$(SPPh$_2$CH$_2$PPh$_2$S). A brief mention has been made of the X-ray crystal structure of this; the Pd–S distances are 2.359(1) and 2.340(1) Å [4].

References:

[1] Clark, G. R.; Palenik, G. J. (Inorg. Chem. **9** [1970] 2754/60).
[2] Clark, G. R.; Palenik, G. J.; Meek, D. W. (J. Am. Chem. Soc. **92** [1970] 1077/8).
[3] Meek, D. W.; Nicpon, P. E.; Meek, V. I. (J. Am. Chem. Soc. **92** [1970] 5351/9).
[4] Palenik, G. J.; Mathew, M.; Steffen, W. L.; Beran, G. (J. Am. Chem. Soc. **92** [1975] 1059/66).

13.11 Substituted Thiocyanato Complexes with Tertiary Phosphites

trans-Pd(SCN)$_2$[P(OMe)$_3$]$_2$. This is made by reaction of [Pd(SCN)$_4$]$^{2-}$ with trimethyl phosphite. It is pale yellow, melting point 90°C. The infrared spectrum suggested that the thiocyanate ligands were bonded through the sulphur atoms [1]. It was also made by reaction of PdCl$_2$ and KSCN in CH$_3$OH with PF$_3$ (the expected product, Pd(SCN)$_2$(PF$_3$)$_2$, was not obtained). In this case a mixture of isomers (–N and –S bonded) was formed, both having a trans geometry. On the basis of infrared evidence the –S bonded complex is first formed (v_{CN} = 2131 cm^{-1} in the infrared) isomerising in solution to the N-bonded form (v_{CN} = 2099 and 2092 cm^{-1} in dimethylformamide). The integrated intensities of the bonds of the latter were measured [1].

cis-Pd(NCS)$_2$(P(OMe)$_3$)$_2$. For this the infrared spectra show v_{CN} at 2125 and 2088 cm^{-1} [9].

Pd(SCN)(NCS)[P(OMe)$_3$]$_2$. For this infrared spectra show v_{CN} at 2103 and 2082 cm^{-1} [9].

cis-Pd(NCS)$_2$[P(OEt)$_3$]$_2$. This is made by reaction of [Pd(SCN)$_4$]$^{2-}$ and P(OEt)$_3$ at room temperature [3, 4] or from [Pd(SCN)$_4$]$^{2-}$ and P(OEt)$_3$ in CH$_3$CN at −10°C [5]. The complex is likely to contain cis thiocyanate ligands bound via sulphur [4]. It was also identified as an intermediate in the [Pd(POEt)$_3$)$_4$]$^{2+}$–NCS$^-$ reaction by infrared spectroscopy [5].

The infrared spectrum of this in CH$_3$CN solution (v_{CN} = 2115, 2080, 2055 cm^{-1}, v_{CS} = 780 cm^{-1}) suggested the predominance of an N-bonded isomer in this solution. ^{1}H NMR data were also recorded [5].

There is evidence from infrared data (2100 to 2000 cm^{-1}, reproduced in paper) that the N-bonded Pd(NCS)$_2$[P(OEt)$_3$]$_2$ forms, in benzene solution and on standing, a thiocyanato-bridged dimer [Pd(SCN)$_2$[P(OEt)$_3$]]$_2$ [6].

[Pd(SCN)(P(OEt$_2$)O[P(OEt)$_2$OH)]$_2$. This is made from K$_2$[Pd(SCN)$_4$] in C$_2$H$_5$OH with P(OEt)$_3$, and can also be made from K$_2$[Pd(SCN)$_4$] and P(OEt)$_2$(OH) [4]. With thiourea (tu) it gives Pd(NCS)(tu)$_2$P(OEt)$_2$OH [8].

[Pd(NCS)(tu)$_3$]P(OEt)$_2$OH. This is made by reaction of thiourea (tu) with [Pd(SCN)P(OEt)$_2$O-[P(OEt)$_2$OH]$_2$ [8].

Pd(NCS)$_2$[PPh(OEt)$_2$]$_2$. This is made by reaction of [Pd(SCN)$_4$]$^{2-}$ with PPh(OEt)$_2$ in CH$_3$CN at −10°C. The infrared spectrum in CH$_3$CN (v_{CN} = 2115, 2080, 2055 cm^{-1}; v_{CS} = 750 cm^{-1}) suggests that the N-bonded isomer predominates in this solvent. ^{1}H NMR data were also recorded [5].

Pd(SCN)$_2$[P(OEt)$_3$]$_2$. This species is apparently formed when solutions of Pd(NCS)$_2$-[P(OEt)$_3$]$_2$ are allowed to stand; at a high P(OEt)$_3$ concentration the formation of this dimer is inhibited. There is thought to be an SCN bridge in the complex [7].

Pd(SCN)$_2$[P(OPh)$_3$]$_2$. This is made by reaction of [Pd(SCN)$_4$]$^{2-}$ with P(OPh)$_3$ in CH$_3$CN at −10°C [5], or from PdCl$_2$(P(OPh$_3$)$_3$)$_2$ and SCN$^-$ [9].

The X-ray crystal structure shows the crystals to be monoclinic, space group P2$_1$/c-C$^5_{2h}$, Z = 4; a = 9.922(6), b = 10.096(8), c = 19.334(14) Å; β = 108.46(5)°, measured density 1.54, calculated density 1.534 g/cm^3. The Pd–P distance is 2.312(1) Å and the Pd–S distance to the S-bonded thiocyanato groups is 2.352(2) Å [9].

The infrared spectrum shows v_{CN} at 2117 cm^{-1}; in CH$_3$CN solution it is likely that the S-bonded isomer predominates; v_{CN} = 2110, 2080, 2045 cm^{-1}; v_{CS} = 745 cm^{-1}. ^{1}H NMR data were also recorded [5].

342

References:

[1] Troitskaya, A. D.; Orlova, I. A. (Zh. Obshch. Khim. **45** [1975] 951; J. Gen. Chem. [USSR] **45** [1975] 938).

[2] Burmeister, S. L.; Gysling, H. J. (Inorg. Chim. Acta **1** [1967] 100/4).

[3] Gogolyukhina, L. F.; Troitskaya, A. D.; Levshina, G. A. (Zh. Obshch. Khim. **44** [1974] 223; J. Gen. Chem. [USSR] **44** [1974] 214/5).

[4] Gogolyukhina, L. F.; Levshina, G. A.; Troitskaya, A. D. (Tr. Kaz. Khim. Tekhnol. Inst. **52** [1973] 22/7; C.A. **80** [1974] No. 103320).

[5] Sodnomov, B. G.; Polovnyak, V. K.; Troitskaya, A. D.; Rusetskii, O. I. (Koord. Khim. **6** [1980] 1061/3; Soviet J. Coord. Chem. **6** [1980] 534/5).

[6] Troitskaya, A. D.; Sentemov, V. V.; Gogolyukhina, L. F.; Antropova, E. I. (Zh. Obshch. Khim. **46** [1976] 1094/5; J. Gen. Chem. [USSR] **46** [1976] 1086/7).

[7] Troitskaya, A. D.; Sentemov, V. V.; Sadakova, G. P.; Alparova, M. V. (Zh. Obshch. Khim. **50** [1980] 706/7).

[8] Gogolyukhina, L. F.; Levshina, G. A.; Troitskaya, A. D. (Tr. Kaz. Khim. Tekhnol. Inst. No. 52 [1973] 28/31; C.A. **80** [1974] No. 103322).

[9] Jacobson, S.; Wong, Y. S.; Chieh, P. C.; Carty, A. J. (J. Chem. Soc. Chem. Commun. **1974** 520/1).

14 Palladium and Arsenic

General References:

Ward, R., Structural Chemistry of Condensed Systems of Transition Metals, MTP [Med. Tech. Publ. Co.] Intern. Rev. Sci. Inorg. Chem. Ser. One **5** [1972] 93/114.

Suchet, J. P., Crystal Chemistry and Semiconduction in Transition-Metal Binary Compounds, Academic, New York 1971.

Hulliger, F., Crystals Chemistry of the Chalcogenides and Pnictides of the Transition Elements, Struct. Bonding [Berlin] **4** [1968] 83/229.

14.1 The Pd–As System

Phase diagrams for the Pd/As systems have been constructed from thermal and X-ray data and from thermal analysis results [1, 2, 4]. There is considerable uncertainty as to the composition, melting points etc. of several phases and eutectics, see below.

A phase formed peritectically at 770°C and tentatively designated as Pd_7As [1] was subsequently identified from single-crystal X-ray data as Pd_5As [2]. However, a later report mentions formation of palladium arsenides including Pd_7As during TEM analysis of Pd/amorphous C specimens on gold grids containing arsenic impurities [3].

Seven intermediate phases were found in the Pd–As system, see **Fig. 87**. As the exact position of the liquidus curve in the region of Pd_5As is uncertain. This phase is not shown in the diagram. The compound PdAs does not exist [2].

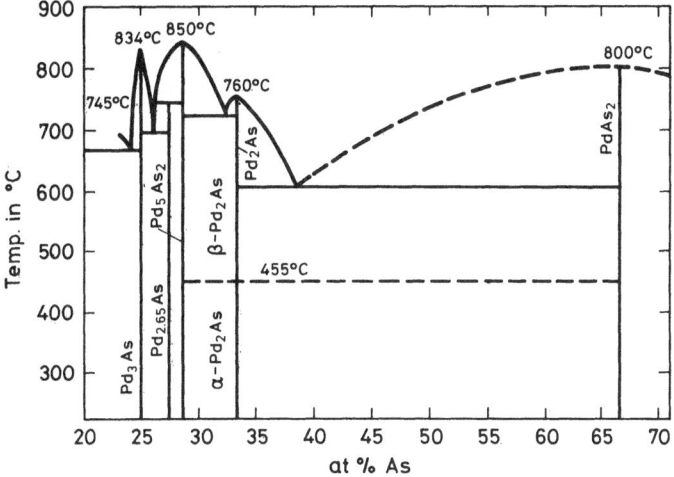

Fig. 87. Pd–As partial diagram.

The existence of the following phases has been claimed:

Pd_7As(?)	Pd_5As(?)	Pd_3As	$Pd_{2.65}As$	Pd_5As_2	$Pd_2As\,(\alpha,\beta)$	$PdAs_2$
m.p. 770°C [1] [2]		c 834°C [2] p 820°C [1]	p 745°C [2]	c 850°C [2] c 860°C [1]	c 760°C, tr 455°C [2] p 728°C, tr 485°C [1]	c 800°C [2]

m.p. = melting point, c = congruent, p = peritectic, tr = transition.

344

A eutectic at 20 at% As, melting at 715°C, was claimed by [1].

The vaporisation thermodynamics of $PdAs_x$ (liquid) (x = 1 to 0.2) have been studied by mass spectrometry combined with Knudsen effusion to determine the variation of arsenic and palladium activity with composition and temperature. Composition variations of the palladium and arsenic activities at 1600 K and of the partial enthalpies of vaporisation between 1400 and 1700 K in the Pd–As system are plotted in the paper. The thermodynamic properties of the Pd/As system are virtually unchanged by added carbon or tungsten but display a positive deviation from ideality on addition of boron. Data are presented in tabular and graphical form [4].

References:

[1] Raub, C. J.; Webb, G. W. (J. Less-Common Metals **5** [1963] 271/7).
[2] Saini, G. S.; Calvert, L. D.; Heyding, R. D.; Taylor, J. B. (Can. J. Chem. **42** [1964] 620/9).
[3] Smith, B. C. (Conf. Ser. Inst. Phys. [London] No. 78 [1986] 147/50; C.A. **104** [1986] No. 234 849).
[4] Storms, E. K.; Szklarz, E. G. (J. Less-Common Metals **136** [1987] 61/73).

14.2 Binary Compounds

Pd_5As. Well formed plate-like crystals have been obtained by chloride vapour transport using a charge of composition $Pd_{2.7}As$ and a temperature gradient of 500 to 650°C for 6 d. The Pd_5As structure is c-face-centred monoclinic with $a = 5.51_4$, $b = 7.72_5$, $c = 8.42_7 \pm 0.01$ Å, $\beta = 99° \pm 5'$, $Z = 4$; $D_{exp} = 11.33$, $D_{calc} = 11.37$ g/cm³ [2]. A more recent report describes the Pd_5As structure as monoclinic, space group $C2-C_2^3$ with $a = 5.520(1)$, $b = 7.739(1)$, $c = 8.426(1)$ Å, $\beta = 98.99(1)°$. X-ray data are tabulated and the structure is illustrated in the paper [4].

Pd_3As occurs in nature as the mineral arsenopalladinite [5]. It has been prepared by heating the elements at 700°C in a vycor tube [6]. Formation of Pd_3As in black powder form has been achieved by dropwise addition of $PdCl_2(CH_3CN)_2$ in toluene solution to molten potassium and arsenic at 200°C [3]. A body-centred tetragonal Fe_3P type structure (space group $I\overline{4}-S_4^2$ [2,6]) has been reported with $a = 9.98_6$, $c = 4.83_6$ Å [7], or $a = 9.95_1$, $b = 4.81_7 \pm 0.01$ Å [6]. A more recent completely indexed powder pattern (data tabulated in paper) gave $a = 9.974 \pm 0.002$, $c = 4.822 \pm 0.002$ Å, $c/a = 0.4834$, $Z = 8$; $D_{calc} = 10.91$, $d_{obs} = 10.88$ g/cm³ [2]. Pd_3As does not superconduct above 0.3 K [1].

$Pd_8As_3(Pd_{2.65}As)$. This phase has been detected in Debye-Scherrer patterns of well-annealed Pd/As alloys containing 27 to 28 at% As. A stoichiometric mixture of Pd and As (27.4 at% As) gave a pure single phase. The powder pattern has been recorded but not indexed, attempts to grow single crystals by vapour transport failed [2]. More recently a new mineral, stillwaterite, with the composition Pd_8As_3 has been described. The crystals are trigonal, space group $P\overline{3}-C_{3i}^1$ or $P3-C_3^1$, with $a = 7.399(4)$, $c = 10.311(15)$ Å. X-ray powder data for synthetic Pd_8As_3 and stillwaterite are tabulated in the paper [8].

$Pd_5As_2(Pd_{2.5}As)$. Polycrystalline samples have been prepared by heating together stoichiometric amounts of palladium and arsenic in a sealed tube, then annealing the crushed bead in vacuo at 600°C. Single-crystal fragments have been separated from an ingot obtained by slow directional cooling of a stoichiometric melt.

Single-crystal X-ray diffraction data establish that Pd_5As_2 displays polytypism with $a = 7.31 \pm 0.01$ Å, $c = 13.7$, 10.34, 27.48 and 96.2 Å, corresponding to a single-layer repeat distance of 3.43 Å on the c-axis. The first cell has hexagonal symmetry, space group $P6_322\text{-}D_6^6$, the last three are all trigonal, space group $P\bar{3}m1\text{-}D_{3d}^3$, $P3m1\text{-}C_{3v}^1$, or $P321\text{-}D_3^2$. The Guinier pattern of polycrystalline Pd_5As_2 can be indexed on a hexagonal cell with a = 7.320 ± 0.002, c = 96 Å. The powder diffraction pattern with the hexagonal axes a = 7.320, c = 27.523 Å, Z = 12, D_{exp} = 10.64, D_{calc} = 10.72 g/cm^3 was compared with those of other M_5X_2 compounds [9]. The presence of the phase has been confirmed during X-ray diffraction studies on Pd–Sn–As and Pd–Pb–As phase diagrams at 500°C [10]. Pd_5As_2 has been detected in zechstein rocks of the Fore-Sudetic monocline [11], in exocontact rocks of the Talnakh intrusive rock [12], and in Oktyabr'skoe deposits [13]. An unnamed mineral Pd_5As_2 with orthorhombic symmetry and lattice constants a = 11.261(4), b = 3.857(1), c = 11.346(5) Å has been described, X-ray powder data are tabulated in the paper [8]. A nickel containing phase $(Pd,Ni)_5As_2$ has also been identified in mineral deposits [13]. Pd_5As_2 is a superconductor below 0.46 K [1].

Pd_2As (see "Palladium" 1942, p. 306). Low temperature, $\alpha\text{-}Pd_2As$, and high temperature, $\beta\text{-}Pd_2As$, forms have been described, the transition temperature is given as 455°C [2], see also phase diagram, p. 343.

The low temperature $\alpha\text{-}Pd_2As$ is obtained by annealing a specimen of stoichiometric composition at 450°C. Guinier X-ray photographs indexed by Ito's method and subsequently reduced to a monoclinic cell gave a = 9.24, b = 8.47, c = 10.45 (all ± 0.01) Å, β = 94°, diffraction symbol 12/m1P, Z = 18; D_{calc} = 10.54, D_{exp} = 10.13 g/cm^3 [2]. $\alpha\text{-}Pd_2As$ is a superconductor below 0.6 K [1].

A sample prepared by heating $Pd_{66.6}As_{33.4}$ for 4 d at 700°C, then annealing at 440°C was indexed on the basis of an orthorhombic unit cell, space group $C_{2v}^{12}\text{-}Cmc2_1$, with a = 3.2$_{45}$, b = 16.8$_{44}$, c = 6.5$_{76}$ Å, Z = 8. The structure is reproduced in the paper, powder and Weissenberg data are tabulated [14]. $\alpha\text{-}Pd_2As$ has been found in exocontact rocks [12] and vein disseminated pentlandite-chalcopyrite ores of Talnakh intrusive rock. Samples are nonmagnetic with a metallic lustre, electron microprobe analysis indicates the presence of Ag (3.23 wt%) and Au (1.38 wt%) isomorphously replacing palladium. The crystals have monoclinic symmetry, space group $C_{2h}^1\text{-}P2/m$ with a = 9.25, b = 8.47, c = 10.44 Å, β = 94° [16].

The high temperature $\beta\text{-}Pd_2As$ has a C22(Fe$_2$P) structure, space group $P\bar{6}2m\text{-}D_{3h}^3$ with a = 6.650(3), c = 3.583(3) Å, Z = 3; D_{calc} = 10.44, D_{exp} = 10.33 g/cm^3. X-ray powder data are tabulated in the paper. Cell parameters determined from single crystal photographs are a = 6.64 ± 0.015, c = 3.58 ± 0.01 Å [2]. Another source gives a = 6.62, c = 3.60 Å [15]. Superconductivity occurs below 1.70 K [1].

Phase diagrams for Pd–Sn–As and Pd–Pb–As systems at 500°C have been investigated by X-ray diffraction, Pd_2As dissolves almost no Pb and ≦ 20% Sn [10]. The palladium in $\beta\text{-}Pd_2As$ is easily substituted by Ni up to the composition PdNiAs [17].

Pd_3As_2. This phase has been observed during electron difraction studies on the Pd/As system, it has orthorhombic symmetry, possible space groups Ammm, A222, and A2mm, with a = 6.58, b = 3.25, c = 17.28 Å [18]. The presence of Pd_3As_2 in Zechstein rocks of the Fore-Sudetic monocline has been detected by microprobe analysis [11].

References:

[1] Raub, C. J.; Webb, G. W. (J. Less-Common Metals **5** [1963] 271/7).
[2] Saini, G. S.; Calvert, L. D.; Heyding, R. D.; Taylor, J. B. (Can. J. Chem. **42** [1964] 620/9).
[3] Carturan, G.; Cocco, G.; Semenato, D. (React. Solids **1** [1985] 31/42).
[4] Matković, T.; Schubert, K. (J. Less-Common Metals **58** [1978] P1/P6).
[5] Claringbull, F.; Hey, M. H. (Mineral Soc. Notice No. 94 [1954] 1004).
[6] Heyding, R. D.; Calvert, L. D. (Can. J. Chem. **39** [1961] 955/7).

[7] Schubert, K.; Bhan, S.; Burkhardt, W.; Gohle, R.; Meissner, H. G.; Pötzschke, M.; Stolz, E. (Naturw. Rundschau **13** [1960] 303).

[8] Cabri, L. J.; LaFlamme, J. H. G.; Stewart, J. M.; Rowland, J. F.; Chen, T. T. (Can. Mineralogist **13** [1975] 321/35).

[9] Saini, G. S.; Calvert, L. D.; Taylor, J. B. (Can. J. Chem. **42** [1964] 1511/7).

[10] El-Boragy, M.; Issa, M. A.; El-Hiti, A. S. (Dirasat Maj. Kulliyat Al-Tarbiyah Jami'at Al-Riyad No. 2 [1978] 81/90; C.A. **91** [1979] No. 179634).

[11] Kucha, H. (Mineral. Pol. **6** [1975] 87/92; C.A. **87** [1977] No. 55893).

[12] Begizov, V. D.; Sluzhenikin, S. F. (Tr. Tsentr. Nauchno-Issled. Geologorazved. Inst. Tsvetn. Blagorodn. Met. **122** [1976] 93/7, 112/6; C.A. **89** [1978] No. 27766).

[13] Evstigneeva, T. L.; Genkin, A. D. (Geokhim. Mineral. **1980** 114/20; C.A. **94** [1981] No. 68758).

[14] Bälz, U.; Schubert, K. (J. Less-Common Metals **19** [1969] 300/4).

[15] Schubert, K.; Frank, K.; Gohle, R.; Maldonado, A.; Meissner, H. G.; Raman, A.; Rossteutscher, W. (Naturwissenschaften **50** [1963] 41/2).

[16] Begizov, V. C.; Meshchankina, V. I.; Dubakina, L. S. (Zap. Vses. Mineral. Obshchestva **103** [1974] 104/7; C.A. **81** [1974] No. 80262).

[17] El-Boragy, M.; Ellner, M.; Schubert, K. (Z. Metallk. **75** [1984] 82/5).

[18] Nedorubova, N. G.; Khar'kin, V. S. (Uch. Zap. Volgograd. Gos. Pedagog. Inst. No. 29 [1970] 19/26; C.A. **76** [1972] No. 38458).

PdAs(?). A phase of this composition formed by dissolving palladium in $PdAs_2$ has been shown by an X-ray powder photograph to have the same lattice constant (5.982 ± 0.002 Å) as the parent $PdAs_2$ [1]. A later paper concludes that the compound PdAs does not exist [2]. However, more recently a microprobe analysis of palladium minerals in Zechstein rocks has produced evidence for the presence of PdAs [3].

$PdAs_2$ (see "Palladium" 1942, p. 306). This compound has been obtained by heating stoichiometric mixtures of the elements in sealed tubes at 700°C for 24 h [2,4] or at 650°C for 30 d [5], then annealing the product at temperatures up to 850°C [2]. Debye-Scherrer patterns confirm a cubic C2 (pyrite) structure, space group T_h^6-Pa3 with a = 5.983 ± 0.001 Å [4]. Another report gives a = 5.98 Å [6]. A single crystal diffraction study gave a = 5.98_7 Å, Z = 4; D_{calc} = 7.93, D_{X-ray} = 7.62 g/cm³ [2]. A later determination gave a = 5.9855 ± 0.0005 Å with Pd–As = 2.498 ± 0.002 Å and As–As = $2.42_6 \pm 0.02$ Å [5]. Finally an electron diffraction study gave a = 5.98 Å [7].

$PdAs_2$ is metallic in behaviour [8,9], plots of resistivity and magnetic susceptibility versus temperature (75 to 600 K) are reproduced [9]. Values for $\chi_m \cdot 10^3$ at room temperature and liquid N_2 temperature are -45 and -50 cgs/mol¹, respectively [8]. Substitutional solid solution limits have been determined for group VIII transition metal atoms in $PdAs_2$ pyrite phase [9].

The standard entropy [10] and heat capacity [11] of $PdAs_2$ at 298 K have been calculated.

The presence of $PdAs_2$ in Zechstein rocks has been detected by microprobe analysis [3]. Several papers report on the formation of $PdAs_2$ during interfacial reactions between palladium metal and gallium arsenide [12 to 15]. The mutual solubility of $PdAs_2$ with other pyrite-type di-pnictides of palladium, platinum and gold has been investigated [16].

References:

[1] Geller, S.; Matthias, B. T. (J. Phys. Chem. Solids **4** [1958] 156/7).

[2] Saini, G. S.; Calvert, L. D.; Heyding, R. D.; Taylor, J. B. (Can. J. Chem. **42** [1964] 620/9).

[3] Kucha, H. (Mineral. Pol. **6** [1975] 87/92; C.A. **87** [1977] No. 55893).

[4] Heyding, R. D.; Calvert, L. D. (Can. J. Chem. **39** [1961] 955/7).

[5] Furuseth, S.; Selte, K.; Kjekshus, A. (Acta Chem. Scand. **19** [1965] 735/41).

[6] Raub, C. J.; Webb, G. W. (J. Less-Common Metals **5** [1963] 271/7).

[7] Nedorubova, N. G.; Khar'kin, V. S. (Uch. Zap. Volgograd. Gos. Pedagog. Inst. No. 29 [1970] 19/26; C.A. **76** [1972] No. 38458).

[8] Hulliger, F. (Nature **200** [1963] 1064/5).

[9] Bennett, S. L.; Heyding, R. D. (Can. J. Chem. **44** [1966] 3017/30).

[10] Rtskhiladze, V. G.; Tsagareishvili, D. Sh.; Agladze, I. I.; Rtskhiladze, D. Sh. (Soobshch. Akad. Nauk Gruz. SSR **102** [1981] 657/60; C.A. **95** [1981] No. 193263).

[11] Rtskhiladze, V. G.; Tsacareishvili, D. Sh.; Agladze, I. I.; Rtskhiladze, D. Sh. (Soobshch. Akad. Nauk Gruz. SSR **103** [1981] 129/31; C.A. **96** [1982] No. 12231).

[12] Romanova, I. D.; Maksimova, N. K.; Potakhova, L. Yu.; Yakubenya, M. P.; Yanovskii V. P. (Poverkhnost No. **1** [1984] 106/9; C.A. **100** [1984] No. 112949).

[13] Zeng, X. F.; Chung D. D. L. (J. Vac. Sci. Technol. **21** [1982] 611/4).

[14] Oustry, A.; Caumont, M.; Escaut, A.; Martinez, A.; Toprasertpong, B. (Thin Solid Films **79** [1981] 241/6).

[15] Olowolafe, J. O.; Ho, P. S.; Hovel, H. J.; Lewis, J. E.; Woodall, J. M. (J. Appl. Phys. **50** [1979] 955/62).

[16] Furuseth, S.; Selte, K.; Kjekshus A. (Acta. Chem. Scand. **21** [1967] 527/36).

14.3 Ternary Compounds

PdAsS. This metallic cobaltite-type phase prepared by an unspecified method has a cell constant a = 5.949 Å [1].

Pd_4As_2S has been shown by microprobe analysis to be present in zechstein rocks [2].

PdAsSe. This metallic cobaltite-type phase, prepared by an unspecified method, has a cell constant a = 6.092 Å [1]. Values of $\psi_M \cdot 10^6$ (c.g.s.) are +14 and −5 at 80 and 295 K, respectively [3]. Conductivity is normal down to 1.2 K [4].

$Pd_{70}As_8Te_{22}$. Guinier powder potographs give lattice constants a = 7.55_5, c = 13.88_4 Å.

A phase diagram for the Pd–As–Te system is reproduced in the paper [5].

K_2PdAs_2 is prepared by heating stoichiometric amounts of the elements together at 900°C under argon. It forms air and moisture sensitive, copper-coloured crystals. X-ray investigations show the crystals to be orthorhombic, space group Cmcm-D_{2h}^{17} with a = 6.536(2), b = 14.121(5), c = 6.025(4) Å, Z = 4; D_{x-ray} = 4.04, D_{calc} = 4.02 g/cm³. The structure contains infinite Pd–As_2 zigzag ribbons [6].

APd_2As_2 (A = Ca, Sr, Ba). These compounds are obtained as steel-grey powders by heating together stoichiometric amounts of the elements at 1000°C for 10 h, then twice tempering at 900°C for 15 h (A = Ca, Sr), or at 800°C for 10 h, then twice tempering at 1000°C for 15 h (A = Ba). They crystallise in a $CeMg_2Si_2$-type structure with a = 4.299(1), c = 10.102(2) Å (A = Ca); a = 4.383(1), c = 10.179(2) Å (A = Sr) and a = 4.346, c = 5.758 Å (A = Ba). Hand-picked crystals of $BaPd_2As_2$ display a second tetragonal structure, space group I4/mmm-D_{4h}^{17}, with a = 4.487(1), c = 20.635(2) Å. Both $BaPd_2As_2$ structures are illustrated and interatomic distances for all four APd_2As_2 structures are tabulated in the paper [7].

References:

[1] Hulliger, F. (Nature **198** [1963] 382/3).
[2] Kucha, H. (Mineral. Pol. **6** [1975] 87/9).
[3] Hulliger, F. (Helv. Phys. Acta **35** [1962] 535/7).
[4] Hulliger, F.; Müller, J. (Phys. Letters **5** [1963] 226/7).
[5] El-Boragy, M.; Schubert, K. (Z. Metallk. **62** [1971] 314/23).
[6] Schuster, H.-U.; Rozsa, S. (Z. Naturforsch. **34b** [1979] 1167/8).
[7] Mewis, A. (Z. Naturforsch. **39b** [1984] 713/20).

14.4 Complexes Containing Thiocyanate and Tertiary Arsines

Pd(SCN)$_2$(AsPr$_3^n$)$_2$. The ν_{CN} frequency is at 2115 cm^{-1} [1].

Pd$_2$(SCN)$_4$(AsPr$_3^n$)$_2$. The infrared spectrum shows the ν_{CN} of the terminal SCN ligand at 2120 cm^{-1} and that of the bridging ligand at 2154 cm^{-1} [1].

Pd$_2$(SCN)$_2$Cl$_2$(AsPr$_3^n$)$_2$. The infrared spectrum shows the ν_{CN} of the bridging SCN$^-$ ligand at 2154 cm^{-1} [1].

Pd(SCN)$_2$(AsBu$_3^n$)$_2$. This is made by reaction of K$_2$[Pd(SCN)$_4$] in aqueous ethanol with the arsine [2].

The infrared spectrum shows ν_{CN} at 2115 cm^{-1} [2], ν_{CS} at 2113 cm^{-1} [1] and δ_{NCS} at 452 and 422 cm^{-1}; the infrared spectra in the ν_{CS} region for the solid and the melt from 1900 to 2400 cm^{-1} were reproduced in the paper. On fusion of the complex it appears, from the infrared data in this region, that partial isomerisation to the N-bonded isomer occurs [2].

Pd(NCS)$_2$(AsBu$_3^n$)$_2$. This is apparently formed by fusion of Pd(SCN)$_2$(AsBu$_3^n$)$_2$ [2] or by reaction of K$_2$[Pd(SCN)$_4$] with the arsine in aqueous ethanol [3].

The infrared spectrum of the solid showed ν_{CN} at 2111 cm^{-1} [3], 2115 cm^{-1} [2] or 2113 cm^{-1} [1, 16], ν_{CS} at 844 [3] or 846 cm^{-1} [2]; the δ_{NCS} band is at 452 and 422 cm^{-1} [2].

Pd$_2$(SCN)$_4$(AsBu$_3^n$)$_2$. The infrared spectrum shows ν_{CN} of the terminal SCN$^-$ ligand at 2120 cm^{-1} and that of the bridging SCN$^-$ ligand at 2153 cm^{-1} [1].

Pd$_2$(SCN)$_2$Cl$_2$(AsBu$_3^n$)$_2$. The infrared spectrum shows ν_{CN} of the bridging ligand at 2154 cm^{-1} [1].

Pd(SCN)$_2$(AsPh$_3$)$_2$. This orange-yellow material, melting at 195°C after a colour change on heating to bright yellow at 130°C, is made from K$_2$[Pd(SCN)$_4$] and AsPh$_3$ in a C$_2$H$_5$OH–H$_2$O–(C$_2$H$_5$)$_2$O mixture [2, 3, 4] or from K$_2$[Pd(SCN)$_4$] and the ligand in ethanol [8, 13]. A similar method but using Na$_2$[Pd(SCN)$_4$] has also been described [5].

The molar conductance in dimethylformamide is 13.3 $\Omega^{-1} \cdot$ cm^2 [3, 4].

The infrared spectrum of the solid shows ν_{CN} at 2120 [2] or 2119 cm^{-1} [4, 5]; ν_{CS} at 834 cm^{-1} [2] and δ_{SCN} at 417 [4, 5] or 418 cm^{-1} [2]. The Pd–S stretch is at 311 [6] or at 306 cm^{-1} [16]. Infrared spectra of solutions of Pd(SCN)$_2$(AsPh$_3$)$_2$ and Pd(NCS)$_2$(AsPh$_3$)$_2$ in benzene were identical (ν_{CN} = 2160, 2120, 2090 cm^{-1}) suggesting isomerisation and the presence of bridging thiocyanate [3]. The infrared spectrum of the complex in benzene with and without added Ph$_3$As was reproduced in the paper (2000 to 2200 cm^{-1}). In dimethylformamide and in $(CH_3)_2SO$ solutions it appears that the thiocyanate ligand retains its S-bonded mode in the complex ($\nu_{CN} = 2119$ cm^{-1}; spectra from 2000 to 2200 cm^{-1} of the complex in benzene with and without added KSCN reproduced in paper). In benzene, CCl_4, cyclopentanone, cyclo-hexanone, nitrobenzene and 2-butanone the infrared spectra suggested the presence of complex equilibria involving –N, –S and bridged thiocyanato isomers. The electronic spectrum (300 to 400 nm, reproduced in paper) was measured of the complex in CCl_4 with and without added KSCN [7]. The effect of solvent on the nature of the bonding in $Pd(SCN)_2(AsPh_3)_2$ has been discussed [12].

The electronic spectrum (300 to 550 nm) of solid $Pd(SCN)_2(AsPh_3)_2$ differed from that of $Pd(NCS)_2(AsPh_3)_2$; in $CHCl_3$ the solution spectra were identical [3]. The electronic spectrum in $(CH_3)_2SO$ was also recorded, as was that in CCl_4 with and without added $AsPh_3$ (CCl_4 spectrum recorded in paper, 300 to 450 nm) [7]. The ^1H NMR shift for the complex in CH_3CN in the presence of the shift reagent Eu(fod)$_3$ (fod = 7,7-dimethyl-1,1,1,2,2,3,3-heptafluoro-octane-4,6-dionato) differs from those of the –N bonded isomer [10].

The dipole moment in benzene is 3.8 D, but no conclusion as to stereochemistry or mode of bonding could be deduced from this value [3]. The $E_{1/2}$ for reduction of the complex in dimethylformamide is 0.32 V; dissociation constants were calculated for loss of thiocyanate ligand from the complex in CH_3CN and $CHCl_3$ [13].

On heating to 156°C for 30 min the N-bonded isomer is formed [3]. Linkage isomerisation to the N-bonded isomer occurs at 145 to 160°C; at 220°C $Pd(NCS)_2(OAsPh_3)_2$ is formed. Thermo-gravimetric and differential thermal analysis studies were also made [8]. A kinetic study of the $Pd(SCN)_2(AsPh_3)_2$–$Pd(NCS)_2(AsPh_3)_2$ linkage isomerisation process in the solid state has been made. The activation energy is ~ 320 kJ/mol [9].

$Pd(NCS)_2(AsPh_3)_2$. This bright yellow N-bonded isomer melts at 195°C and is made by heating $Pd(SCN)_2(AsPh_3)_2$ to 156°C for 30 min [3]. It is formed from $Pd(SCN)_2(AsPh_3)_2$ from 145 to 160°C [8], or from $Pd(SCN)_2(AsPh_3)_2$ in solution [4].

The molar conductance in dimethylformamide is 13.8 $\Omega^{-1} \cdot$ cm^2 [3, 4].

The infrared spectrum of the solid shows ν_{CN} at 2090 [3], 2089 [4], 2085 [2] or 2088 cm^{-1} [5]; ν_{CS} at 854 [1, 2, 13], 853 [5] or 861 cm^{-1} [3]. The Pd–N stretch is at 265 cm^{-1} [9] but could not be identified in another study [11]. Band intensities were measured for ν_{CN} and ν_{CS} [5]. Infrared spectra in benzene showed bands (ν_{CN} at 2160, 2120 and 2090 cm^{-1}) identical with those found for $Pd(SCN)_2(AsPh_3)_2$ in solution [3]. In pyridine, acetone, acetonitrile, benzonitrile and adipo-nitrile ν_{CN} bands at 2120 to 2113 cm^{-1} were observed suggesting that $Pd(SCN)_2(AsPh_3)_2$ was the main solute species present [7]; in solutions in benzene [3, 10], CCl_4, $CHCl_3$, CH_2Cl_2, cyclo-pentanone, cyclohexanone, nitrobenzene, 2-butanone and 3-pentanone a number of bands between 2170 and 2082 cm^{-1} were observed, suggesting that a complex mixture of isomers of –N, –S and bridged thiocyanate isomers was present [7]. Electronic spectra were measured of the solid (300 to 550 nm) [3] and in $CHCl_3$ [3, 7], CH_3CN, C_6H_6, CCl_4 and CH_2Cl_2 [7]. The effect of solvent on the mode of bonding in $Pd(NCS)_2(AsPh_3)_2$ has been discussed [12].

The ^1H NMR chemical shift in CS_2 differs from those of the –S bonded isomer in the presence of Eu(fod)$_3$ (fod = 7,7-dimethyl-1,1,1,2,2,3,3-heptafluoro-octane-4,6-dionato) [10].

The dipole moment in benzene is 3.6 D, close to that (3.8) observed for the S-bonded isomer [3].

A kinetic study of the $Pd(SCN)_2(AsPh_3)_2$–$Pd(NCS)_2(AsPh_3)_2$ linkage isomerisation process in the solid state was carried out; the activation energy is ~ 320 kJ/mol [9].

On heating the compound, $Pd(NCS)_2(OAsPh_3)_2$ is formed at 200°C. The thermogravimetry and differential thermal analysis (DTA) of the system was measured [8].

References:

[1] Chatt, J.; Duncanson, L. A. (Nature **178** [1956] 997/8).
[2] Sabatini, A.; Bertini, I. (Inorg. Chem. **4** [1965] 1665/7).
[3] Burmeister, J. L.; Basolo, F. (Inorg. Chem. **3** [1964] 1587/93).
[4] Basolo, F.; Burmeister, J. L.; Poë, A. J. (J. Am. Chem. Soc. **85** [1963] 1700/1).
[5] Miezis, A. (Acta Chem. Scand. **27** [1973] 3746/60).
[6] Keller, R. N.; Johnson, N. B.; Westmoreland, L. L. (J. Am. Chem. Soc. **90** [1968] 2729/30).
[7] Burmeister, J. L.; Hassel, R. L.; Phelan, R. J. (Inorg. Chem. **10** [1971] 2032/8).
[8] Sedova, G. N.; Kirillova, M. A. (Zh. Neorgan. Khim. **32** [1987] 1510/2; Russ. J. Inorg. Chem. **32** [1987] 905/6).
[9] Wakita, H.; Kinoshita, S. (Fukuoka Daigaku Rigaku Shuho **17** [1987] 35/7; C.A. **107** [1987] No. 162624).
[10] Anderson, S. J.; Norbury, A. H. (J. Chem. Soc. Chem. Commun. **1975** 48/9).

[11] Goodgame, D. M. L.; Malerbi, B. W. (Spectrochim. Acta A **24** [1968] 1254/5).
[12] Burmeister, J. L.; Hassel, R. L.; Phelan, R. J. (Chem. Commun. **1970** 679/80).
[13] Tsytsyktueva, L. A.; Polovnyak, V. K.; Gazizov, K. K.; Gazizova, D. M. (Zh. Neorgan. Khim. **31** [1986] 2146/8; Russ. J. Inorg. Chem. **31** [1986] 1237/8).

14.5 Complexes Containing Thiocyanate and Chelating Arsines

Pd(SCN)₂(Ph₂As(CH₂)ₙAsPh₂) (n=1 to 3). These are made by reaction of the ligand with K₂[Pd(SCN)₄] in ethanol.

The $E_{1/2}$ reduction in dimethylformamide occurs at 0.25 V; dissociation constants for the complexes in CH_3CN and $CHCl_3$ were calculated [1].

Pd(SCN)₂(As–S). This dimethyl-o-methylthiophenylarsine (As–S) complex is made by reaction of the ligand with Li(NCS) and K₂[PdCl₄]. It is orange-yellow [2, 3].

The ν_{CN} frequency is 2094 cm⁻¹ [4]. The molecular conductance in nitrobenzene is $1 \times 10^{-4} \ \Omega^{-1} \cdot cm^2$ [2, 3].

Pd(SCN)₂(AsPSC₃₀H₂₄). Reaction of Na₂[PdCl₄] with NaSCN in 1-butanol with diphenyl-(o-diphenylarsinophenyl)phosphine sulphide gave the orange complex. The molar conductance in CH_3CN at 25°C was $<1 \ \Omega^{-1} \cdot cm^2$ [5].

The infrared spectra of the solid showed ν_{CN} at 2119 and 2110 cm⁻¹, suggesting the presence of S-bonded thiocyanate [5, 6]. In CH_2Cl_2, ν_{CN} appears at 2116 cm⁻¹; the band intensity was measured. The infrared spectrum of the solid (2000 to 2200 cm⁻¹) was reproduced in the paper [6]. The electronic spectra in CH_2Cl_2 solution and also of the solid were measured and were similar, suggesting that the solute retained the structure of the solid [5].

Pd(SCN)(NCS)(AsPC₃₀H₂₄). This is the tentative formulation given to the yellow material obtained by dissolving [Pd(AsPC₃₀H₂₄)][PdCl₄] in hot dimethylformamide (AsPC₃₀H₂₄ is diphenyl(o-diphenylarsino)phosphine). It can also be made by heating [Pd(AsPC₃₀H₂₄)][PdCl₄] to 220°C for 45 min.

The complex is monomeric in $CHCl_3$ and a nonconductor in CH_3CN [5]. The molar conductance in CH_3CN is $8.3 \ \Omega^{-1} \cdot cm^2$; it melts at 253 to 253.5°C [5].

The infrared spectrum of the solid shows ν_{CN} at 2117 cm⁻¹, indicative of an S-bonded thiocyanate ligand and also at 2058 cm⁻¹ indicative of an N-bonded thiocyanate group (spectra 2000 to 2200 cm^{-1}, reproduced in paper) [5]. The infrared spectrum of a solution in CH$_2$Cl$_2$ also shows ν_{CN} at 2118 and 2085 cm^{-1} [9]; the intensities were measured. Infrared spectra of the solid and the solution in CH$_2$Cl$_2$ were reproduced in the paper (2000 to 2200 cm^{-1}) [6]. The electronic spectra of the solid and of the solution in CH$_2$Cl$_2$ were measured [5].

Pd(SCN)$_2$(As$_2$C$_{26}$H$_{24}$). This yellow material is obtained from o-(Ph$_2$As)$_2$C$_6$H$_4$ and [Pd(SCN)$_4$]$^{2-}$ in ethanol.

The infrared spectrum of the solid shows ν_{CN} at 2116 and 2112 cm^{-1}. In CH$_2$Cl$_2$, however, two bands at 2119 and 2086 cm^{-1} are seen (intensities measured) which could result from Pd(SCN)(NCS)(As$_2$C$_{26}$H$_{24}$). Electronic spectra were recorded for the solid and its solution in CH$_2$Cl$_2$ [6].

Pd(SCN)(NCS)(As$_2$C$_{26}$H$_{24}$). This may be formed when Pd(SCN)$_2$(As$_2$C$_{26}$H$_{24}$) is dissolved in CH$_2$Cl$_2$ [6].

References:

[1] Tsytsyktueva, L. A.; Polovnyak, V. K.; Gazizov, K. K.; Gazizova, D. M. (Zh. Neorgan. Khim. **31** [1986] 2146/8; Russ. J. Inorg. Chem. **31** [1986] 1237/8).
[2] Livingstone, S. E. (J. Chem. Soc. **1958** 4222/6).
[3] Livingstone, S. E. (Chem. Ind. [London] **1957** 143/5).
[4] Chiswell, B.; Livingstone, S. E. (J. Chem. Soc. **1959** 2931/6).
[5] Nicpon, P.; Meek, D. W. (Inorg. Chem. **6** [1967] 145/9).
[6] Meek, D. W.; Nicpon, P. E.; Meek, V. I. (J. Am. Chem. Soc. **92** [1970] 5351/9).

15 Palladium and Antimony

Palladium Thioantimonide PdSbS is a cobaltite-type metallic phase; $a = 6.185$ Å.

Reference:

Hulliger, F. (Nature **198** [1963] 382/3).

Substituted Thiocyanato Complexes with Tertiary and Chelating Stibines

Pd(SCN)$_2$(SbPh$_3$)$_2$. This is made by reaction of K$_2$[Pd(SCN)$_4$] in ethanol with SbPh$_3$ [1, 6].

The infrared spectrum of the solid shows v_{CN} at 2119 and 2115 [1, 2] or 2093 cm^{-1} [3]; the Pd–S stretch is at 290 cm^{-1} [7]. The intensity of the v_{CN} band was measured [3]. S-bonding of the thiocyanate ligand was deduced on the basis of the infrared data [1, 2, 3]. In dimethylformamide, (CH$_3$)$_2$SO, acetone and CH$_3$CN solutions v_{CN} appeared at 2110 cm^{-1} suggesting retention of the S-bonded structure for the solute, but in C$_6$H$_6$, CCl$_4$, CHCl$_3$, cyclopentanone and cyclohexanone, bands at 2165 to 2085 cm^{-1} were observed, suggesting that complex equilibria involving –N, –S and bridged thiocyanate bonded isomers were present in such solutions. The electronic absorption spectra in C$_6$H$_6$, CCl$_4$ and CHCl$_3$ were measured [5].

Reduction in dimethylformamide at $E_{1/2} = 0.27$ V occurs, and dissociation constants were measured in CHCl$_3$ and CH$_3$CN for the complex [6].

Pd(NCS)(SNC)(o-C$_6$H$_4$PPh$_2$SbPh$_2$). This is made from Na$_2$[Pd(SCN)$_4$] and (o-diphenylphosphinophenyl)diphenylstibine in dimethylformamide. It is deep yellow, melting at 189°C. The molar conductance in nitrobenzene is 2 $\Omega^{-1} \cdot$ cm^2. The infrared spectrum of the solid has bands at 2110 and 2078 cm^{-1}, shifting to 2115 and 2100 cm^{-1} in CH$_2$Cl$_2$ solution. The presence of both –N and –S bonded thiocyanate was suggested [4].

[Pd(NCS)(o-C$_6$H$_4$PPh$_2$SbPh$_2$)$_2$]NCS. This is made from Na$_2$[Pd(SCN)$_4$] and (o-diphenylphosphinophenyl)diphenylstibine in refluxing dimethylformamide. It is orange-red, melting at 185°C. The molar conductance in C$_6$H$_3$NO$_2$ is 101 $\Omega^{-1} \cdot$ cm^2. The infrared spectrum shows v_{CN} at 2090 and 2060 cm^{-1} in the solid and at 2085 cm^{-1} in CH$_2$Cl$_2$ solution; the electronic absorption spectrum was also measured [4].

Pd(SCN)$_2$(o-C$_6$H$_4$AsPh$_2$SbPh$_2$). This is made from Na$_2$[Pd(SCN)$_4$] in refluxing dimethylformamide with (o-diphenylarsinophenyl)diphenylstibine. It is orange, melting at 203°C, and its molar conductance in nitrobenzene is <1.0 $\Omega^{-1} \cdot$ cm^2. The infrared spectrum shows v_{CN} at 2100 cm^{-1} for the solid and at 2112 and 2108 cm^{-1} in CH$_2$Cl$_2$ solution. It was suggested that the thiocyanate ligands are S-bonded to the metal atoms. The electronic absorption spectrum was also measured [4].

References:

[1] Burmeister, J. L.; Basolo, F. (Inorg. Chem. **3** [1964] 1587/93).
[2] Sabatini, A.; Bertini, I. (Inorg. Chem. **4** [1965] 1665/7).
[3] Miezis, A. (Acta Chem. Scand. **27** [1973] 3746/60).
[4] Levason, W.; McAuliffe, C.A. (Inorg. Chim. Acta **16** [1976] 167/72).
[5] Burmeister, J. L.; Hassel, R. L.; Phelan, R. J. (Inorg. Chem. **10** [1971] 2032/8).
[6] Tsytsyktueva, L. A.; Polovnyak, V. K.; Gazizov, K. K.; Gazizova, D. M. (Zh. Neorgan. Khim. **31** [1986] 2146/8; Russ. J. Inorg. Chem. **31** [1986] 1237/8).
[7] Keller, R. N.; Johnson, N. B.; Westmoreland, L. L. (J. Am. Chem. Soc. **90** [1968] 2729/30).

Physical Constants and Conversion Factors

Avogadro constant N_A (or L) = 6.02214×10^{23} mol^{-1}

Faraday constant   F = 9.64853×10^4 C/mol

molar gas constant   R = 8.31451 J·mol^{-1}·K^{-1}

molar volume (ideal gas) V_m = 2.24141×10^1 L/mol
(273.15 K, 101325 Pa)

Planck constant   h = 6.62608×10^{-34} J·s

elementary charge e = 1.60218×10^{-19} C

electron mass m_e = 9.10939×10^{-31} kg

proton mass m_p = 1.67262×10^{-27} kg

1 kg = 2.205 pounds

1 m = 3.937×10^1 inches = 3.281 feet

1 m^3 = 2.642×10^2 gallons (U.S.)

1 m^3 = 2.200×10^2 gallons (Imperial)

Force	N	dyn	kp
1 N	1	10^5	1.019716×10^{-1}
1 dyn	10^{-5}	1	1.019716×10^{-6}
1 kp	9.80665	9.80665×10^5	1

Pressure	Pa	bar	kp/m²	at	atm	Torr	lb/in²
1 Pa = 1N/m²	1	10^{-5}	1.019716×10^{-1}	1.019716×10^{-5}	9.86923×10^{-6}	7.50062×10^{-3}	1.450378×10^{-4}
1 bar = 10^6 dyn/cm²	10^5	1	1.019716×10^4	1.019716	9.86923×10^{-1}	7.50062×10^2	1.450378×10^1
1 kp/m² = 1 mm H_2O	9.80665	9.80665×10^{-5}	1	10^{-4}	9.67841×10^{-5}	7.35559×10^{-2}	1.422335×10^{-3}
1 at (technical)	9.80665×10^4	9.80665×10^{-1}	10^4	1	9.67841×10^{-1}	7.35559×10^2	1.422335×10^1
1 atm = 760 Torr	1.01325×10^5	1.01325	1.033227×10^4	1.033227	1	7.60×10^2	1.469595×10^1
1 Torr = 1 mmHg	1.333224×10^2	1.333224×10^{-3}	1.359510×10^1	1.359510×10^{-3}	1.315789×10^{-3}	1	1.933678×10^{-2}
1 lb/in² = 1 psi	6.89476×10^3	6.89476×10^{-2}	7.03069×10^2	7.03069×10^{-2}	6.80460×10^{-2}	5.17149×10^1	1

Work, Energy, Heat

	J	kW·h	kcal	Btu	eV
1 J = 1 W·s = 1 N·m = 10^7 erg	1	2.778×10^{-7}	2.39006×10^{-4}	9.4781×10^{-4}	6.242×10^{18}
1 kW·h	3.6×10^6	1	8.604×10^2	3.41214×10^3	2.247×10^{25}
1 kcal	4.1840×10^3	1.1622×10^{-3}	1	3.96566	2.6117×10^{22}
1 Btu (British thermal unit)	1.05506×10^3	2.93071×10^{-4}	2.5164×10^{-1}	1	6.5858×10^{21}
1 eV	1.602×10^{-7}	4.450×10^{-14}	3.8289×10^{-11}	1.51840×10^{-10}	1

$1\,\text{cm}^{-1} = 1.239842 \times 10^{-4}\,\text{eV}$

$1\,\text{hartree} = 27.2114\,\text{eV}$

$1\,\text{Hz} = 4.135669 \times 10^{-15}\,\text{eV}$

$1\,\text{eV} \triangleq 23.0578\,\text{kcal/mol}$

Power

	kW	hp	$\text{kp} \cdot \text{m} \cdot \text{s}^{-1}$	kcal/s
1 kW = 10^3 J	1	1.35962	1.01972×10^2	2.39006×10^{-1}
1 hp (horsepower, metric)	7.3550×10^{-1}	1	7.5×10^1	1.7579×10^{-1}
$1\,\text{kp} \cdot \text{m} \cdot \text{s}^{-1}$	9.80665×10^{-3}	1.333×10^{-2}	1	2.34384×10^{-3}
1 kcal/s	4.1840	5.6886	4.26650×10^2	1

References:

International Union of Pure and Applied Chemistry, Manual of Symbols and Terminology for Physicochemical Quantities and Units, Pergamon, London 1979; Pure Appl. Chem. **51** [1979] 1/41.

The International System of Units (SI), National Bureau of Standards Spec. Publ. 330 [1972].

Landolt-Börnstein, 6th Ed., Vol. II, Pt. 1, 1971, pp. 1/14.

ISO Standards Handbook 2, Units of Measurement, 2nd Ed., Geneva 1982.

Cohen, E. R., Taylor, B. N., Codata Bulletin No. 63, Pergamon, Oxford 1986.

Key to the Gmelin System
of Elements and Compounds

System Number	Symbol	Element
1		Noble Gases
2	H	Hydrogen
3	O	Oxygen
4	N	Nitrogen
5	F	Fluorine
6	**Cl**	**Chlorine**
7	Br	Bromine
8	I	Iodine
8a	At	Astatine
9	S	Sulfur
10	Se	Selenium
11	Te	Tellurium
12	Po	Polonium
13	B	Boron
14	C	Carbon
15	Si	Silicon
16	P	Phosphorus
17	As	Arsenic
18	Sb	Antimony
19	Bi	Bismuth
20	Li	Lithium
21	Na	Sodium
22	K	Potassium
23	NH_4	Ammonium
24	Rb	Rubidium
25	Cs	Caesium
25a	Fr	Francium
26	Be	Beryllium
27	Mg	Magnesium
28	Ca	Calcium
29	Sr	Strontium
30	Ba	Barium
31	Ra	Radium
32	**Zn**	**Zinc**
33	Cd	Cadmium
34	Hg	Mercury
35	Al	Aluminium
36	Ga	Gallium

System Number	Symbol	Element
37	In	Indium
38	Tl	Thallium
39	Sc, Y La–Lu	Rare Earth Elements
40	Ac	Actinium
41	Ti	Titanium
42	Zr	Zirconium
43	Hf	Hafnium
44	Th	Thorium
45	Ge	Germanium
46	Sn	Tin
47	Pb	Lead
48	V	Vanadium
49	Nb	Niobium
50	Ta	Tantalum
51	Pa	Protactinium
52	**Cr**	**Chromium**
53	Mo	Molybdenum
54	W	Tungsten
55	U	Uranium
56	Mn	Manganese
57	Ni	Nickel
58	Co	Cobalt
59	Fe	Iron
60	Cu	Copper
61	Ag	Silver
62	Au	Gold
63	Ru	Ruthenium
64	Rh	Rhodium
65	Pd	Palladium
66	Os	Osmium
67	Ir	Iridium
68	Pt	Platinum
69	Tc	Technetium[1]
70	Re	Rhenium
71	Np,Pu...	Transuranium Elements

HCl · CrCl$_2$ · ZnCrO$_4$ · ZnCl$_2$

Material presented under each Gmelin System Number includes all information concerning the element(s) listed for that number plus the compounds with elements of lower System Number.

For example, zinc (System Number 32) as well as all zinc compounds with elements numbered from 1 to 31 are classified under number 32.

[1] A Gmelin volume titled "Masurium" was published with this System Number in 1941.

A Periodic Table of the Elements with the Gmelin System Numbers is given on the Inside Front Cover